17th International School on Condensed Matter Physics (ISCMP) : Open Problems in Condensed Matter Physics, Biomedical Physics and Their Applications 2012

Journal of Physics: Conference Series Volume 398

Varna, Bulgaria
2-7 September 2012

ISBN: 978-1-62276-864-6
ISSN: 1742-6588

Printed from e-media with permission by:

Curran Associates, Inc.
57 Morehouse Lane
Red Hook, NY 12571

CURRAN ASSOCIATES INC.
proceedings
.com

Some format issues inherent in the e-media version may also appear in this print version.

Printed by Curran Associates, Inc. (2013)

For permission requests, please contact the Institute of Physics
at the address below.

Institute of Physics
Dirac House, Temple Back
Bristol BS1 6BE UK

Phone: 44 1 17 929 7481
Fax: 44 1 17 920 0979

techtracking@iop.org

Additional copies of this publication are available from:

Curran Associates, Inc.
57 Morehouse Lane
Red Hook, NY 12571 USA
Phone: 845-758-0400
Fax: 845-758-2634
Email: curran@proceedings.com
Web: www.proceedings.com

Papers

PREFACE

We are pleased to introduce the Proceedings of the 17th International School on Condensed Matter Physics: Open Problems in Condensed Matter Physics, Biomedical Physics and their Applications, organized by the Institute of Solid State Physics of the Bulgarian Academy of Sciences. The Chairman of the School was Professor Alexander G Petrov. Like prior events, the School took place in the beautiful Black Sea resort of Saints Constantine and Helena near Varna, going back to the refurbished facilities of the Panorama hotel. Participants from 17 different countries delivered 31 invited lecturers and 78 posters, contributing through three sessions of poster presentations. Papers submitted to the Proceedings were refereed according to the high standards of the *Journal of Physics: Conference Series* and the accepted papers illustrate the diversity and the high level of the contributions.

Not least significant factor for the success of the 17 ISCMP was the social program, both the organized events (Welcome and Farewell Parties) and the variety of pleasant local restaurants and beaches. Visits to the Archaeological Museum (rich in valuable gold treasures of the ancient Thracian culture) and to the famous rock monastery Aladja were organized for the participants from the Varna Municipality.

These Proceedings are published for the second time by the *Journal of Physics: Conference Series*. We are grateful to the Journal's staff for supporting this idea.

The Committee decided that the next event will take place again in Saints Constantine and Helena, 1–5 September 2014. It will be entitled: Challenges of the Nanoscale Science: Theory, Materials and Applications.

Doriana Dimova-Malinovska, Diana Nesheva, Emilia Pecheva, Alexander G Petrov and Marina T Primatarowa
Editors

All papers published in this volume of *Journal of Physics: Conference Series* have been peer reviewed through processes administered by the proceedings Editors. Reviews were conducted by expert referees to the professional and scientific standards expected of a proceedings journal published by IOP Publishing.

Wide field nanometric materials analysis by diffraction limited far field optical nanoscopy

P C Montgomery, D Montaner, F Anstotz and B Serio
Institut d'Electronique du Solide et des Systèmes (InESS),
UDS-CNRS, UMR 7163, 23 rue du Loess, 67037 Strasbourg, France.

E-mail: paul.montgomery@unistra.fr

Abstract. In the development of nanomaterials and biomaterials, new characterization techniques are required that overcome the challenges presented by the increasing dimensional ratio between the different entities to be studied and the growing complexity introduced by the use of heterogeneous materials and technologies. Diffraction limited far field optical nanoscopy techniques are receiving growing interest because of their capacity to detect nm structures over very large fields and at high speed. In this paper we present a classification scheme of optical nanoscopy techniques and in particular highlight four categories of far field diffraction limited techniques based on increasing the contrast, measuring the phase, using deconvolution or using nano-markers. We thus demonstrate that by increasing the power of detectability, observability or measurability, a wealth of information concerning nanometric structures becomes available even though all the lateral details may not be resolved. These techniques conserve all the advantages of classical imaging such as real time or high speed measurement, large quantities of useful data, non-invasiveness, non-destructiveness and ease of use. Such a summary will be useful in stimulating the search for new solutions to certain difficult problems in nano-characterization of nanomaterials and biomaterials over wide fields.

1. Introduction

In considering one of the key enabling technologies of the last 50 years, microelectronics, it can be noted that research in this area has been marked from its conception by the growth in the gap between the size of the basic element and that of the finished system. If this ratio was a factor of 10 at the beginning of the 1960's, it will approach 10^6 in the next few years. For example, in today's microprocessors, the gate length of a transistor is of the order of tens of nm while the overall size of the processor is more than 1 cm^2. In future components and systems based on nanotechnologies and biotechnologies, this ratio will continue to increase by several orders of magnitude. In addition, components and systems will grow in complexity since they will contain several different technologies as is already the case in microsystems (MEMS, MOEMS) and Systems on Chips (SOC). These two factors, the dimensional ratio problem and the complexity, present two significant challenges in the field of future nano-characterization to make these new technologies viable.

In the case of biomaterials, for example in layers of hydroxyapatite (the mineral part of bones and teeth), morphological, chemical, optical and mechanical information is necessary at several scales, from the nm level of the basic crystals to the mm level for the functional material [1]. In living organisms such as cells, the details of interest vary from the nm level for molecules, to 10 nm for larger proteins and up to several tens of nm for vesicles and this dynamically in solution.

In terms of high resolution characterization techniques, electron microscopy and near field microscopy provide certain solutions for imaging the basic elements at the nm level, but as with all techniques, while being very good in certain applications their limitations make them unsuitable for others. For example, in electron microscopy, the electron beam and the vacuum or near-vacuum conditions can be destructive for many types of samples. In near field microscopy, point by point scanning limits the measurement bandwidth and the field of measurement and the physical presence of the tip leads to measurement biases and a restriction to surface or near surface characterization.

Far field optical microscopy has received renewed interest in recent years for several reasons, not the least being the innate high bandpass due to the parallel sampling of imaging that gives the capacity for real time analysis. The development of super-resolution techniques also makes it possible to go well beyond the limits of diffraction. Finally, there is the realization that far field imaging can in fact be used extensively for extracting important information from nano-structures and thus being able to solve certain scientific problems in the "nano-world". These different approaches mentioned, i.e. near field, far field and super-resolution, generally fall under the term of nanoscopy.

One thing that can be remarked when approaching this new field of nanoscopy is that there are so many different techniques that exist that it can be quite confusing for the materials specialist who is looking for an appropriate characterization technique. There are several different ways of classifying nanoscopy techniques. In figure 1 we propose one particular classification scheme that can help in better distinguishing one technique from another. It is neither exhaustive nor definitive, but merely useful. A first level of classification concerns the distance at which the optical information is obtained, in the far field, with an imaging objective, or in the near field with a physical probe placed in the nanometric vicinity of the surface such as SNOM (scanning near-field optical microscopy).

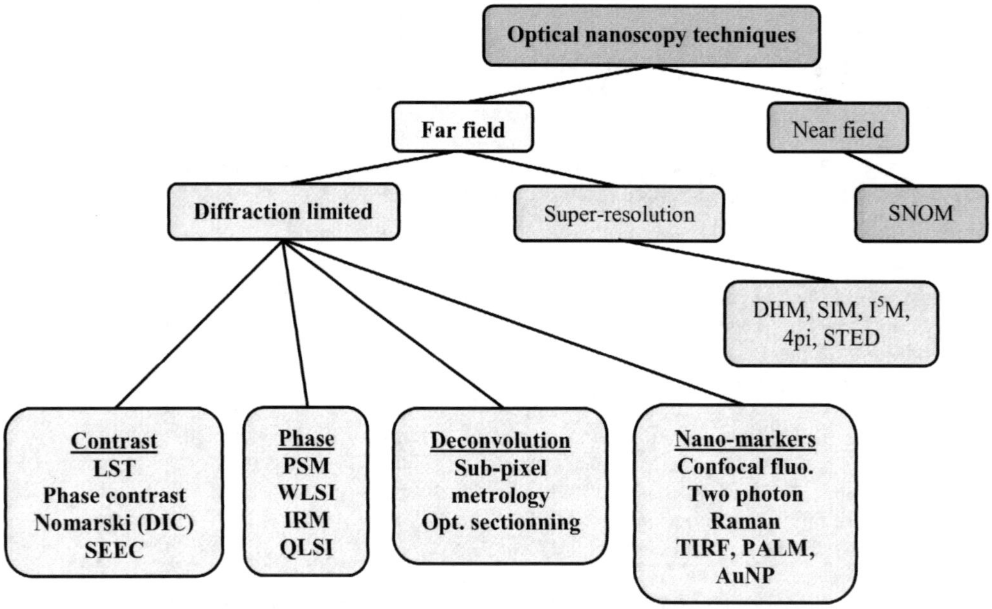

Figure 1. The four sub-divisions of diffraction limited far field techniques (marked in blue) in the context of a global classification scheme for nanoscopy techniques

Amongst the far field techniques, a second level of classification can be made concerning the lateral resolution attained. The lateral resolution is either limited by the effects of diffraction or, by the super resolution techniques used to go beyond diffraction limited imaging. Resolving lateral details below the wavelength of the light used while remaining in far field conditions, can be achieved by diffraction tomography (or DHM, digital holographic microscopy), SIM (structured illumination microscopy), I5M, 4pi or STED (stimulated emission depletion). These techniques are of particular interest in the analysis of biological samples where a lateral resolution of 20 nm to 100 nm is required.

The second category of far field techniques concerns those that are diffraction limited, which can be sub-divided into four categories according to the method used to give the nanometric sensitivity:

1. Increasing the contrast by means of the illumination, the phase or the polarisation.
2. Measuring the phase by interferometry.
3. Using deconvolution in sub-pixel metrology techniques.
4. Using nano-markers such as fluorochromes or gold nano-particles.

It is this particular category of far field, diffraction limited techniques (marked in blue in figure 1) that is the subject of this paper. While being diffraction limited and unable to resolve all the lateral details, they can nonetheless be used to extract a number of different types of important information from nano-structures while conserving all the advantages of classical imaging [2]. For example, characterization can be performed over wide fields of hundreds of µm and even mm, at a very high rate, often allowing real time measurement and the production of a large quantity of data. The use of an optical probe is also non-invasive, non-destructive and non-toxic for living organisms.

It is worth underlining that even though the origins of several far field, diffraction limited techniques described date from over twenty years ago or even more, it is their "updating" using the combination of various physical principles and improved cameras and image processing techniques that make them so useful in nano-characterization. We pointed this out in several papers at the beginning of the 1990's [2-4] but it is only in more recent years that the idea is becoming more popular. Both well-known and not so well-known techniques are worth another look at in the context of today's nano- and biomaterials since they could provide some elegant solutions for certain problems in nano-characterization. Such a review could also stimulate the development of new instrumentation.

In the four sections that follow, we therefore present a summary of some of the results of far field, diffraction limited nanoscopy using contrast, phase measurement, deconvolution and nano-markers. We present some of the principles and some example results to demonstrate the type of information that can be obtained in nano-characterization.

2. Nanoscopy using high contrast for detection of nm structures

When the size of an object under a conventional optical microscope is slightly smaller than the Rayleigh limit, the intensity decreases rapidly to the point of the object becoming no longer visible. This can be described in terms of Rayleigh scattering, in which the object size $d \ll \lambda$ and the intensity of the scattered light decreases as a function of d^6. But since scattering is uniform in all directions, one way of making nano-particles visible in the far field is simply to increase the contrast. For images containing small features that are present on a large uniform background, the contrast is given by the Weber contrast, C_W:

$$C_W = \frac{(I_{Max} - I_{Min})}{(I_{Min})}$$

where I_{Max} is the maximum intensity (intensity of the features) and I_{Min} is the minimum intensity (background intensity). For an 8 bit image depth, C_W therefore varies from 1 to 255. In other types of images in which the structures have equivalent bright and dark features, the contrast is given by the Michelson contrast, C_M, which varies between 0 and 1 and is defined by:

$$C_M = \frac{(I_{Max} - I_{Min})}{(I_{Max} + I_{Min})}$$

2.1. Dark field microscopy

In classical dark field microscopy [2, 3] the sample contrast comes from light scattered by the sample using a dark field illumination device by illuminating the sample with highly inclined light rays that are not collected by the objective lens and generally do not form part of the image. To illustrate the effects of contrast on particle visibility, the results in figure 2 show dark field images of microprecipitates in crystals of GaAs using two different types of near IR illumination. In this case the microprecipitates are visible (figure 2(a)) because they have a size near to the resolution limit of about 1 µm. But the contrast of the particles is fairly low (C_W = 1 to 2) since the background intensity is far from zero (I_{Min} = 20 to 80 grey levels). While the direct illuminating light rays do not enter the objective lens, light diffused from scatterers above and below the observation plane does. Even with image processing, the contrast is not sufficient to observe nano-particles.

On the other hand, it should be pointed out that even with low contrast conditions, observation of µm sized microprecipitates can actually reveal important structural information about nano-structures. The image in figure 2(a) is in fact the addition of three images taken at different depths, which after image processing and colour coding give a 3D impression of the orientation of the distribution of the particles (figure 2(b)), which actually decorate a series of looped dislocations which are atomic in size. The reason the dislocation paths become observable is because not only they are decorated by "native-markers", but they are also extended beyond the lateral resolution limit of the microscope. This example could therefore have been used in section 5 to illustrate the use of nano-markers.

A novel way of considerably reducing the background intensity in dark field microscopy is by using a thin beam of laser illumination at 90° to the direction of observation in the laser scanning tomography (LST) technique. For example, by focusing a near IR (λ = 1 µm) laser beam on the cleaved face of a crystal of GaAs using LST, the cloud of 10 nm sized microprecipitates appearing after annealing, becomes clearly visible, as in figure 2(c) [3, 4]. In this case, the nanoparticles act like point sources that are visible individually if their density is less than 10^9 to 10^{10} cm^{-3} or as a continuous background level if their density is greater. While the technique works well in bulk materials, it is not suitable for analyzing defects in thin layers since it is limited to an observation plane that is more than 5 µm from the surface.

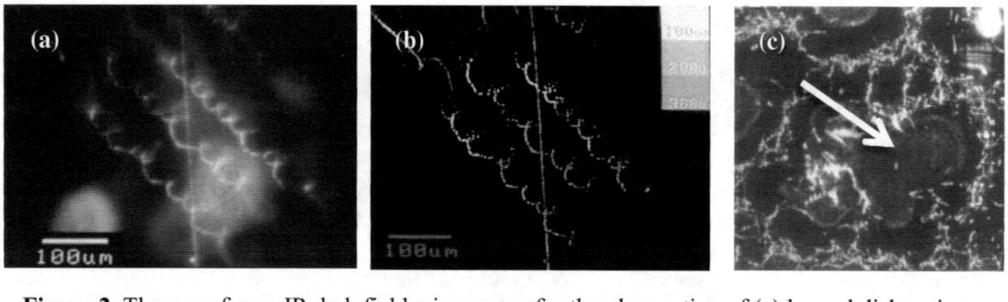

Figure 2. The use of near IR dark field microscopy for the observation of (a) looped dislocations decorated with microprecipitates in GaAs, (b) 3D observation of (a) after image processing (P. Montgomery 1990) and (c) cloud of 10 nm nano-particles (indicated by arrow) in GaAs using LST (1 x1 mm) (M. Castagné & P. Gall-Borrut, CEM2, Univ. Montpellier II, 1990)

Other useful properties of Rayleigh scatterers are that they are sensitive to wavelength, the scattering intensity varying with $(1/\lambda)^4$ and they can be sensitive to the orientation of polarization of the illumination beam when assimilated to dipoles. The use of several illumination wavelengths [5] and the effects of polarization in more recent improvements of the LST technique allow the determination of the shape and nature of the different nanoparticles present. In fluorescence microscopy, the orientation of molecules in the far field has been observed due to the asymmetric polarization response of the Airy disc [6].

2.2. Preferential chemical etching and Nomarski (DIC)

A powerful although destructive technique for revealing nano-defects in the sub-surface of crystals is by the use of preferential chemical etching, which produces variations in roughness according to the defects present. Nomarski microscopy (or DIC, differential interference microscopy,) can be used to improve the contrast of the edge details, making use of the high sensitivity of interferometry in a sheared image pair of the surface. The results in figure 3(a) show a DIC image of a surface of GaAs that has been etched to a depth of 0.6 μm [7]. What can be seen is a dislocation parallel to the surface that is decorated with micro-defects that is known as the Cottrel atmosphere. In this way, the etching reveals the presence of nanometric and even extended atomic sized defects that are amplified by DIC microscopy. The relief can also be quantified using phase stepping microscopy (see section 3.2).

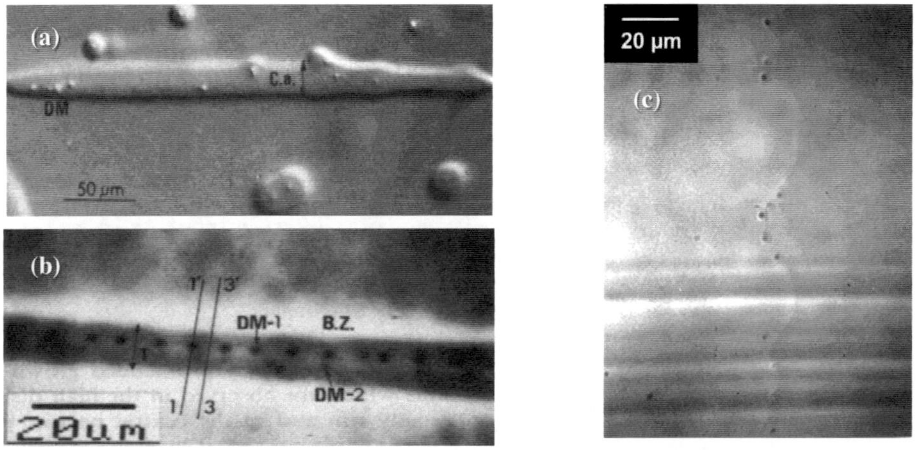

Figure 3. Atomic sized dislocation paths made visible in GaAs by (a) Normarski after chemical etching (J. Weyher, 1990), (b) near IR phase contrast microscopy (P.C. Montgomery, CEM2, Univ. Montpellier II, 1990) and (c) near IR phase contrast microscopy with image integration (P.C. Montgomery, CEM2, Univ. Montpellier II, 1989)

2.3. Phase contrast microscopy

A non-destructive way of observing the same type of defects described previously is to increase the contrast of small variations in the refractive index within the crystal using near IR phase contrast microscopy (figure 3(b)) [7], revealing clearly the extended Cottrel atmosphere around the decorated dislocation. Measurement of the intensities shows a contrast $C_M = 0.1$. By using image integration to reduce camera noise, near IR phase contrast microscopy was used to produce the exceptional results in figure 3(c) in which despite the extremely low contrast $C_M = 0.01$, the observation of the dislocation path was possible. The extended atomic sized defect is made visible by the associated surrounding strain field in the tin doped GaAs crystal [8], the "pinning" effect of the dislocation path by microprecipitates being clearly visible.

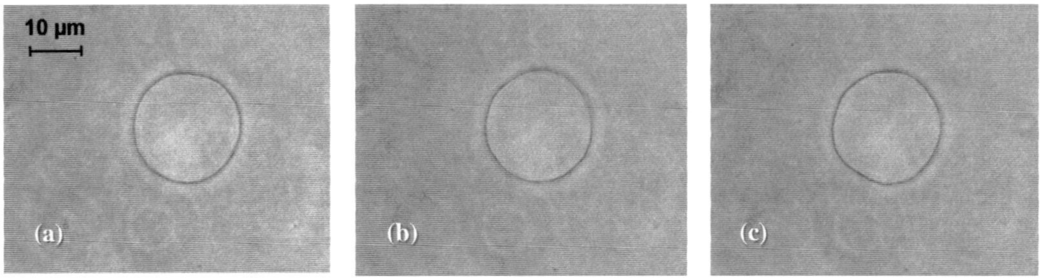

Figure 4. A series of strobed images ($\delta t = 1$ s) using phase contrast microscopy of a giant vesicle showing shape fluctuations (V. Vitkova, Laboratory of Liquid Crystals, ISSP, BAS, Sofia, Bulgaria).

A more recent application of phase contrast microscopy in nano-characterization, is in the measurement of the elastic properties of lipid membranes through the study of thermal shape fluctuations of giant vesicles [9]. Although the thickness of the membranes of the unilamellar giant vesicles (10 µm radius) is less than 5 nm, because their refractive index is about 10% higher than that of the aqueous surroundings, the edge of the sphere becomes clearly visible. The thermal fluctuations are very complex, being the sum of many different vibration modes. To avoid blurring, the addition of stroboscopic illumination using a xenon flashlamp makes it possible to freeze the movement. The strobed images (4 µs pulse duration) in figure 4 taken at intervals of 1 s show clearly such fluctuations. Strobing from 0.1 to 1 MHz is possible. The system has been used to successfully estimate the friction between the monolayers comprising the bilayer of the membrane, as well as the bending elasticity modulus of the blocked exchange of molecules between the monolayers [9].

2.4. Surface Enhanced Ellipsometric Contrast microscopy (SEEC)
Another way of increasing the contrast is by using polarised light. The SEEC (surface enhanced ellipsometric contrast, commercially known as Sarfus) technique, uses dedicated substrates specially prepared with anti-reflection coatings that are designed for use in crossed polarised light to decrease the background intensity (I_{Min}) and thus increase the contrast by a factor of 10 to 100 [10]. The presence of nm layers or nanoparticles changes the state of the polarization, producing contrast and thus become visible. The results in figure 5 show an image from a video (figure 5(a)) of the crystallization of a 9 nm thick layer of polystyrene-b-polyethylene oxide (PS-b-PEO) diblock copolymers. In figure 5(b) can be seen bundles of DWNT (double walled carbon nanotubes) that are individually 9 nm in diameter. The technique is finding use in a growing number of applications in nano-characterization, such as for example in the study of nano-dots, nanowires, graphene, crystallites, thin films, polymers, membranes, polyelectrolytes, soft lithography, biochips and live cells.

Figure 5. The use of the SEEC technique to observe (a) the crystallization of a 9 nm layer of PS-PEO (G. Reiter, ICSI, Mulhouse, France) and bundles of DWNT carbon nanotubes (E. Flahaut, CIRIMAT, Toulouse, France) © Nanolane

3. Nanoscopy using phase for nm roughness measurement
The second family of far field diffraction limited nanoscopy techniques is that based on the measurement of the phase of the light reflected from a surface. Comparison of the wavefront of the light from the sample with that from a flat reference surface results in interference fringes and the conversion of the phase into intensity that allows the measurement of nm surface roughness and shape.

3.1. Interference Reflection Microscopy (IRM)
A simple interferometric setup exists for studying the behaviour of live cells on microscope slides in IRM (interference reflection microscopy, or RICM, reflection interference contrast microscopy).

Viewing through the glass substrate, interference is produced between the substrate surface and the cell membrane, producing fringes that can be quantified so as to measure the distance between the membrane and the substrate surface in adhesion studies. The high sensitivity of interferometry in the "fluctuation contrast" technique, also allows the detection of nm variations due to bunches of ligand-receptor links within the substratum [11].

3.2. Phase Stepping Microscopy (PSM)

By optimizing the optical configuration of the interferometer in a dedicated objective, the fringe contrast, C_M, which is also known as the fringe modulation or visibility in interferometry, can have quite a high value between neighbouring bright and dark fringes, corresponding to a height difference of $\lambda/4$, or roughly 170 nm in visible light. Thus even with an 8 bit image depth, measurement of intermediate intensities of fringes in combination with optimized phase stepping algorithms thus allows the measurement of nanometric or better surface roughness [2, 4, 14]. PSM (phase stepping microscopy), CSI (coherence scanning interferometry, also known as WLSI, white light scanning interferometry) are used for measuring a wide variety of surfaces such as semiconductors, metals, polymers, etc [1, 4]. Surfaces can be profiled in real time using 4D interference microscopy [12].

As mentioned in section 2.2, PSM is also used to quantify the surface roughness after preferential chemical etching, such as the hillocks and valleys on tin doped GaAs using a DSL (dilute Sirtl-like) solution in figure 6(a) [13] and on HgI_2 using a KI solution in figure 6(b) [14] and to associate them to the different types of nano-defects in the crystals over very wide fields of hundreds of µm or several mm. The etching "amplifies" the presence of the nano-defects laterally and the high axial sensitivity of the PSM technique allows the roughness to be quantified and the defects to be catalogued.

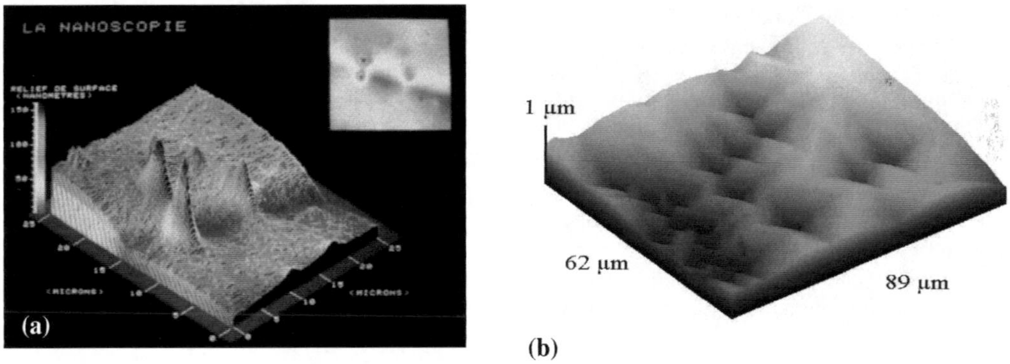

Figure 6. The use of PSM for measuring, identifying and cataloguing the different types of defects in microelectronic crystals after preferential chemical etching in (a) Tin doped GaAs using a DSL solution (P. Montgomery, J. Weyher, CEM2, Université Montpellier II, Montpellier) and (b) HgI_2 using a KI solution (InESS, Strasbourg)

Another example of the usefulness of PSM even though it is diffraction limited, is in the calibration of the laser fluence of an excimer laser (λ = 308 nm) for the annealing of 80 nm thick amorphous Si layers for flat panel displays [15]. To achieve this, a single measurement of the 3D roughness is made of a sample placed at the edge of the laser impact which has a "top hat" intensity profile and therefore has a linear variation of the laser fluence with distance along one of the axes (figure 7(a)). A profile of the roughness along this axis shows the variation in roughness as a function of the laser fluence (empty diamonds in figure 7(b)). The two zones to avoid are visible, zone B, in which the grain size is too small and the "SLG" (super lateral growth) region, in which the grain size is maximum but irregular. The optimal zone is at A, with a large and constant grain size and providing the optimal laser fluence required.

A comparison with the measurements made with AFM (black diamonds in figure 7(b)) shows that those made by PSM are smaller. This difference is due to the fact that the grains of poly-Si vary in lateral size from 0.1 μm to 1 μm, which is near to the lateral resolution limit of 0.4 μm of the microscope. But it can be noted that even though the roughness is undersampled by PSM, the general shape of the line profile across the SLG region is maintained. This application is a good demonstration of the use of PSM as a rapid technique over large areas in a type of combinatorial analysis. The optimal parameter (the laser fluence) is identified due to the variation of another parameter (the roughness), not between different samples but between different positions on a single sample that has been judiciously placed at the edge of the laser beam and has taken only a few seconds to measure.

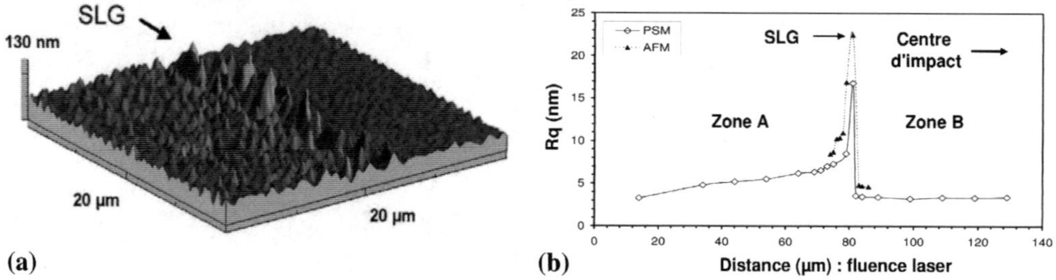

(a) **(b)**

Figure 7. PSM used in combinatorial analysis to determine the optimal zone of laser fluence in the crystallization of poly-Si in flat screen processing (a) PSM roughness near the edge of the laser impact and (b) Comparison of the roughness measurements between PSM/AFM (InESS, Strasbourg).

A more recent variation of the PSM technique, QLSI (quadriwave lateral shearing interferometry), allows the measurement of phase variations in a single image and therefore higher speed analysis. A high resolution wavefront sensor based on a 2D hole grating (modified Hartman mask) in front of the CCD is used in combination with FT analysis. The technique is particularly useful in the study of cell membrane dynamics and intracellular activities due to the high sensitivity that reveals very thin or small structures such as the lamellipodia surrounding cells [16].

4. Nanoscopy by deconvolution for nm positioning

In the case of simple structures consisting of small particles, straight edges or regular structures, a third family of far field nanoscopy techniques exists using deconvolution, based on the knowledge of the optical transfer function of the optical system for sub-pixel or sub-voxel metrology.

4.1. Edge and particle positioning

White light interferometry was used in early sub-pixel techniques for helping to better measure the positions of structural edges in integrated circuits, with a lateral measurement uncertainty in position of 10-20 nm [17]. In LST, analysing the 3D distribution of microprecipitates inside semiconductor crystals at a sub-micron scale can be carried out by measuring the intensity response of the particles as a function of either the Gaussian beam profile of the laser beam or the $sinc^2$ longitudinal PSF of the optical system, giving a measurement uncertainty of 0.1 μm along the optical axis [4].

4.2. Position referenced microscopy

A more recent version of sub-pixel metrology has been developed for the precise repositioning of live cells (figure 8(a) and (b)) using a carefully designed reference mask consisting of a pseudo-periodic pattern placed just below the sample (figure 8(c) and (d)) [18]. With a lateral measurement uncertainty of 20 nm, this technique allows the absolute reference positioning of live specimens (figure 8(d)) and enables for example the following of the evolution of the absorption of apoptotic bodies by fibroblasts (figure 8(b)) during long periods of tens of hours that are removed from a controlled environment for imaging on the microscope at regular intervals.

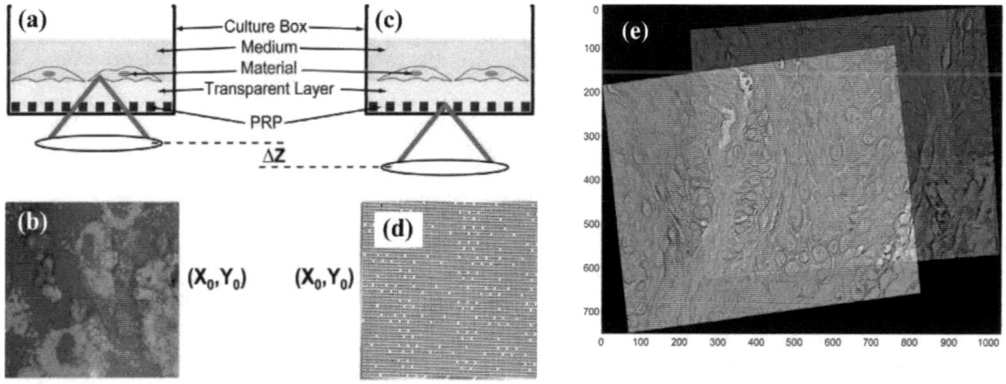

Figure 8. Position referenced microscopy : (a) Focused on the cells, (b) Confocal image of fibroblasts (red) in the process of absorbing apoptotic bodies (green), (c) Focused on the pseudo-periodic reference structure; (d) The pseudo periodic reference structure and (e) Alignment of two successive images following correction (P. Sandoz, J.A. Galéano, FEMTO-ST, Besançon, France)

5. Nanoscopy using nano-markers for nm measurement of position or structure

The fourth and last group of techniques considered is by using nano-markers to study nano-structures. Apart from "native-markers" such as microprecipitates decorating dislocations in crystals, mentioned in section 2.1, the use of "foreign-markers" such as fluorochromes and gold nano-particles are finding extensive use in the study of nano-structures.

5.1. Total Internal Reflection Microscopy (TIRF)

While different fluorochromes are used to target many different types of nano-structures today in fluorescence imaging, the contrast can be enhanced further by reducing the background intensity (I_{Min}) using total internal reflection at an interface (figure 9(a)), or TIRF (total internal reflection microscopy) [19].

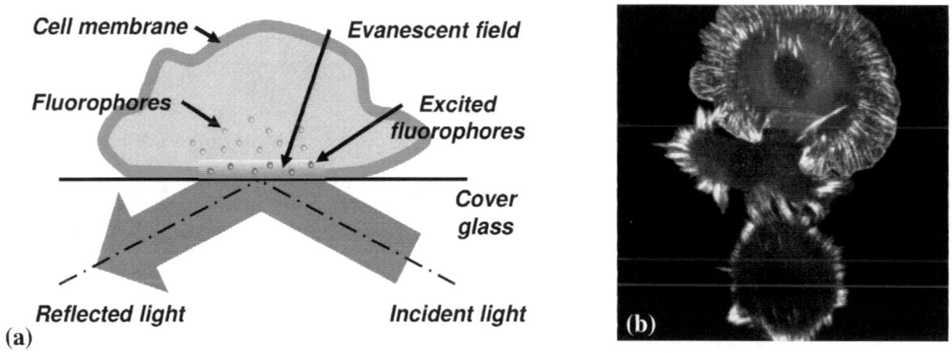

Figure 9. Total Internal Reflection Microscopy : (a) Layout for illumination in TIRF and (b) Study of membrane adhesion (© www.leica-microsystems.com)

The evanescent field excites only the fluorophores that are attached to structures within a depth of about 100 nm. Detailed images within the region of the cell membrane (figure 9(b)) are thus accessible, opening up the possibilities of studying dynamic processes such as cell membrane adhesion, or the movement of nanometric vesicles. The study of individual molecules is possible with stochastic techniques such as PALM (photo-activated localization microscopy) and STORM (stochastic optical reconstruction microscopy).

5.2. Gold nanoparticles (AuNP)

Finally, in place of fluorophores, nano-particles of gold (AuNP) can also be used as useful markers, becoming point sources that are not resolved but are observable. They can thus be used in combination with two photon microscopy to map out the nanometric structure of collagen fibres using single 5 nm AuNP that are moved along the fibre [20].

6. Conclusions

In this paper we have underlined the growing interest in a certain class of far field optical nanoscopy techniques that are limited by diffraction but nonetheless allow the detection, observation or measurement of different types of information of structures at a nanometric scale. Different techniques exist to reveal such information, such as by increasing the contrast, by measuring the phase, by using deconvolution techniques, or by employing nano-markers such as fluorophores or gold nano-particles. By keeping a wide field, it is thus possible to observe the distribution of nano-defects inside crystals, to measure nanometric surface roughness, to study and measure nano-layers, to perform nanometric sub-pixel metrology and to reveal nanometric structures and dynamic functions in living cells.

7. Acknowledgements

Thanks are extended to V. Vitkova (BAS, Sofia, Bulgaria), N. Médard (Nanolane, France), R. Serra (Phasics, Palaiseau, France), P. Sandoz (FEMTO-ST, Besançon, France), and J.M. Ellens (Leica Microsystems, Nanterre, France) for permission to use their material in this paper.

8. References

[1] Pecheva E, Montgomery P C, Montaner D and Pramatarova L 2007 *Langmuir* **23** 3912
[2] Montgomery P 1990 *Nanotechnology* **1** 54
[3] Fillard J P, Montgomery P C, Gall P, Castagne M and Bonnaffe J 1990 *J. Cryst. Growth* **103** 109
[4] Fillard J P 1996 *Near Field Optics and Nanoscopy* (Singapore: World Scientific) p 197
[5] Ma M, Nango N, Ogawa T, Watanabe M and Eguchi M 2000 *J. Crystal Growth* **208** 282
[6] Beversluis M R, Novotny L and Stranick S J 2006 *Optics Express* **14 7** 2650
[7] Weyher J L and Montgomery P C 1990 *J. Cryst. Growth* **106** 476
[8] Montgomery P C and Fillard J P 1989 *Electron. Lett.* **25** 2 89
[9] Genova J, Zheliaskova A, Vitkova V and Mitov M D 2009 *J. Optoelectronics & Advanced Materials* **11** 9 1222
[10] Aussere D and Valignat M P 2007 *Optics Express* **15** 13 8329
[11] Atilgan E and Ovryn B 2011 *Biomedical Optics Express* **2** 8 2417
[12] Montgomery P, Anstotz F, Montaner D, Pramatarova L and Pecheva E 2010 *J. Phys. Conf. Ser.* **253** 012017
[13] Montgomery P C and Weyher J 2007 L'érosion nanoscopique *Voir L'Invisible* ed ECRIN, (Sophia-Antipolis : Omniscience) p 60
[14] Ponpon J P, Montgomery P C, Sieskind M and Amann M 2000 *Appl. Surf. Sci.* **165** 233
[15] Benatmane A, Montgomery P C, Fogarassy E and Zahorsky D 2003 *Appl. Surf. Sci.* **208-209** 189
[16] Bon P, Maucor G, Wattelier B and Monneret S 2009 *Optics Express* **17** 15 13080
[17] Dockrey J W and Hendricks D 1989 *Proc. Int. Conf. on Integrated Circuit Metrology, Inspection and Process Control III (San Diego, USA, 19 July 1989) (Proc. SPIE* vol 1087*)* ed K M Monahan (Bellingham: SPIE) p 120
[18] Galeano J A, Sandoz P, Gaiffe E, Launay S, Robert L, Jacquot M, Hirchaud F, Pretet J L and Mougin C 2011 *Biomedical Optics Express* **2** 5 1307
[19] Chung E, Kim D and So P T C 2006 *Optics Letters* **31** 7 945
[20] Chen B, Estrada L, C Hellriegel C and Gratton E 2011 *Biomedical Optics Express* **2** 3 511

Optical thin film metrology for optoelectronics

Peter Petrik[1,2,3]

[1] Fraunhofer Institute for Integrated Systems and Device Technology, Schottkystrasse 10, 91058 Erlangen, Germany

[2] Research Centre for Natural Sciences - Institute for Technical Physics and Materials Science, Konkoly Thege Rd. 29-33, 1121 Budapest, Hungary

[3] Doctoral School of Molecular- and Nanotechnologies, Faculty of Information Technology, University of Pannonia, Egyetem u. 10, Veszprem, H-8200, Hungary

E-mail: peter.petrik@ttk.mta.hu

Abstract. The manufacturing of optoelectronic thin films is of key importance, because it underpins a significant number of industries. The aim of the European joint research project for optoelectronic thin film characterization (IND07) in the European Metrology Research Programme of EURAMET is to develop optical and X-ray metrologies for the assessment of quality as well as key parameters of relevant materials and layer systems. This work is intended to be a step towards the establishment of validated reference metrologies for the reliable characterization, and the development of calibrated reference samples with well-defined and controlled parameters. In a recent comprehensive study (including XPS, AES, GD-OES, GD-MS, SNMS, SIMS, Raman, SE, RBS, ERDA, GIXRD), Abou-Ras et al. (Microscopy and Microanalysis 17 [2011] 728) demonstrated that most characterization techniques have limitations and bottle-necks, and the agreement of the measurement results in terms of accurate, absolute values is not as perfect as one would expect. This paper focuses on optical characterization techniques, laying emphasis on hardware and model development, which determine the kind and number of parameters that can be measured, as well as their accuracy. Some examples will be discussed including optical techniques and materials for photovoltaics, biosensors and waveguides.

1. Introduction

Although the development of optoelectronic materials strongly depends on accurate and reliable metrologies, it has recently been shown that in spite of high sensitivities, large deviations (exceeding the error bars) can be obtained between the results of elemental depth profiling performed by different methods (see figure 1 from Ref. [1]). This fact puts emphasis on proper modeling, calibration, and verification for all metrologies. While the agreement between the profiles shown in figure 1 are far from being perfect, the situation is even worse taking into account that the total amount of Ga over the whole layer has been normalized to a reference value determined by X-ray fluorescence.

Most of the methods taking part in the study are summarized in Table 1. In the first group, reliable optical models have to be found for the proper depth profiling. In the second group, the sputter rate has to be calibrated - e.g. by reference measurements with ellipsometry. The last group using cross sections consists of the mostly direct methods. However, the agreement is not perfect in this case either. Though depth resolutions of 1 nm or even below (in case of ellipsometry) can be reached, lateral resolution is a problem for the first two groups of Table

Journal of Physics: Conference Series **398** (2012) 012002 doi:10.1088/1742-6596/398/1/012002

1. Even for ellipsometry, the best lateral resolutions are in the range of approximately 50 μm. The sensitivity is in most cases at or below 1 %. However, as the comparison in figure 1 reveals, the accuracy of (and agreement between) depth profiles is worse than these specified values for most of the methods.

From the huge field of metrologies optical techniques will be in the focus of this short review. Examples of development of instrumentation and modeling for photovoltaics and liquid crystal modulated waveguide sensorics will be presented.

Table 1. Measurement techniques for the determination of elemental depth distribution. The listed parameters are typical values used in the study of Abou-Ras et al. [1]. RBS: Rutherford Backscattering Spectrometry; ERDA: Elastic Recoil Detection Analysis; GIXRD: Grazing Incidence X-Ray Diffraction; SNMS: Sputtered Neutral Mass Spectrometry; SIMS: Secondary Ion Mass Spectrometry; XPS: X-Ray Photoelectron Spectrometry; AES: Auger Electron Spectrometry; GD-OES: Glow-Discharge Optical Emission Spectrometry; TEM-EDX: Energy-Dispersive X-Ray Spectrometry in a Transmission Electron Microscope.

Method	Depth resolution (nm)	Sensitivity (at. %)
Depth profiling by modeling		
Ellipsometry	1	0.2-2
RBS	10	1
ERDA	10	10^{-4}
GIXRD	100	1
Depth profiling by sputtering		
SNMS	1	0.05
SIMS	4	$10^{-7} - 10^{-3}$
XPS	1-10	0.1
AES	10	0.3
GD-OES	3-100	$10^{-5} - 10^{-3}$
Raman depth profiling	100	1
Depth profiling using cross section		
Scanning Auger	1	3
TEM-EDX	Specimen thickness	0.5
SEM-EDX	Few 100	3

2. Instrumentation

Probably the most significant research field within optoelectronics is photovoltaics. Low cost and effective manufacturing of solar panels requires large area processing, most of the steps being thin film deposition. Taking advantage of its sensitivity, speed and non-destructive manner, ellipsometry is an attractive tool for in line monitoring of thin film properties.

The research group of M. Fried at the MFA developed a patented mapping ellipsometry concept for large area thin film characterizations [2–6]. The basic idea is the use of a divergent light source. Figure 2 shows the construction (left), the schematics (middle) and the beam path (right) of the mapping ellipsometer. This way, a large area can be illuminated, limited only by the geometrical constraints of the deposition or monitoring chamber. In our application at the Center for Photovoltaics Innovation and Commercialization at the University of Toledo this area is several dm^2. This new version applies a spherical mirror to collect light to the detector.

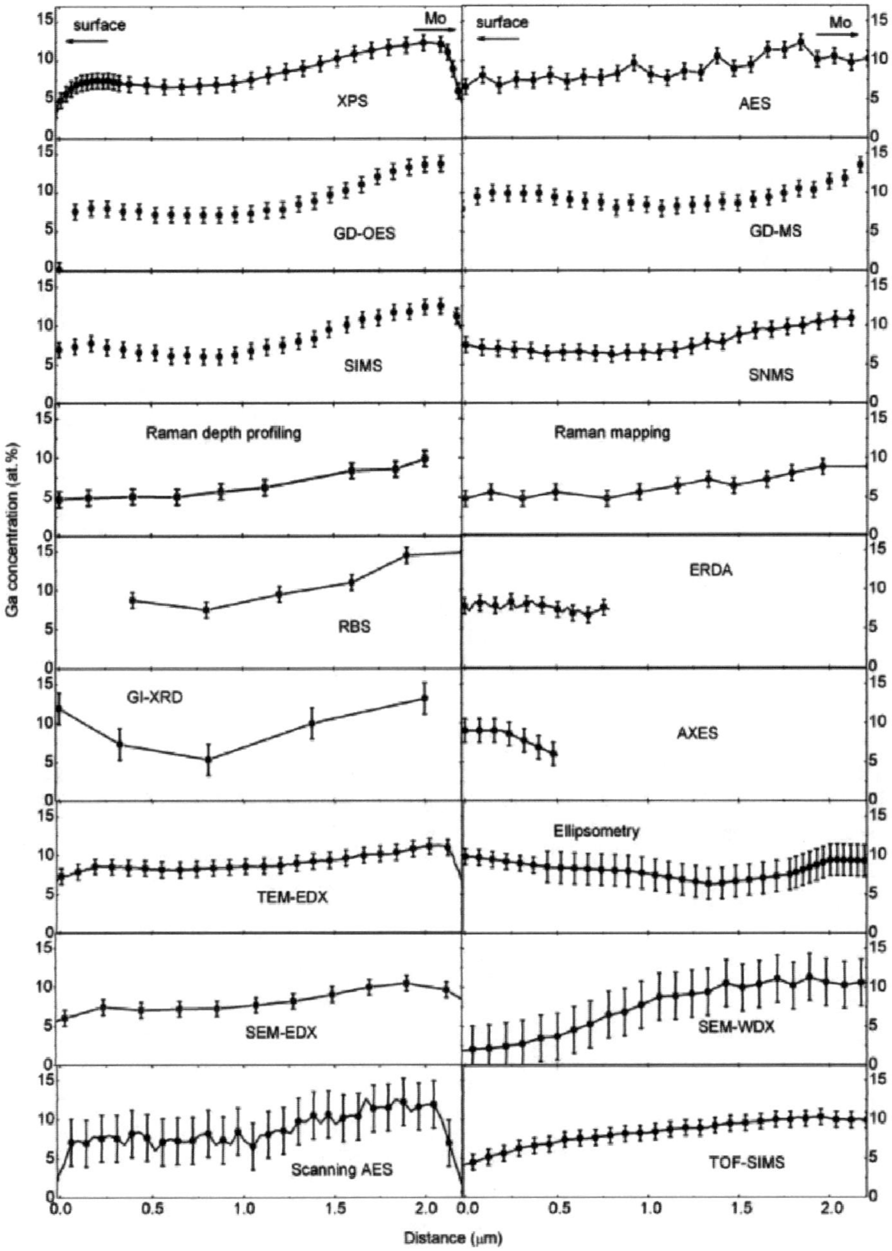

Figure 1. Ga distributions in $Cu(In,Ga)Se_2$ measured by different techniques. The surface is at distance zero. The lines are guides for the eyes. The error bars are estimated individually for each technique. Note that the differences between the different techniques are in a lot of cases larger than the error bars [1]. The used techniques are summarized in Table 1, except for GD-MS (Glow-Discharge Mass Spectrometry), AXES (Angle-Dependent Soft X-Ray Emission Spectroscopy), SEM-WDX (SEM-Wavelength-Dispersive X-ray Spectrometry), and TOF-SIMS (Time-of-Flight SIMS). [Reprinted from Microscopy and Microanalysis 17, Abou-Ras et al., Comprehensive comparison of various techniques for the analysis of elemental distributions in thin films, 728 (2011). Copyright (2011) Microscopy Society of America.]

Journal of Physics: Conference Series **398** (2012) 012002 doi:10.1088/1742-6596/398/1/012002

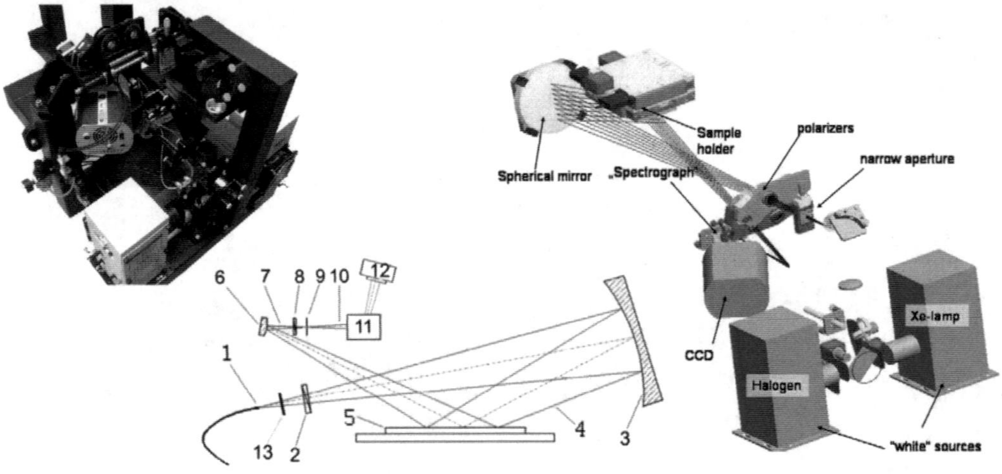

Figure 2. Expanded beam mapping ellipsometry. The prototype, located in the Center for Photovoltaics Innovation and Commercialization at University of Toledo (Ohio) can be seen on the left-hand side. The working principle and the main components are shown in the middle ([1] pont source; [2] polarizer; [3] spherical mirror; [4] non collimated beam; [5] sample; [6] cylindrical mirror; [7] corrected beam; [8] analyzer; [9] pinhole; [10] beam after pinhole; [11] corrector-disperser optics; [12] ccd detector; [13] rectangular (narrow) aperture), whereas the exploded drawing on the right-hand side shows the beam-guiding in the chamber and the main components of the arrangement (see [4]).

The image on the CCD is created by a pin hole. In this spectroscopic version one index of the CCD is the spectral information and the other index is the spatial information along a line. 2D images can be taken by moving the sample perpendicular to that line. Note that in a roll-to-roll production this movement is already part of the manufacturing process. This allows an easy integration of the tool as an in line process monitor [7].

Following the first demonstration of the concept [8], there has been a remarkable progress in the field of lable-free optical biosensor development [9–11]. These sensors utilize the fact that the propagation of light in waveguides sensitively depend on the refractive index difference at the waveguide surface (i.e. between the waveguide and the ambient - the latter being e.g. a protein solution). The light is coupled into the waveguide from the substrate, using a grating. The coupling angle sensitively depends on the above mentioned difference, which allows the detection of refractive index changes down to 10^{-5}. Consequently, also proteins and other objects adsorbed to the surface can be detected with high sensitivity. In the research group of R. Horvath a new sensor concept has been developed that avoids using moving parts by measuring the coupling angle (see e.g. Ref. [12]), but also makes the sensor highly integrable [13, 14]. A schematic drawing of the device is shown in figure 3. The tool utilizes two coupling gratings, whereas one of the light beams is modulated using a liquid crystal modulator. The phase shift sensitively depends on the refractive index, i.e. the number of molecules attached to the surface of the waveguide.

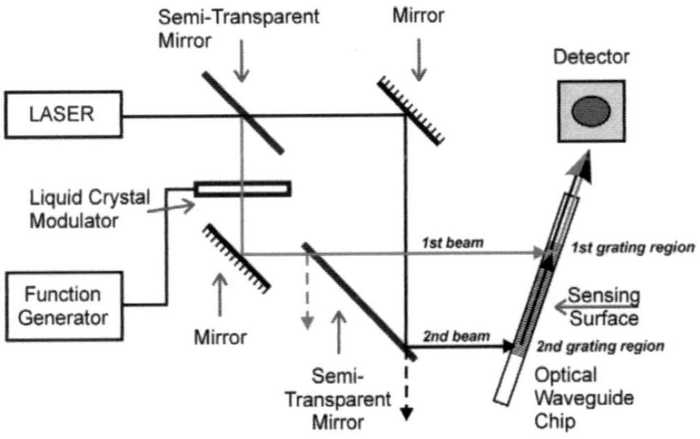

Figure 3. Grating coupled interferometer. The gray and black lines represent the reference and measurement arms of the interferometer, respectively [14]. [Reprinted from Applied Physics B 97, Kozma et al., Grating coupled interferometry for optical sensing, 5 (2009). Copyright (2009) Springer Verlag.]

3. Evaluation and modeling

The fairly large deviation between the metrologies shown in figure 1 is to a great extent a question of proper modeling. A large part of optical techniques is indirect in a sense that derived layer properties can only be determined if a suitable and reliable optical model exists for the structure. The test, whether a model is acceptable is not straightforward, although a couple of rules can be set to create reliable models (see Ref. [15] or page 2 of Ref. [16]).

The optical modeling of optoelectronic thin films is usually a complex problem for at least two reasons: (i) the electronic band structure and the related dielectric function is strongly dependent upon the preparation conditions, which rules out almost completely the use of reference dielectric functions, hence requires the application of analytical models; (ii) thin films created by the most usual (mostly deposition-related) techniques are only in exceptional cases uniform to the scale of precision of the most sensitive optical techniques like ellipsometry. In most cases at least the surface nanoroughness and an interface layer to the substrate have to be taken into account [16–19]. It can be shown that the optical modeling of one of the most frequently applied transparent conductive oxide (TCO) materials, Indium-Tin Oxide (ITO) is also a complex process [20–27], in which special care has to be taken for the modeling of vertical non-uniformity [28]. Most significant improvement was obtained by introducing a surface roughness layer as well as upper and bottom sublayers [28].

ZnO is also a key material for optoelectronic applications [29], used e.g. as TCO in photovoltaics to substitute ITO. The measurement of its optical properties is still a challenging task [30, 31]. A typical dielectric function of a low-resistance ZnO film is shown in figure 4 as a composition of different oscillators corresponding to excitonic and other electronic transitions [32]. The significance of the interpretation of optical spectra lies with the fact that the optical properties (especially at photon energies around the band gap) correlate with crucial electronic properties as e.g. the specific resistance [30, 33]. Figure 5 reveals a fairly good correlation between the optically measured exciton strength and the specific resistance. An even better correlation was shown by Hwang et al. between the band gap (also studied in Ref. [30]) and the carrier concentration for sputter deposited and annealed ZnO:Al films [33].

A possible strategy for the model-independent measurement of the dielectric function of ZnO

Journal of Physics: Conference Series **398** (2012) 012002 doi:10.1088/1742-6596/398/1/012002

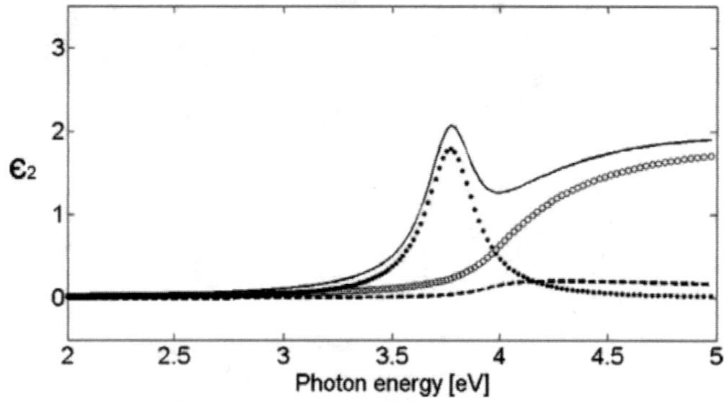

Figure 4. Typical dielectric function of a sputter-deposited ZnO:Al sample with low specific resistance from Ref. [30]. Solid line is the sum of the individual oscillators shown by the circled, dotted, and dashed lines. The dotted and dashed lines show the discrete-exciton and the continuum-exciton oscillators, respectively (see further details in Ref. [30]). [Reprinted from Applied Surface Science 255, Major et al., Optical and electrical characterization of aluminium doped ZnO layers, 8907 (2009). Copyright (2009) Elsevier.]

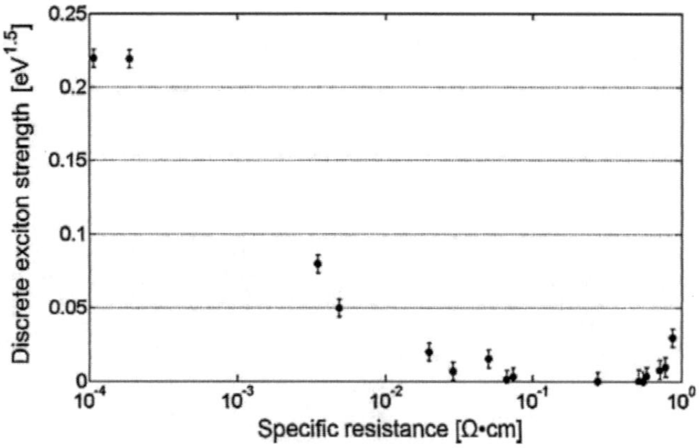

Figure 5. Discrete exciton strength parameter as the function of specific resistance [30]. [Reprinted from Applied Surface Science 255, Major et al., Optical and electrical characterization of aluminium doped ZnO layers, 8907 (2009). Copyright (2009) Elsevier.]

in a wide spectral range (as for most non-zero gap materials) is to determine the layer thickness by fitting the spectra in a relatively narrow transparent wavelength range. Under this conditions the Cauchy model can usually be applied, neglecting the absorption (i.e. only the real part of the refractive index will be fitted using a polynomial with fit parameters A, B, and C as follows: $n = A + B/\lambda^2 + C/\lambda^4$, where λ denotes the wavelength). As soon as the thickness is known, the real and imaginary parts of the refractive index (n and k) or dielectric function (ϵ_1 and ϵ_2) can be determined for each wavelength independently, because two ellipsometric angles (Ψ and Δ) are measured at each angle of incidence, whereas the number of unknown model parameters is only two: n and k. This leads to a model-independent point-by-point determination of the

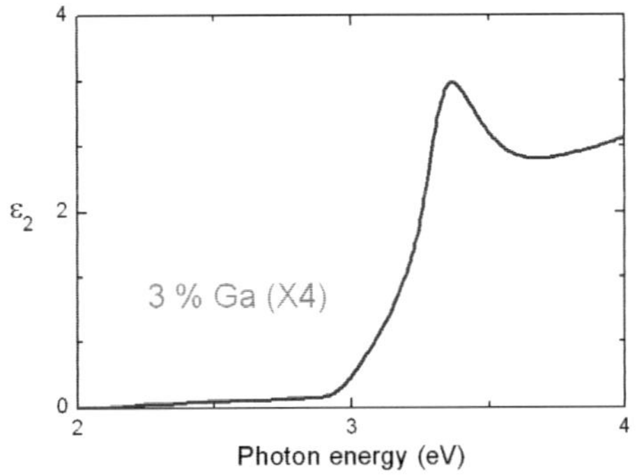

Figure 6. Imaginary part of the dielectric function of a sputter-deposited GaInZnO layer with nominally 3 and 6 % Ga and In contents, respectively, using point-by-point determination based on the layer thickness fitted in the transparent spectral region.

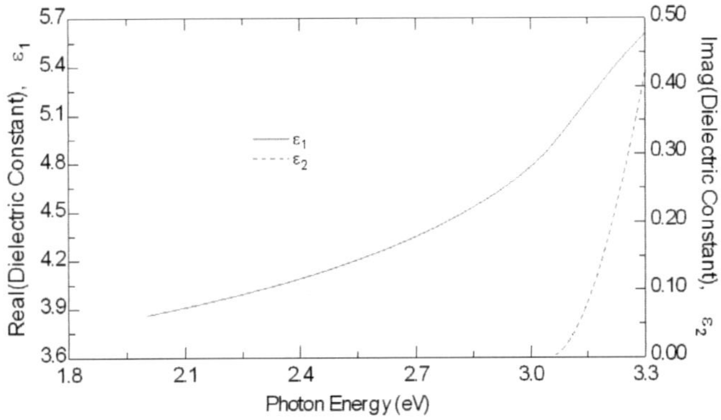

Figure 7. Typical dielectric function of a sputter-deposited GaInZnO film evaluated using the Tauc-Lorentz model.

dielectric function as shown in figure 6. Note the good agreement between the point-by-point (figure 6) and the oscillator (figure 4) approach. By checking the effect of whether the proper layer thickness was chosen, or most importantly, the effect of layer thickness to the results, the ϵ_2 spectrum of figure 6 was found to be stable. In case of Si oscillations may appear when using a wrong layer thickness [34].

To analyze the band gap region, (or more specifically: to obtain numerical gap parameters) the Tauc-Lorentz model can also be used, though in a limited spectral range (see figure 7). This model can be used to fit the decay of ϵ_2 at the band edge, assuming a quadratic behavior. The equation of the Tauc-Lorentz model can be obtained by multiplying the quadratic near-edge Tauc function [35, 36] by a Lorentz oscillator [37], which finally takes the form $k = A(E - E_g)^2/(E^2 - BE + C)$, where E denotes the photon energy, A, E_g, B, and C are the amplitude, gap, strength, and broadening parameters. From this model the energy of the band gap and the amplitude of the excitonic oscillator at the band egde can directly be determined.

Waveguides are basic building blocks of potential integrated optic amplifiers and lasers, also applied in biosensors mentioned above. Ion implantation is a tool with remarkable capabilities

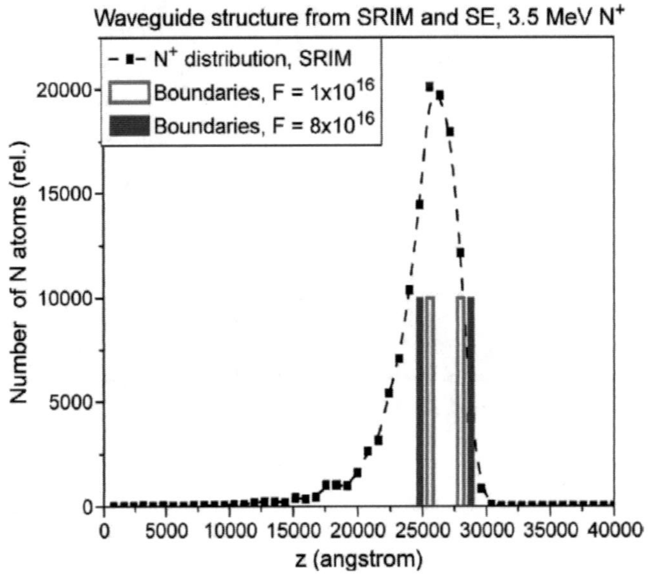

Figure 8. Depth position of the waveguiding layer determined by ellipsometry and SRIM (Stopping and Range of Ions in Matter) [39]. [Reprinted from IEEE Photonics Journal 4, Bányász et al., MeV Energy N^+-Implanted Planar Optical Waveguides in Er-Doped Tungsten-Tellurite Glass Operating at 1.55 μm, 721 (2012). Copyright (2012) IEEE.]

also for waveguide fabrication [38, 39]. Waveguides prepared by 3.5-MeV N^+ implantation into Er:Te glasses have been demonstrated by Banyasz et al. [39], characterized by M-line spectroscopy and ellipsometry. A fluence-dependent change of refractive index in the range of 0.04-0.09 have been revealed by ellipsometry in a waveguide layer approximately 2.5 micron below the implanted Er:Te glass surface. A three-layer model, consisting of (1) a surface nanoroughness layer, (2) a non-damaged surface layer (with a thickness of about 2.5 micron) and (3) a buried waveguide layer in the stopping region of the ions (as verified by the SRIM [Stopping and Range of Ions in Matter] program [40]) was used with Cauchy parametrizations in layers (2) and (3). The surface roughness layer was an effective medium mixture of the ambient and the underlying non-damaged Er:Te glass with a volume fraction fixed at 50 %. The position of the buried waveguide layer determined by ellipsometry is depicted in figure 8 together with the N^+ distribution calculated by SRIM.

4. Conclusions

It was shown in this brief review that modeling, calibration and verification is still a crucial issue for reliable metrology. Examples have been presented for the improvement of instrumentation and modeling for a couple of optical techniques and materials. The demand for the characterization of layers and structures with ever increasing complexity requires the construction of complicated optical models with numerous fit parameters. However, the importance of the verification of results has to be pointed out.

Acknowledgments
Support from the European Community's Seventh Framework Program, European Metrology Research Program (EMRP), ERA-NET Plus, under Grant Agreement No. 217257 (the EMRP is jointly funded by the EMRP participating countries within EURAMET and the European Union) as well as from the National Development Agency grant TMOP-4.2.2/B-10/1-2010-0025 and OTKA grant Nr. K81842 is greatly acknowledged.

References

[1] Abou-Ras D, Caballero R, Fischer C H, Kaufmann C A, Lauermann I, Mainz R, Mönig H, Schöpke A, Stephan C, Streeck C, Schorr S, Eicke A, Döbeli M, Gade B, Hinrichs J, Nunney T, Dijkstra H, Hoffmann V, Klemm D, Efimova V, Bergmaier A, Dollinger G, Wirth T, Unger W, Rockett A A, Perez-Rodriguez A, Alvarez-Garcia J, Izquierdo-Roca V, Schmid T, Choi P P, Müller M, Bertram F, Christen J, Khatri H, Collins R W, Marsillac S and Kötschau I 2011 *Microsc. Microanal.* **17** 728
[2] Juhasz G, Horvath Z, Major C, Petrik P, Polgar O and Fried M 2008 *Phys. Stat. Solidi C* **5** 1081
[3] Major C, Juhasz G, Petrik P, Horvath Z, Polgar O and Fried M 2009 *Vacuum* **84** 119
[4] Fried M, Juhasz G, Major C, Petrik P, Polgár O, Horváth Z and Nutsch A 2011 *MRS Proceedings* **1323** 157
[5] Fried M, Juhasz G, Major C, Nemeth A, Petrik P, Polgar O, Salupo C, Dahal L R and Collins R W 2011 *Thin Solid Films* **519** 2730
[6] Nemeth A, Attygalle D, Dahal L R, Aryal P, Huang Z, Salupo C, Petrik P, Juhasz G, Major C, Polgar O, Fried M, Pecz B and Collins R W 2011 *MRS Proceedings* **1321** 267
[7] Fried M, Juhasz G, Major C, Nemeth O P A, Petrik P, Dahal L R, Huang Z, Attygalle D, Shan A, Chen J, Podraza N J and Collins R W Fast imaging/mapping spectroscopic ellipsometry for large area samples unpublished
[8] Tiefenthaler K and Lukosz W 1985 *Thin Solid Films* **126** 205
[9] Ramsden J J 1993 *Phys. Rev. Lett.* **71** 295
[10] Horvath R, Pedersen H, Skivesen N, Selmeczi D and Larsen N B 2003 *Opt. Lett.* **28** 1233
[11] Horvath R, Cottier K, Pedersen H C and Ramsden J J 2008 *Biosens. Bioelectron.* **24** 799
[12] Horvath R, Skivesen N and Pedersen H C 2004 *Appl. Phys. Lett.* **84** 4044
[13] Cottier K patent published on September 18, 2008 Integrated optical sensor wO2008110026
[14] Kozma P, Hamori A, Cottier K, Kurunczi S and Horvath R 2009 *Appl. Phys. B* **97** 5
[15] Aspnes D E 1982 *Thin Solid Films* **89** 249
[16] Petrik P 2012 *Characterization of polysilicon thin films using in situ and ex situ spectroscopic ellipsometry* Ph.D. thesis Technical University of Budapest http://petrik.ellipsometry.hu/petrik_phd.pdf
[17] Petrik P, Lohner T, Fried M, Biró L P, Khánh N Q, Gyulai J, Lehnert W, Schneider C and Ryssel H 2000 *J. Appl. Phys.* **87** 1734
[18] Petrik P, Fried M, Vazsonyi E, Basa P, Lohner T, Kozma P and Makkai Z 2009 *J. Appl. Phys.* **105** 024908
[19] Petrik P 2012 *Nanocrystals - Synthesis, Characterization and Applications* ed Neralla S (http://dx.doi.org/10.5772/48732: InTech)
[20] Fukarek W and Kersten H 1994 *J. Vac. Sci. Technol. A* **12** 523
[21] Woollam J A, McGahan W A and Johs B 1994 *Thin Solid Films* **241** 44
[22] Synowicki R 2002 *Thin Solid Films* **20** 37
[23] Dobrowolski J, Li L and Hilfiker J 1999 *Applied Optics* **38** 4891
[24] Losurdo M, Giangregorio M, Capezzuto P, Bruno G, Rosa R D, Roca F, Summonte C, Plá J and Rizzoli R 2002 *J. Vac. Sci. Technol.* **20** 37
[25] Jung Y S 2004 *Thin Solid Films* **467** 36
[26] Kim J K, Chhajed S, Schubert M F, Schubert E F, Fischer A J, Crawford M H, Cho J, Kim H and Sone C 2008 *Advanced Materials* **20** 801
[27] Tejo-Cruz C, Mendoza-Galván A, López-Beltrán A M and Garcia-Jimenez M 2009 *Thin Solid Films* **517** 4615
[28] Lohner T, Kumar K J, Petrik P, Subrahmanyam A and Bársony I Analyses of optical thickness and refractive index of magnetron sputtered ito thin films: comparison of reflectometry and variable angle spectroscopic ellipsometry techniques unpublished
[29] Özgür U, Alivov Y I, Liu C, Teke A, Reshchikov M A, Dogan S, Avrutin V, Cho S J and Morkoc H 2005 *J. Appl. Phys.* **98** 041301
[30] Major C, Nemeth A, Radnoczi G, Czigany Z, Fried M, Labadi Z and Barsony I 2009 *Appl. Surf. Sci.* **255** 8907
[31] Gupta R K, Cavas M and Yakuphanoglu F 2012 *Spectrochimica Acta Part A: Molecular and Biomolecular Spectroscopy* **95** 107

[32] Yoshikawa H and Adachi S 1997 *Japanese Journal of Applied Physics* **36** 6237
[33] Hwang Y, Kim H and Um Y 2012 *Current Applied Physics* **currently in press**
[34] Aspnes D E, Studna A A and Kinsbron E 1984 *Phys. Rev. B* **29** 768
[35] Tauc J, Grigorovici R and Vancu A 1966 *Phys. Status Solidi* **15** 627
[36] Tauc J 1969 *Optical Properties of Solids* ed Nudelman S and Mitra S S (New York: Plenum)
[37] G E Jellison J and Modine F A 1996 *Appl. Phys. Lett.* **69** 371
[38] Banyasz I, Berneschi S, Khanh N Q, Lohner T, Fried M, Petrik P, Zolnai Z, Lengyel K, Peter A, Watterich A, Nunzi-Conti G, Pelli S and Righini G C 2010 *IOP Conference Series: Materials Science and Engineering* **15** 012027
[39] Banyasz I, Berneschi S, Brenci M B M, Fried M, Khanh N Q, Lohner T, Conti G N, Pelli S, Petrik P, Righini G C, Speghini A, Watterich A and Zolnai Z 2012 *IEEE Photonics Journal* **4** 721
[40] Ziegler J F 2004 *Nucl. Instr. Meth. Phys. Res. B* **219** 27 http://www.srim.org

Journal of Physics: Conference Series **398** (2012) 012003 doi:10.1088/1742-6596/398/1/012003

Common mistakes in luminescence analysis

Y Wang[1] and P D Townsend[2]
[1] School of Science, China University of Geosciences, Beijing, China 100083
[2] University of Sussex, Brighton, BN1 9QH, UK

E-mail: pdtownsend@gmail.com

Abstract. Luminescence techniques are powerful and sensitive probes to study imperfections, impurities and modifications of insulating materials. They are used in a wide range of disciplines from condensed matter physics to archaeology and mineralogy and the methods have developed over nearly a century. Early equipment was often not quantitative and data were collected in formats that were difficult to process and manipulate, and so signals were frequently presented in terms of the initial signals without corrections for equipment spectral sensitivity. Unfortunately not only did this distort the information but often it resulted in incorrect interpretations. Further, the incorrect data handling has persisted into modern usage both by physicists and those in other fields who merely use luminescence as a sensitive technique. Several main types of problem are considered. These include temperature errors in thermoluminescence dosimetry; subtleties in the signal intensity corrections for the responses of both the spectrometer and detectors; grating polarization effects; sample anisotropy; and common errors in spectral deconvolution, especially failure to transform from wavelength to energy plots.

1. Introduction

Luminescence signals provide information on relaxation processes in both inorganic and biological materials and the photon energy of the transition is primarily defined by the electronic structure around the emission site. Subtle variations in the structure will then influence the transition energy and excited state lifetime. Such variations have been successfully used to track changes over volumes as large as 50 neighbouring shells. Luminescence is therefore a powerful probe of defect structures in insulators, as well as responding to impurities, phase changes and distortions such as those caused by stress or nanoparticle inclusions. Equally, the changes induced by local distortion make luminescence a useful probe to distinguish between healthy and diseased biological material and it is used in the emerging field of Optical Biopsy.

Luminescence signals have been studied for more than a century with techniques from simple visual observation to black and white or colour photography. By the 1960s more quantitative detectors, such as photomultiplier tubes became available. However the spectral information was generally displayed on a chart recording as a monochromator swept through the wavelength range of the system. Such raw data were then published and used as the basis for discussion and interpretation of the luminescence. For some applications this was acceptable, for example in mineralogical applications changes in spectra revealed different component materials and the presence of rare earth ions were easily identified by characteristic line spectra. It must be recalled that in the 1960s attempts to remove the background dark current, and then scale the signals to correct for the transmission efficiency of the monochromator and sensitivity of the PM tube involved tedious manual processing (N.B. on line computer processing was not available).

The sensitivity of luminescence was still attractive and widely used both by physicists and those from other disciplines who were expert in their own fields, but not necessarily in luminescence. In this context there are many examples of the problems from those based in semiconductor science and materials modifications. This situation has unfortunately resulted in many errors in signal processing and discussion which have persisted to the present day. In part this is because data are compared with earlier work and in part because different communities of research areas handle signals in similar manners, often with identical recording equipment. Overall, systematic errors have become established. Three examples of these problems will now be discussed.

2. Thermoluminescence dosimetry

Radiation dosimetry evolved in the 1950s for health reasons for those working in the nuclear industries. Rather than use photographic film badges it was found that many materials (such as LiF with added impurities) could both respond similarly to body tissue and also offer an extremely sensitive and reliable radiation dosimeter. The science of thermoluminescence (TL) is that in an insulating material exposure to ionising radiation releases electrons which can become trapped at impurity/defect sites. Subsequently the material can be heated and during the heating the electrons are released, move through the material to more stable sites, and finally emit luminescence as they drop into the lower energy states. For dosimetry this is excellent as the total emitted signal is proportional to the original irradiation dose. Radiation dose measurements by TL are valuable not only for personnel dosimetry but also for archaeological dating of materials which had been heated at some time in their history (i.e. to offer a time zero). Consequently dosimetry signals from pottery can be linked to age, and the information is helpful both for historic reasons and for sensing fakes or later copies of a particular style of ceramic.

Experimentally the signal rises through a peak (called a glow curve) as the rate of charge release increases exponentially with sample temperature, but the signal then falls as the number of trapped charges become exhausted [1]. The peak temperature Tm is related to the activation energy or trapping (E) and the attempt to escape frequency (υ), with other factors influencing the detailed shape of the glow peak. In order to see a weak signal against the background noise of the PM tube detector it became fashionable to use very high heating rates. The temperature that was measured was taken with a thermocouple on a heater strip and this was controlled to give a rapid linear heating ramp. The dosimeters are by definition insulating materials, and hence poor thermal conductors. This means there is a changing temperature gradient dT between the heater strip and the emitting face of the dosimeter. For *dosimetry* this error between the true sample temperature and the heater is irrelevant as one is only concerned with the integrated signal. Further, most research groups tend to use very similar equipment so there is the same systematic error between different workers in the field and there is no confusion in discussions within the community.

The dT problem becomes apparent once one attempts to move to a more scientific analysis of the data, as the error dT will change with heating rate (β) and the entire curve shape is distorted. Therefore the dosimetry community may discuss a feature at say 250°C but, as recorded, their estimates of E and υ will be wrong, as will be temperature of the glow peak. More critically is that comparison of a glow peak temperature with literature data obtained via other techniques, such as isothermal annealing of absorption or ESR signals, will not match. Hence neither the TL experts, nor those in other fields, will realise they are discussing the same process. Dosimetry heating rates can run to more than 100 degrees per second so the dT values are very significant. Figure 1 indicates the scale of the errors that can accrue. It contrasts the glow peak for a theoretical glow curve, where the true surface temperature is rising at 50 degrees per second, with the measured curve where the signals are plotted in terms of the heater temperature and there is a typical thermal temperature lag developing. Even for this modest heating rate the recorded signal is wrong by some 50 degrees.

As already mentioned, this systematic error is not critical for dosimetry, but it is a major error in terms of modelling of the processes. Nevertheless many journal articles continue to appear with modelling based on the heater temperature, despite a number of publications which have given detailed indications of the errors involved, or empirical suggestions as to how to minimise them [2-5].

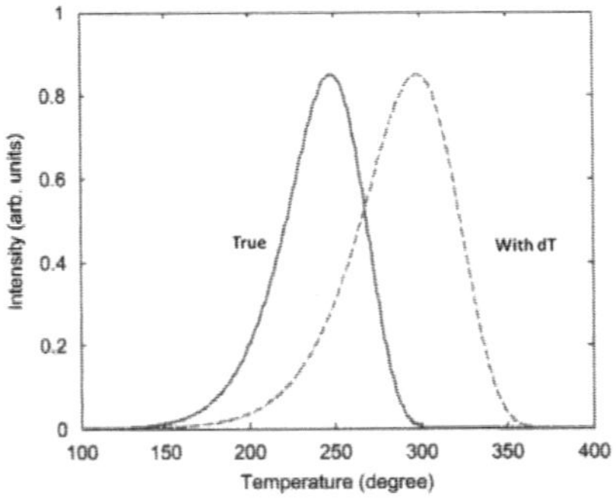

Figure 1. The solid line shows a true glow curve of a sample heated at 50 degrees per second and the dashed line is typical of recorded data based on the thermocouple temperature of the heater strip. The thermal gradient across the insulating sample introduces a variable lag of dT which is ~50 degrees at the peak value in this example.

3. Errors in corrections of the instrument sensitivity

3.1 Diffraction grating effects

The initial step in correcting the spectral data is to remove the dark current or other sample independent signals that are produced by the detector. Signal to noise can be excellent for strong signals and PM tubes but the linear dynamic range of CCD detectors is often far less than the range for the PM tubes. The next step is to correct the measured spectral intensity for distortions introduced by the transmission characteristics of any optics, filters and polarizers etc in the system. Polarizers are often used in data recorded from anisotropic crystals but their spectral response is rarely flat across the spectra. Similarly with diffraction grating monochromators (and spectrometers) one must avoid the problems of second or third order diffraction and so blocking filters are required at longer wavelengths to remove higher order spectra. Note for a grating the dispersion is given by $n\lambda = 2d \sin\theta$ so at a nominal setting of say 800 nm the unfiltered signal will contain not only 800 nm light but also second order from 400 nm and third order from 266 nm etc. Since the grating performance is far better at 400 nm than at 800 nm (as seen in figure 2). Similarly with PM tubes the efficiency of detecting the second order 400 nm signal will be superior to that of detection of the real 800 nm signal. In one sensitive system used in Sussex the overall efficiency at 400 nm was 200 times greater than at 800 nm so the second order terms swamped the true signals unless filters were added.

For steady persistent signals there is no difficulty in acquiring spectra with and without the second order filters but in measurement of transient luminescence signals, or those that are time dependent, this is a serious problem. In one of the Sussex systems designed to record the emission spectra during thermoluminescence the problem was avoided by the use of a pair of diffraction grating spectrometers, each with its own photon imaging PM tube [6]. This has an additional benefit that the red region could use a diffraction grating blazed for long wavelengths (i.e. giving higher efficiency) and a second order filter was permanently installed on the red system.

Intensity correction of the signal requires knowledge of the spectral response of both the monochromator and detector. Often the calibration is made for the package but, as will be mentioned, this can be deceptive if the signals are polarized. Typical grating sensitivities are sketched in figure 2 [7, 8]. The transmission efficiency is strongly dependent on whether the light is polarized parallel to the rulings on the grating (P type) or normal to them (S type). For nominally unpolarized light the net sensitivity is a mixture of the two responses. For any system that is intended to record the full spectral range from say 200 to 900 nm the peak will always be near 400 nm, and sensitivity falls steeply in the UV and less rapidly at the red end of the spectrum.

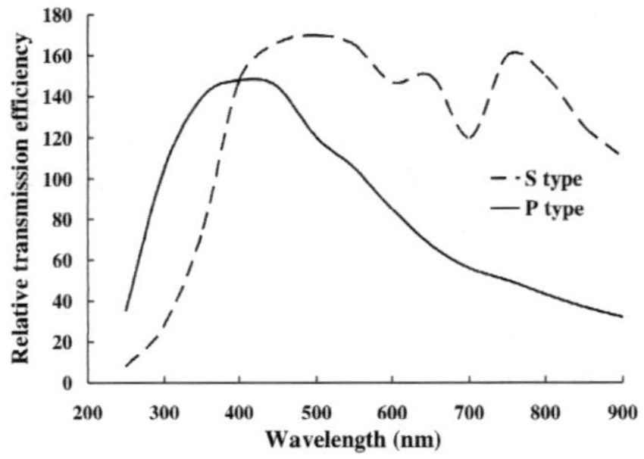

Figure 2. Relative transmission efficiency of a diffraction grating spectrometer for light polarized parallel (P) and perpendicular (S) to the grating lines.

For luminescence studies of anisotropic crystalline material figure 2 indicates there is an unexpected problem. If the different emission bands have a polarized component then across the range of the spectrum the transmission characteristics differ by a factor of 8. This strongly distorts the relative intensities of the components. Some experimentalists introduce a polarizer in front of the spectrometer to select one view of the emission, but because in many configurations (e.g. in a cryostat) it is not possible to rotate the sample. The temptation is to rotate the polarizer by 90 degrees to record the other polarized view. Quite clearly this gives a completely different response and generates false data. Correct solutions would be to select a polarizer orientation, and have a sensitivity calibration for this setting, and then rotate the sample. A simpler option is to choose the polarizer orientation relative to the sample and use an optical fibre link which randomises the polarized signal before it enters the spectrometer. This is preferable as the same grating response applies to each polarizer setting as well as for an unpolarized input.

An additional word of caution is that for some anisotropic materials the emission not only depends on the face that is viewed but also on the axis along which it has been excited. Examples of such phenomena exist for ion implanted crystals.

3.2 Spectral response of the detector

For a wide spectral coverage of luminescence the preferred PM tube has invariably been a multi-alkali photocathode (termed S20). The response of such tubes can be quite variable at the long wavelength end of the spectrum so they need individual calibration [9, 10]. They may also vary with age and, more critically, with exposure to light. The sensitivity changes are apparent in variations in the dark current and in monochromatic systems the red response can be enhanced by exposure to short wavelength light. Hence a spectral scan from short to long wavelengths can differ from scans made in the reverse direction. In recent years various ways of coupling the light to the PM tube have offered enhanced red performance [10-13]. Some of these same enhancements can be achieved with S20 cathode photon imaging tubes as are used in spectrometer systems.

For spectrometer systems there are geometric advantages in using CCD detector arrays, not least because they have greater red coverage. Once again they are very variable and depend on the manufacturer, whether or not they are designed to be illuminated on the front or rear face, and features such as anti-reflective coatings to enhance performance. In general they have a limited dynamic range, so caution is needed if the spectrum has a mixture of weak and strong signals. Coated, or thinned detectors, may display thin film interference effects which can appear as real sample data. This is problematic if the samples are thin films which may also generate signal oscillations with wavelength. For time resolved measurements of luminescence the CCD response is considerably slower than from a PM tube and PM tubes are compatible with lock-in amplifier detection.

The main feature to note is that it is essential to derive a spectral calibration curve for the entire system in order to correctly process the raw data. Over a wide spectral range this implies the use of a detection system which includes long wavelength data that have been filtered to remove high order spectral overlap with the true long wavelength signal. Not all systems automatically include this feature and for equipment used as a general tool (e.g. not by specialists in luminescence) this can (and does) generate misleading data.

4. Errors in data processing

The ubiquitous use of diffraction gratings means that the emission spectra are obtained as intensity as a function of wavelength. The monochromator has a fixed bandwidth entrance slit so the collected data after processing and correcting for the system response appear as a display of intensity versus wavelength $\{I(\lambda)d\lambda$ versus $\lambda\}$. If there are many overlapping feature then some peak deconvolution is necessary. Unfortunately many people using luminescence fail to recognise that such deconvolution cannot be made with the wavelength data. Instead the signals *must* be transformed into an energy plot of $\{I(E)dE$ versus $E\}$. There are basically two common errors in this attempt at signal deconvolution. The first is to feed the wavelength data into a curve fitting package but unfortunately the analysis is not just incorrect in terms of the physics, but it is totally misleading. The second incorrect approach is to transform only the wavelength axis into photon energy ($E = hc/\lambda$). Such a transformation is immediately apparent as the intensities of the peaks are the same as in the wavelength view of the signals. The correct route requires adjustment of the intensity data from a fixed $d\lambda$ to a fixed dE. Mathematically this is trivial as $dE = -hc/\lambda^2$. However the λ^2 term has a major influence on the intensities. Less obvious is that the energy of the peak taken from a wavelength plot does not match the peak energy of the correctly transformed data. More obvious is that the intensity changes can strongly influence our discussions as to which are the major features.

The normal assumption is that luminescence processes of emission sites in solids result in Gaussian or Lorentzian line shapes [14-16]. Figure 3 therefore shows how a set of symmetric Gaussian bands (in energy) would have appeared on the wavelength version of the data. In wavelength plots the peak shifts as a function of full width half maximum and the value is only correct as the bandwidth approaches zero. This is of course why such errors are not obvious when recording line spectra, as from rare earth ions. The current example has considered a red emission band centred at 2.0eV, since for the longer wavelength signals one can sense more clearly why the apparent wavelength data are so sensitive to bandwidth. In this example the change from a narrow to a wide FWHM shifts the wavelength peak by nearly 200 nm!

To emphasise the scale of the changes that have occurred along the processing route figure 4 displays the data sets that would have been recorded by a grating spectrometer with either a red sensitive PM tube or two different types of CCD detector. Figure 4 is the view that would appear before the signals were corrected for the responses of the collection systems. Whilst it is evident that with UV sensitive detectors there are probably 4 emission bands present, it is very tempting to assume that with the CCD detector there are probably some minor bands in addition to the 4 main ones. It is also clear that the PM and CCD do not define the peak values at the same wavelengths. Certainly one could not assume that after the spectrometer corrections and a transformation into an energy plot the spectrum is actually comprised of four Gaussian features of equal intensity of equal bandwidth (0.8 eV) centred at 1.50. 2.03, 4.16 and 5.50 eV. In true wavelength terms these peaks are near 826, 438, 299 and 225 nm.

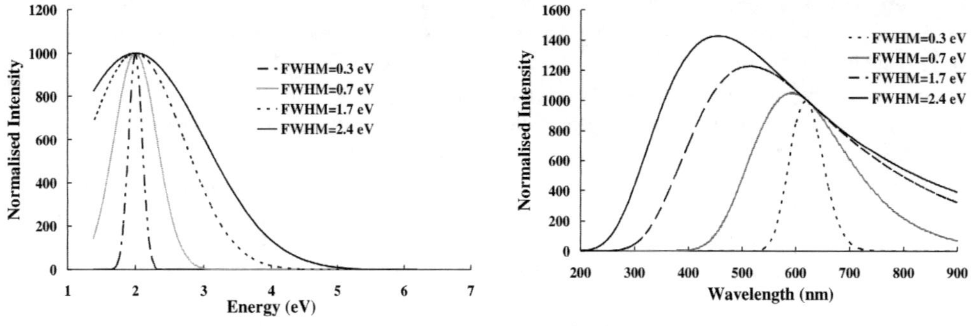

Figure 3. Equivalent wavelength and energy plots of a set of Gaussian luminescence bands centred at the same photon energy but different FWHM values

To emphasise the scale of the changes that have occurred along the processing route figure 4 displays the data sets that would have been recorded by a grating spectrometer with either a red sensitive PM tube or two different types of CCD detector. Figure 4 is the view that would appear before the signals were corrected for the responses of the collection systems. Whilst it is evident that with UV sensitive detectors there are probably 4 emission bands present, it is very tempting to assume that with the CCD detector there are probably some minor bands in addition to the 4 main ones. It is also clear that the PM and CCD do not define the peak values at the same wavelengths. Certainly one could not assume that after the spectrometer corrections and a transformation into an energy plot the spectrum is actually comprised of four Gaussian features of equal intensity of equal bandwidth (0.8 eV) centred at 1.50. 2.03, 4.16 and 5.50 eV. In true wavelength terms these peaks are near 826, 438, 299 and 225 nm.

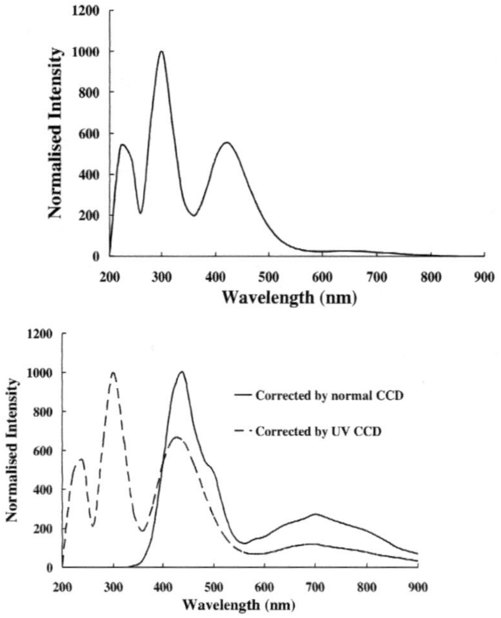

Figure 4. The spectra that would initially be collected by a spectrometer and PM tube (left) and the same signal as sensed by two different types of CCD detector (right). Note that after spectrometer correction and

transformation to an energy plot there are just 4 Gaussian emission bands of equal intensity and bandwidth.

5. Final comments

The examples presented here emphasise that because of the long history of luminescence techniques much of the early literature did not take into account all the factors which cause signal distortion as the data were collected. Further, many of the users of luminescence are experts in different disciplines and so the luminescence technology is used merely as a tool without a full appreciation of the problems. Overall this has resulted in a literature in which many signal processing errors are either not appreciated or not mentioned in the publications. It is then difficult to guess whether system corrections have been applied and, for energy plots, if both intensity and energy axes have been transformed from the wavelength recorded data. Unfortunately the current literature of luminescence data, even in physics based areas such as semiconductor physics, is rarely explicit in the type of corrections which have been made, and the trend is to show and discuss data presented on a wavelength axis. The reader must then guess which, or if, detector corrections were applied. For the examples used in this article it is apparent that such problems lead to errors in the data collection as well as subsequent signal processing and interpretation. One ongoing problem is that publications often show data in ways which match, or can be compared with earlier articles. There is thus inertia in attempting to move all the luminescence literature towards clearly stated and properly corrected signals.

References

[1] McKeever SWS 1985 *Thermoluminescence of solids* (Cambridge: Cambridge University Press)
[2] Betts D S, Couturier L, Khayrat A H, Luff B J and Townsend P D 1993 *J. Phys. D: Appl. Phys* **26** 843
[3] Kitis G and Tuyn J W N 1998 J. Phys. D: Appl. Phys **31** 2065; ibid 1999 *Rad. Prot. Dosim.* **84** 371
[4] Stadtmann H, Delgado A and Gómez-Ros J M 2002 *Radiation Protection Dosimetry* **101** 141
[5] Ege A, Wang Y and Townsend P D 2007 *Nucl Insts and Methods* A **576** 411
[6] Luff B J and Townsend P D 1993 *Meas. Sci Tech* **4** 65
[7] Palmer C and Loewen E 2005 *Diffraction Grating Handboo*k (New York: Newport Corp)
[8] Ditchburn R W 1967 *Light Chapt 6* (London: Blackie)
[9] Marmonier C 2002 *Photomultiplier Tubes - principles and applications* (Brive: Photonis)
[10] Townsend P D 2003 *Contemporary Physics* **44** 17
[11] Townsend P D, Downey R, Harmer S W, Wang Y, Cormack A, McAlpine R and Bauer T 2006 *J. Phys. D: Appl. Phys* **39** 1525
[12] Townsend PD, Valberg L, Momchilov N, Harmer SW, Downey R and Cormack AJ 2008 *J. Phys. D: Appl. Phys* **41** 185504
[13] Harmer SW and Townsend PD 2010 *J. Phys. D: Appl.Phys.* **4**, 415101
[14] Curie D 1963 *Luminescence in crystals* (London: Methuen)
[15] Henderson B, Wertz JE 1968 *Advances in Physics* **17** 749
[16] Lubchenko AF 1956 *Ukr Fiz Zh.* **1** 265

On the orientational ordering of water and finite size of molecules in the mean-field description of the electric double layer - a mini review

A Iglič[1,2] **and E Gongadze**[1,2]

[1] Laboratory of Biophysics, Faculty of Electrical Engineering, University of Ljubljana, Tržaška 25, SI-1000 Ljubljana, Slovenia

[2] Laboratory of Clinical Biophysics, Chair of Orthopaedics, Faculty of Medicine, University of Ljubljana, Zaloška 9, SI-1000 Ljubljana, Slovenia

E-mail: `ales.iglic@fe.uni-lj.si`

Abstract. The distribution of ions in an electrolyte solution close to a charged surface is determined by competition between electrostatic interactions and the entropy of the ions. The diffuse electric double layer (EDL) created influences the overall electrostatic interaction of the charged surface with its environment. Different approaches were introduced through the years to overcome some of the limitations of classical EDL Poisson-Boltzmann models and to better capture the complex phenomena of the EDL. The core of these models is the spatial variation of the relative permittivity, achieved by taking into account the spatial and orientational distribution of water molecules and the excluded volume effect.

1. Introduction

The contact between a charged surface and an electrolyte solution implies a characteristic spatial and orientational distribution of ions and water molecules which is reflected in the formation of the electric double layer (EDL) [1–6]. Study of the electric double layer began in 1879 with Hermann von Helmholtz, but today it still remains a scientific challenge. Helmholtz treated the double layer as a simple capacitor, assuming that the surface charge density is neutralized by the counterions located at a distance equal to their hydrated radius. Gouy [1] and Chapman [1, 2] considered the thermal motion of ions and pictured a diffuse double layer composed of ions of opposite charge (counterions) attracted to the surface, and ions of the same charge (co-ions) repelled from it. Ions are embedded in a dielectric continuum while the electric potential is subject to the Poisson-Boltzmann (PB) differential equation. Within the so-called Poisson-Boltzmann theory [1–3], the ions in an electrolyte solution are treated as dimensionless and a uniform permittivity of the electrolyte solution is assumed. Surfaces are considered to be uniformly charged (see also Refs. [3–8]).

The Stern model [9] was the first attempt to incorporate the finite size of ions in EDL theory by combining the Helmholtz [10] and Gouy-Chapman [1, 2] models Furthermore, in 1942, Bikermann introduced the first complete modified Poisson-Boltzmann (modified Gouy-Chapman) model with consideration of steric effects in the electrolyte phase [11]. The tendency was continued by Grimley and Mott [12, 13], and Freise [14] who included the excluded volume effect with a pressure-dependent potential. Wicke and Eigen [15] used a thermodynamic approach, multiplying the numerical density of ions by a factor containing the number of vacant sites. Also more recently, the finite size of particles has been incorporated into the electric double layer theory in different ways [16], for example by using a lattice statistics model [17] or by including surface charge correlations where ions and solvent molecules were treated as hard spheres [5]. The properties of the electric double layer may thus be strongly influenced by the ordering of water molecules [7, 8, 18–22] and their depletion in the region of the electric double layer [23]. Outhwaite developed a modified Poisson-Boltzmann (PB) theory of the electrical double layer composed of a mixture of hard spheres with point dipoles and finite sized ions. The Langevin expression for the permittivity appeared as the first closure, while the second closure led to the Onsager expression for the relative permittivity [18].

It was shown that close to the charged surface, water dipole ordering in the saturation regime and depletion of water molecules may result in a strong local decrease of permittivity [20, 21, 23, 24].

In this work EDL mean-field theories of electrolyte solution in contact with a charged surface are reviewed. It is indicated that consideration of the orientational ordering of water molecules near a highly charged (metal) surface, the finite size of molecules, electronic polarizability and cavity field of water molecules is essential to better describe the physical properties of the EDL at high values of the surface charge density [23].

2. Gongadze-Iglič model

The effective dipole moment of a water molecule should be known before a satisfactory statistical mechanical study of water and aqueous solutions is possible [25–27]. The dipole moment of a water molecule in liquid water differs from that of an isolated water molecule because each water molecule is further polarized and orientationally perturbed by the surrounding water molecules [25–28].

In the past treatment of the cavity and the correlations between water dipoles in the Onsager [29], Kirkwood [30] and Fröhlich models [25] were limited to the case of small electric field strengths. Generalization of the Kirkwood-Onsager-Fröhlich theory in the saturation regime was performed later by Booth [26]. However, Booth's model does *not consider the excluded volume effect* in an electrolyte solution near a charged surface.

In order to develop an integrating framework to clarify the factors influencing the relative permittivity, in this review a generalisation of the Langevin Poisson-Boltzmann (LPB) model [22] is presented. The model takes into account the cavity field [25] (but not the structural correlations between water dipoles), as well as the finite size of ions [8, 23] in the saturation regime (an important consideration for an electrolyte solution in contact with a highly charged surface).

In the model electronic polarization is taken into account by assuming that the point-like rigid (permanent) dipole is embedded in the centre of a sphere with a volume equal to the average volume of a water molecule in the electrolyte solution (figure 1). The permittivity of the sphere is taken to be n^2, where $n = 1.33$ is the optical refractive index of water. The relative (effective) permittivity of the electrolyte solution (ε_r) is then expressed as:

$$\varepsilon_r(\mathbf{r}) = n^2 + \frac{|\mathbf{P}|}{\varepsilon_0 E}, \tag{1}$$

where $\mathbf{P}$ is the polarization vector due to the net orientation of permanent point-like water dipoles having dipole moment $\mathbf{p}$. The external dipole moment ($\mathbf{p}_e$) of a point-like dipole at the center of the sphere with permittivity n^2 can then be expressed in the form [25] $\mathbf{p}_e = 3\mathbf{p}/(2 + n^2)$ whence it follows:

$$\mathbf{p} = \frac{2 + n^2}{3}\,\mathbf{p}_e \ . \tag{2}$$

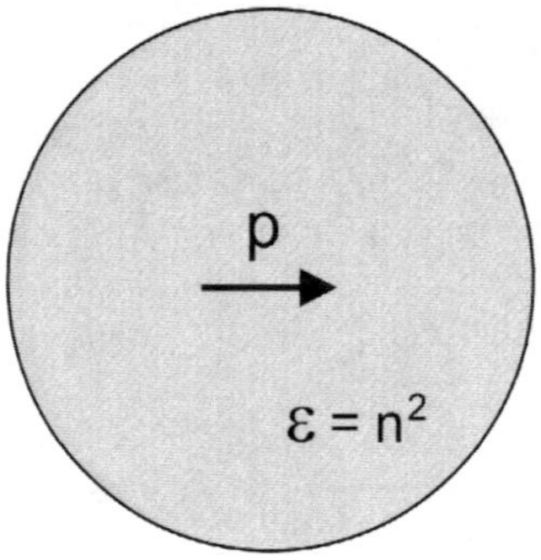

Figure 1. In the model a single water molecule is considered as a sphere with permittivity $n^2 = 1.33^2$ and a point-like rigid (permanent) dipole with dipole moment $\mathbf{p}$ at the centre of the sphere. Here $n = 1.33$ is the optical refractive index of water.

The short range interactions between dipoles are neglected. The local electric field strength at the centre of the sphere at the location of the permanent (rigid) point-like dipole (figure 2) is [25] :

$$\mathbf{E}_c = \frac{3\,\varepsilon_r}{2\,\varepsilon_r + n^2}\,\mathbf{E} \ , \tag{3}$$

where the expression represents the field inside a spherical cavity with dielectric permittivity n^2 embedded in a medium with permittivity ε_r. In the following, Eq. (3) is simplified to the form (strictly valid for $\varepsilon_r \gg n^2$ only): $\mathbf{E}_c = \frac{3}{2}\,\mathbf{E}$. The energy of the point-like dipole $\mathbf{p}$ in the local field $\mathbf{E}_c$ may then be written as:

$$W_i = -\mathbf{p}\cdot\mathbf{E}_c = \gamma\, p_o\, E\cos(\omega) \ , \tag{4}$$

where p_0 is the magnitude of the dipole moment $\mathbf{p}_e$, ω is the angle between the dipole moment vector $\mathbf{p}$ and the vector $-\mathbf{E}$ and γ is [23]:

$$\gamma = \frac{3}{2}\left(\frac{2+n^2}{3}\right) \ . \tag{5}$$

The polarization $P(x) = n_w(x)\,\left\langle \mathbf{p}(x,\omega)\right\rangle_B$ is given by [23]:

$$P(x) = n_w(x)\left(\frac{2+n^2}{3}\right) p_0 \left\langle \cos(\omega)\right\rangle_B = -\,n_w(x)\left(\frac{2+n^2}{3}\right) p_0\, \mathcal{L}\left(\gamma p_0 E\beta\right) \ , \tag{6}$$

where we took into account the well known result (see for example [25]):

$$\left\langle \cos(\omega)\right\rangle_B = \frac{\int_0^\pi \cos\omega\, \mathcal{P}(x,\omega)\, 2\,\pi\, \sin\omega d\omega}{\int_0^\pi \mathcal{P}(x,\,\omega)\, 2\,\pi\, \sin\omega d\omega} = -\,\mathcal{L}\left(\gamma p_0 E\beta\right) \ , \tag{7}$$

and $\beta = 1/kT$, where kT is the thermal energy. Since $\sigma < 0$, the projection of the polarization vector $\mathbf{P}$ on the x-axis points in the direction from the bulk to the charged surface and $P(x)$ is considered negative.

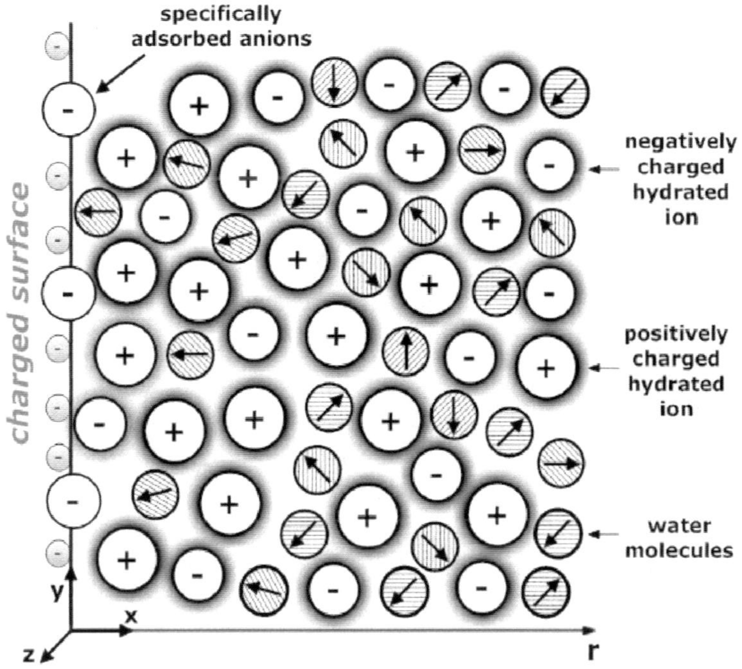

Figure 2. Schematic figure of the EDL near a negatively charged planar metal surface. The water dipoles in the vicinity of the charged surface are partially oriented towards the surface (adapted from [23, 37]).

Since in the bulk solution the number densities of water molecules (n_{0w}), counterions (n_0) and co-ions (n_0) are constant, their number densities can be expressed in a simple way by calculating the corresponding probabilities that a single lattice site in the bulk solution is occupied by one of the three kind of particles in the electrolyte solution (counterions, co-ions and water molecules) [8, 33, 34]. However, in the vicinity of a charged surface the number densities of ions and water molecules are influenced by the charged surface (figure 2), so the probabilities that a single lattice site is occupied by a particle of one of the three kinds should be corrected by the corresponding Boltzmann factors, leading to ion and water dipole distribution functions in the form [23]:

$$n_+(x) = n_s \frac{n_0 e^{-e_0\phi\beta}}{n_0 e^{e_0\phi\beta} + n_0 e^{-e_0\phi\beta} + n_{0w}\left\langle e^{-\gamma p_o E\beta\cos(\omega)}\right\rangle_\omega}, \tag{8}$$

$$n_-(x) = n_s \frac{n_0 e^{e_0\phi\beta}}{n_0 e^{e_0\phi\beta} + n_0 e^{-e_0\phi\omega} + n_{0w}\left\langle e^{-\gamma p_o E\beta\cos(\omega)}\right\rangle_\omega}, \tag{9}$$

$$n_w(x) = n_s \frac{n_{0w}\left\langle e^{-\gamma p_o E\beta\cos(\omega)}\right\rangle_\omega}{n_0 e^{e_0\phi\beta} + n_0 e^{-e_0\phi\omega} + n_{0w}\left\langle e^{-\gamma p_o E\beta\cos(\omega)}\right\rangle_\omega}, \tag{10}$$

where n_s is the number density of lattice sites as defined above and

$$\left\langle e^{-\gamma p_o E\beta\cos(\omega)}\right\rangle_\omega = \frac{2\pi \int_\pi^0 d(\cos\omega)\, e^{-\gamma p_o E\beta\cos(\omega)}}{4\pi} = \frac{\sinh(\gamma p_0 E\beta)}{\gamma p_0 E\beta}. \tag{11}$$

is the dipole Boltzmann factor after rotational averaging over all possible angles ω. Eqs. (8)-(11) can be rewritten as [23]:

$$n_+(x) = n_0\, e^{-e_0\phi\beta}\, \frac{n_s}{\mathcal{D}(\phi,E)} \ , \tag{12}$$

$$n_-(x) = n_0\, e^{e_0\phi\beta}\, \frac{n_s}{\mathcal{D}(\phi,E)} \ , \tag{13}$$

$$n_w(x) = \frac{n_{0w}\, n_s}{\mathcal{D}(\phi,E)}\, \frac{\sinh(\gamma p_0 E\beta)}{\gamma p_0 E\beta} \ . \tag{14}$$

where $\mathcal{D}(\phi,E) = 2n_0 \cosh(e_0\phi\beta) + \frac{n_{0w}}{\gamma p_0 E\beta}\sinh(\gamma p_0 E\beta)$. Combining Eqs. (6) and (14) gives the polarization in the form:

$$P(x) = -\left(\frac{2+n^2}{3}\right) \frac{p_0\, n_{0w}\, n_s}{\mathcal{D}(\phi,E)}\, \frac{\mathcal{L}(\gamma p_0 E\beta)}{\gamma p_0 E\beta}\, \sinh(\gamma p_0 E\beta) \ . \tag{15}$$

Using the definition of the function $\mathcal{F}(u)$: $\mathcal{F}(u) = \mathcal{L}(u)\,(\sinh u/u)$ Eq. (15) transforms into:

$$P = -\, p_0\, n_{0w}\, n_s \left(\frac{2+n^2}{3}\right) \frac{\mathcal{F}(\gamma p_0 E\beta)}{\mathcal{D}(\phi,E)} \ . \tag{16}$$

Combining Eqs. (1) and (16) yields the relative (effective) permittivity [23] :

$$\varepsilon_r(x) = n^2 + \ n_{0w}\, n_s\, \frac{p_0}{\varepsilon_0}\left(\frac{2+n^2}{3}\right) \frac{\mathcal{F}(\gamma p_0 E\beta)}{\mathcal{D}(\phi,E)\, E} \ . \tag{17}$$

Using the above expression for $\varepsilon_r(x)$, we can then write the Poisson equation in the form [8, 23]:

$$\nabla \cdot [\varepsilon_0\, \varepsilon_r(x) \nabla\phi(x)] = -\rho_{free}(x) \ , \tag{18}$$

where $\rho_{free}(x)$ is the macroscopic (net) volume charge density of co-ions and counterions (see also Eqs. (12) and (13)) [23] :

$$\rho_{free}(x) = e_0\, n_+(x) - e_0\, n_-(x) = -2\, e_0\, n_s n_0\, \frac{\sinh(e_0\phi\beta)}{\mathcal{D}(\phi,E)} \ . \tag{19}$$

The boundary conditions are [23]:

$$\nabla\phi\,(x=0) = -\frac{\sigma\, \mathbf{n}}{\varepsilon_0\, \varepsilon_r(x=0)} \ , \quad \phi\,(x\to\infty) = 0 \ . \tag{20}$$

In the limit of vanishing electric field strength ($E \longrightarrow 0$) and zero potential ($\phi \longrightarrow 0$), the above derived expression for relative permittivity (Eq. (17)) gives the well-known Onsager expression for permittivity:

$$\varepsilon_r \cong n^2 + \left(\frac{2+n^2}{3}\right)^2 \frac{n_{0w} {p_0}^2 \beta}{2\,\varepsilon_0} \ . \tag{21}$$

In the above derived expression for the relative (effective) permitivitty (Eq. (17)), the value of the dipole moment $p_0 = 3.1\,\mathrm{D}$ (Debye is $3.336 \cdot 10^{-30}\,\mathrm{C/m}$) predicts a bulk permittivity $\varepsilon_r = 78.5$. This value is considerably smaller than the corresponding value in previous similar

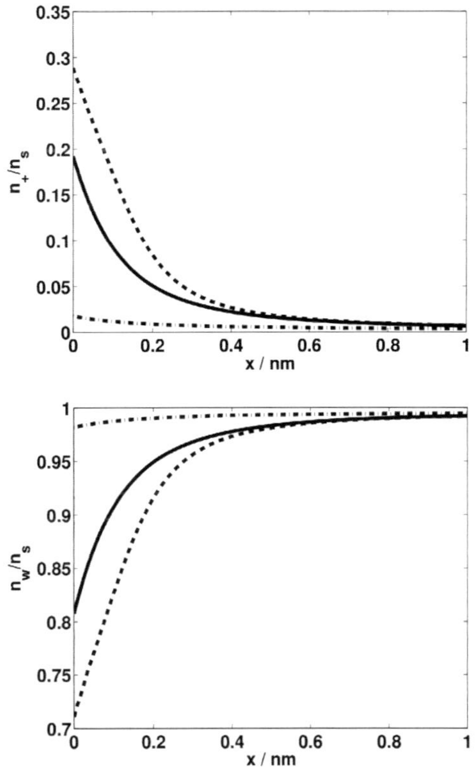

Figure 3. The relative number density of counter ions (n_+/n_s) and water dipoles (n_w/n_s) (calculated using Eqs. (12) and Eq. (14) as a function of distance from a planar charged surface x calculated for three values of surface charge density: $\sigma = -0.1\,\mathrm{As/m^2}$ (dashed-dotted line), $\sigma = -0.2\,\mathrm{As/m^2}$ (full line) and $\sigma = -0.3\,\mathrm{As/m^2}$ (dashed line) (from [23]). Values of parameters assumed: dipole moment of water $p_0 = 3.1\,D$, bulk concentration of salt $n_0/N_A = 0.15\,\mathrm{mol/l}$, optical refractive index $n = 1.33$, bulk concentration of water $n_{ow}/N_A = 55\,\mathrm{mol/l}$, where N_A is the Avogadro number.

models ($p_0 = 4.79\,\mathrm{D}$) [22, 31, 32] which did not take into account the cavity field and electronic polarizability of water molecules.

Figures 3 and 4 show the calculated spatial dependence of the relative number density of counter ions (n_+/n_s), water dipoles (n_w/n_s) and $\varepsilon_r(x)$ within the finite sized Gongadze-Iglič (GI) model in planar geometry for three values of the surface charge density σ.

It can be seen in figure 3 that close to the charged plane there may be a considerable excluded volume effect on the density profile of the counterions and on the solvent molecules. The concentration of counterions there is comparable to the concentration of solvent lattice sites, so that the concentration of solvent lattice sites deviates significantly from its value far from the charged surface.

The ratio between the concentration of the counterions near the charged plane and the bulk counter-ion concentration as a function of the surface charge density σ is higher for dimensionless ions than for ions of finite size. The discrepancy between the results for dimensionless ions and for ions of finite size grows with increasing $|\sigma|$ [35]. The deviation can be attributed to the steric

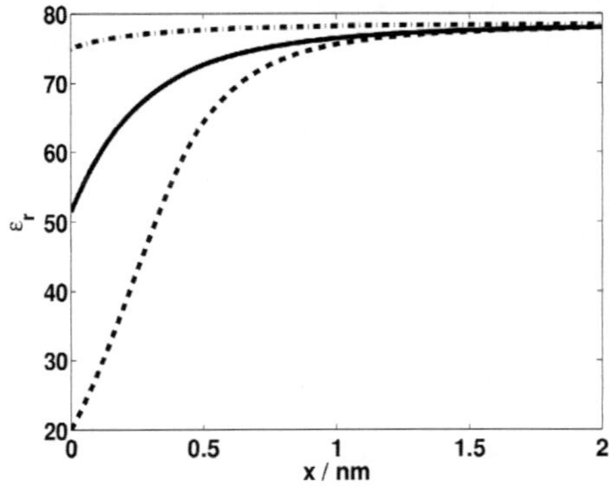

Figure 4. Relative permittivity ε_r (Eq. (17)) as a function of distance from a planar charged surface x (from [23]). Three values of surface charge density were considered: $\sigma = -0.1\,\mathrm{As/m^2}$ (dashed-dotted line), $\sigma = -0.2\,\mathrm{As/m^2}$ (full line) and $\sigma = -0.3\,\mathrm{As/m^2}$ (dashed line). Eq. (18) was solved numerically taking into account the boundary conditions as described in [23]. Values of the model parameters are the same as in the previous figure.

effect of counterions and solvent molecules in the small region in the vicinity of the charged surface. The counter-ion concentration profile shows a rapid decrease for small lattice constant and a plateau region near the charged plane for large lattice constant. For large counterions we can therefore distinguish between two regions within the electric double layer: the saturated layer dominated by steric repulsion and the diffuse layer extending into the solution.

The decrease of $\varepsilon_r(x)$ towards the charged surface is pronounced with increasing σ (figure 4) and is a consequence of the increasing depletion of water molecules near the charged surface (due to the excluded volume effect as a consequence of accumulation of counterions near the charged surface) (figure 4) and increased orientational ordering of water dipoles (saturation effect). Comparison between the predictions of the GI model [23] and the Langevin Poisson-Boltzmann (LPB) model model [8, 22] which does not take into account the finite size of molecules (i.e. assumes a constant density of water molecules), shows a stronger decrease of relative permittivity of the electrolyte solution near the highly charged surface stronger in the GI model than in the LPB model, mainly due to depletion of water molecules in the vicinity of the charged surface in the GI model (see figure 3).

3. Langevin-Bikerman model

In order to differentiate between the influence of the finite size of ions and the influence of the cavity field on the relative permittivity near the charged surface, the equations of the above described Gongadze-Iglič (GI) [23] model may be written in the limit of $\gamma \to 1$ and $n \to 1$ [8] to get [8, 36] :

$$\nabla \cdot \left[\varepsilon_0 \, \varepsilon_r(x) \nabla \phi(x) \right] = -\rho_{free}(x) \quad , \tag{22}$$

where $\rho_{free}(x)$ is the macroscopic (net) volume charge density of co-ions and counterions in the Langevin-Bikerman (LB) model :

$$\rho_{free}(x) = -2\, e_0 \, n_s n_0 \, \frac{\sinh\left(e_0 \phi \beta\right)}{\mathcal{H}(\phi, E)} \quad , \tag{23}$$

while $\varepsilon_r(x)$ is the relative permittivity in the LB model [31, 32, 36] :

$$\varepsilon_r(x) = 1 + n_{0w} n_s \frac{p_0}{\varepsilon_0} \frac{\mathcal{F}(p_0 E\beta)}{E\,\mathcal{H}(\phi, E)} , \tag{24}$$

where $\mathcal{H}(\phi, E) = 2n_0 \cosh(e_0\phi\beta) + \frac{n_{0w}}{p_0 E\beta} \sinh(p_0 E\beta)$.

Comparison between the space dependency of the relative permittivity within the GI model and within its limiting LB model for $\gamma \to 1$ and $n \to 1$ shows that consideration of the cavity field and electronic polarizability of water molecules makes the reduction of permittivity of the electrolyte solution near the charged surface stronger in the GI model [37]. More important, in the LB model (i.e. limit of $\gamma \to 1$ and $n \to 1$ of the GI model) the value $p_0 = 4.79\,D$ [31, 32, 36] (similarly as in the LPB model [8, 22]) should be used in order to get $\varepsilon_r(x \to \infty) = 78.5$.

4. Bikerman model

In the limit of $p_0 \to 0$ the particle distribution functions (12)-(14) transform into Fermi-Dirac-like Bikerman distribution functions [11–15, 17, 35, 38, 39] :

$$n_+(x) = \frac{n_0 n_s}{n_{0w}} \frac{e^{-e_0\phi\beta}}{1 + (2n_0/n_{0w})\cosh(e_0\phi\beta)} , \tag{25}$$

$$n_-(x) = \frac{n_0 n_s}{n_{0w}} \frac{e^{e_0\phi\beta}}{1 + (2n_0/n_{0w})\cosh(e_0\phi\beta)} , \tag{26}$$

$$n_w(x) = \frac{n_s}{1 + (2n_0/n_{0w})\cosh(e_0\phi\beta)} , \tag{27}$$

while Eq. (18) transforms into the Bikerman equation [11–15, 17, 35, 38, 39] :

$$\nabla \cdot [\varepsilon_0\, \varepsilon_r \nabla\phi(x)] = -\rho_{free}(x) \tag{28}$$

where we made the transformation $\varepsilon_0 \to \varepsilon_r\varepsilon_0$ with $\varepsilon_r = 78.5$, while $\rho_{free}(x)$ is defined by Eq.(23). Eq. (25) predicts a Fermi-Dirac-like distribution for counterions if the lattice constant a is large enough. For higher values of the surface charge density ($|\sigma|$), the counter-ion density saturates close to the charged surface to its close packing value, while the well-known Gouy-Chapman model [1–3] predicts unreasonably high values beyond the close-packing values (see also [11, 17, 35]).

5. Gouy-Chapman model

In the limit of a very dilute solution everywhere in the electrolyte solution:

$$\sum_{j=+,-} n_j(x) \ll n_w(x) , \tag{29}$$

and by taking into account the approximation $n_{0w} \simeq n_s$, we can neglects the second term in the denominator of Eqs. (25) and (26), so the particle distribution function (25) and (26) transform into Boltzmann distribution functions within the Gouy-Chapman model [1–4, 7, 40]:

$$n_+(x) = n_0\, e^{-e_0\phi\beta} , \tag{30}$$

$$n_-(x) = n_0\, e^{e_0\phi\beta} , \tag{31}$$

while the number density of water molecules (Eq.(27)) becomes constant [17, 35] :

$$n_w(x) = n_{0w} . \tag{32}$$

The Bikerman Eq.(28) transforms into the Poisson-Botzmann equation within the Gouy-Chapman model [1–3]:

$$\nabla \cdot [\varepsilon_0\, \varepsilon_r \nabla\phi(x)] = 2e_0 n_0 \sinh(e_0\phi\beta) . \tag{33}$$

6. Discussion and Conclusions

In theoretical models of the EDL, the effect of a polarizing environment can be captured by introducing the cavity field [25, 29, 30]. In the past, treatments of the cavity field in the Onsager [29], Kirkwood [30] and Fröhlich [25] models were limited to the case of small electric field strengths far away from the saturation limit [41]. Later a generalization in the saturation regime was performed [23, 26, 41].

In this review it was shown that considering the cavity field of a single water molecule and the finite size of the ions result in a more realistic description of the decrease of the permittivity near the charged surface [8, 23]. The corresponding analytical expression for the spatial dependence of relative permittivity of an electrolyte solution near the charged surface is given in [8, 23, 34].

The dipole moment of an isolated water monomer is 1.855 D, while the dipole moment of a water molecule in bulk liquid water [27] is around 2.4 - 2.6 D [28, 42]. By considering the cavity field at the centre of a water molecule and its electronic polarizability [25], a magnitude of the external water dipole moment around 3 D was suggested within the mean-field lattice statistics approach [23] in accordance with predictions of previous theoretical treatments within the lattice model approach [27, 42]. Neglecting the cavity field and the electronic polarizability within lattice statistics theoretical models leads to much higher predicted values of the water external dipole moment around 5 D [22, 31].

To ascertain whether the described Gongadze-Iglič (GI) mean-field model (which includes orientational ordering of water, cavity field, electronic polarizability of water and finite size of molecules) has improved the agreement between theory and experiments with respect to the classical Gouy-Chapman model, one should compare the measured and the predicted values of electric potential and differential capacitance of the EDL in both models. Within the GI model the predicted values of the electric potential at higher surface charge densities σ are substantially more negative than the corresponding values within the Gouy-Chapman model (see Ref. [33] and references therein).

Within the Gouy-Chapman model, we can estimate the electric potential ϕ_0 at the surface (of the electrode for example) by the using Grahame equation [20]:

$$\sigma \;=\; \sqrt{8\, n_0\, \varepsilon_0 \varepsilon_r/\beta} \cdot \sinh\left(\frac{e_0\, \beta\, \phi_0}{2}\right) \;, \tag{34}$$

where ϕ_0 is the surface potential, i.e. $\phi(x=0)$. The corresponding Gouy-Chapman differential capacitance is [20]:

$$C_{GC} = \frac{d\sigma}{d\phi_0} = \sqrt{2\, e_0^2\, \beta\, n_0\, \varepsilon_0 \varepsilon_r} \cdot \cosh\left(\frac{e_0\, \beta\, \phi_0}{2}\right) . \tag{35}$$

The Gouy-Chapman model provides relatively good predictions for monovalent salts at concentrations below 0.2 mol/l in aqueous solutions and magnitudes of surface potentials below $50-80 mV$ [20]. However, by increasing the surface charge density (i.e. increasing the magnitude of ϕ_0) the differential capacitance calculated within the GI model decreases. By applying a voltage of up to 1V, the differential capacitance drops to very small values in accordance with experimental results (see also Refs. [20, 21, 33]). Obviously, the GI model also works efficiently at higher surface charge densities or potentials, where the classical Gouy-Chapman model completely fails as the differential capacitance C_{GC} (unlike the experimental results) strongly and monotonously increases with increasing ϕ_0.

On the contrary, in the GI model, in accordance with the experimental results [20, 43], the differential capacitance decreases with increasing ϕ_0 after reaching its maximum. The predicted values of differential capacitance at high values of ϕ_0 are much smaller than the corresponding values calculated using the empirical formula for $\varepsilon(x)$ [21]. Moreover, if the Gongadze-Iglič (GI) model is modified by also taking into account the distance of closest approach [8], the predicted values of differential capacitance would closely approach to the experimental values [43].

Acknowledgement(s)
This work was supported by Slovenian Research Agency (ARRS) grants J3-2120, J1-4109, J1-4136, J3-4108 and P2-0232.

References
[1] Gouy M G 1910 *J. Physique* (France) **9** 457
[2] Chapman D L 1913 *Philos. Mag.* **25** 475.
[3] McLaughlin S 1989 *Ann. Rev. Biophys. Chem.* **18** 113
[4] Cevc G 1990 *Biochim. Biophys. Acta* **1031** 311
[5] Lamperski S and Outhwaite C W 2002 *Langmuir* **18** 3423
[6] Bivas I and Ermakov Y A 2007 *Elasticity and Electrostatics of Amphiphilic Layers* (Advances in Planar Lipid Bilayers and Liposomes vol 5) ed A. Leitmannova Liu (Amsterdam: Elsevier) chapter 11 pp 313-343
[7] Israelachvili J N 1997 *Intermolecular and Surface Forces* (London, Academic Press)
[8] Gongadze E, van Rienen U, Kralj-Iglič V and Iglič A 2012 *Comput. Meth. Biomech. Biomed. Eng.* doi: 10.1080/10255842.2011.624769
[9] Stern O 1924 *Zeitschrift fur Elektrochemie* **30** 508
[10] Helmholtz H 1879 *Ann. Phys.* **243** 337
[11] Bikerman J J 1942 *Phil. Mag.* **33** 384
[12] Grimley T B and Mott N F 1947 *Discuss. Faraday Soc.* **1** 3
[13] Grimley T B 1950 *Proc. R. Soc. Lond. A* **201** 40
[14] Freise V 1952 *Z. Elektrochem.* **56** 822
[15] Eigen M and Wicke E 1954 *J. Phys. Chem.* **58** 702
[16] Bhuiyan L B and Outhwaite C W 2009 *Coll. Int. Sci.* **331** 543
[17] Kralj-Iglič V and Iglič A 1996 *J. Phys.* II France **6** 477.
[18] Outhwaite C W 1983 *Mol. Phys.* **48** 599
[19] Gruen D W R and Marčelja S 1983 *J. Chem. Soc., Faraday Trans.* II. **79** 225
[20] Butt H J, Graf K and Kappl M 2003 *Physics and Chemistry of Interfaces* (Weinheim, Wiley-VCH Verlag)
[21] Bazant M Z, Kilic M S, Storey B and Ajdari A 2009 *Adv. Colloid Interface Sci.* **152** 48
[22] Gongadze E, van Rienen U, Kralj-Iglič V and Iglič A 2011 *Gen. Physiol. Biophys.* **30** 130
[23] Gongadze E and Iglič A 2012 *Bioelectrochemistry* **87** 199
[24] Outhwaite C W 1976 *Mol. Phys.* **31** 1345
[25] Fröhlich H 1964 *Theory of Dielectrics* (Oxford, Clarendon Press)
[26] Booth F 1951 *J. Chem. Phys.* **19** 391
[27] Adams D J 1981 *Nature* **293** 447
[28] Dill K A and Bromberg S 2003 *Molecular Driving Forces* (New York and London, Garland Science)
[29] Onsager L 1951 *J. Am. Chem. Soc.* **58** 1486
[30] Kirkwood J G 1939 *J. Chem. Phys.* **7** 911
[31] Abrashkin A, Andelman D and Orland H 2007 *Phys. Rev. Lett.* **99** 077801
[32] Iglič A, Gongadze E and Bohinc K 2010 *Bioelectrochemistry* **79** 223
[33] Gongadze E 2011 *Dr. Ing. Thesis* (Rostock, University of Rostock)
[34] Gongadze E, van Rienen U, Kralj-Iglič V and Iglič A 2011 *Cell Mol Biol Lett.* **16** 576
[35] Iglič A and Kralj-Iglič V 1994 *Electrotechnical Rev.* **61** 127
[36] Gongadze E, Kabaso D, Bauer S and Slivnik T, Schmuki P, van Rienen U and Iglič A 2011 *Int. J. Nanomed.* **6** 1801
[37] Gongadze E and Iglič A 2012 *Bulg. J. Phys.* **39** 12
[38] Dutta M and Sengupta M 1954 *Proc. Natl. Inst. Scif. India* **20** 1
[39] Wiegel F W and Strating P 1993 *Mod. Phys. Lett. B* **7** 483
[40] Evans D F and Wennerström H 1994 *The Colloidal Domain* (New York, Wiley-VHC)
[41] Szalai I, Nagy S and Dietrich S 2009 *J. Chem. Phys.* **131** 154905.
[42] Gregory J K, Clary D C, Liu K, Brown M G and Saykally R J 1997 *Science* **275** 814
[43] Nebe B, Lüthen F, Lange R and Beck U 2007 *Macromol. Biosci.* **7** 567

Hydroxyapatite synthesis on solid surfaces using a biological approach

A Wang[2], J Mei[2], Y Y Tse[2], IP Jones[2] and RL Sammons[1]

[1]University of Birmingham School of Dentistry, Birmingham B4 6NN, UK
[2]University of Birmingham School of Metallurgy and Materials, Birmingham, B15 2TT, UK

E.mail: r.l.sammons@bham.ac.uk

Abstract. Many naturally occurring mineralisation processes yield hydroxyapatite (HA) or related salts, but biological routes to calcification have not generally been exploited for production of hydroxyapatite for clinical and industrial applications. *Serratia sp.* NCIMB 40259 is a non-pathogenic Gram-negative bacterium which is capable of growing as a biofilm on many surfaces and can be used to form HA coatings on a variety of polymeric and metallic materials, including titanium. Here we review previous work and report the results of more recent studies on the influence of titanium compositional and surface properties on *Serratia* adherence and proliferation and biomineralisation on commercially pure titanium (cp Ti) discs and a Ti mesh. Bacterial adherence was equivalent on cpTi and Ti6Al4V, and biofilms formed on both rough and mirror-polished cpTi surfaces. Embedded alumina particles and alkali treatment did not noticeably alter the precipitation of *Serratia* HA, nor the structure of the coating in comparison with non-treated substrates. Coatings were retained after sintering at 800°C in argon, although the original curved plate-like crystals changed to nano-scale β-tricalcium phosphate particles. A phosphorous-rich diffusion zone formed at the coating-titanium interface. Bacterial mineralisation may have applications as a method for producing coatings on implants in non load-bearing sites, and non-clinical applications where a high surface area is the major concern.

1. Introduction

Biomineralisation occurs in nature in many forms, some of which are beneficial to humans and some detrimental. In the body, biomineralisation leads to the formation of the skeleton and teeth but pathogenic calcification can also occur in arteries, in skin, on heart valves, as dental calculus, as kidney stones and as encrustations blocking medical devices such as urinary catheters [1, 2]. Although many of these naturally occurring processes yield hydroxyapatite (HA) or related salts, biological routes have not generally been exploited for production of HA for clinical and industrial applications. In this paper we briefly review our previous work on HA formation by *Serratia* and then describe the results of more recent investigations [3] using this approach to form a HA coating on titanium substrata, starting with biofilms. Finally we briefly discuss an alternative method of producing HA crystals and coatings using *Serratia* not as a biofilm but as planktonic cells in suspension.

2. Biomineralisation by Serratia: Background

Serratia sp. NCIMB 40259 was originally isolated from heavy-metal ion contaminated soil where it survives by removing heavy metal contaminants e.g. cadmium from its surroundings as metal phosphates [4]. Initially identified as a species of *Citrobacter*, it was reclassified as a species of *Serratia* in 2002[5].

The mechanism by which metal phosphates are formed by this species of *Serratia* involves an acid phosphatase enzyme, located in the bacterial cell wall, within the periplasmic space, and on hair-like projections called fimbriae [6]. The enzyme cleaves organic phosphates liberating inorganic phosphate, which combines with metal ions to produce metal phosphates which precipitate within the extracellular polymeric matrix EPM) surrounding the cells. The EPM provides both the nucleation sites for crystal formation and restricts crystal growth [7,8,9].

When *Serratia* are presented in the laboratory with calcium ions from calcium chloride and an organic phosphate (glycerol-2-phosphate), nanoscale crystals of Ca-deficient hydroxyapatite are produced. The Ca/P ratio varies depending on the mineralisation conditions: Mineralisation at pH 8.6 yields calcium deficient HA which may either sinter (at 1200° C) to β-TCP [10] or remain as Ca-deficient HA, depending on the initial Ca/P ratio, the pH and/or the presence of sodium citrate [10]. As in many biological apatites, the absence of the typical HA O-H stretching peak in the FTIR spectrum suggests that the HA crystal lattice is defective [11]. Na^+ and Cl^- ions are present at about 1% in the original crystals and remain as impurities in the final sintered product [12]. A porous scaffold material that supports the growth of osteoblasts in culture was made by culturing the bacteria on polyethylene sponge, allowing biomineral formation and then sintering the mineral-encrusted biofilm to destroy the bacteria and the sponge template leaving only the mineral "skeleton" that replicates the original structure of the biofilm, with a very high surface area [11,12]. By presenting the bacteria with a mixture of CaCl2 and SrCl2 in the biomineralisation solution, co-crystals are formed containing Ca and Sr in the crystal lattice in amounts corresponding to the supplied concentrations [2]. Such materials could be useful as bone-substitute materials including applications in the repair of osteoporotic bone. On the other hand, as a procedure for coating solid surfaces with HA it may be a useful non-line-of-sight method [13], whilst its high surface area makes it potentially advantageous for waste water treatment for removal of phosphates [14] or radionuclides [15,16].

3. Influence of titanium properties on Serratia adherence and proliferation

3.1 Chemical composition

Serratia attached equally well to commercially pure (cpTi) and TiAl6V4 alloy [16]. There was no significant difference in proliferation on these after 24h. The bacteria proliferated significantly better on pure vanadium than aluminium (figure 1).

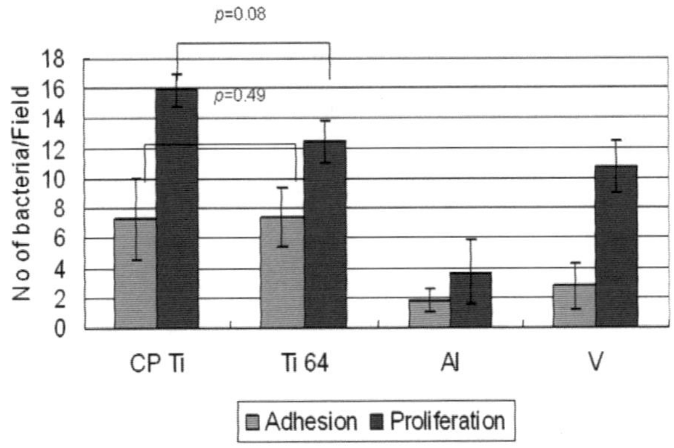

Figure 1. *Serratia* attachment to cpTi, TiAl6V4, pure Al and Vanadium mirror polished surfaces. Attachment time was 4h in phosphate buffered saline and proliferation time 24h in carbon-limited minimal medium [9]. Bacteria on each substrate were counted using secondary electron imaging (JEOL 7000, Japan) in ten fields chosen at random from each metal disc, with three discs of each material.

3.2 Influence of grain boundaries and crystal orientation

Using a combination of back-scattered electron imaging to show grain boundaries, and electron back-scattered diffraction to show crystal orientation on polished cpTi, it was found that neither grain boundaries nor grain orientation affected *Serratia* adherence [16].

3.3 Influence of surface topography on biofilm formation

Biofilm formed on a range of model dental implant surfaces, as well as on Al_2O_3 grit-blasted and mirror polished cpTi shown in figure 2a. There was no influence of surface roughness (ra value) but some indication that surface hydrophobicity might be more important, there being a negative correlation between static water contact angle (figure 2b) and protein after 7 days in culture with T1, T3, T4, T6 (figure 2c). However further work must be done to confirm this.

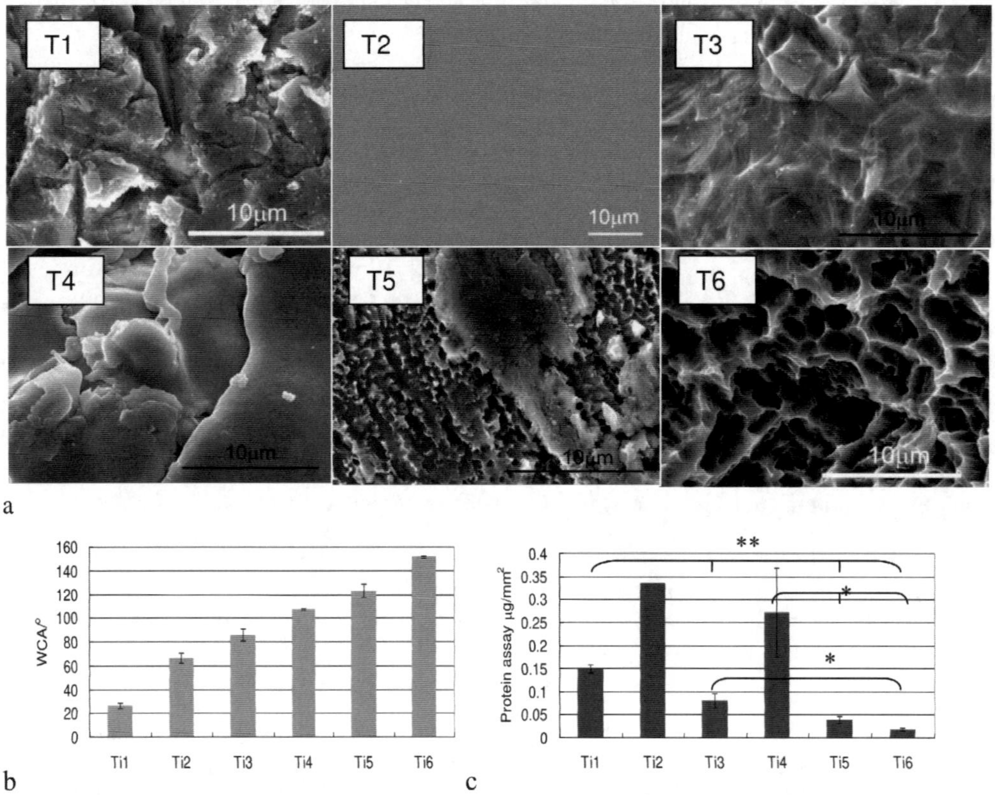

Figure 2: a) cpTi surfaces used for comparison of bacterial proliferation; b) hydrophobicity of the surfaces as measured by water contact angle; c) bacterial load estimated by a protein assay after 7 days in continuous culture in lactose-limiting medium. * indicates statistical significance in t test; n = 3; P = <0.05*; <0.01**.

According to the protein assay and SEM images, biofilm formed on the mirror polished surface (T2) at least as well as on the rougher ones. The bacteria formed densely-packed colonies on T2 and on T6 they formed vertical stacks and palisades in the troughs, with more randomly aligned cells on the ridges (figure 3).

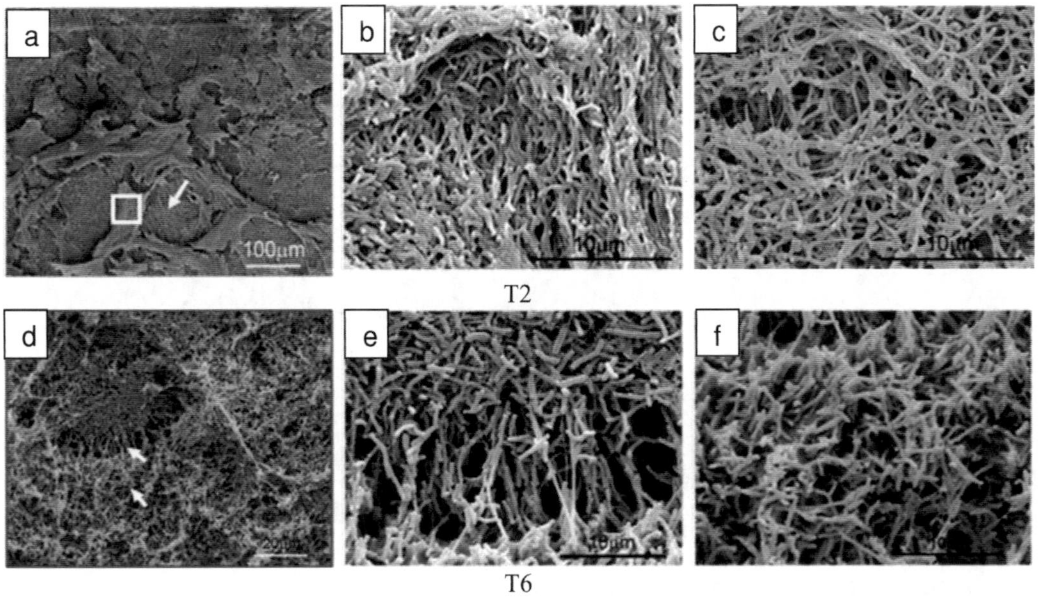

Figure 3: Biofilm formation on T2 (a-c, mirror polished) and T6 (d-f, grit-blasted acid etched) surfaces shown in Figure 2. On T2a the boxed area is shown more highly magnified in b and the arrowed site in c. On T6 the upper arrowed site (a trough) is shown in b and the lower arrowed site (a ridge) in c. Size bars = a, 100 µm; d, 20 µm ; b,c,e,f, 10µm.

4. Biomineralisation on grit-blasted cpTi

4.1 Structure of the coating before heat-treatment

When biomineralisation occurs, individual cells become coated with an envelope of clusters of needle or plate-shaped crystals. A few uncoated bacterial cells can also be seen amongst the coated ones (figure 4a). A cross-section through the mineralised biofilm shows the porosity of the structure and the thickness of the mineral envelope around each cell (figure 4 b,c). In addition, clumps of cells are surrounded by shells of mineral approximately 0.5µm thick. Selected area diffraction confirmed the presence of HA and the Ca/P ratio (EDX) was 1.57±0.14.

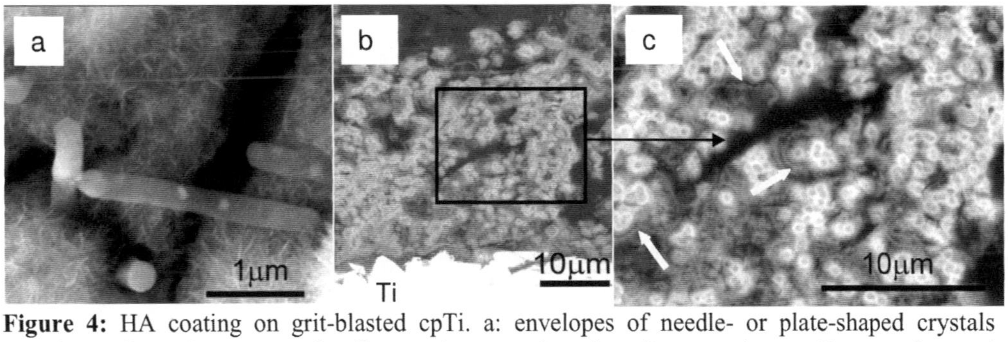

Figure 4: HA coating on grit-blasted cpTi. a: envelopes of needle- or plate-shaped crystals covering cells; a few uncoated cells are also seen. b and c: Cross-section c: Clumps of coated individual cells are surrounded by shells of mineral (arrows).

4.2 Structure of the coating after heat treatment

To burn off the biomass and sinter the crystal coating samples were subjected to the following heating regime: ramp at 5° C/min to 550° C, dwell for 3h, ramp at 5° C/min to 800° C for 10h and then cool at 5° C/min. Figure 5 shows the overall appearance of the coating. Cracks are apparent in the surface. Higher magnification revealed that the coating consisted of spheres of nano-particles with a Ca/P ratio of 1.34±0.04, consistent with β-tricalcium phosphate. A transition layer was formed at the interface with the substrate (figure 5).

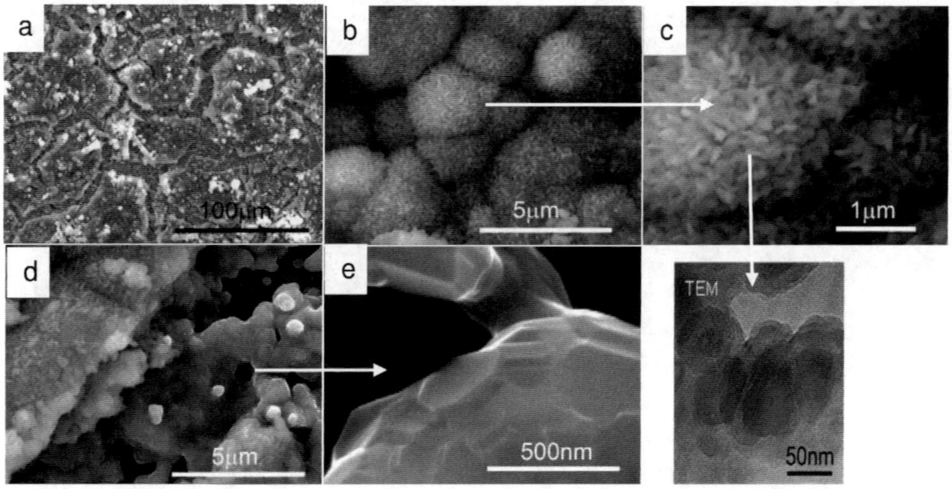

Figure 5: Heated HA coating on cpTi. a: overall appearance of the coating (scale bar = 100µm). Higher magnification images in b and c show clusters of nanoparticles and TEM appearance below. d and e: a transition layer at the interface between the coating and the substrate (figure 6 below).

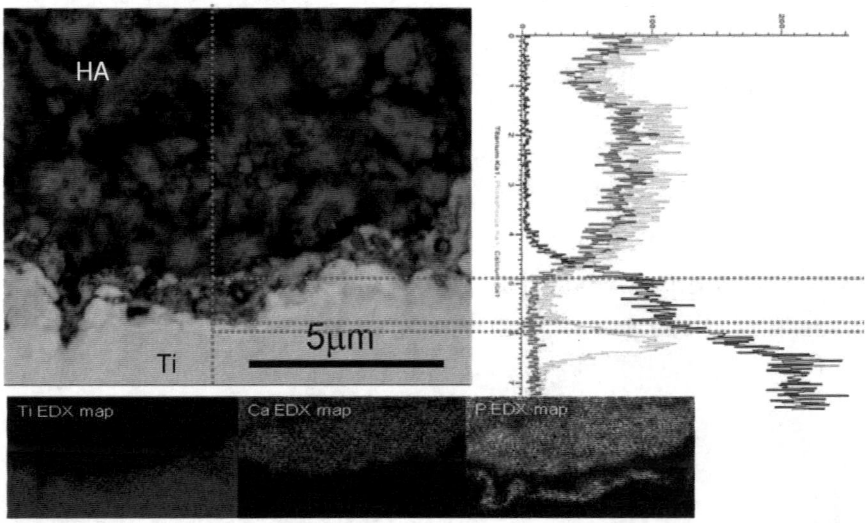

Figure 6: Cross-section of the interface between the HA coating and the cpTi substrate after heating. A line scan shows migration of P from a P-deficient zone into the substrate, confirmed by EDX maps.

5. Biomineralisaion on grit blasted titanium treated with sodium hydroxide

The interface between the HA coating and cp Ti which had been grit-blasted with Al_2O_3 and then treated with 5M sodium hydroxide to form a sodium titanate layer was investigated. Focused ion beam milling was used to produce a cross-section 150nm wide. Imaging revealed a layered structure with continuity between the sodium titanate and the HA crystals (figure 7). The sodium titanate layer consisted of a denser lower region and a "fibrous" upper region interdigitating with HA crystals. Embedded Al_2O_3 particles were surrounded by sodium titanate and similarly covered with HA.

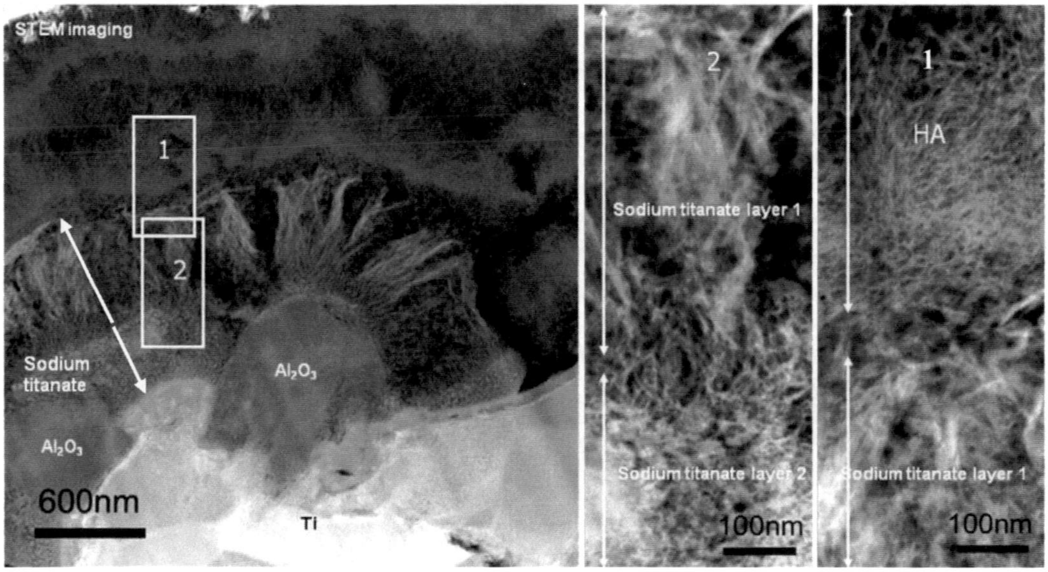

Figure 7. Interface between sodium titanate and HA on grit blasted cpTi. Cross-sections were produced by focused ion beam milling in a Jeol7000 FEG SEM and imaged using TEM in scanning mode (STEM).

6. Biomineralisation of a titanium mesh

Serratia are 1-2μm long and possess flagellae that enable them to swim in suspension and theoretically thereby enter cavities and channels that are wider than a few microns. However build up of the biofilm on the outside of a porous structure might prevent bacterial infiltration into the interior. To investigate this, a Ti mesh made from 50 μm diameter wires with a thickness of about 1mm was used as a substrate (figure 8) The meshes were incubated in *Serratia* culture for three days to allow bacterial adhesion and proliferation and then loaded into a bioreactor perfused with mineralisation solution at a flow rate of 1.5 ml/h for nine days. Under these conditions each wire of the mesh was coated with nanoscale crystals, even in the interior. Following heating (550°C for 3 hours & 800 °C for 5 hours in an argon furnace) the coating remained on the surface. EDS mapping confirmed the presence of large crystals of calcium phosphate between the wires and a line scan of the interface showed migration of phosphate from the coating into the Ti in a diffusion zone as with the Ti sustrate described above (figure 8).

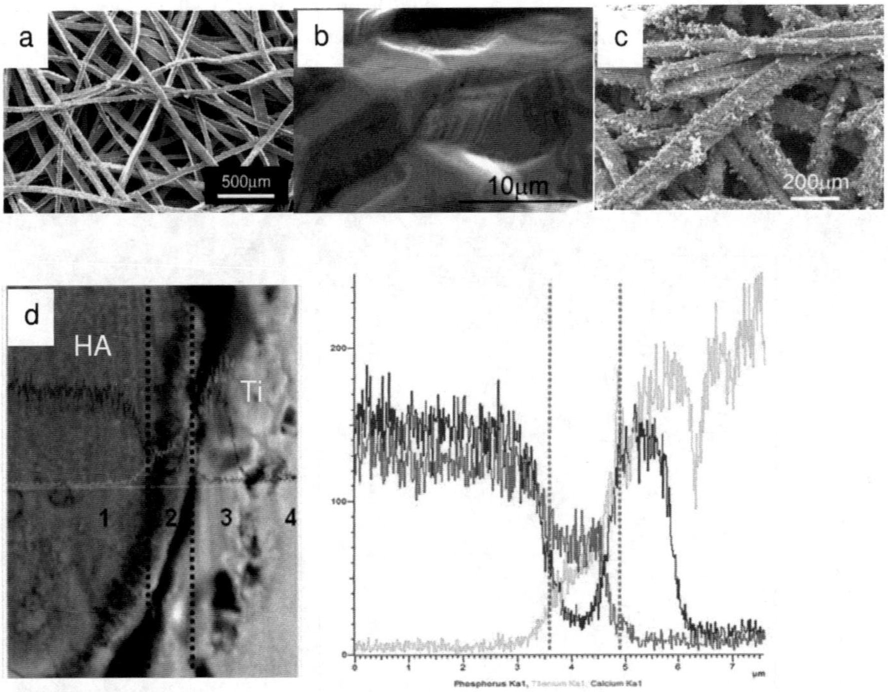

Figure 8. Biomineralisation on Ti wire mesh. a and b Appearance of the uncoated mesh and individual wire surface showing Ti crystal structure. c: wires after biomineralisation and heating, showing retention of the calcium phosphate crystals. d: Cross section of the interface between the wire and adjacent HA crystals. An EDX line map from left to right of the image shows migration of P into the titanium substrate.

7. Use of planktonic *Serratia* to promote calcium phosphate precipitation

Hydroxyapatite crystals may develop on Ti, which has been treated with 5M NaOH to produce a sodium titanate surface, when it is placed in simulated body fluid containing calcium and phosphate ions, by an ion exchange mechanism [17]. The ability of planktonic (free swimming) *Serratia* bacteria, harvested from the biofermenter outflow, to promote crystal formation on alkali-treated Ti in biomineralisation solution was tested and the results shown in figure 9. Detonated nanodiamonds (DND) were added to some samples as these have previously been shown to promote HA nucleation under these conditions [18]. However DND appeared to inhibit crystal formation or growth to some extent, giving a lower yield of precipitated crystals. All the titanium discs were coated with crystals although the clusters of crystals formed in the presence of nanodiamonds appeared more consolidated. Further work is necessary to elucidate whether biomineralisation by *Serratia* bacteria in suspension involves the action of the acid phosphatase as in the intitial stages following biofilm formation or whether crystal nucleation is simply promoted by components of the bacterial cell wall or extracellular matrix.

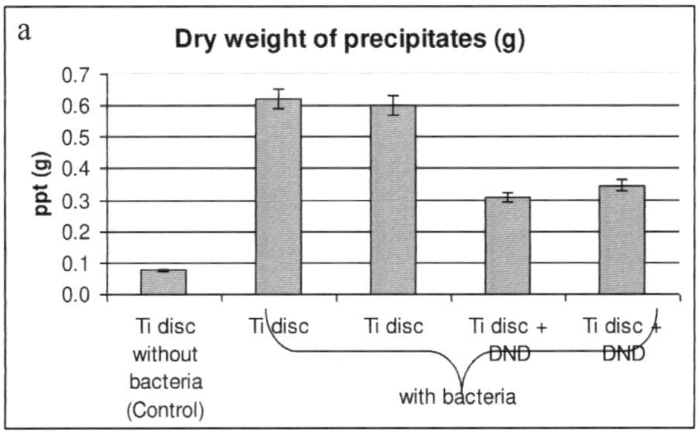

Figure 9. Biomineralisation promoted in mineralisation medium by planktonic *Serratia* in suspension and on alkali-treated Ti discs placed in the same container. a: A lower yield of crystals was obtained as a precipitate in the presence of DND; b, c: back-scattered electron images of the precipitate on the titanium discs revealed a denser particulate structure with DND (c).

8. Conclusion

Coating titanium discs with HA using bacterial biomineralisation may be superior to other methods when surface area is essential but physical strength is not so important, such as implants in non-load bearing sites and as catalyst carriers, or for water purification. The coating structure is not determined by the topography of the substrate, but the structure of the biofilm, yielding an extremely high surface area with nano-, micro- and macro-scale roughness. This unique structure and the nano-scale calcium-deficient HA may be beneficial *in vivo* to facilitate tissue mineralisation and bone ingrowth whilst its high surface area may find applications for drug release.

9. Acknowledgements

A.Wang acknowledges an overseas research scholarship from the University of Birmingham.

10. References

[1] Ciftiçioğlu N, McKay D S 2010 *Pediatr Res.* **67**: 490

[2] Sammons R, Wang, A, Thackray A, Yong P, Kuboki Y, Ametani A and Macaskie L 2010 *Nano Biomedicine* **2** 70

[3] Wang A 2011, PhD thesis, University of Birmingham, UK

[4] Macaskie LE, Dean A C R 1984 *J Gen Microbiol* **130** 53

[5] Pattanapipitpaisal P et al 2002 *Environ Technol* **23** 731

[6] Bonthrone K M, Quarmby J, Hewitt C J, Allan V J M, Paterson-Beedle M, Kennedy J F and Macaskie L E 2000 *Environ Technol* **21** 123

[7] Jeong B C, Hawes C, Bonthrone K M and Macaskie L E 1997 *Microbiol* **143** 2497

[8] Finlay J A, Allan V J M, Conner A, Callow M E, Basnakova G, Macaskie L E 1999 *Biotechnol Bioengineering* **63** 87

[9] Macaskie L E, Bonthrone K M, Yong P, Goddard D T 2000 *Microbiol-UK* **146** 1855

[10] Thackray A C, Sammons R L, Macaskie L E, Yong P, Lugg H, Marquis P M. 2004 *J Mater Sci Mater Med* **15** 403

[11] Medina Ledo H, Thackray A C, Jones I P, Marquis P M, Macaskie LE, Sammons R L. 2008 *J Mater Sci Mater Med* **19** 3419
[12] Sammons R L, Thackray A C, Medina-Ledo H, Marquis P M, Jones I P, Yong P, Macaskie LE 2007 *J Phys Conf Ser* **93**, 012048
[13] Macaskie L E, Yong P, Paterson-Beedle M, Thackray A C, Marquis P M, Sammons R L, Nott K P, Hall L D 2005 *J Biotechnol* 118 187
[14] Yong P, Macaskie LE, Sammons RL, Marquis PM, 2004 Biotechnol Lett 26: 1723-1730
[15] Paterson-Beedle M, Macaskie LE, Lee C H, Hriljac J A, Jee K Y, Kim W H 2006 *Hydrometallurgy* **83** 141
[16] Handley-Sidhu S, Renshaw J C, Yong P, Kerley R, Macaskie L E 2011 *Biotechnol Letts* **33** 79.
[17] Takadama H, Kim H-M, Kokubo T, Nakamura T 2001 *J. Biomed Mater Res* **55** 185
[18] Pramatarova L, Radeva E, Pecheva E, Hikov T, Krasteva N, Dimitrova R, Mitev D, MontgomeryP, Sammons R and Altankov G 2011 The advantages of polymer composites with detonation nanodiamond particles for medical applications *On Biomimetics* ch. 14, ed. L Pramatarova (Croatia; InTech Publications) pp 297-320

Modelling of two-and four-terminal thin-film silicon tandem solar cells

S Reynolds[1] and V Smirnov[2]
[1] School of Engineering, Physics and Mathematics, University of Dundee, Dundee DD1 4HN, United Kingdom.
[2] IEK-5 Photovoltaik, Forschungszentrum Jülich, D-52425 Jülich, Germany.

E-mail: s.z.reynolds@dundee.ac.uk

Abstract. We have performed semi-empirical modelling of amorphous silicon/microcrystalline silicon thin-film solar cells in tandem optical configuration, under series (two-terminal) and independent (four-terminal) electrical connection. The four-terminal connection relaxes the constraint of current matching between the cells. Computer simulations indicate an increase in maximum initial efficiency from 10.0% (two-terminal) to 10.8% (four-terminal). Following degradation of the amorphous silicon top-cell the figures are 8.5% and 9.8% respectively, with an optimum top-cell absorber layer thickness in four-terminal connection of 150 nm. Changes in solar spectral quality have been simulated by applying weighted spectra, and improvements in efficiency favouring four-terminal connection occur under red- and blue-rich conditions. However, these typically convey only a small fraction of the total annual insolation, and gains of 5% or less in annual electrical output are predicted. Possible increased optical and electrical losses due to additional contact layers in four-terminal connection are not taken into account.

1. Introduction

Tandem solar cells comprising amorphous silicon (a-Si:H) top cell/microcrystalline silicon (μc-Si:H) bottom cell enable higher efficiencies to be achieved in comparison with equivalent single-junction cells due primarily to reduced photocarrier thermalisation loss resulting from more effective spectral matching [1]. An additional benefit is that the a-Si:H top-cell absorber layer may be made thinner than in a single cell, resulting in improved resistance to light-induced degradation [2]. Given the compatibility of the deposition processes for each sub-cell, these advantages have led to sustained scientific and commercial interest in such 'micromorph' cells and modules over the past decade [3-5]. The great majority of tandem cells employ series electrical connection of sub-cells to give a *two-terminal* (2T) device, which results naturally from the deposition process. 2T connection imposes an important boundary condition - that the current in both sub-cells must remain equal. To achieve optimal power output under these conditions requires precise current matching, *e.g.* by careful adjustment of absorber-layer thickness. However, if the sub-cells are electrically decoupled in a *four-terminal* (4T) device then their operating points are independently controllable, ensuring maximum power transfer at all times.

A number of operational advantages may in principle be gained by 4T connection. Since the a-Si:H top-cell need no longer support the generation current of the μc-Si:H bottom-cell, it can be thinner and will be more stable as a result. The 'intermediate tunnel junction' between the sub-cells and its attendant resistive losses is obviated. Variations in service conditions such as changes in intensity and spectral quality of the incident light will not unbalance the cell. For example, sunlight from a sun low in the sky contains a greater proportion of red light. Assuming the tandem cell has been optimized for AM1.5, this will pass through the top-cell and generate most current in the bottom-cell. Thus in 2T connection the power deliverable to an external load under these conditions would be strongly limited by the top-cell. Similarly, if the light contains a larger blue fraction then the tandem becomes bottom-cell limited. In 4T connection no such limitation exists, and provided the maximum-power point of each component cell is tracked independently the electrical matching remains optimal.

Despite these potential advantages, little practical interest has been shown until recently in the development of 4T micromorph cells. Madan [6] has described a range of experiments on prototype 4T structures, and outlined possible production pathways. Kondo et al [7] used a 4T structure to measure the component currents in a tandem cell thereby enabling a more detailed analysis. Dadouche et al [8] have conducted a computer simulation study of 4T cells. They used a numerical solution of the device equations for each cell, enabling parameters such as defect densities and top-cell layer thickness to be adjusted. Their simulations are carried out for the case of a polymorphous silicon top-cell. They conclude, in common with Madan [6], that a thin top-cell leads to optimal performance in a degraded 4T micromorph, i.e. one in which this sub-cell has a high defect density. A study of 4T connected modules was recently performed by Hudansky *et al* [9], who found significant improvements in initial efficiency over commercial micromorph modules, under both standard test conditions and real outdoor conditions. Curiously, similar stabilised efficiencies following degradation were obtained, contrary to expectations.

Here we present a device simulation procedure rather different to that of Dadouche et al [8]. It is based on a semi-empirical approach, and makes use of suitably-scaled experimental *J-V* curves. The power delivered in each connection for a given top-cell thickness is evaluated, in both un-degraded and light-soaked states. This enables the optimum thickness in both states to be determined for 2T and 4T connections. We have also simulated the effects of varying solar spectral quality over typical outdoor conditions, by including a weighting function applied to the AM1.5 spectrum coupled with statistical data on the irradiance distribution *vs.* average photon energy. The potential benefits of the 4T connection are identified and discussed in the context of previous studies.

2. Modelling

The model assumes a series optical structure, in which the top cell, whose thickness may be varied, filters the light absorbed by the bottom cell, which remains constant in thickness throughout. This is shown schematically in figure 1. No account is taken of any optical or electrical losses arising from additional contact layer(s) required for 4T connection. In order to determine the short-circuit currents, the product of the quantum efficiency (*QE*) and the AM1.5 spectrum, filtered in the case of the bottom cell, is integrated over the entire wavelength range, separately for each cell. These photocurrents are

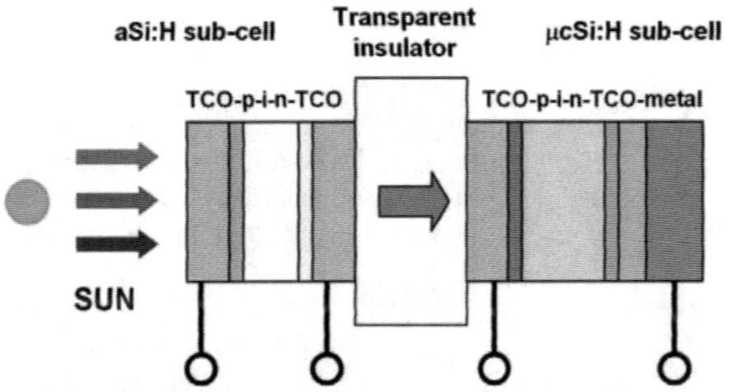

Figure 1. Principle of four-terminal tandem solar cell

then used to scale representative *J-V* curves of good-quality un-degraded a-Si:H and μc-Si:H solar cells along the current density axis, thereby providing an approximation to the characteristics of the component cells. It is accepted that this cannot provide an exact match to the true characteristic, as alongside a general increase in current density J_{SC}, both fill-factor *FF* and open-circuit voltage V_{OC} generally decrease with increasing cell thickness (and are also wavelength-dependent). However, over the i-layer thickness range of interest, the increase in J_{SC} is the dominant factor. To obtain the 2T *J-V* curve, the scaled characteristics are solved simultaneously by graphical means for a range of equal currents, and from this, the maximum power under 2T connection may be calculated. For the 4T case, the unrestricted maximum powers for both cells are calculated, and then added together to obtain the total power.

The *QE* curves used for the top- and bottom-cells are shown in figures 2 and 3 respectively. In the case of the top-cell, the curves are generated by taking a thick (1 μm) un-degraded cell, and then multiplying by the optical absorption curve for the relevant thickness of a-Si:H. This artifice essentially assumes the short-wavelength *QE* is thickness-independent, and the long-wavelength *QE* is not limited by extraction, only by generation of carriers. These curves are then used to scale the *J-V* characteristics in figures 4 and 5, in the manner described above.

The effect of top-cell photo-degradation is included in our simulation in a simple way. The data of Klaver *et al* [2] and Platz *et al* [10] reveal that while all three solar cell parameters J_{SC}, V_{OC} and *FF* are reduced by photo-degradation, *FF* is most strongly affected. For an i-layer thickness of 600 nm or less, J_{SC} falls by less than 10% on 1000 hours exposure to AM1.5, and V_{OC} by less than 3%. However, *FF* falls roughly linearly with thickness, by up to 30%. The *FF* data from those sources are re-drawn in figure 6. In order to model the *FF* degradation, the output from the a-Si:H cell in 2T and 4T configurations is reduced in proportion to the ratio of the degraded to the un-degraded *FF* value at the appropriate thickness.

The influence of solar spectral variations is included in the simulations by means of weighted AM1.5 spectra. This broadly matches a range of insolation conditions, such as those reported by Krishnan *et al* [11]. The weight is set by a linear function, and the spectrum quantified in terms of its average photon energy (*APE*) calculated over the range 350 to 1050 nm. This range is commonly adopted in work on silicon solar cells, and in this context the *APE* of the AM1.5 spectrum is 1.88 eV. The total power between these limits was normalised to a constant value for all simulations, enabling direct calculation of the conversion efficiency. Typical spectra obtained are shown in figure 7. The integration procedure is then applied to obtain the short circuit currents, the *J-V* characteristics scaled, and then solved numerically for 2T connection as described above. In these simulations, for simplicity

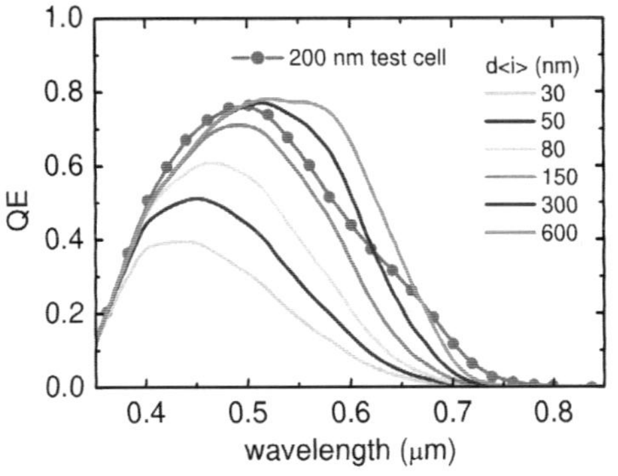

Figure 2. *QE* vs. i-layer thickness d<i> for a-Si:H top cell. See text for how thickness variation is obtained.

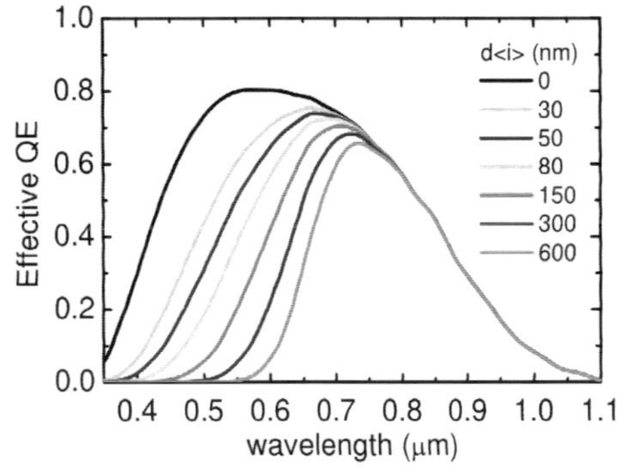

Figure 3. Effective *QE* for μc-Si:H bottom cell of thickness 1.1 μm, including filtering by a-Si:H top-cell of thickness d<i>.

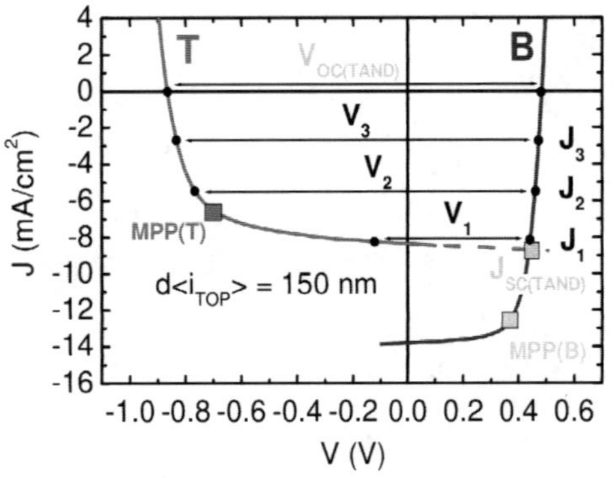

Figure 4. Construction used to obtain the 2T tandem-cell characteristic from top- and bottom-cells (T and B), evaluated at points 1, 2 and 3. The maximum-power points for the separate cells (4T connection) are marked.

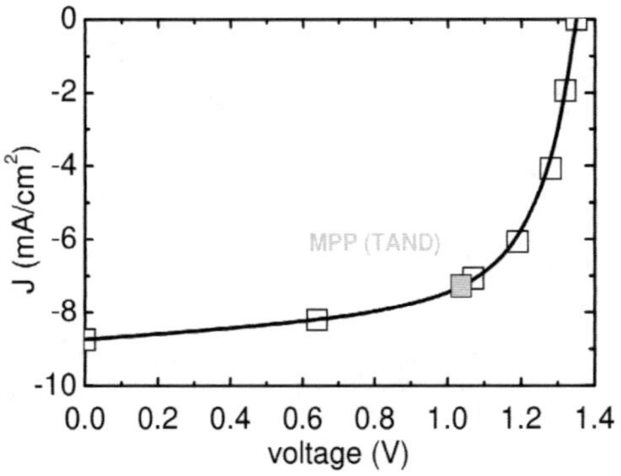

Figure 5. Tandem cell (2T connection) J-V obtained from construction illustrated in figure 4.

we are using the un-degraded J-V characteristic of the top-cell and maintain a constant top-cell thickness of 300 nm.

The power delivered under 2T and 4T connections may be estimated by asserting that the fractional changes in open-circuit voltage and fill-factor are substantially smaller than those in the current, and may thus be neglected. The V_{OC} values used in our simulations were 0.90 V and 0.52 V for a-Si:H and μc-Si:H sub-cells respectively, and 1.42 V for the tandem cell. FF was assumed constant at 70% throughout.

3. Results and discussion

3.1. Effect of top-cell thickness on 2T and 4T performance

The power produced by each cell under 4T conditions *vs.* top-cell thickness is shown in figure 8. In the un-degraded state, the top cell power increases with increasing thickness. The curve gradually levels off, whereas in practice the output reduces above 400 nm due to recombination loss exceeding the gain

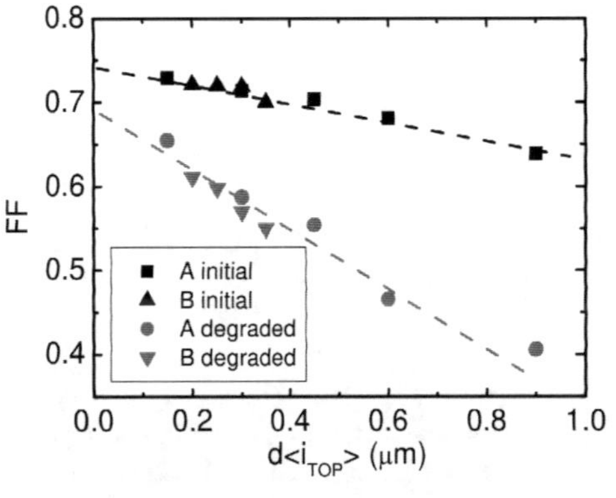

Figure 6. Variation in fill-factor with i-layer thickness in initial and degraded states for typical a-Si:H cells. A and B denote data taken from references [2] and [10], respectively.

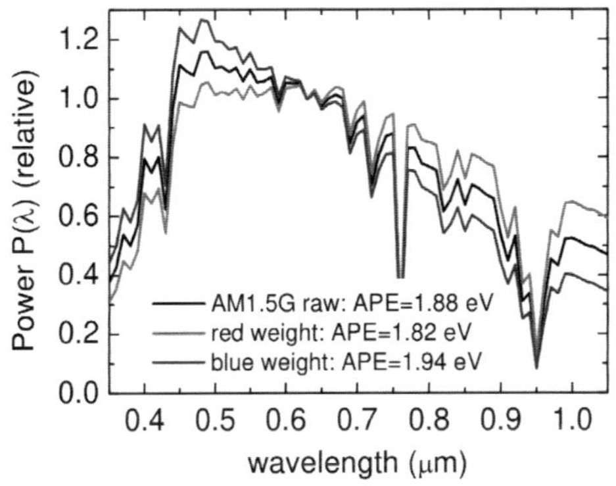

Figure 7. Weighted solar spectra generated using linear filter function. Spectra are normalised to contain equal power between 350 and 1050 nm.

from increased optical absorption. This feature is illustrated in the simulations of Dadouche *et al* [8]. In the degraded state, the inclusion of the fill-factor term results in a peak in output from the top-cell for a thickness of 250-300 nm. The bottom-cell output falls rapidly as the top-cell thickness increases up to 100 nm, reflecting strong top-cell absorption in the blue/green portion of the spectrum. Above 300 nm, the bottom-cell output continues to fall as progressively more of the visible spectrum is absorbed in the top-cell, but does so more gradually as the deep red/IR spectral content lies below the a-Si:H band gap.

The efficiency of the 2T and 4T connected tandems, in initial and degraded states, is shown in figure 9, as a function of top-cell thickness. In the initial state, the efficiencies of both 2T and 4T connections continue to rise up to 600 nm top-cell thickness. The 2T efficiency falls as the thickness is further increased and the system becomes bottom-cell limited (not shown on this scale). The 4T efficiency tends towards a plateau. If there is no additional recombination loss in the top-cell with increasing thickness, then the highest 4T tandem efficiency will be achieved with a thick top-cell, because carriers created close to the band edges, by band-gap energy photons, undergo the minimum thermalisation loss, and will thus generate more electrical energy when absorbed in the top-cell than the bottom cell. In practice, recombination increases as the internal electric field falls, which reduces *FF* [2], and the 4T curve will reach a maximum dependent on the intrinsic defect density in the un-degraded top-cell. We note that the 4T connection (10.8%) yields a higher efficiency than the 2T (10.0%) connection since in the former case both cells operate at maximum power transfer.

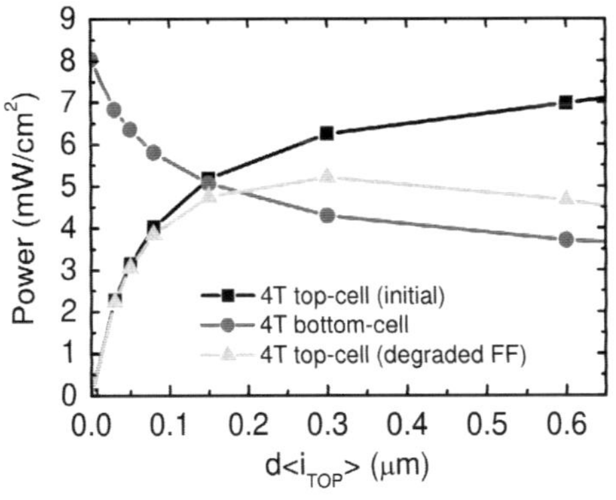

Figure 8. Power delivered by top- and bottom-cells in independent connection (4T). It is assumed that the bottom-cell parameters do not degrade.

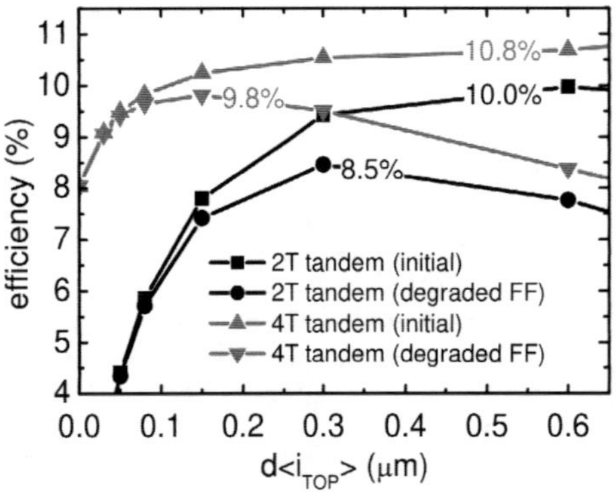

Figure 9. Summary of conversion efficiency *vs.* top-cell thickness for 2T and 4T tandem connections, under initial and degraded top-cell conditions.

For i-layer thicknesses > 80 nm, degradation reduces the power produced by a given top-cell, in either configuration. The 2T optimized efficiency of 8.5% is reached for a top-cell thickness around 350 nm. However, the 4T peak efficiency of 9.8% occurs at 100-150 nm. There is a higher carrier thermalisation loss than in the former case, since more higher-energy photons will pass through the thinner top-cell to the lower bandgap bottom-cell, but this is more than recouped by reduced Joule heating in the top-cell by the bottom cell due to the current mismatch. Under these conditions, the model predicts that the top-cell delivers some 45% of the total tandem power in 4T, compared with 65% in 2T configuration.

Overall, we predict, in the degraded state, a significant benefit of the 4T connection over the 2T connection – an improvement in absolute photovoltaic efficiency of some 1.3%, and a more stable output. However, the percentage figures quoted here should be taken as indicative, rather than absolute, since they depend on the quality of the 'initial' *J-V* characteristics and the approximations inherent in this model.

We now compare our results with those of Dadouche *et al* [8] in more detail, beginning with the component cells (4T connection, figure 8). While the initial and degraded top-cell powers show very similar dependence on top-cell thickness, the bottom-cell power contribution is rather smaller in their case, even though the bottom cell in their model is slightly thicker (1.5 μm against 1.1 μm). At a top-cell thickness of 100 nm, our model predicts a bottom-cell contribution of 5.5 mW/cm^2, compared with 3 mW/cm^2. The figures are 4 mW/cm^2 and 2.5 mW/cm^2 at 400 nm thickness. Comparison with experimental data [12] supports the view that our bottom-cell currents are too large. This is probably because we do not take account of reflections, scattering, absorption in extra TCO layers etc.

Comparing the total output in 2T and 4T connections (figure 9) with the equivalent figures in Dadouche *et al* [8], there is reasonable overall agreement between the curves, although some differences in detail exist. The main discrepancies are: (i) In 2T connection our model predicts little degradation for top-cell thicknesses below 150 nm, rising to a 25% relative power loss at 600 nm. In their model, the relative loss is around 15%, and is independent of top-cell thickness between 50 and 500 nm. (ii) In 4T connection, for a degraded top-cell, we observe a clear maximum in efficiency around 150 nm, whereas in their study it is almost independent of top-cell thickness. (iii) For a degraded cell, our model predicts an increase in optimized efficiency from 8.5% (2T) to 9.8% (4T), a 1.3% improvement, against a prediction of only 0.2% in their model. At present we have no clear explanation for these discrepancies.

3.2. Effect of solar spectral distribution on 2T and 4T performance

The simulated efficiency for a tandem cell comprising an un-degraded top-cell of thickness 300 nm coupled with a 1.1 μm bottom-cell, under 2T and 4T connection over a range of spectral conditions is

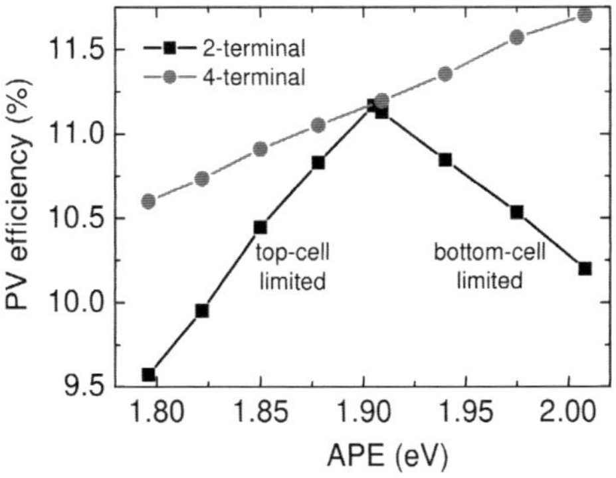

Figure 10. Simulation of photovoltaic efficiency for 2T and 4T connected tandem cells, *vs.* average photon energy *APE*.

presented in figure 10. Under 2T connection, the tandem is limited by the top-cell current under red-rich conditions, and by the bottom-cell current under blue-rich conditions. The optimum current-match for this simulation is obtained with a slightly blue-rich spectrum relative to AM1.5 – at an *APE* of 1.91 eV as against 1.88 eV. Under 4T connection, current matching is no longer an issue, and a progressive improvement in efficiency is obtained with increasing blue content. This is because an increasing proportion of photons have an energy above the a-Si:H bandgap, and the electron-hole pair thus generated by these photons is subject to a smaller thermalisation loss than in the μc-Si:H cell.

However, figure 10 does not convey the true picture, since it does not take into account the prevalence of red-rich or blue-rich conditions over a period of time. In order to predict the likely improvement obtainable outdoors, some statistical information is required. Figure 11 shows a histogram of irradiance distribution (*vs. APE*) over a three-year cycle after Minemoto et al [13], re-scaled so that the peak irradiance lies at zero *APE*. By means of this distribution we have estimated the annual energy output, by summing the electrical energy produced by each histogram element, and then calculating an 'annual average' efficiency by dividing by the total irradiance. The result is shown in figure 12. This plot shows the effect of shifting the mean value of the distribution shown in figure 11 relative to the matched condition in 2T connection (at *APE* = 1.91 eV in this case). There is little change in the 4T response, since the underlying efficiency rises monotonically with *APE*. However in comparison to figure 10, the 2T response shows a lower peak in annual efficiency, some 0.3% below the corresponding 2T value. This is because the 'wings' of the distribution are converted at a lower efficiency than the centre. The 2T response falls either side of the peak because the cell is no longer matched to the *APE* at the centre of the distribution. This indicates that a 2T module

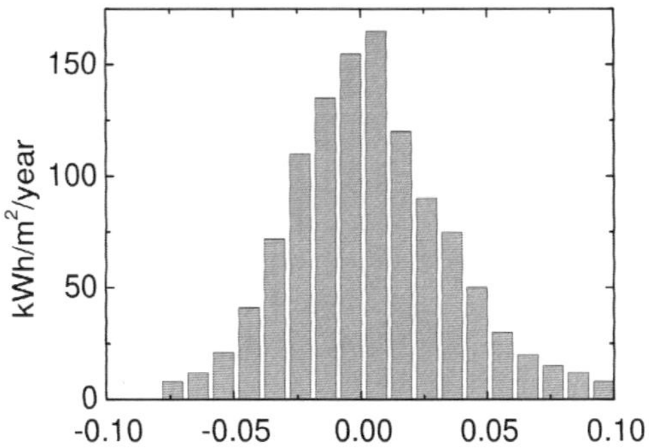

Figure 11. Histogram showing solar irradiance distribution over three annual cycles [13]. 90% lies within ± 0.05 eV of the mean *APE*.

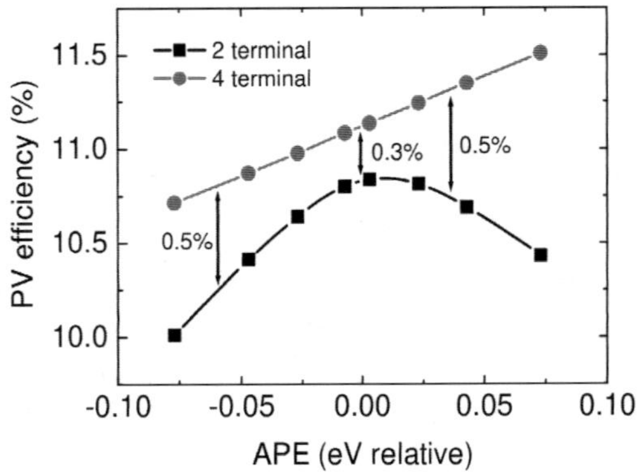

Figure 12. Annual-averaged efficiencies of 2T and 4T connected tandem cells vs. position of irradiance peak (figure 11) relative to 2T current-matched *APE*.

matched to AM1.5, but not to the *local* outdoor conditions, may incur losses in annual efficiency of around 0.5%. Further work is required to determine the prevalence of local variations in *APE* but a preliminary investigation suggests a range of 1.86 – 1.92 eV.

Ulbrich *et al* [14] recently demonstrated, by applying blue or IR bias light from an LED in addition to the AM1.5 spectrum, that the current-matching condition for a two-terminal tandem cell coincides with a *minimum* in the fill factor, and that maximum power is obtained at a point where the top-cell short-circuit current is around 5% greater than that of the bottom cell. Taking this into account modifies the efficiency curve for the 2T tandem cell (figure 10), making the reduction in efficiency more gradual. As a consequence, the 2T annual average efficiency is improved, and the relative benefit of 4T connection is less pronounced. We are currently investigating the potentially important consequences of this, and will report our findings in a subsequent publication.

4. Conclusions

The performance of two-terminal (series connected) and four-terminal (independent) amorphous silicon/microcrystalline silicon tandem solar cells has been modelled using a semi-empirical approach, in which the short-circuit current obtained by integration of the product of the quantum efficiency curve and the solar spectrum is used to scale the *J-V* curves. The current-voltage curve for the series-connected cells is then obtained by graphical solution. In order to calculate efficiencies it is assumed that (i) the open-circuit voltage and fill-factor for all cells are insensitive to spectral effects and thus remain constant throughout, (ii) in the case of the two-terminal cell the circuit current is the lesser of the two sub-cell photo-generated currents.

The model predicts improved overall efficiency in four-terminal connection, for both as-deposited and light-degraded cells, by 0.8% and 1.3% respectively, under AM1.5 illumination. Relaxation of the current-matching boundary condition enables a thinner top-cell than is optimal in two-terminal connection to be used, which reduces both the degree of degradation and the fraction of the total power delivered by the top-cell. An optimal top-cell absorber layer of around 150 nm (with a 1.1 μm bottom-cell) is indicated.

A filter function is introduced which enables the spectrum to be varied in a broadly realistic manner over the range typically encountered outdoors (average photon energy between 1.8 and 2.0 eV). The two-terminal tandem cell shows a peak in efficiency of 11.2% at *APE* = 1.90 eV, where its top- and bottom-cell currents are matched, but falls steeply to 9.6% and 10.4% at 1.8 and 2.0 eV respectively. In contrast, the four-terminal connection, in which both cells are operated at the maximum power point, shows a monotonic increase in efficiency, from 10.6% to 11.7%.

A more realistic long-term (annual-average) efficiency is calculated based on a typical annual irradiance *vs. APE* distribution. From this we estimate that four-terminal connection provides an improvement in absolute efficiency of at least 0.3%, but could exceed 0.5% (or an increase of 5% in annual energy production) under a realistic degree of mismatch.

Our model does not take into account coupling and losses between the sub-cells. In two-terminal connection ohmic loss can result from the intermediate tunnel junction, whereas in the four- terminal case there may be increased optical loss due to the need for thicker TCO layer(s) at the intermediate contact to support the current extracted.

Future plans include development of a model based on the ASA program suite [15-17], which fully takes into account the electrical and optical device properties, and may be calibrated against a series of top- and bottom-cells. This offers greater flexibility to examine the effect of varying the full range of design parameters, including additional TCO contacts, and allows refinements such as an intermediate reflector to be considered.

Acknowledgments
The authors are grateful to Andreas Lambertz for measurement of solar cell characteristics at short notice, and to Kaining Ding for helpful discussions and his assistance in developing ASA code.

References

[1] Green M A 2006, *Third Generation Photovoltaics: Advanced Solar Energy Conversion* (Berlin: Springer) chapter 5
[2] Klaver A and van Swaaij R A C M M 2008 *Sol. En. Mater. Sol. Cells* **92** 50
[3] Shah A V, Schade H, Vanecek M, Meier J, Vallat-Sauvain E, Wyrsch N, Kroll U, Droz C and Bailat J 2004 *Prog. Photovolt: Res. Appl.* **12** 113
[4] van den Donker M N, Gordijn A, Stiebig H, Finger F, Rech B, Stannowski B, Bartl R, Hamers E A G, Schlatmann R and Jongerden G J 2007 *Sol. En. Mater. Sol. Cells* **91** 572
[5] Smirnov V, Das C, Melle T, Lambertz A, Hülsbeck M, Carius R and Finger F 2009 *Mater. Sci. Eng. B* **159** 44
[6] Madan A, *Materials and Manufacturing Processes* 2007 **22** 412
[7] Kondo M, Nagasaki S, Miyahara H, Matsui T, Fujibayashi T, Sato A and Matsuda A 2003 *Proc. 3rd World PVSEC, Vols. A-C* p 2779
[8] Dadouche F, Béthoux O, Gueunier-Farret M E, Johnson E V, Roca I Cabarrocas P, Marchand C and Kleider J-P 2011 *EPJ Photovoltaics* **2** 20301
[9] Hudanski L, Kasouit S, Francke L, Damon-Lacoste J, Besnier J-F, Roschek T, Ahmed K, Zhao L, Cassagne V and Vermeersch M 2011 *37th IEEE PVSC* (Seattle WA, 19-24 June 2011) p 605
[10] Platz R, Hof Ch, Fischer D, Meier J and Shah A 1998 *Sol. En. Mater. Sol. Cells* **53** 1
[11] Krishnan P, Schüttauf J W A, van der Werf C H M, Houshyani Hassanzadeh B, van Sark W G J H M and Schropp R E I 2009 *Sol. En. Mater. Sol. Cells* **93** 691
[12] Schicho S, Hrunski D, van Aubel R and Gordijn A 2010 *Prog. Photovolt: Res. Appl.* **18** 83
[13] Minemoto T, Fukushige S and Takakura H 2009 *Sol. En. Mater. Sol. Cells* **93** 1062
[14] Ulbrich C, Zahren C, Noll J, Blank B, Gerber A, Gordijn A, Rau U 2011 *26th EUPVSEC* (Hamburg, 5-9 September 2011) 302
[15] Zeman M, Willemen J A, Vosteen L L A, Tao G and Metselaar J W 1997 *Sol. En. Mater. Sol. Cells* **46** 81
[16] Pieters B E, Krc J and Zeman M 2006 *4th IEEE WCPEC* p 1513
[17] Ding K, Kirchartz T, Pieters B E, Ulbrich C, Ermes A M, Schicho S, Lambertz A, Carius R and Rau U 2011 *Sol. En. Mater. Sol. Cells* **96** 3318

Some results on the germanium telluride density of states

C Longeaud*[1], J Luckas[1,2] and M Wuttig[2]

[1] Laboratoire de Génie Electrique de Paris (UMR 8507), Plateau de Moulon, 11 rue Joliot Curie, 91190 Gif sur Yvette, France
[2] I. Physikalisches Institut IA, RWTH Aachen University, 52056 Aachen, Germany

*E-mail: longeaud@lgep.supelec.fr

Abstract. Germanium telluride (GeTe) is one of the most studied phase change materials. Surprisingly, only little is known about the distribution of states (DOS) in its band gap. In this paper we investigate both experimentally and theoretically the amorphous GeTe DOS and we propose a model for this DOS as well as orders of magnitude of some of the transport parameters of this material. The DOS we propose is subsequently compared to those suggested to explain the threshold switching and the resistance drift phenomena.

1. Introduction

Phase change materials (PCM) exhibit amazing properties that make them excellent candidates for rewritable data storage. Though already included in everyday devices, improving their potentialities is still a matter of research. Intense studies aim at data storage using multiple resistance levels created by different amorphous volumes within phase change memory devices. However, it was observed that drift of these resistance levels may eventually lead to a loss of information if they come too close. Some theories propose that this drift originates from an evolution with time of the density of states (DOS) of the PCM band gap [1, 2]. Another phenomenon is still a matter of debate: the threshold switching and the drift of the threshold voltage. Again the models proposed are based on the DOS of the PCM band gap [3] and on its evolution with time or annealing of the material [4].

Germanium telluride (GeTe) is one of the most studied PCM but the states distribution in its band gap remains speculative. In its amorphous state (a-GeTe) most authors agree on the presence of band tail states but their extent in the gap is only roughly estimated. Additionally, deeper states may be present but their position in energy and distributions are unknown [5, 6]. To investigate the PCM density of states the modulated photocurrent technique (MPC) seems to be well adapted [7]. Indeed, we have recently probed the a-GeTe DOS by means of this technique [8] and shown that the 'standard' treatment of the MPC data [9, 10] should be modified in order to take into account the evolution of the band gap width with respect to temperature [11]. In this paper, we investigate the a-GeTe DOS both experimentally and by means of numerical calculations, trying to reproduce evolutions with temperature *T* of the dark and photo conductivity, of the MPC spectra and variations with photon energy of the absorption coefficient measured on this material.

After a brief description of the samples and the experiments we used to characterize them we present experimental results. We insist on the modulated photocurrent technique and show how it is possible to account for the widening of the a-GeTe band gap with decreasing temperature to derive a refined DOS spectroscopy. Then we use a numerical simulation to try to reproduce the experimental results by a proper choice of transport parameters and DOS. We detail the method we used to estimate the extended states hole mobility μ_p and the equivalent density of states N_v at the valence band edge E_v. From the results of these numerical calculations we propose a model for the a-GeTe DOS. We detail the different states distributions as well as their capture coefficients. From this model, we expect to get a better understanding of the DOS in a-GeTe and, further, to investigate on the possible link(s) between the respective evolutions of resistivity, threshold switching and DOS. We finally look at the agreements of this model of DOS with some models found in the literature.

2. Samples and experiments

GeTe amorphous thin films, 100-200 nm thick, were deposited on glass substrates by means of a LS 320 von Ardenne Sputter system operating in constant 20–25 W power mode, 20 s.c.c.m. Ar flow and with stoichiometric targets of 99.99 % purity. For Fourier Transform Infrared (FTIR) spectroscopy in reflectance mode the glass was covered with a 200 nm thick aluminium film before a-GeTe deposition. For electrical measurements two parallel rectangular Al ohmic electrodes of 8 mm length and 2 mm apart were deposited on top of the phase change films deposited on glass.

Steady-state dark and photo- conductivity (SSPC) were measured each 10 K on samples fitted onto the cold finger of a dynamically pumped cryostat for temperatures in the range 100-300 K. A constant flux (F_{dc} = 10^{16} cm^{-2}s^{-1}) of infrared light (850 nm) was used for photoconductivity measurements. After reaching the measurement temperature under dark conditions, we waited 10 min for the temperature to stabilize and took the average value of 20 acquisitions, one each second, to extract the dark conductivity. Then the light was turned on and we waited one more minute under light before recording again 20 values, the average of which leads to the estimate of photoconductivity. The error was estimated to be less than 5 %. The same light source was used to perform MPC experiments to determine the DOS distribution in the band gap of these films. MPC measurements were achieved each 20 K or each 5 K for $100 \leq T \leq 300$ K. The amplitude of the modulated flux was 4 times lower than the mean dc flux and the modulation frequencies at a given temperature were in the range 12 Hz – 40 kHz. For each couple (temperature, frequency) the measured phase shift and modulus of the alternating current were the average of 20 acquisitions by a lock-in amplifier and the error could be estimated to be of the order of 10%. From FTIR measurements, we deduced the evolution of the band gap with T. Reflectance measurements were achieved for $4 \leq T \leq 300$ K with photon energies E from 0.04 to 1 eV. The absorption coefficient $\alpha(E)$ was deduced from the analysis of the reflectance pattern by means of a Tauc-Lorentz oscillator model proposed by the software SCOUT [12]. The band gap was defined as the energy value for which $\alpha(E)$=10^4 cm^{-1} known as the E_{04}. In addition, Seebeck measurements were performed from room temperature to 220 K giving the position of the Fermi level E_f with respect to E_v. Finally, for some samples, photo-deflection spectroscopy (PDS) experiments [13, 14] were achieved at room temperature and the curves were calibrated with the absorption coefficient.

3. Experimental results

Figure 1 displays a typical Arrhenius plot of the dark conductivity and total conductivity under illumination -that we shall call photoconductivity in the following- of an a-GeTe film. These films presented very little difference between dark and photo current from 300 K down to 200 K. The dark conductivity at T = 300 K was of the order of σ_{dark} = 5x10^{-4} S/cm and, from the plot of figure 1 we deduced that σ_{dark} had an activation energy $E_a \approx 0.31$-0.32 eV for $210 \leq T \leq 300$ K. For temperatures below 200 K the temperature variations of the dark conductivity clearly depart from a straight line and this departure may be interpreted as the onset of a hopping conduction as we shall see latter.

Figure 2 displays a PDS spectrum normalized to the absorption coefficient for which three zones can be observed. At low energy the absorption coefficient varies exponentially with energy and we can define an Urbach energy E_u of the order of 63 meV. This linear variation is followed by a parabolic variation and, further, by a saturation when all the photons are absorbed by the layer.

Although the samples were not highly photoconductive over a large range of temperature, it was possible to perform MPC experiments from 300 K down to 100 K. It was shown [9, 10] that from the measurement of the modulus of the photoconductivity $|\sigma_{ac}|$, resulting from modulated generation rate G_{ac}, and of ϕ, its phase shift referred to the excitation, it is possible to achieve a DOS spectroscopy deducing the variations with energy of the quantity NC_p/μ_p, N being the density of states at an energy E_ω and C_p the capture coefficient of the states trapping free carriers (holes in a-GeTe, hence the index p), from the equations

$$\frac{N(E_\omega)C_p}{\mu_p} = \frac{2}{\pi k_B T} qG_{ac} \frac{\sin\phi}{|\sigma_{ac}|} \tag{1}$$

with

$$E_\omega - E_v = k_B T \ln\left(\frac{C_p N_v}{\omega}\right) \tag{2}$$

k_B being the Boltzmann constant, ω the angular frequency of the modulation and q the electron charge.

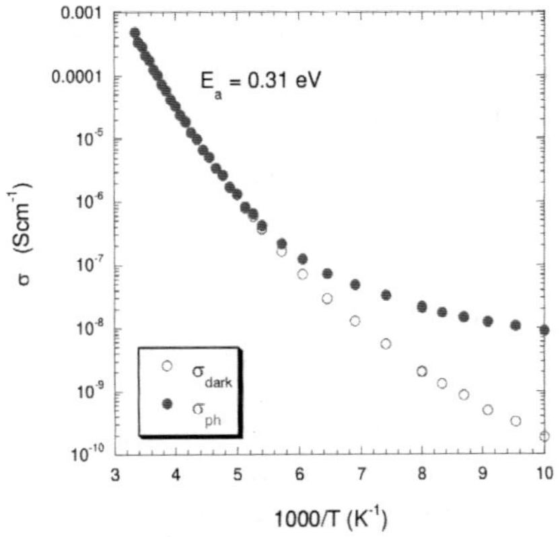

Figure 1. Typical variations with T of dark and photo conductivities observed on an a-GeTe film.

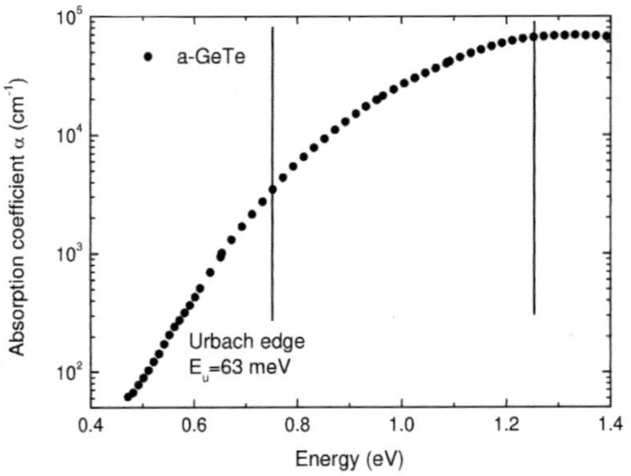

Figure 2. Typical PDS spectrum normalized to the absorption coefficient observed on the a-GeTe films (symbols). The vertical lines define three zones: an exponential variation (left), a parabolic variation (middle) and a saturation of the absorption coefficient (right).

The reader may note that all the quantities on the right hand side of Eq. (1) are known. Therefore, by varying experimentally T and ω, Eqs. (1)-(2) offer the possibility to scan the DOS of the material provided C_p, μ_p and N_v are known. One quantity of particular importance is the attempt to escape frequency $\nu = C_pN_v$ for it fixes the energy scale of the DOS spectroscopy (see Eq. (2)).

The symbols of figure 3 display a first scan of the density of states of an a-GeTe film. Each symbol corresponds to a given temperature and, at this temperature, the deepest symbol in energy corresponds to the lower frequency (12 Hz) and the highest to the higher frequency (40 kHz). For this plot the attempt to escape frequency was fixed at $\nu = 10^{12}$ s^{-1}. The upper envelope of these symbols describes a DOS made of deep states and of a valence band tail (VBT). It can be seen that all the sets of symbols fit onto this envelope for the points obtained at high frequency and depart from it for the points obtained at low frequency. The choice of ν was not done at random but precisely to obtain this behaviour. Indeed, it was shown [10] that the departure of the points from the upper envelope could be due to the influence of the mean dc flux needed to achieve the modulation. This mean dc flux gives rise to a recombination zone in the gap, the extent in energy of which depends on its intensity: the larger the flux, the larger the extent. Eq. (1) being valid only for trapping states, when applied to experimental data controlled by the recombination region, the deepest states probed at a given temperature, it leads to this departure from the upper envelope that describes the trapping states outside the recombination region. The estimate of ν was done in order to have the upper points of a spectrum obtained at a given T matching with the upper points of a spectrum obtained at the nearest temperatures at least for the VBT states. For lower values of ν the spectra would cross one another and for higher values they would not match at all. This procedure is one of the possibilities to estimate the ν values that are appropriate to describe the valence band tail states.

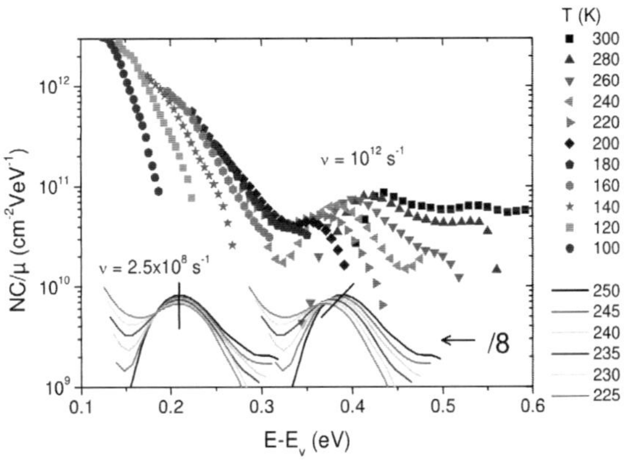

Figure 3. NC/μ spectroscopy achieved from MPC measurements from 300 to 100 K each 20 K steps (symbols). The bump around 0.4 eV has been probed more accurately with MPC measurements in the range 250-225 K each 5 K (Lines). Its energy scaling was done with two different values of ν and it was plotted after a division by 8 of NC/μ to give a clear overview of the different distributions.

As far as deep states are concerned it can be seen that MPC spectra reveal a bump around 0.4 eV. A better investigation of this bump was achieved by varying the temperature in 5 K steps for $225 \leq T \leq 250$ K. The spectra are plotted by full lines in figure 1 for two different values of ν: 10^{12} s^{-1} for the spectra on the right hand side of figure 3 and $2.5\text{x}10^{8}$ s^{-1} for the spectra on the left hand side. To have a clear overview, the NC/μ values for these spectra have been divided by 8 as shown in the figure. A well defined bell shape is described by the different spectra. In addition, with $\nu = 2.5\text{x}10^{8}$ s^{-1} the maxima of the spectra are located at the same energy which is not the case with $\nu = 10^{12}$ s^{-1} as indicated by the short full line segments. Obviously, if this bell shape distribution corresponds to a peak of states its maximum must stay at the same energy position independently of the experimental temperature. This suggests that the proper ν associated to these states is of the order of 10^{8} s^{-1} and it is a wrong value of ν used in Eq. (2) that leads to the distribution of deep states described by symbols around 0.4 eV in the first scan we made of the DOS. The alignment of the maxima of several MPC spectra is another possibility to estimate the attempt to escape frequency of the probed states [15].

The two procedures described above to find estimates of ν can be refined if one takes into account the fact that the band gap of a-GeTe varies with temperature. Indeed, FTIR measurements at different temperatures showed that the a-GeTe band gap varies with T following the equation

$$E_{04}(T) = E_{04}(T=0\ K) - \xi T^{2} \qquad (3)$$

ξ being of the order of $\xi \approx 1.3\text{x}10^{-6}$ eV/K^2. It means that between 300 K and 100 K the gap width increases by 0.1 eV, a rather large value. In addition, Seebeck measurements showed that this evolution of the gap was essentially due to a movement of the valence band edge toward a fixed conduction band edge. Therefore, Eq. (2) has to and can be corrected to take account of the gap shift with T. Two types of corrections can be proposed and are detailed in Ref. 11. To correct the MPC energy scaling given by Eq. (2), and to define the same energy origin for all the MPC spectra taking account that the conduction band edge does not move with temperature in agreement with the Seebeck measurements, we may consider that the shrinkage of the gap results in a modification of the trap energies proportional to their positions at 0 K, i.e. states near the valence band edge shift by the full correction term $-\xi T^{2}$, mid-gap states only by $-0.5\ \xi T^{2}$ and the energy positions of the states close to the conduction band are almost unchanged. In this case the correction is made pro rata to the energy position of the traps and one has

$$\Delta E = k_B T \ln\left(\nu/\omega\right) - \xi T^2\left(1 - \frac{\Delta E}{E_{04}}\right) \tag{4}$$

giving

$$\Delta E = \frac{k_B T \ln\left(\nu/\omega\right) - \xi T^2}{1 - \xi T^2 / E_{04}} \tag{5}$$

that leads to the spectra displayed in figure 4.

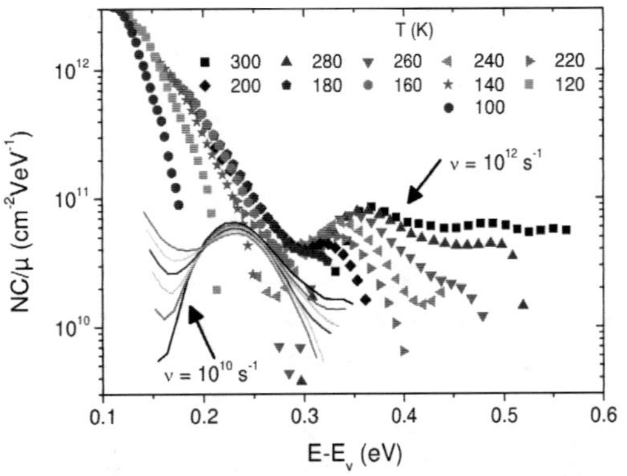

Figure 4. *NC/μ* distributions of states taking account of a pro rata correction to the spectra displayed in figure 3.

It can be seen in figure 4 that the VBT is still well described using $\nu = 10^{12}$ s^{-1} all the parts of the spectra obtained at high frequency fitting well from one spectrum to the other. However, for temperatures below 140 K the spectra clearly depart from the upper envelope as if the mean dc flux used to perform the experiment was too high and a wide recombination region was needed to reach equilibrium conditions. This possibility has to be considered but the VBT increasing exponentially the recombination zone created by the mean dc flux does not need to extend widely to create enough recombining states. The reader may also note that the disagreement of the spectra with the upper envelope increases with decreasing temperature. The reason of this increasing disagreement has to be probably found in an increasing influence of the hopping conduction with decreasing temperature. Indeed, we have shown that hopping conduction could be responsible for such departures of the MPC spectra from the upper envelope [16]. We would like to underline that the temperature range in which hopping seems to have a large influence on the MPC spectra, say below 160 K, corresponds to the temperature range for which dark conductivity is not following anymore an activated behaviour in the Arrhenius plot of the dark conductivity (See figure 1).

As far as deep states are concerned the bell shape distribution is still present but we had to increase the ν value up to $\nu = 10^{10}$ s^{-1} to align the maxima of the different spectra at the same energy of ≈ 0.24 eV. For the deepest states (0.5 eV above E_v) we lack information to estimate the ν value, for none of the procedures described above seem to be applicable in this case. That is why the MPC spectra have been plotted with the same ν as for the VBT states though the true value is probably different.

To investigate more precisely on this distribution of states we have developed a numerical calculation that can be used to reproduce the experimental results. The basic idea of this approach is not to reproduce the experimental data point after point but rather to underline the main trends observed experimentally.

4. Numerical calculation

The numerical calculation that we have developed can reproduce the behaviours of many techniques among which the SSPC, the MPC and the variation with photon energy of the absorption coefficient. The program, DeOSt, is available on the web site of the LGEP [17]. The material parameters (band gap width, mobilities, gap states distribution, capture coefficients, etc..) and the 'experimental' parameters (temperature, flux, frequency of the modulation,...) can be defined by the user. The program calculates the data corresponding to the chosen experiment and these data can be treated as 'experimental' results subsequently and compared with real experimental results. The number of adjustable parameters is rather large. Therefore, one may introduce as many experimental data as possible to minimize the number of unknowns. For instance one can introduce the band gap width, choose the valence band tail width, fix the Fermi level position, or some ν values from the experimental data. The other parameters are adjustable in order to eventually reproduce the general experimental trends observed in SSPC, MPC and PDS but these adjustments cannot be achieved randomly since we have to fit the results of three independent experiments.

From the SSPC measurements we deduce an order of magnitude of the dark Fermi level position above the valence band edge from the activation energy of dark conductivity $E_f - E_v \approx 0.31$ eV. Considering that an expression of the conductivity σ is

$$\sigma = q\mu_p N_v \exp[-(E_f - E_v)/k_B T] \tag{6}$$

we can deduce a good order of magnitude of the product $\mu_p N_v$ at room temperature using also the σ value at 300 K, $\sigma = 5\text{x}10^{-4}$ S/cm, to finally end with $\mu_p N_v \approx 10^{21}$ cm^{-1}V^{-1}s^{-1}. We can combine these results with those of the MPC experiment. This technique reveals a valence band tail decreasing exponentially from the valence band edge following the equation

$$N(E) = N(E_v)\exp[-(E_v - E)/E_{vbt}] \tag{7}$$

We can then deduce an order of magnitude of the characteristic energy of the VBT, $E_{vbt} \approx 30$-33 meV. The best matching we have for the different MPC spectra describing the VBT is given by $\nu = C_p N_v = 10^{12}$ s^{-1}, from which we deduce an estimate of the ratio $C_p/\mu_p \approx 10^{-9}$ cm/V. The estimate of NC/μ from Eq. (1) is completely determined by experimental data. For instance, if we consider the data obtained for $E - E_v = 0.2$ eV, $NC/\mu \approx 5\times10^{11}$ cm^{-2}eV^{-1}V, it is possible to estimate $N(E_v) \approx 2\times10^{23}$ cm^{-3}eV^{-1} from Eq. (7) assuming $E_{vbt} \approx 33$ meV. Finally, if we suppose that $N_v \approx k_B T N(E_v)$, we end with $N_v \approx 5\times10^{21}$ cm^{-3}, $\mu_p \approx 0.2$ cm^{2}V^{-1}s^{-1} and $C_p \approx 2\times10^{-10}$ cm^{3}s^{-1}.

The PDS spectrum is also a help for the determination of the DOS distribution. Indeed, in the photon energy range 0.5-0.8 eV the variations of the absorption coefficient describe an Urbach tail with a characteristic energy of 63 meV. Since the absorption coefficient can be defined as the convolution product of all the transitions from full states below the Fermi level toward empty states above it, it means that, assuming that the conduction band tail (CBT) is also varying exponentially, the characteristic energy of the CBT is much larger than that of the VBT, and of the order of 60 meV.

Nevertheless, the DOS distribution does not consist only in band tails for, if it was the case, the dark Fermi level would be pinned at their intersection, i. e. 0.25 eV above E_v as shown by the calculation. This value is too small to account for the experimental activation energy and, in addition, MPC has revealed deep states distributions among which a peak of states around 0.25 eV. This peak of states is most probably of donor type since acceptor states would push the Fermi level toward the valence band edge. It is well described by the MPC spectra with an attempt to escape frequency of $\nu = 10^{10}$ s^{-1}, that is a capture coefficient of the order of $C_p \approx 2\times10^{-12}$ cm^{3}s^{-1} and from the NC/μ value we can deduce $N_{max} \approx 10^{22}$ cm^{-3}eV^{-1}.

Figure. 5(a) displays a first draft of a DOS that can be drawn from the experimental results. The band gap width has been set at 0.75 eV to obtain a good match between the calculated and experimental PDS spectra. Figure 5(b) displays the MPC spectra that can be calculated from the DOS introduced in the numerical calculation. The valence band tail and the bump around 0.4 eV are well reproduced but the deepest states found experimentally above 0.4 eV are not reproduced by the calculation.

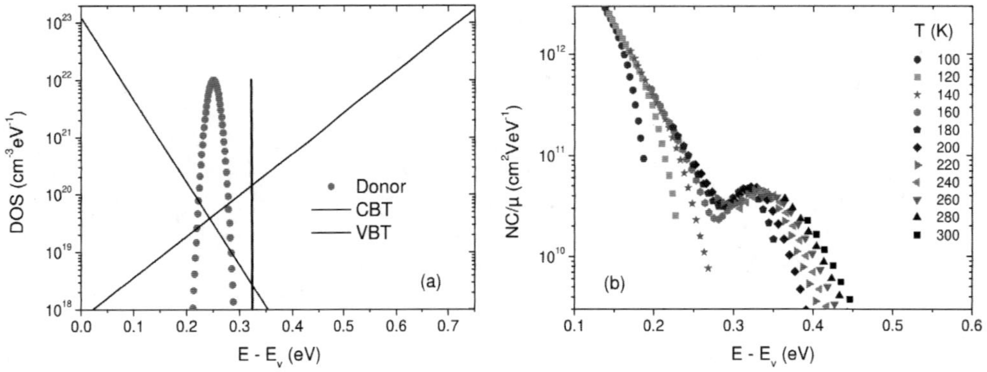

Figure 5. (a) density of states deduced from the experiments the vertical line indicates the position of the dark Fermi level (b) MPC spectra calculated from the DOS displayed in (a).

Hence, this DOS does not bring satisfying results. In addition, the calculation showed that the dark Fermi level is not fixed and moves to different positions when the temperature varies, the result being a dark conductivity with an activation energy of 0.27 eV. Deeper states have to be added to fix the dark Fermi level position and give a contribution to the MPC spectra above say 0.45 eV.

The only way to reproduce a contribution of a DOS distribution to the MPC spectra for energies above 0.45 eV is to add at least one peak of states above the Fermi level. These states must be of acceptor type otherwise the Fermi level will move up toward the conduction band. The determination of their parameters (density and capture cross-section) is rather difficult because being located above the dark Fermi level and close to it they will not be pure trapping states as they can play a role in the recombination of carriers. They may give a contribution to the MPC data but for such states the application of Eq. (1) is theoretically not possible since it applies only to trapping states. Nevertheless, as we have done experimentally we can apply Eq.(1) to the calculated data and play with the parameters of these states to eventually get a result as close to the experiment as possible.

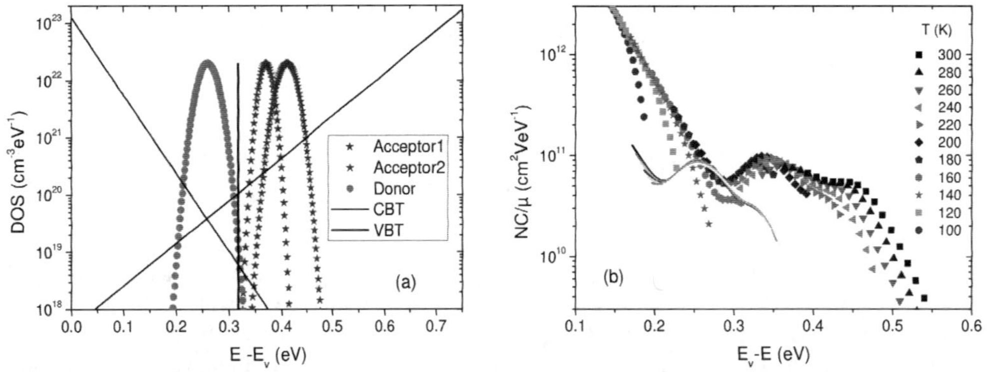

Figure 6. (a) Density of states introduced into the numerical calculation. The dark Fermi level position is indicated by the vertical line. (b) MPC spectra obtained from the DOS in (a) plotted with $\nu = 10^{12}$ s^{-1} (symbols) and with $\nu = 10^{10}$ s^{-1} (lines). Same temperatures, same symbols and colors as figure 4.

Actually, to extend the MPC spectra towards high energy (deepest states) we had to add two distributions of acceptor states above the dark Fermi level. Figures 6(a) and 6(b) display the DOS introduced in the calculation and the corresponding MPC spectra, calculated for the same temperature as the experimental ones, respectively. The similarity between figure 6(b) and figure 4 is rather good though we have not reproduced the deepest states observed experimentally for energies of 0.55 eV and above. For this purpose we should have introduced more states above the Fermi level but, considering the lack of information concerning them, it would have been very speculative.

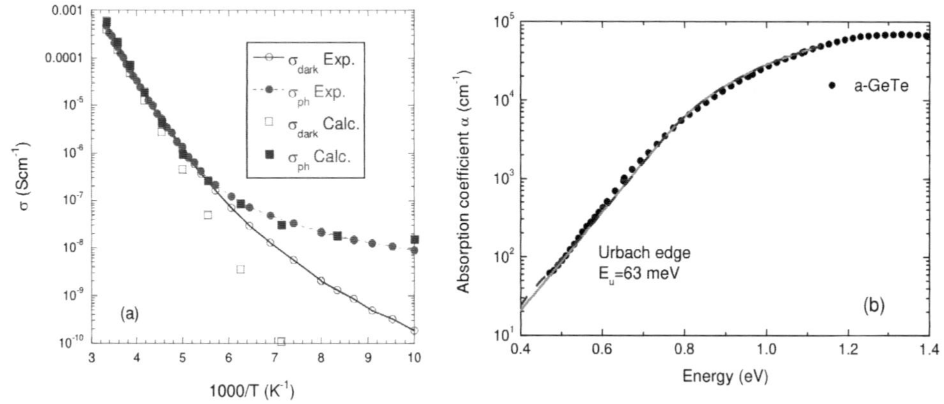

Figure 7. (a) Plot of dark and photo conductivity, experimental and calculated with the DOS of figures 6(a). (b) Plot of the absorption spectra, experimental (symbols), calculated with only the band tails (full line) and including the deep states (dashed line).

With such a distribution of states we can also reproduce variations of the photoconductivity on the whole range of investigated temperatures and those of dark conductivity for $T \geq 210$ K (See figure 7(a)), the disagreement at low temperatures originating from the fact that our calculation does not take hopping transport into account. We have also reproduced the PDS spectrum (See figure 7(b)). For this purpose we have chosen different values for the optical matrix elements C_{opt} of the band tails and the deep states. Taking only the band tails into account, with a VBT extent of 33 meV and a CBT extent of 60 meV and $C_{opt} \approx 8\times10^{-22}$ cm$^{-5/2}$eV, leads to the full line curve matching the PDS spectrum in figure 7(b). We could not choose the same optical elements for the deep states otherwise they would have given rise to a large bump around 0.6 eV. Choosing optical elements 40 times lower than for the band tails leads to the fit displayed by a dashed line in figure 7(b). Nevertheless, these values for the optical elements are only estimates from our calculation and should be confirmed experimentally.

Before concluding we would like to add a few words on the distributions of states that we propose.

5. Discussion

The reader may note that the band tails we propose are in contradiction with those suggested by Bahl and Chopra [5, 6] that had imagined a CBT narrower than the VBT. However, they stressed that their proposal was only schematic because of the lack of information at the time they wrote their communication. In addition, a-GeTe is p-type and if the band tail states were distributed as they proposed the Fermi level would be pushed toward the conduction band, leading to n-type material. The tail states distributions may vary slightly from one sample to the other but MPC experiment has always revealed a rather narrow valence band tail distribution and the PDS results suggest a large conduction band tail to account for the large Urbach edge.

Concerning the deep states, we have modelled them by essentially two distributions of states on each side of the dark Fermi level, donor states below E_f and acceptor states above. These states were chosen to be monovalent whereas multivalent states with negative correlation energy have been proposed as for instance in the valence alternating pair (VAP) model [18]. However, this VAP model, proposed for lone pair semiconductors as chalcogenide materials, has been recently questioned in the case of tellurides such as $Ge_2Sb_2Te_5$ [19]. That is why we have preferred to model the density of deep states by monovalent states that are easier to handle and to understand, particularly in terms of recombination paths, and with which we can describe rather accurately the transport behaviours we have observed experimentally.

We would like to underline that the distribution of states we propose is rather close to the one proposed by Ielmini *et al.* [3] to explain the threshold switching phenomenon. These authors have suggested that the threshold switching phenomenon originates from the presence of two deep levels of states. Assuming that the electrons are the majority carriers, their description is that the deeper level is a reservoir that may fill the upper level, close to the band, if the applied field is high enough to lower the barrier between states of the deeper level and states of the upper one. Our model of DOS includes these possible two levels, one of the main differences being that the majority carriers would be holes in our case but the process could be exactly the same.

However, we have found that the a-GeTe DOS distribution includes a large valence band tail that is not taken into account in Ielmini's model. Besides, experimental data suggests that hopping is predominant at low temperature. Our simulation cannot deal with hopping in the case where the states have different capture cross-sections since the calculation time can rapidly becomes too long, but we have made a test with the same capture cross section for all the states and we have reproduced the same trends as those observed experimentally. We are then convinced that hopping is present, masked at room temperature by multiple trapping transport but becomes significant say below 160 K. A large literature on hopping transport in band tails can be found in which the concept of transport level is salient. Under 'normal' conditions, low temperature and low field, this transport level is far from the band edge. However, it was demonstrated that the transport level in band tail can be pushed as close as 50 meV from the band if the field is large enough [20]. Playing with the equations proposed by Cleve *et al.* it is easy to find that, with the VBT we have found, a field of 200 kV/cm, a reasonable value for a threshold field, would push the Fermi level to 75 meV ($3k_BT$ at room temperature) above the valence band. Hopping in band tail and evolution of the transport level with the applied field could be then an alternative to the Ielmini's model.

Another point that we can deal with is the resistance drift phenomenon. This behaviour is usually attributed to an evolution of the band gap and/or band gap states with aging. Our model can reproduce this behaviour easily. On the one hand, an opening of the band gap without a modification of the DOS, as suggested by Boniardi *et al.* [21] and Pirovano *et al.* [22] for $Ge_2Sb_2Te_5$, will lead to a decrease of conductivity since the Fermi level is pinned by the deep states, and on the other hand, a decrease (increase) of the acceptor (donor) states will move E_f toward the conduction band resulting in conductivity decrease reproducing the resistance drift phenomenon.

This discussion shows that the DOS distribution we propose neither rejects nor validates one particular model or the other. The DOS we proposed seems to be slightly more refined that any of the models mentioned above, gathering them in a whole picture. We think that, based on this DOS, the existing models describing the threshold switching and the resistance drift phenomena could be refined or some alternative explanation may be proposed if needed. We have already started SSPC, PDS and MPC measurements on aged samples and the results of these experiments should be helpful for future understanding of the complete transport mechanisms in PCM materials.

6. Conclusion

We have investigated on the density of states and transport parameters of a-GeTe by means of three different experiments: dark and photo-conductivity, PDS and MPC. We have underlined that the 'standard' treatment of MPC had to be refined to account for the evolution of the band gap width with temperature, and suggested a new treatment of the MPC data. From the experimental results of these three techniques we have shown that some of the a-GeTe transport parameters could be deduced.

Introducing these parameters into a numerical simulation we have reproduced all the main trends of the three experiments with a model of DOS distribution made essentially of two band tails, one donor deep states below the Fermi level and acceptor deep states above the Fermi level.

We have compared the DOS we propose with some of the models of the literature and found that it does not contradict any of them, providing a more elaborate picture of the states distribution instead since the features of this DOS can explain the threshold switching and the resistance drift phenomena. Experimental investigations on the DOS of aged or annealed a-GeTe should provide, by comparison with the as deposited state, a better comprehension of these two puzzling characteristics of PCM materials [23].

Acknowledgements
This work was partly supported by a Procope project (#21870VB) funded by Egide. Many thanks to R. Carius and J. Klomfass for the PDS measurements and D. Krebbs for valuable discussions.

References

[1] Pirovano A *et al.* 2004 *IEDM Tech. Dig.* **51** 714
[2] Boniardi M *et al.* 2009 *J. Appl. Phys.* **105** 084506
[3] Ielmini D and Zhang Y 2007 *J. Appl. Phys.* **102** 054517
[4] Karpov I V, Mitra M, Kau D, Spadini G, Kryukov Y A and Karpov V G 2007 *J. Appl. Phys.* **102** 124503
[5] Bahl S K and Chopra K L 1970 *J. Appl. Phys.* **41** 2196
[6] Bahl S K and Chopra K L 1969 *J. Appl. Phys.* **40** 4940.
[7] Oheda H 1981 *J. Appl. Phys.* **52** 6693
[8] Luckas J, Krebs D, Salinga M, Wuttig M and Longeaud C 2010 *Phys. Status Solidi C* **7** 852
[9] Brüggemann R, Main C, Berkin J and Reynolds S 1990 *Philos. Mag. B* **62** 29
[10] Longeaud C and Kleider J P 1992 *Phys. Rev. B* **45** 11672
[11] Luckas J, Kremers S, Krebs D, Salinga M, Wuttig M and Longeaud C 2011 *J. Appl. Phys.* **110** 013719
[12] Theiss W, http://mtheiss.com/.
[13] Boccara A C, Fournier D and Badoz J 1980 *Appl. Phys. Lett.* **36** 130
[14] Jackson W B, Amer N M, Boccara A C and Fournier D 1981 *Applied Optics* **20** 1333
[15] Longeaud C, Kleider J P, Kaminski P, Koslowski R, Pawlowski M and Cwirko J 1999 *Semicond. Science Technol.* **14** 747
[16] Longeaud C and Tobbeche S 2009 *J. Phys.: Condens. Matter* **21** 045508
[17] Longeaud C http://www.lgep.supelec.fr/index.php?page=scm-logiciels
[18] Kastner M, Adler D and Fritzshe H 1976 *Phys. Rev. Lett.* **37** 1504
[19] Robertson J, Xiong K and Peacock P W 2007 *Thin Solid Films* **515** 7538
[20] Cleve B, Hartenstein B., Baranovskii S D, Scheidler M, Thomas P and Baessler H, 1995 *Phys. Rev. B* **51** 16705
[21] Boniardi M, Redaelli A, Pirovano A, Tortorelli I, Ielmini D and Pellizer F 2009 *J. App. Phys.* **105** 084506
[22] Pirovano A, Lacaita A L, Pellizzer F, Kostylev S A, Benvenuti A and Bez R 2004 *IEEE Transaction on Electronic Devices* **51** 714.
[23] Luckas J, Rausch P, Krebs D, Zalden P, Boltz J, Jost P, Raty J-Y, Salinga M, Klomfaß J, Carius R, Longeaud C and Wuttig M 2012 Frontiers in Electronic Materials, Wiley-VCH, ISBN 978-3-527-41191-7

Surface Energy, Surface Debye Temperature and Specific Heat of Nanocrystals

M Michailov and I Avramov

Institute of Physical Chemistry, Bulgarian Academy of Sciences,
"Acad. G. Bonchev" Str., Block 11, 1113 Sofia, Bulgaria

E-mail: mike@ipc.bas.bg

Abstract. The present study provides insight into the role of surface free energy and surface Debye temperatures in the specific heat of nanocrystals. In the case of small phases, consisted of comparable number of bulk and surface atoms, the variation of heat capacity with respect to that of infinitely large phase is significant. This effect, not accounted for in the Debye model is due to the substantial impact of interface energy of clusters. Here, in the framework of classical Gibbs thermodynamics, we introduce surface heat capacity that accounts for surface free energy contribution. Defined as a difference, between surface and bulk terms, the excess heat capacity reveals temperature gap and size limits of nanoclusters where the contribution of surface energy is essential. Considering the role of surface Debye temperatures, we evaluate the deviation of specific heat for nanoclusters having different surface atomic density and surface orientation. Being in agreement with recent experimental findings, the present model opens up a way to predict basic, size-dependent thermodynamic properties of nanocomposite systems.

1. Introduction

The nanocomposite materials consisted of atomic and molecular clusters demonstrate peculiar properties, tightly dependent on the size of their building particles [1-3]. Being of great technological importance nanocrystalline glass-ceramics, ferromagnetic materials, nanostructured dielectrics, nanotubes, nanofluids and nanodroplets have attracted special attention in the last decade [4-7]. The principal point of this attention is focused on the problem how to predict and control basic physical properties of materials. Hence, the problem is associated with detailed knowledge of thermodynamic behaviour of these nanoscale systems. On that physical background, in the present study we analyse the impact of surface free energy and surface Debye temperatures on the specific heat of nanocrystals.

Experimental evidence of size-dependent heat capacity has been recently reported for metallic nanoclusters and nanodielectrics [8-10]. Despite its clear physical background, the size-dependent effect, not taken into account in the classical Einstein-Debye theory, is still subject to limited number of theoretical treatment. In a previous study [11], confined to the Einstein heat capacity model and the Lindeman model of melting of small atomic clusters, we have shown significant deviation of specific heat for nano-sized crystals. In the present paper, implementing alternative approach, we extend and refine the classical Debye model in the case of finite size phases [11,12]. Here, in view of the role of dividing surfaces in Gibbs thermodynamics, [13], we account for the impact of the surface free energy and surface Debye temperatures on the specific heat of nanoclusters. The present model considers the heat capacity of n-atomic cluster with n_B bulk and, n_S

surface atoms, having total number of atoms $n = n_B + n_S$. It involves the cluster heat capacity C_V as an additive function of bulk C_V^B and the surface C_V^S , terms, i.e.

$$C_V\left(\frac{n_S}{n}\right) = \frac{n_B}{n} C_V^B + \frac{n_S}{n} C_V^S \tag{1}$$

where $\frac{n_B}{n}$ and $\frac{n_S}{n}$ are the relative number of surface and bulk atoms in the cluster. Since $\frac{n_B}{n} + \frac{n_S}{n} = 1$, the heat capacity can be expressed by the relative number $\frac{n_S}{n}$ of surface atoms. Let us point out that significant variable for the tiny nanoscale effects is the relative number of surface atoms related to the total number of atoms in the system and therefore in our further treatment we will consider the value $C_V\left(\frac{n_S}{n}\right)$. Hence, Eq. (1) transforms to:

$$C_V\left(\frac{n_S}{n}\right) = \frac{nC_V^B + n_S\left(C_V^S - C_V^B\right)}{n} \tag{2}$$

The above equation contains the important quantity:

$$\Delta C_V = \frac{n_S}{n}\left(C_V^S - C_V^B\right) \tag{3}$$

defined as a cluster excess heat capacity compared to that the large system. Here, C_V^B relates to infinitely large phase where the contribution of dividing surface is missing. In the following sections of the present paper, we will attest that the value of ΔC_V is significant for the thermodynamic properties of nanocrystals.

2. Excess Heat Capacity and Surface Energy

According to classical thermodynamics [13,14], the atomic heat capacity C_V could be determined by the second derivative of the Helmholtz thermodynamic potential *F*, i.e.

$$C_V = -\frac{T}{n}\left(\frac{\partial^2 F}{\partial T^2}\right)_V \tag{4}$$

For pure substances and large phases $F = n\mu$ where μ is the chemical potential. In that case, C_V^B related to one atom in the bulk is:

$$C_V^B = -T\left(\frac{\partial^2 \mu}{\partial T^2}\right)_V \tag{5}$$

The above expression does not account for the role of the interface. In the case of finite size system where the contribution of surface free energy could be significant, the exact Helmholtz potential [14] has the form:

$$F(n) = n\mu + \sum_i \sigma_i A_i \tag{6}$$

σ_i being the specific surface energy of ith crystal face and A_i is the respective surface area. In isotropic approximation, the last term in Eq. 6 representing the surface free energy turns to $F(n) = n\mu + n_S\sigma$ where σ is the surface energy related to the area of single atom. Here the total number of surface atoms $n_s = \sum_i n_{s_i}$, $n_S\sigma = \sum_i n_{s_i}\sigma_i$, and therefore σ is considered as an effective or average surface energy of the nanocrystal. It is important to note that the model presented here is developed in constrain of the classical capillary theory, i.e. σ does not depend on the size of the cluster. Additional refining that accounts for variation of surface energy with the system size does not change the basic conclusions of our approach. The Eq. 6 allows evaluating independently the excess contribution of surface atoms to heat capacity. Hence, the size dependence of C_V according to Eq. 4 and Eq. 6 will be expressed through the relative number of surface atoms $\frac{n_S}{n}$.

$$C_V\left(\frac{n_S}{n}\right) = -T\left[\left(\frac{\partial^2 \mu}{\partial T^2}\right)_V + \sum_i \frac{n_{S,i}}{n}\left(\frac{\partial^2 \sigma_i}{\partial T^2}\right)_V\right] \tag{7}$$

It is clearly seen that equation (7) consists of bulk, $C_V^B = -T\left[\left(\frac{\partial^2 \mu}{\partial T^2}\right)_V\right]$, and of surface, $C_V^S\left(\frac{n_s}{n}\right) = -T\sum_i \frac{n_{S_i}}{n}\left(\frac{\partial^2 \sigma_i}{\partial T^2}\right)_V$, terms. Therefore it determines the excess heat capacity ΔC_V, already defined in Eq. 3. Evidently, this excess is due to the surface contribution. The substitution of $C_V^S\left(\frac{n_S}{n}\right)$ and C_V^B from Eq. 7 into Eq. 3 yields:

$$\Delta C_V = -T\sum_i \frac{n_{S_i}}{n}\left(\frac{\partial^2 \sigma_i}{\partial T^2}\right)_V \tag{8}$$

This expression reveals an important relation between the heat capacity, relative number of the surface atoms $\frac{n_S}{n}$ and surface energy σ. This dependence will be discussed in details in Section 4 of the present article.

3. The Impact of Surface Debye Temperature

The next step in our approach relates to non-thermodynamic analysis of heat capacity for finite-sized system, i.e. atomic cluster in terms of the classical Debye model [19]. In a previous study we have already treat a similar problem [11,12]. Extensive experimental studies reveal a larger mean-square displacement of surface atoms compared to that of the bulk [15,16]. The creation of a free surface and broken symmetry leads to significant reduction of the total forces affecting atomic vibrations, both normal and parallel to the surface. Therefore, the acoustic phonon spectrum of small system is perturbed and the specific heat of cluster with comparable number of surface and bulk atoms is expected to differ from that of the large system [17,18]. This divergence will be evaluated in the frame of Debye theory. Following the Debye model, the basic equation for C_V is expressed through the Debye function $F_D\left(\frac{T}{T_D}\right)$ and corresponding Debye temperature $T_D = \frac{h\nu_D}{k_B}$ as:

$$C_V^B = 3k_B\, F_D\left(\frac{T}{T_D}\right) \tag{9}$$

where k_B is the Boltzmann constant. Let us point that classical Debye model relates to ensemble of harmonic quantum oscillators for infinitely large phase. In the case of finite size system (nanocluster), the phase is consisted of $n = n_B + n_S$ atoms, with bulk ν_D and surface ν_S Debye frequencies and respectively aligned Debye temperatures T_D, $T_D^{S_i}$. According to Eq. 1, the heat capacity $C_V\left(\frac{n_s}{n}\right)$ is considered as an additive function of bulk and surface terms and therefore from Eq. 9 it follows:

$$C_V\left(\frac{n_s}{n}\right) = 3k_B\left[F_D\left(\frac{T}{T_D}\right) + \left[\sum_i \frac{n_{S_i}}{n} F_D\left(\frac{T}{T_D^{S_i}}\right) - \frac{n_s}{n} F_D\left(\frac{T}{T_D}\right)\right]\right] \tag{10}$$

The above general expression for $C_V\left(\frac{n_s}{n}\right)$ holds the essential feature to distinguish between bulk and surface Debye temperature contribution. It is worthwhile to stress here that being a subject of large number of studies, both T_D and T_D^S are experimentally assessable quantities [15]. Therefore, the validity of Eq. 10 could be a subject to relatively simple experimental verification.

4. Surface Energy and Surface Debye Temperature

We have derived two physically independent theoretical expressions for $C_V\left(\frac{n_s}{n}\right)$ in the case of small, finite size phases. The first expression, Eq. 7, follows from classical Gibbs thermodynamics [13,14] while the second, Eq. 10, follows from Debye theory [19]. The formal similarity between Eq. 7 and Eq. 10 is conspicuous. Both are consisted of bulk and surface terms, and therefore the following set of relations can be drawn:

$$\left(\frac{\partial^2 \mu}{\partial T^2}\right)_V = -\frac{3k_B}{T} F_D\left(\frac{T}{T_D}\right) \tag{11}$$

and

$$\left(\frac{\partial^2 \sigma}{\partial T^2}\right)_V = -\frac{3k_B}{T}\left[F_D\left(\frac{T}{T_D^s}\right) - F_D\left(\frac{T}{T_D}\right)\right] \tag{12}$$

Here, Eq. 11 results from the bulk terms similarity and Eq. 12 results from the surface terms similarity. The integration of Eq. 12 give:

$$\sigma = \sigma_o - \int_0^T \left(\int_0^T \frac{3k_B}{T} \left[F_D\left(\frac{T}{T_D^{s_i}}\right) - F_D\left(\frac{T}{T_D}\right) \right] dT \right) dT \tag{13}$$

where σ_o is the surface energy at 0 K. This equation satisfies the requirement, $\left(\frac{\partial \sigma}{\partial T}\right)_V \Big|_{T\to 0} \to 0$, imposed by the third law of thermodynamics. The Eq. 13 has two very important characteristics. First, it demonstrates the functional dependence of surface free energy on temperature and second, it reveals for the first time the relation between surface free energy σ and surface Debye temperature $T_D^{s_i}$. Moreover, since $T_D^{s_i}$ depends on the crystallographic orientation of the surface [15], the expression of σ in Eq. 13 accounts for the symmetry and atomic density of the respective crystal faces.

Following the model suggested here, we can also derive an expression for the excess heat capacity, identical to Eq.3 but now in terms of Debye model for finite size
phases. After substitution of Eq. 10 into Eq. 3, $\Delta C_V\left(\frac{n_S}{n}\right)$ turns to:

$$\Delta C_V\left(\frac{n_S}{n}\right) = 3k_B \left[\sum_i \frac{n_{S_i}}{n} F_D\left(\frac{T}{T_D^{s_i}}\right) - \frac{n_S}{n} F_D\left(\frac{T}{T_D}\right) \right] \tag{14}$$

It is essential to point that for nanocrystals the quantity $C_V\left(\frac{n_S}{n}\right)$ strongly depends on the absolute value of surface Debye temperature. Hence, a new contribution, not accounted for in the classical Debye model appears. This expression will be applied to evaluate the relative increase of specific heat when the size of nanoclusters decreases. It will also trace out the temperature gap where the size-effects are significant.

5. Discussion

According to Gibbs thermodynamics, the substantial difference between large phases and nanoclusters originates from the role of surface. The broken symmetry at the interface, resulting in significant reduction of the local forces, strongly affects the atomic vibrations of surface atoms. Following the experimental findings, the mean-square displacements of surface atoms are larger compared to these in the bulk [15,16]. A clear evidence is the ratio $\frac{T_D^s}{T_D} \approx 0.6 \pm 0.1$, calculated from the existing diffraction data (TEAS, LEED and RHEED) on surface and bulk Debye temperatures given in Ref. [15]. The pointed difference between surface and bulk vibration energies causes deviation of heat capacity of finite sized nanocluster from that of infinitely large phase. This expected divergence is manifested in figure 1. The temperature variation of $C_V(n)$ for clusters with variable size (number of atoms) is calculated according to Eq. 10. The parameters taken are: $T_D = 400K$, $T_D^s = 0.6T_D$, $n = 1000$ (upper dotted curve), $n = 27000$ (middle dashed curve) and $n = \infty$ (lower solid curve). The presented results reveal significant contribution of surface layer to the heat capacity of small phases. Nanocluster of 1000 atoms has almost 20% greater C_V than the corresponding bulk phase.

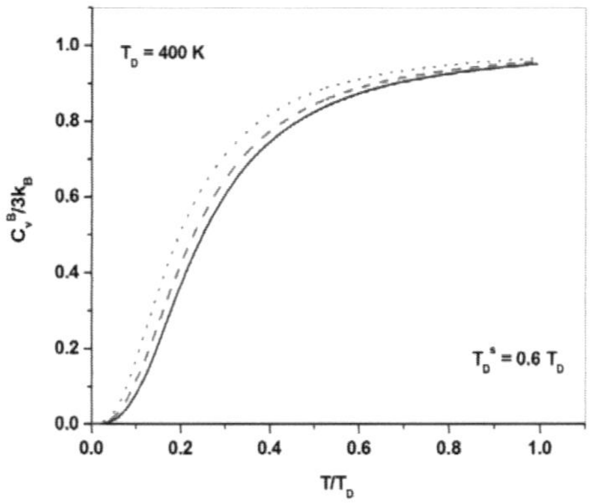

Figure 1. Heat capacity, C_V^B, of infinitely large ($n \to \infty$) homogeneous phase (solid curve) compared to that of phases consisted of n-atomic clusters according to Eq. (10) (T_D=400 K; T_D^s=0.6T_D). The upper dotted curve reflects the heat capacity behaviour of 1000-atomic nanocluster (nanocrystal 10x10x10 atoms); the dashed line in between corresponds to 27000-atomic nanocluster (30x30x30 atoms).

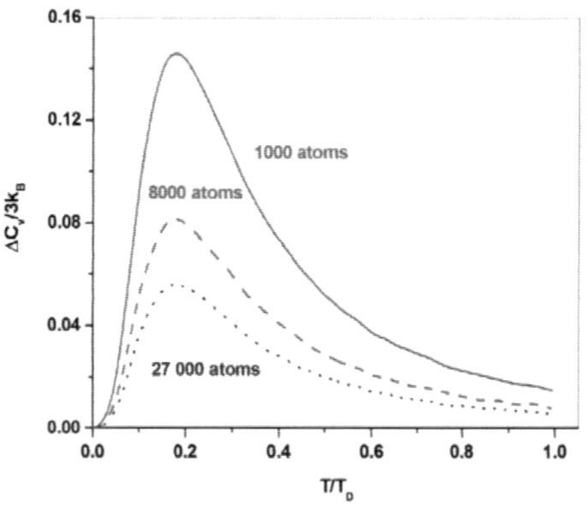

Figure 2. The excess heat capacity, $\Delta C_V\left(\frac{n_S}{n}\right)$, of phase consisted of n-atomic clusters (n = 1000 solid line, n = 8000, dashed line and n = 27000 dotted line) as a function of T. At high T the limit $\Delta C_V\left(\frac{n_S}{n}\right) \to 0$ where both infinitely large homogeneous phase and phase consisting of nanoclusters satisfy the Dulong and Petit 3R limit.

It is important to note here, that observed size-dependent behaviour of $C_V\left(\frac{n_s}{n}\right)$ is consistent with the classical thermodynamic limit, i.e. at high temperatures both infinitely large homogeneous phase and phase consisted of *n-atomic* nanoclusters satisfy the Dulong and Petit $3k_B$ condition. Similar course has the temperature dependence of $\Delta C_V\left(\frac{n_S}{n}\right)$, figure 2. The value of $\Delta C_V\left(\frac{n_S}{n}\right)$ define the range where the surface contribution is essential. As it is seen on figure 2, the increase of excess heat capacity is clearly pronounced in immense temperature gap $0.1 < \frac{T}{T_D} < 0.8$. For cluster with size 10x10x10 atoms at $\frac{T}{T_D} \approx 0.2$ the heat capacity is 14% higher than the infinitely large bulk phase. The observed enhancement of heat capacity is in agreement with recent experimental findings [21-23], for example 40 nm Fe nanoclusters [9,10]. According to the present model, this originates of the reduced Debye temperature at the surface, $T_D^s < T_D$, accounted for by Eq. 14. The dependence of the relative increase of heat capacity, $\frac{\Delta C_V}{C_V^B}$, on the number of atoms n is presented in figure 3.

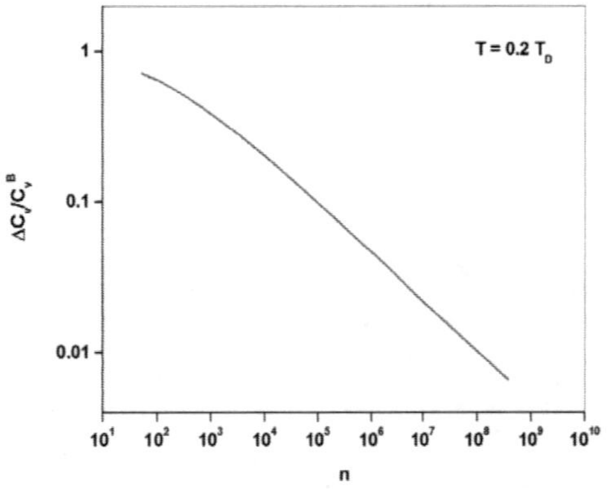

Figure 3. The dependence of the relative increase of heat capacity on the number of atoms n at T = 0.2 T_D. The impact of surface term in Eq. 10, i.e. the deviation of $C_V\left(\frac{n_S}{n}\right)$ from C_V^B is about 40% for clusters in the range of 10^3 atoms (as example: nanocrystal with cubic shape with 10 atoms in edges).

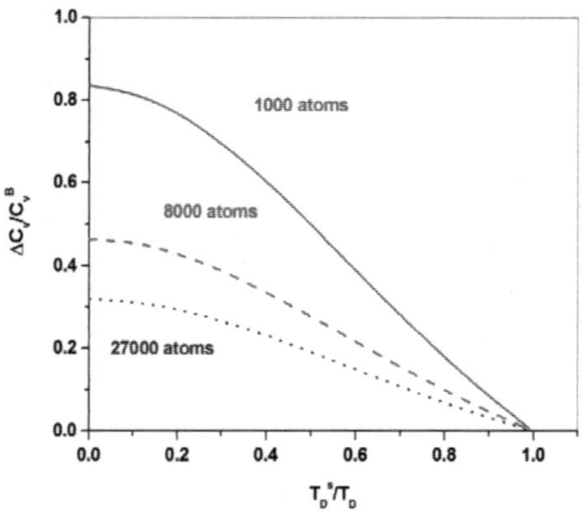

Figure 4. Anisotropy dependence of surface excess heat capacity of nanocrystals build up of 1000, 8000 and 27000 atoms. At fixed cluster size, the crystal surface orientations with lower Debye temperature have larger contribution to the heat capacity. The observed effect is more pronounced for smaller nanocrystals (upper curve).

The variation of $C_V\left(\frac{n_s}{n}\right)$ is shown at T = 0.2 T_D where the observed deviation is well-defined, as it is manifested in figure 1 and figure 2. The relative increase remains significant even for $n = 10^5$ atoms. At that size, the variation of the heat capacity of n-atomic clusters is larger than 3% with respect to infinitely large phase. The important divergence $\Delta C_V\left(\frac{n_S}{n}\right) = C_V\left(\frac{n_S}{n}\right) - C_V^B \geq 0.01 C_V^B$ holds for n $\leq 4\text{x}10^6$ atoms, size corresponding to cubic nanocrystal with 150 atoms in edge.

Having established that the excess heat capacity tightly depends on the number of atoms building the cluster, figure 3, we now explore how the specific heat correlates with the surface Debye temperature. Obviously, the surface phonon modes depend on the symmetry and atomic density of the crystal surface. Therefore, being relevant to the energy of atomic vibrations i.e. to the surface Debye temperature of the corresponding crystal face, the surface excess heat capacity is expected to be anisotropy dependent. The curves shown in figure 4 reveal this variation at different clusters sizes (nanocrystals build up of 1000, 8000 and 27000 atoms) when the surface Debye temperature changes. Each value of T_D^s could be assigned to a given imaginary surface with specific crystallographic orientation, for example, made-up nanocrystal entirely faceted by (111) or (100) or (110) or other high index vicinal surfaces. It is seen that crystal faces with lower Debye temperature (larger atomic vibrations) have more pronounced impact to the surface heat capacity. The steeper slope of the upper curve ($n = 1000$ atoms) suggests stronger effect at smaller clusters. Apparently, when T_D^s approaches to the bulk T_D, the value of ΔC_V tends to zero which is a sign of similar surface and bulk phonon modes.

The final issue of the present study is addressed to the relation between surface free energy and surface Debye temperature. As it follows from Eq. 13, σ correlates with the crystal anisotropy through the surface Debye temperature of specific crystal faces [12]. The integration constant σ_o in Eq. 13 can be evaluated from the existing computational and experimental data for surface energy at fixed temperature [24]. Let us note that σ_o being the surface energy of crystal at absolute zero is not universal constant and depends on specific substance and crystallographic orientation. A typical example, manifesting the temperature behaviour of surface energy is given in figure 5 for Cu(111) and Cu(100) crystal faces.

The values of parameters $T_D = 322\ K$, $T_D^s = 210$ K for <100>, $T_D^s = 244\,K$ for <111> $\sigma_o = 0.92[eV/atom]; \sigma_o = 0.71\ [eV/atom]$ are taken from Ref. [15, 24].

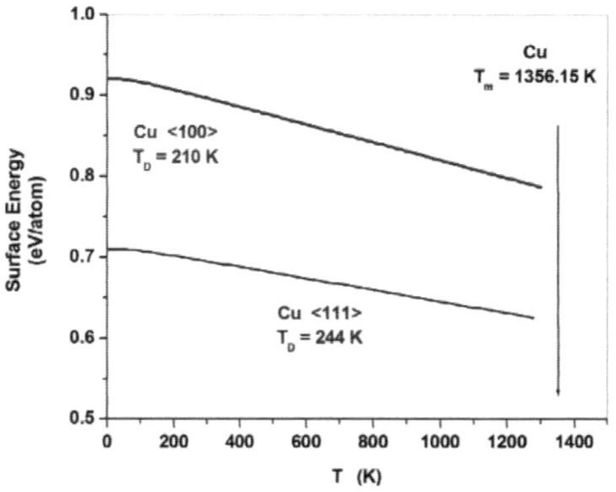

Figure 5. Temperature dependence of the surface free energy of Cu crystal, according Eq. 13. Depending on the atomic density and crystal anisotropy, the surfaces with orientations <111> (lower curve) and <100> (upper curve) have different free energy at 0 K. T_m indicates the bulk melting point of Cu.

It is seen that the temperature trend of surface energy is anisotropy dependent and correlates with the surface Debye temperature of the specific crystal face. This result opens up a way, the large number of existing experimental data (see Ref 15, 16) for T_D^s and T_D to be applied directly for evaluation the surface energy of specific crystal faces. For very small facetted crystals the number of atoms along the edges and vertices is becoming comparable to the number of atoms on the surfaces. So, additional terms should be taken into account. However, this refinement does not change the principal conclusions of the model and it is not a subject to the present study [20].

6. Conclusion

This study reveals the role of surface free energy in the specific heat of small finite size phases. The approach is developed in the framework of classical Gibbs thermodynamics and Debye theory of heat capacity. The presented model introduces an excess heat capacity accounted for by the difference of surface and bulk Debye temperatures. On that physical background, we can predict the heat capacity behaviour of nanoclusters as a function of their size (number building atoms). We found that specific heat of nanocomposite systems can be enlarged by: (i) decreasing the size of nanocrystals and (ii) increasing the relative part of high index surfaces that facet the nanocrystal. Making a bridge between Gibbs free energy and Debye temperatures, the suggested model opens up a way for evaluation of the surface energy of crystals directly from accessible experimental data for T_D^s and T_D .

Acknowledgments
The authors thank Dimo Kashchiev (Institute of Physical Chemistry, Sofia) for valuable discussions and critical reading of the manuscript.

References

[1] Baletto F and Ferando R 2005 *Rev. Mod. Phys.* **77** 371
[2] Mey Q S and Lu K 2007 *Progress in Material Science* **52** 1175
[3] Zhu Y F, Lian J S and Jiang Q 2009 *Journal of Phys. Chem.* C **113** (39) 16896-16900; see also: Zhu Y F, Zheng W and Jiang Q 2009 *Appl. Phys. Lett.* **95** 083110
[4] Kong L B, Zhang T S, Ma J and Boey F 2008 *Progress in Materials Science* **53** (2) 207
[5] Hakuta Y, Hayashi H and Arai K 2003 Current Opinion in: *Solid State and Materials Science* **7** (4-5) 341
[6] Zebli B, Vieyra H A and Carmeli I 2009 *Phys. Rev. B 79, 205402*
[7] Starace A K, Gomez J, Wang J, Pradhan S and Glatzmaier G 2011 *J. Appl. Phys*. **110**, 124323 (2011); http://dx.doi.org/10.1063/1.3672685
[8] Wang L, Tan Z, Meng S, Liang D and Li G 2001 *J. Nanoparticle Res.* **3** 483
[9] Wang T M and Lia J 1995 *Solid State Commun.* **94** 201
[10] Bai H, Luo J, Jin D and Sun J 1996 *J. Appl. Phys.* **79** 361
[11] Avramov I and Michailov M 2008 *J. Phys.: Condensed Matter* **20** 295224
[12] Michailov M and Avramov I 2010 *Solid State Phenomena* **159** 171-174
[13] Gibbs J W 1962 *Thermodynamics*, Dover Publications, Inc. New York, vol. **I**
[14] Rusanov A I 1967 *Fazovye Ravnovesiya i Poverkhnostnye Yavleniya*, Khimiya, Leningrad.
[15] Van Hove M A, Weinberg W and Chan C-M 1986 *Low Energy Electron Diffraction,* Springer Series in Surface Science **6**, ed. G. Ertle, Springer, p. 45
[16] Beauvais C, Rouxel D, Michailov M and Mutaftschiev B 1995 *Surf. Sci.* **324** 1
[17] Couchman P R and Karasz F E 1977 *Physics Letters* **62A** 59
[18] Sadaiyandi K 2009 *Materials Chemistry and Physics* **115** 703
[19] Debye P, 1912 *Ann. Physik* **39** 789; see also: Ashcroft N W and Mermin N D 1981 *Solid State Physics*, Cornell University, Sounders College Philadelphia, Holt-Saunders Japan, Tokyo, p. 458
[20] Michailov M and Avramov I 2012 *(in preparation)*
[21] Lu K, Wang J and Wei W D 1992 *J. Phys. D: Appl. Phys.* **25** 808
[22] Hu X, Wang G, Wu W, Jiang P and Zi J 2001 *J. Phys.: Condens. Matter* **13** L835
[23] Wang L, Tan Z, Meng S, Druzhinina A, Varushchenko R and Li G 2001 *J. Non-Cryst. Solids* **296** 139
[24] Vitos L, Ruban A V, Skriver H L and Kollar J 1998 *Surf. Sci.* **411** 186

The Hubbard model and optics: reflectivity of $1D$ and $2D$ systems

Vladan Celebonovic
Institute of Physics,Univ.of Belgrade,Pregrevica 118,11080 Zemun-Belgrade,Serbia
E-mail: vladan@ipb.ac.rs

Abstract. The Hubbard model (HM) was proposed around the middle of the last century as a model of the behavior of correlated electrons. It has so far been solved only in the $1D$ case. The HM is used in condensed matter physics in work on high T_c superconductors, organic conductors, graphene,...It is known that reflectivity is related to the electrical conductivity, and both quantities can be measured.

This paper is divided into several sections. It begins with a brief introduction to the HM, and the links between the reflectivity and conductivity of a material with its parameters. In the second and third parts,the explicit calculations of the reflectivity of a 1D and 2D square rectangular lattices will be presented. Expressions for the reflectivity thus obtained are tied to the material parameters via previous results of the author on the conductivity of a 1D lattice. We attempt to find a set of material parameters in both cases of lattices, for which the reflectivity becomes close to zero, or equal to an arbitrary constant. If such a set of material parameters could be found, it could have interesting applications in material science such as invisibility .

1. Introduction

1.1. The Hubbard model

The theoretical framework which is now called "The Hubbard model" (HM) is the achievement of the British physicist John Hubbard ($1931-1980$). Despite the fame and wide applicability of his model, only a couple of biographies of Hubbard are easily accessible. One of them is posted at `http://theor.jinr.ru/~kuzemsky/hubbio2.html`. He was a brilliant scientist and a very discrete person.

Around the middle of the last century, the problem of the metal-insulator (MI) transition was high on the list of open problems of condensed matter physics. A certain amount of success in analyzing the MI transition was achieved in the early days of quantum mechanics ([1] and references given there). A more general theory of MI transitions was formulated towards the middle of the last century [2]. The first microscopic theory of the (MI) transition has been proposed in [3], and this marked the start of the development of the (HM).

At first glance, the HM seems to be extremely simple. The Hamiltonian by which it is defined contains only two terms: the "free" kinetic term H_0 and the "interaction" term H_I

$$H = H_0 + H_I \tag{1}$$

Complexity quickly becomes visible when this Hamiltonian is written completely. In two dimensions,using the second quantisation formalism, it has the form:

$$H = -\sum_{ij}\sum_{\sigma} t_{ij} c^{+}_{i\sigma} c_{j\sigma} + \frac{1}{2}\sum_{ijkl}\sum_{\sigma\sigma'} < ij|v|kl > c^{+}_{i\sigma} c^{+}_{j\sigma'} c_{l\sigma'} c_{k\sigma} \tag{2}$$

In this expression,latin indices denote sites on a lattice, σ is the spin index, and symbols of the form $c^{+}_{i\sigma}$ denote second quantisation operators creating an electron with spin σ at a lattice site i. The hopping integral between sites i and j is denoted by t_{ij}. Finally,v is the Coulomb potential. Performing any kind of a calculation with such a Hamiltonian would be prohibitively complicated. However, Hubbard introduced two simplifying assumptions which take into account the essential physics and ease the mathematics. They can be expressed as:

$$t_{ij} = \begin{cases} t & (i,j) \quad nn \\ 0 & otherwise \end{cases}$$

and

$$< ij\,|v|\,kl >= \begin{cases} U & (i = j = k = l) \\ 0 & (otherwise) \end{cases}$$

where nn denotes nearest neighbors.

In spite of the fact that it was proposed nearly 60 years ago, the HM has so far been solved analytically only in the $1D$ case [4]. The interest of this model for the condensed matter community can be appreciated by the number of papers which use it in some way. A search at `http://prola.aps.org` performed at the time of this writing (end of July 2012) with the keyword "Hubbard model" has given 1981 paper containing those words in the title or abstract published in Phys.Rev.Lett. and Phys.Rev.B in the period 2000 - 2012.

Knowledge of the Hamiltonian of a system opens up the possibility of calculating its transport properties. One of the methods for such calculations is the so called "memory function" method, developed in the 1970s as a consequence of previous work of Kubo. Within this method, the conductivity is determined by the following equations:

$$\chi_{AB}(\omega) =<< A; B >>= -i\int_0^{\infty} \exp i\omega t < [A(t), B(0)] > dt \tag{3}$$

where $\mathbf{A} = \mathbf{B} = [\mathbf{j}, \mathbf{H}]$, $\mathbf{j}$ denotes the current operator, and

$$\sigma(\omega) = i\frac{\omega_P^2}{4\pi\omega}[1 - \frac{\chi_\omega}{\chi_0}] \tag{4}$$

The symbol ω_P denotes the plasma frequency, which is given by $\omega_P^2 = 4\pi e^2 n/m$, and where e, n, m are the electron charge, number density and mass respectvely. The current operator is defined as follows [5]:

$$\mathbf{j} = \frac{\partial}{\partial t}\mathbf{P} = i[\mathbf{H}, \mathbf{P}] \tag{5}$$

where

$$\mathbf{P} = \sum_i \mathbf{R_i} n_i \tag{6}$$

is the polarisation operator. In expression (6) $\mathbf{R_i}$ denotes the position of lattice site i, and n_i is the number of particles at lattice site i. Using the $1D$ Hubbard Hamiltonian and the so-called memory function method, it was shown that the electrical conductivity of a $1D$ system of correlated electrons is given by [6]

$$\sigma_R(\omega) = (1/2\chi_0)(\omega_P^2/\pi)[\omega^2 - (bt)^2]^{-1}(Ut/N^2)^2 \times S \tag{7}$$

and the symbol S denotes the following function

$$\begin{aligned} S = 42.49916 \times (1 + \exp(\beta(-\mu - 2t)))^{-2} + 78.2557 \times \\ (1 + \exp(\beta(-\mu + 2t\cos(1 + \pi))))^{-2} + (bt/(\omega_0 + bt)) \times \\ (4.53316 \times (1 + \exp(\beta(-\mu - 2t)))^{-2} + \\ 24.6448(1 + \exp(\beta(-\mu + 2t\cos(1 + \pi)))))^{-2}) \end{aligned} \tag{8}$$

where $b = -4 \times [1 + \cos(1 - \pi)] = -1.83879$, μ denotes the chemical potential of the electron gas on a $1D$ lattice determined in [6], N is the number of lattice sites. The immaginary part of the conductivity is given by

$$\sigma_I = \frac{\omega_P^2}{4\pi\omega}\left(1 - \frac{\chi_R}{\chi_0}\right) \tag{9}$$

and

$$\chi_R(\omega) \cong [128U^2t^3cos^2((1-2\pi)/2)]/[1 + \exp(\beta(-\mu - 2t)))^2N^4(\omega + 2bt)^2)] \tag{10}$$

1.2. Relation with optics

According to standard optics [8] the propagation of an electromagnetic signal through a medium is defined by the frequency dependent dielectric function $\epsilon(\omega)$ and the refractive index $N(\omega) = [\epsilon(\omega)]^{1/2}$. Both can be expressed as complex functions :

$$\epsilon(\omega) = \epsilon_R(\omega) + i\epsilon_I(\omega) \tag{11}$$

and

$$N(\omega) = n(\omega) + iK(\omega) \tag{12}$$

where $K(\omega)$ is the extinction coefficient. The real and imaginary components of the dielectric function can be expressed as

$$\begin{aligned} \epsilon_R(\omega) = n^2 - K^2 \\ \epsilon_I(\omega) = 2nK \end{aligned} \tag{13}$$

The dielectric function $\epsilon(\omega)$ is related to the susceptibility $\chi(\omega)$ by: $\epsilon = 1 + 4\pi\chi$ and the reflectivity $R(\omega)$ is defined as the ratio

$$R(\omega) = \frac{(n-1)^2 + K^2}{(n+1)^2 + K^2} \tag{14}$$

Our aim is to introduce various material parameters into this definition; the only way to achieve it is to use the electrical conductivity, which is itself a function of material parameters.

The conductivity can be expressed as a complex function, and it is related to the dielectric function by

$$\epsilon_R(\omega) + i\epsilon_I(\omega) = 1 + i\frac{4\pi(\sigma_R(\omega) + i\sigma_I(\omega))}{\omega} \tag{15}$$

Using expressions for the real and imaginary parts of ϵ and σ, some algebra leads to

$$n = \frac{2\pi\sigma_R}{\omega K} \tag{16}$$

and

$$K^4 + K^2(1 - \frac{4\pi\sigma_I}{\omega}) - (\frac{2\pi\sigma_R}{\omega})^2 = 0 \tag{17}$$

Solving for K gives

$$K_{1,2}^2 = \frac{4\pi\omega\sigma_I - \omega^2 \pm [(4\pi\omega\sigma_R)^2 + (\omega^2 - 4\pi\omega\sigma_I)^2]^{1/2}}{2\omega^2} \tag{18}$$

Inserting results for n and K into eq.(10) leads to the final expression for the reflectivity

$$R = \frac{(2\pi\sigma_R)^2 - K\omega[4\pi\sigma_R - K\omega(1+K^2)]}{(2\pi\sigma_R)^2 + K\omega[4\pi\sigma_R + K\omega(1+K^2)]} \tag{19}$$

Using the fact that $(1-ax)/(1+ay) = (1-ax)\sum_{l=0}^{\infty}(-1)^l(ay)^l$ (19) can be transformed into

$$R \cong 1 - \frac{2K\omega}{\pi\sigma_R} + 2\frac{(K\omega)^2}{(\pi\sigma_R)^2} - \frac{3}{2}(\frac{K\omega}{\pi\sigma_R})^3 + ... \tag{20}$$

This result is valid regardless of the dimensionality of the system under study. The remainder of this contribution is devoted to the calculation of the function R for $1D$ and $2D$ systems,using results obtained through the HM.

2. Reflectivity of a 1D lattice

Imagine,as a first example, that the aim is to calculate the reflectivity of a $1D$ nonconducting lattice. This means that $\sigma_R \to 0$. Calculating the ratio K/σ_R for this case,it can be shown that $K/\sigma_R \to \infty$, and this means that R becomes unphysically large. The implication is that this calculation cannot be applied to insulators.

However, if the conductivity of a $1D$ lattice is finite, the situation becomes more interesting. Assusming that the number of lattice sites N and the frequency ω are sufficiently large, eqs.(9) and (10) show that the imaginary part of the conductivity $\sigma_I \to 0$. Taking the square root of eq.(18)and inserting the result into eq.(20), one gets the following expression for the reflectivity of a $1D$ lattice

$$R \approx 1 - \frac{\omega^2}{\pi^2\sigma_R^4} + \frac{[\omega^4 + 16\pi^2\sigma_R^2\omega^2]^{1/2}}{\pi^2\sigma_R^4}\ldots \tag{21}$$

In order to obtain theoretical data meaningfully comparable with experiments, the first step is to obtain values of the conductivity. As an example, figure 1 represents the dependence of σ_R on the temperature T calculated by eqs.(7) and (8) for the following set of HM parameters: $N = 150$,$U = 4t$,$\omega_P = 3U$,$\omega = 0.7U$,$s = 1$,$n = 0.8$,$b = -1.83879$, where various symbols denote: N is the number of lattice sites,s is the value of the lattice constant,U is the Hubbard U, t is the hopping energy,ω_P is the plasma frequency,ω denotes the frequency. The conductivity is normalized to $\sigma_R = 1$ for $t = 0.005$,$T = 116K$ [6].

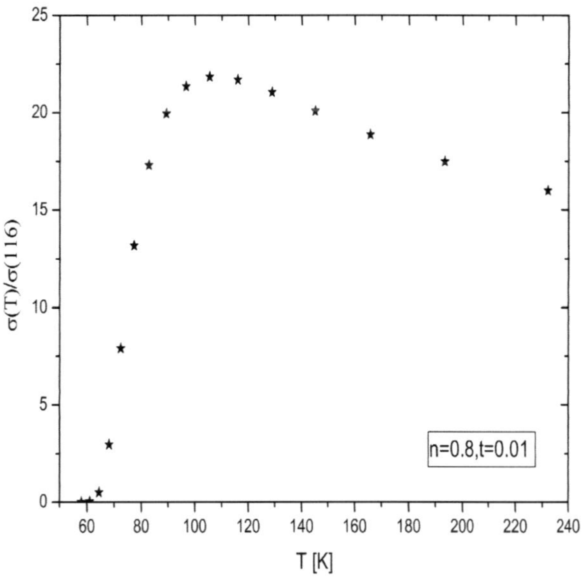

Figure 1. normalized conductivity of a $1D$ lattice

Knowing the function $\sigma_R(T)$, the next step in the calculation is to determine the function $R(T)$. This can be done by inserting the values of the normalized conductivity into the expression for R. For values of the conductivity shown on figure 1, one gets the representation of the curve $R(T)$ shown on figure 2.

Similar curves could be obtained for other choices of the parameters.

3. Reflectivity of a 2D lattice

A more complicated, but also more interesting system, is the rectangular Hubbard plane: a $2D$ rectangular plane made up of two $1D$ lattices intersecting at right angles at mutual distance a. The reflectivity of a Hubbard plane can be calculated by starting from the $2D$ Hubbard Hamiltonian, calculating the electrical conductivity and then the reflectivity. A simpler way to perform this calculation would be to define the conductivity of the lattice as $\sigma_r \rightarrow \sqrt{\sigma_{rx}^2 + \sigma_{ry}^2}$ as in [7], where the suffixes denote the two coordinate axes,and then determine the reflectivity. It is assumed in this definition that the conductivities along the two lattice axes are mutually independent.

Introducing this definition into eq.(20) for the reflectivity, it follows that

$$R \cong 1 - \frac{2K\omega}{\pi\sqrt{\sigma_{rx}^2 + \sigma_{ry}^2}} + 2\left(\frac{\omega K}{\pi\sqrt{\sigma_{rx}^2 + \sigma_{ry}^2}}\right)^2 - \frac{3}{2}\left(\frac{K\omega}{\pi\sqrt{\sigma_{rx}^2 + \sigma_{ry}^2}}\right)^3 + ... \tag{22}$$

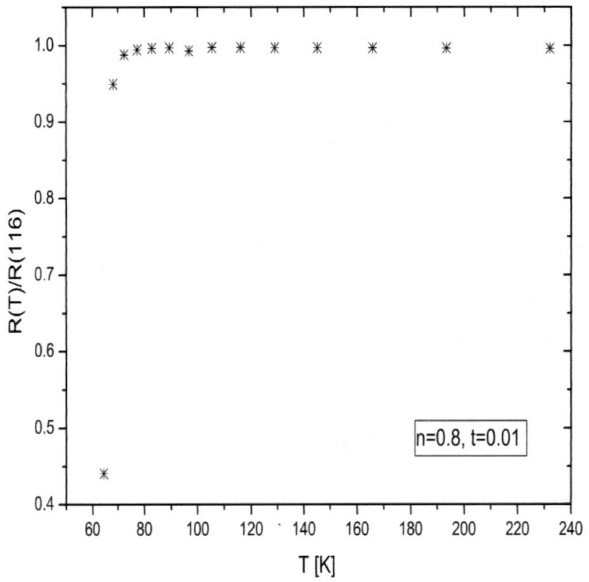

Figure 2. normalized reflectivity of a $1D$ lattice

In the two dimensional case the function K can be defined using the same reasoning for both the real and imaginary components of the conductivity. Taking that $\sigma_{ix} = \sigma_{iy} = 0$ and introducing $\alpha = \sigma_{ry}/\sigma_{rx}$, it follows that

$$K_{1,2}^2 = -\frac{1}{2} \pm \frac{1}{2\omega^2}[(4\pi\omega\sigma_x^2\sqrt{1+\alpha^2})^2 + \omega^4]^{1/2} \quad (23)$$

These two expressions open up the possibility of investigating the reflectivity as a function of various material parameters defined within the HM, but also on the ratio of the conductivities along the lattice axes. It follows from eq.(23) that

$$\frac{K\omega}{\pi\sqrt{\sigma_x^2+\sigma_y^2}} = \frac{K\omega}{\pi\sigma_x\sqrt{1+\alpha^2}} = \frac{\omega}{\pi\sigma_x\sqrt{1+\alpha^2}}[-\frac{1}{2} + (\frac{1}{2\omega^2})[16\pi^2\omega^2\sigma_x^4(1+\alpha^2) + \omega^4]^{1/2}]^{1/2} \quad (24)$$

Take as an example that $\alpha = 0$, which means that the lattice is not conducting in the y direction . In this case, the conductivity and reflectivity of a $2D$ plane are given by the expressions for a $1D$ system. Take as a next example that $\alpha = 0.1$. This means that the conductivity in the x direction is 10 times higher than the value along the y axis. Such a material would be highly anisotropic,and its reflectivity would depend on the temperature in the following way:

Assuming that $\sigma_{ry} = 0.5\sigma_{rx} = \sigma$, that is $\alpha = 1/2$, it follows that

$$\frac{K\omega}{\pi[\sigma_{rx}^2 + \sigma_{ry}^2]^{1/2}} = \frac{2\omega}{\pi\sigma\sqrt{5}}[-\frac{1}{2} + \frac{1}{2\omega^2}[\omega^4 + 20\pi^2\omega^2\sigma^4]^{1/2}]^{1/2} \quad (25)$$

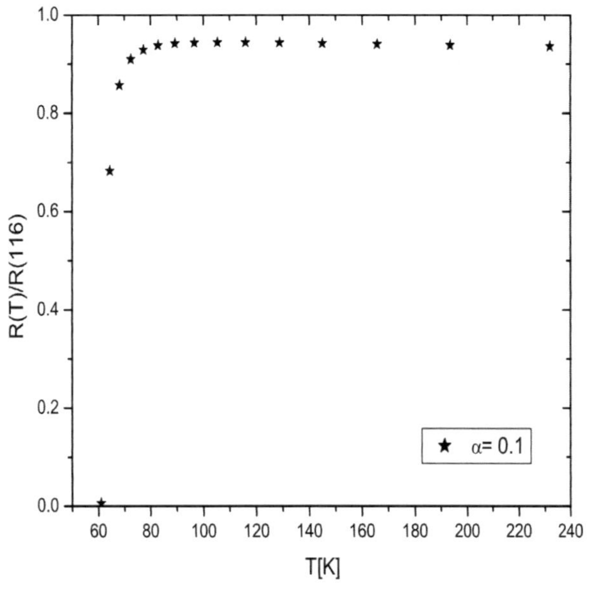

Figure 3. normalized reflectivity of a $2D$ lattice, for $\sigma_{ry} = 0.1\sigma_{rx}$

and the reflectivity follows by inserting this result into eq.(22).

An example of the temperature dependence of the reflectivity of a Hubbard plane is shown in figure 4. The points were calculated for the same values of the HM parameters as used for figure 1,and it was assumed that the ratio of conductivites along the two axes is $1/2$. The results were normalized to $R = 1$ at the point $t = 0.005$ $\beta = 116$.

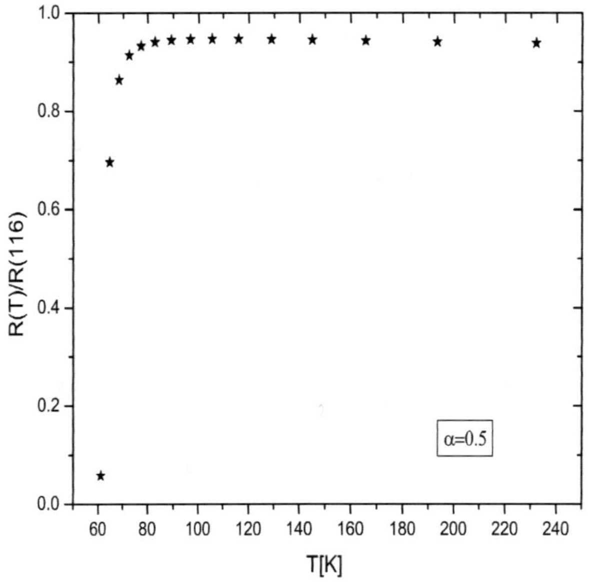

Figure 4. normalized reflectivity of a $2D$ lattice, for $\sigma_{ry} = 0.5\sigma_{rx}$

Another example is shown in figure 5. All the parameters were kept the same, except the ratio of the conductivities, which was increased to $\alpha = 1$.

4. Discussion and conclusions

In this paper we have discussed a seemingly purely theoretical subject - the calculation of reflectivity of $1D$ and $2D$ systems described by the HM. The obvious question is whether our calculation has any link with real material science, or it is just an interesting problem in mathematical physics. The links exist and are important. In general terms,standard electrodynamics shows that electrical conductivity of many physical systems is related to their reflectivity [8]. A logical consequence is that measuring, or calculating, one of these quantities, one can gain information about the other. The $1D$ calculation discussed in the present paper has applications in research on higly anisotropic materials.

Examples of such materials are the so-called Bechgaard salts, whose generic chemical formula is $(TMTSF)_2X$ or transition metal oxides,such as $Tl_2Ru_2O_7$ [9]. Here $TMTSF$ denotes a higly complicated molecule called tetra-meta-thio-seleno-fulvalene,and X is some anion. These materials are characterized by a high degree of anisotropy; the ratio of electrical conductivities along the three crystal axes is typically $\sigma_x : \sigma_y : \sigma_z \cong 1 : 0.01 : 0.001$, which implies that they can be modelled as effectively one dimensional. For recent reviews of some aspects of research on $(TMTSF)_2X$ see, for example [10], [11].

As a simple extension of the calculation for $1D$, we have performed the calculation of reflectivity for a $2D$ rectangular lattice. Purely mathematically, the main point which made

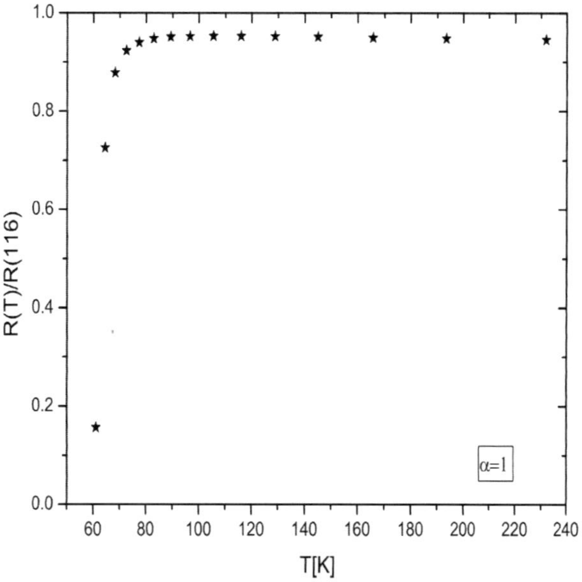

Figure 5. normalized reflectivity of a $2D$ lattice, for $\sigma_{ry} = \sigma_{rx}$

this calculation possible was defining the conductivity of this lattice as $\sigma = \sqrt{\sigma_x^2 + \sigma_y^2}$, where x, y denote the two lattice axes.Previous work [6] has shown that the coductivity of a $1D$ lattice is a function of tunable constants and the temperature to which the specimen is subdued. In the $2D$ case, one more parameter has to be added - the ratio of conductivities along the two axes, which is a pure number ,denoted here by α, characterizing the isotropy of the lattice. No attempt was made to obtain an anaytical expression for the reflectivity of a $2D$ lattice. Instead of such an attempt,the conductivity was calculated for a $1D$ case with the same choice of parameters as in [6], and this result was then applied to the two dimensional case for several values of the ratio of the conductivities along the two axes.

The results are shown on the figures. With all the other parameters kept constant, the reflectivity at sufficiently low temperatures is influenced by the ratio of conductivities along the lattice axes. For the same value of the temperature, the reflectivity is higher for $\alpha = 1$, than for $\alpha = 0.5$. This result should be checked for several more values of α.

The $2D$ part of calculations reported here can find applications in work on conducting materials with $2D$ structure. It may seem from the figures that the behavior of the reflectivity for $1D$ and $2D$ lattices is strikingly similar. This is due to the definition of the conductivity of a $2D$ lattice which was used. However, the factor which regulates the form of the temperature dependence of the reflectivity of a $2D$ lattice is the ratio of conductivities along the two lattice axes.

A $2D$ material in which there is enormous interest at present is of course graphene. However, our results cannot be directly applied to graphene, because it has a honeycomb lattice. On the other hand, it is known that graphene has neglible reflectivity [12]; note figure 3 at $T \cong 60K$. Another $2D$ material with considerable biomedical interest is human skin.Calculating, and even more so measuring its reflectivity could be of high interest in surgery, for example in cases of burning.

Lowering the value of reflectivity of an object, naturally brings into mind the topic of invisibility. Non-radiating objects are visible because they reflect incident light. Objects which do not reflect light, can not be seen - they are invisible. Numerous publications have appeared on the problem of achieving invisibility of various objects by embedding them in different kinds of cloaks (for example [13]). Equations for the reflectivity which we have obtained express it as a function of various material parameters. By inverting this reasoning, it should be possible to choose such values of material parameters which would make its reflectivity arbitrarily close to zero. It should be possible to achieve theoretically, by solving eq.(20). Stating eq.(20) in the form $R(x) = 0$,where $x = K\omega/\pi\sigma_R$ leads to the value $x = 0.744875$. This idea will be pursued further elsewhere.

Acknowledgments

This paper was prepared within the research project 174031 financed by the Ministry of Education and Science of Serbia. The author is grateful to Dr. Steve Reynolds from the University of Dundee for his help in proofreading the paper.

References

[1] Mott N 1982 *Proc.R.Soc.London* **A 382** 1
[2] Mott N 1968 *Rev Mod Phys.* **40** 677
[3] Hubbard J 1963 *Proc R Soc London* **A 276** 238
[4] Lieb E H and Wu F Y 1968 *Phys Rev Lett* **20** 1445
[5] Mahan G D 2000 *Many-Particle Physics,third edition* (New York:Kluwer Academic/Plenum Publishers)p 24
[6] Celebonovic V in 2006,*Trends in Materials Science Research*, Editor Caruta B M, Nova Science Publishers Inc.,New York,pp.245-260
[7] Celebonovic V 2010 *J Phys Conf Ser* **253** 012004
[8] Wallis R F and Balkanski M 1986 *Many-Body Aspects of Solid State Spectroscopy* Amsterdam:North Holland
[9] van den Brink J 2006 *Nature Materials* **5** 427
[10] Jerome D 2012 *J Supercond Nov Magn* **25** 633
[11] Bourbonnais C and Sedeki A 2011 *C R Physique* **12** 532
[12] Nair R R, Blake P , Grigorenko A N, Novoselov K S, Booth T J, Struber T, Peres N M R and Geim A K 2008 *Science* **320** 1308
[13] Zhang B,Luo Y, Liu X and Barbastathis G 2011 *Phys Rev Lett* **106** 033901

Thermal Rupture of Monatomic Metal Nanowires

M Michailov and D Kashchiev

Institute of Physical Chemistry, Bulgarian Academy of Sciences,
"Acad. G. Bonchev" Str., Block 11, 1113 Sofia, Bulgaria

E-mail: mike@ipc.bas.bg

Abstract. The present study deals with the problem of thermal stability and rupture of metal nanowires, considered here as a monatomic chain located on an atomically smooth crystalline substrate. Based on a recently developed Monte Carlo computational model, the simulation results reveal a scenario of nanowire breaking via formation of atomic vacancies. Depending on temperature, the fluctuations in the positions of the nanowire atoms generate wave-shaped nanowire configurations with specific active sites for breaking. This process is followed by formation of one-atom vacancies in the nanowire. Their overgrowth into vacancies of two, three and more atoms leads to permanent nanowire rupture and, hence, to loss of the nanowire electronic and other physical properties. The time evolution of the atomic-scale nanowire morphology and breaking mechanism are discussed in detail. The proposed model of nanowire rupture sheds light on the general problem of thermal stability of artificial atomic structures on crystal surfaces and their application to electronic and other devices with exotic physical features.

1. Introduction

Metal nanowires are of great technological importance due to their exotic physical properties. In the same time these systems excite remarkable academic curiosity because of the physics behind their reduced dimensionality and resulting quantum behavior [1]. Being at the cutting edge of nanoscale science, the theoretical and experimental studies of this problem are generally focused on the issue of "how to create nanowires" [1-3]. Considerably less attention has been paid to the thermal stability and rupture of these structures [4-6]. That is why, the present work deals with the opposite issue of "how a nanowire destroys itself without external influence". Focusing our study on that, we discuss the kinetic behavior of a monatomic metal nanowire (i.e. a chain of single metal atoms) located on an atomically smooth crystal surface. We investigate the nanowire atomic morphology and the probabilities of one- and two-atom vacancy formation in the nanowire.

2. Physical and Computational Model

The physical model in the present study is designed to represent a flat step-free atomically smooth crystal surface with a monatomic chain of the same material placed on it. As a model system we have chosen the Cu(111) surface with chain of 57 aligned Cu atoms on top. The initial system configuration is shown in figure 1a and is that of a perfectly ordered fcc(111) crystal surface with an infinite atomic chain on it because of imposed periodic boundary conditions. The system size in the corresponding space dimensions is $57 \times 33 \times 5$ lattice units.

The computational model is grounded on classic Monte Carlo (MC) sampling with many-body tight-binding second-moment approximation potential between interacting atoms [7]. In this approach, the total cohesive energy E_i of atom *i* in the system is expressed by

$$E_i = E_i^r + E_i^b, \tag{1}$$

E_i^r and E_i^b being, respectively, the repulsive and attractive energies. The first energy is given by [7]

$$E_i^r = \sum_{j, r_{\alpha\beta} < r_c} A_{\alpha\beta} \exp\left[- p_{\alpha\beta}\left(\frac{r_{ij}}{r_{\alpha\beta}^0} - 1 \right) \right] \tag{2}$$

Here α and β denote the chemical nature of atoms, r_{ij} is the distance between atoms *i* and *j*, $r_{\alpha\beta}^0 = (r_{\alpha\alpha} + r_{\beta\beta})/2$ if $\alpha \neq \beta$, $r_{\alpha\alpha}$ and $r_{\beta\beta}$ are the nearest-neighbor atomic distances of the corresponding pure metals, $A_{\alpha\beta}$ and $p_{\alpha\beta}$ are energy and compressibility related free parameters, and r_c is the cutoff distance for the interaction. The energy E_i^b is the band energy obtained in the form of the second-moment approximation of the electron density of states and is given by [7]

$$E_i^b = -\sqrt{\sum_{j, r_{ij} < r_c} \xi_{\alpha\beta}^2 \exp\left[-2q_{\alpha\beta}\left(\frac{r_{ij}}{r_{\alpha\beta}^0} - 1 \right) \right]} \tag{3}$$

where $\xi_{\alpha\beta}$ is an effective hopping integral and $q_{\alpha\beta}$ describes its dependence on the relative interatomic distance. In general, E_i^b can be considered as an expression that represents a sum over the local electronic charge density induced at site *i* from atoms at site *j*. Hence, $\xi_{\alpha\beta}$ is expected to be sensitive to the number of surrounding atoms. The energy calculation is performed over all atoms within a sphere having radius 3 times the Cu nearest-neighbor distance. The values of the hoping integral $\xi_{\alpha\beta}$ and all related parameters $A_{\alpha\beta}$, $p_{\alpha\beta}$, and $q_{\alpha\beta}$ in Eqs. 2 and 3 are taken from [7,8]. These values are consistent with the overall thermodynamic behavior of the system, since they reproduce basic physical characteristics of Cu crystal such as melting point, evaporation energy, elastic constants, etc.

The statistical distribution applied is canonical, i.e. ensemble with constant number of particles at fixed temperature [9]. At equilibrium, the system energy assessed over all atoms by Eq. 1 reaches its minimum and fluctuates around a constant value. Full lattice dynamics of both substrate layer and nanowire ensures complete relaxation of the system. The time evolution of the system is examined by a series of successive snapshots and pair-distribution function analysis. Complete simulation details are described elsewhere [10].

3. Time Evolution and General Scenario of Nanowire Rupture

In general, various parameters such as the temperature, type of atomic bonding to the substrate, crystallographic orientation, lattice mismatch between substrate and nanowire and elastic strain have an impact on the nanowire lifetime and rupture. In the present study, based on MC simulation experiments, we focus our attention exclusively on the significant role of thermal fluctuations. These fluctuations cause changes in the morphology of the atomic chain and result in a loss of essential nanowire physical properties. To that extent, detailed analysis of our model system reveals different consecutive stages in the general scenario of nanowire time evolution, stability and rupture. At constant non-zero temperature, the nanowire tends to minimize its free energy by the thermal fluctuations in the positions of its building atoms. The nanowire develops a wave-shaped atomic morphology, the waves having clearly manifested maxima and minima (figure 1a). At a given critical value of the fluctuation amplitude, the atom thermal energy exceeds the interatomic binding energy and the nanowire breaks down (figure 1b). The breakdown is defined as the formation of a one-atom vacancy.

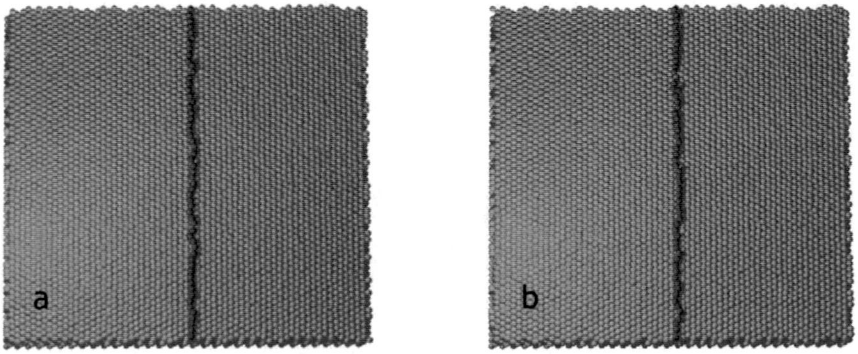

Figure 1. Breaking scenario of monatomic nanowire (red balls) on fcc(111) crystal surface (green balls) at 300 K. **(a)** Formation of active sites on the nanowire as a result of thermal fluctuations in the system. The wave-shaped nanowire geometry leads to weakening of the binding between the atoms at the wave maxima and minima. **(b)** Formation of one-atom vacancy and temporary nanowire rupture provided the nanowire discontinuity is restored via a random filling of the vacancy by an atom belonging to the nanowire.

Detailed time evolution and snapshot analysis suggests that the locus of rupture is the atomic position at the wave maximum or minimum (figure 1). Being considered as an active site, this special position is particularly significant, because it corresponds to weaker bonding of the atom occupying it. As seen in figures 1a and 1b, the nanowire is populated with a number of active sites on which a next generation of one-atom vacancies is very likely. The creation of such a vacancy results in breakdown, for example, of the nanowire conductivity and therefore loss of the respective electronic properties. It is essential to point out that the generation of a one-atom vacancy is not a process that irreversibly changes the nanowire morphology and integrity. Indeed, the vacancy can be filled up again by a randomly walking adatom that has been already detached from the chain. This process has been observed regularly in our MC experiments at sufficiently elevated temperatures. Hence, the nanowire could restore its integrity after a certain time. Since the formation of the first one-atom vacancy does not necessarily lead to permanent rupture, we consider this event as a step preceding the final stage of nanowire destruction.

The probability P_v of appearance of at least one vacancy in the nanowire until time t, defined as:

$$P_v(t) = \frac{N(t)}{N_{tot}} \tag{4}$$

is one of the parameters that characterizes the temporary nanowire rupture at a certain temperature. Here $N(t)$ is the number of nanowires possessing at least one vacancy until time t, and N_{tot} is the total number of nanowires studied. The statistical analysis of over $N_{tot} = 50$ nanowires at temperature $T = 300$ K resulted in the $P_v(t)$ curve shown in figure 2. The existence of two nanowire states is clearly seen in the figure: while at time $t < 200$ MC steps the nanowire is vacancy-free and thus stable, for $t > 200$ MC steps it is already unstable with respect to rupture. Naturally, the value of the time at which the nanowire loses its stability is a temperature function and also depends on the physical properties of materials (atomic bounding with the substrate, lattice misfit, crystallographic orientation, nanowire thickness, etc.).

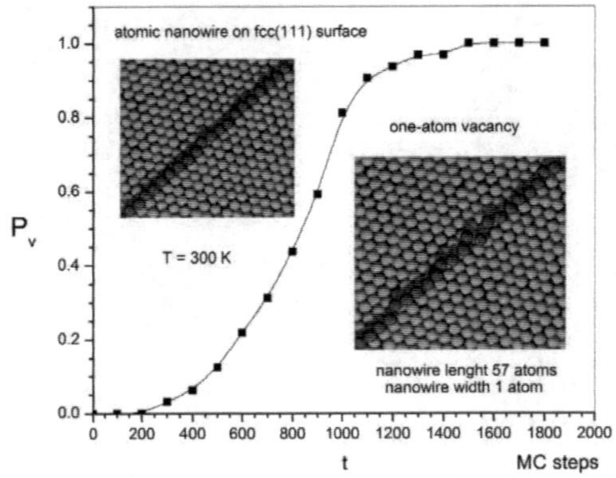

Figure 2. Time dependence of the probability to form at least one atomic vacancy. Initial (left inset) and final (right inset) states of the nanowire atomic morphology are shown.

The later stage of nanowire evolution leads to irreversible breaking. During that stage the system is in a state in which the probability of filling up the nanowire vacancy is practically nil. This state is a result of generation of a larger vacancy containing more than one atomic vacancies. Our simulations showed that for a vacancy built-up even of two atomic vacancies, the event of nanowire integrity recovering is almost unlikely. The process of generation of one- and many-atom vacancies is visualized in figures 3a and 3b, respectively. It is worth pointing out that the formation of a two-atom vacancy is always preceded by the birth of a one-atom vacancy growing into a larger vacancy. The one- and two-atom vacancies shown in figure 3 originate, respectively, from active sites and one-atom vacancies shown in figure 1b. Therefore, both of these act as precursors for nanowire rupture.

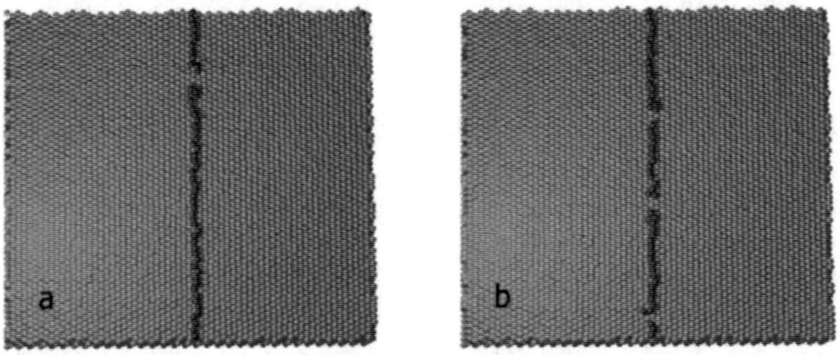

Figure 3. (a) Single two-atom vacancy in a permanently ruptured nanowire;
(b) One- and two-atom vacancies formed consecutively at three different active sites in the nanowire.

The curves in figure 4 depict the time dependences of the probabilities of appearance of at least one atomic vacancy and at least one two-atom vacancy. The shift of the latter with respect to the former reveals that at least one two-atom vacancy is born at a moment (ca. 1700 MC steps) when at least one atomic vacancy is already generated with certainty. The coexistence of one- and many-atom vacancies is seen in figure 3b. Inasmuch as P_v is a temperature function, the shift magnitude opens up a way to control temperature-dependent time intervals for simultaneous generation of differently sized vacancies. This shift relates to the problem of the temperature limits of temporal or permanent break down of the nanowire electronic or other properties.

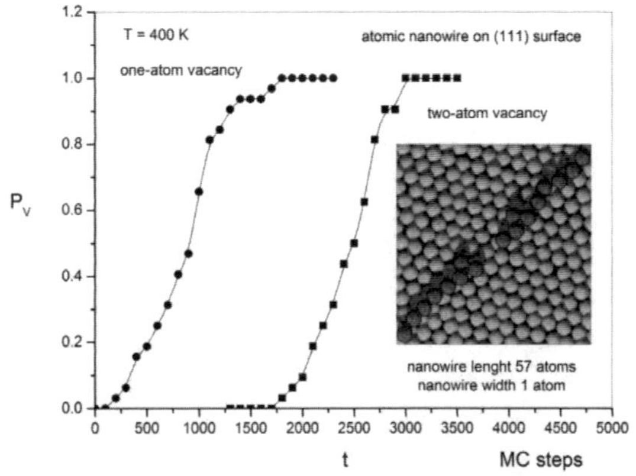

Figure 4. Time dependences of the probabilities to form at least one atomic vacancy and at least one two-atom vacancy. The inset represents an irreversibly growing vacancy.

A similar behavior was observed for nanowires constituted of two, three and more monatomic chains. Since the nanowire is wider, however, the formation of waves is less favorable with respect to that of a monatomic chain. In that case the active sites are local nanowire narrowing due to atomic roughening of the nanowire steps. The nanowire rupture occurs when the narrowing initiated at one of the nanowire steps becomes greater than the nanowire width. This process is manifested in figure 5 which illustrates the time evolution of a nanowire having three monatomic chains.

Figure 5. Consecutive snapshots illustrating the rupturing scenario of a nanowire of three monatomic chains (red balls) on fcc(111) crystal surface (green balls) at 300 K. The rupture occurs at a nanowire narrowing formed at a step of the nanowire.

Detailed snapshot analysis reveals another important effect related to the anisotropy of the substrate surface on which the nanowire is placed [10]. In the case of nanowires with width of more than about four atomic diameters, the nanowire narrowing beginning at B-type steps are greater than those beginning at A-type steps, figure 6. These two types of steps exist because of the specific crystallographic anisotropy of the fcc(111) surface and are illustrated in figure 6 (middle panel). This anisotropy leads to a different step free energy, relaxation ability, and diffusion barrier for displacement of atoms in directions A and B (figure 6) [11]. That is why, the roughing of B-steps costs less energy compared to that of A-steps. Because of symmetry reasons this effect does not exist on the cubic (100) surface. Therefore, the breaking process is not symmetric and depends on the substrate surface orientation through the anisotropy of the surface dynamics in different directions. It is important to note that the observed difference disappears at high enough temperatures where the diffusion energy of atoms belonging to the steps is higher than the energy barrier for perpendicular translation in both directions (see figure 6, right panel).

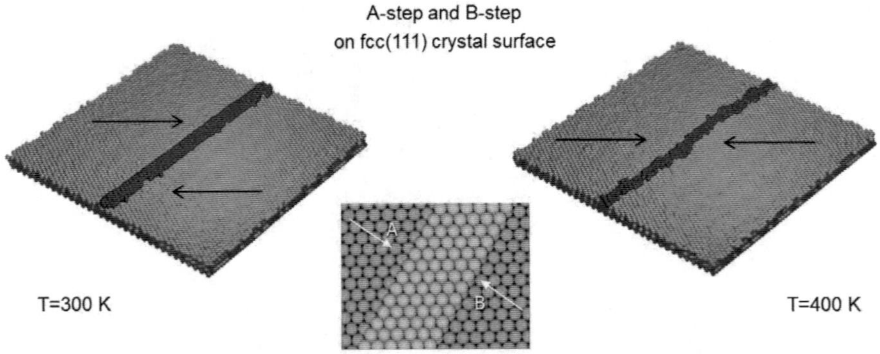

Figure 6. Anisotropy of thermal fluctuations of nanowire on fcc(111) substrate surface. At low temperature, $T = 300$ K, the B-step fluctuations are clearly pronounced in contrast to those of the opposite A-step. Increasing the temperature to $T = 400$ K, this anisotropy disappears because of the higher atom energy. The atomic configurations of A- and B-steps are presented in the middle panel of the figure.

4. Conclusion

The present study provides insight into thermal behaviour and structural evolution of monatomic metal nanowires homoepitaxially grown on crystal surfaces. The physical model considered outlines a general scenario of nanowire rupture as a three-step process of formation of vacancies. The first two steps involve the generation of rupture precursors (active sites and one-atom vacancies), and the last step is the actual formation of irreversibly growing vacancies of two and more atoms. This scenario is grounded on the observation that an atomic vacancy appears only when an atom at an apex of the wave-like nanowire leaves its place by attaching itself to one of the two nanowire steps. It is thus solely at a nanowire apex that an atom is active with respect to atomic vacancy formation. Before becoming active, the atom is unable to vacate its place in the nanowire, i.e. it is passive with respect to such a vacancy formation. Therefore, an atomic vacancy is born in two steps: first, a passive atom becomes an active atom by arriving at a nanowire apex owing to the undulatory motion of the whole nanowire, and second, this already active atom vacates its place by detaching itself from the nanowire. After that and this is the third and final step in the overall process of nanowire rupture, the so-born vacancy grows irreversibly. Detailed theoretical analysis on each step in the presented nanowire breaking model, including temperature dependence of rupture, probability distribution of vacancy cluster generation, lifetime and the impact of lattice mismatch between substrate and nanowire is in progress.

References

[1] Agrait N, Levy-Yeyati A and van Ruitenbeek J 2003 *Phys. Rep.* **377**, 81
[2] Tersoff J and Tromp R 1993 *Phys. Rev. Lett.* **70**, 2782
[3] Tosatti E, Prestipino S, Kostlimeier S, Corso A and Tolla F 2001 *Science* **291**, 288
[4] Jelinek P, Perez R, Ortega J and Flores F 2003 *Phys. Rev B* **68**, 085403
[5] da Silva E Z, da Silva A R J and Fazzio A 2001 *Phys. Rev. Lett.* **87**, 256102
[6] Liu H, Jiang E, Bai H, Wu P, Li Z and Sun C 2008 *Journal of Computational and Theoretical Nanoscience* **5**, 1-4
[7] Cleri F and Rosato V 1993 *Phys. Rev. B* **48**, 22
[8] Michailov M 2009 *Phys. Rev. B* **80**, 035425
[9] Landau D P and Binder K 2000 *A Guide to Monte Carlo Simulations in Statistical Physics*, Cambridge University Press, 2000
[10] Michailov M in: *"Nanophenomena at Surfaces: Fundamentals of Exotic Condensed Matter Properties"* 2011 Springer Series in Surface Science, Ed. M. Michailov, vol. 47, Chapter 6
[11] Giesen M and Ibach H 1999 *Surf. Sci.* **431**, 109

Journal of Physics: Conference Series **398** (2012) 012011 doi:10.1088/1742-6596/398/1/012011

Coherent phonons as a new element of quantum computing and devices

Rusko Ruskov and Charles Tahan

Laboratory for Physical Sciences, 8050 Greenmead Dr., College Park, MD 20740

E-mail: charlie@tahan.com, ruskovr@gmail.com

Abstract. We explore the possibility of strongly coupling semiconductor qubit states to nanomechanical resonators (phonons) in silicon. These systems may be relevant to qubit transduction schemes, as supporting technology for quantum information processing, for qubit characterization, and for quantum-enabled devices. Specifically, we consider systems where cavity phonons can interact with suitable qubit states in the 1-10 GHz (and higher) regime (tunable using strain/electric and/or magnetic fields). These results may be useful for several solid-state devices as well as being of interest to the optomechanics community.

1. Introduction

The purpose of this proceeding paper is to make a simple introduction to the use and appearance of phonons in nanomechanical structures in the context of the emerging field of quantum computing and quantum-enabled device applications. After re-capitulating essential features of cavity Quantum Electrodynamics (cQED)[1, 2], we introduce in an analogous way confined or propagating acoustic phonons and study the possibility to observe their quantum coherent nature in artificially created phononic cavities. Our main observation in further pursuing such an analogy is the finding that single impurities in an ideal crystal environment (e.g. in a silicon nanostructure, see below) may couple strongly to confined phonons, just as a single atom does with an optical cavity. By exploring recent achievements in optomechanics [3, 4], we propose a new way to probe/measure the quantum state of a phononic cavity, by using the phonon non-linearity introduced by an impurity (single donor or acceptor in a silicon phononic bandgap structure). The proposed scheme(s) can have applications related to conversion/transmission of quantum information between distant spin-qubit quantum nodes, using flying photonic qubits as transduction elements, while the qubit-photon interaction will be mediated by strongly coupled phonons.

2. Elements of quantum computing architecture

Quantum computing (QC) is commonly described in terms of qubits or two level systems (TLS) that can be coherently manipulated locally (so-called single qubit gates) or allowed to controllably interact with other TLSs (two-qubit or multi-qubit gates) [5]. Physically, the Hamiltonian (unitary) coupling between two qubits can be achieved by direct qubit-qubit interaction (dipole interaction in case of atoms, capacitive coupling in case of superconducting Josephson qubits, etc.), or by mediating the interaction via coupling to a common cavity (optical or microwave): so-called cavity bus. For a distributed quantum system, when a qubit (or qubits) are placed distantly from each other, a coherent coupling between the distant sub-systems is made via (ideally lossless) quantum channels/waveguides. While ideally, all these elements (qubits or cavities/waveguides) are lossless, in reality they are subject to relaxation/decoherence processes (e.g. due to spontaneous photon emission in case of atomic TLS) or intrinsic/extrinsic cavity losses (due to various photon scattering mechanisms). Thus, in the QC "race" the various physical approaches will complement each other rather than to compete: the photonic (in the optical frequency range, hundred THz) flying qubits will be suitable for long distance quantum state transfer, (due to very low-loss optical channels, even at room temperatures), superconducting Josephson junction (JJ) qubits and cavities (in the microwave range, MHz to GHz) will be suitable for quantum operations (due to potentially high qubit coherence times, very high qubit-cavity coupling), finally, semiconductor spin-qubits will be most suitable for quantum information storage (due to very high coherence times, of order of miliseconds to seconds). The need to inter-couple these various physical systems, eventually using tunable and strong coupling will dictate the use of phonons as mediators, as described below.

3. Cavity QED and circuit QED

Elements of the QC architecture (e.g., an atom strongly coupled to an optical cavity) were first realized in cavity QED [1, 2]. It was proposed to consider such a compound system as a node of a "quantum network", where different nodes are coupled via a photon waveguide and tunable couplings[6]. By "reducing" the atom to a two-level system the atom-cavity coupling comes into play in a famous Jaynes-Cummings Hamiltonian [7], which (in the rotating wave approximation) establishes a coherent exchange of TLS excitation σ^+ to an excitation of the confined optical cavity mode, $b^\dagger_{\boldsymbol{q},\sigma}$:

$$H_g \approx \hbar g_{\boldsymbol{q}} \left(\sigma^+_{ge} b_{\boldsymbol{q},\sigma} + \sigma^-_{ge} b^\dagger_{\boldsymbol{q},\sigma}\right) \tag{1}$$

The so-called strong coupling regime refers to the situation when $g > \{\Gamma_{\mathrm{TLS}}, \kappa_{\mathrm{cav}}\}$, where Γ_{TLS} is the TLS relaxation rate to all modes (including a continuum of modes not confined to the cavity) and κ_{cav} is the leakage rate of the confined cavity mode (which, in the optical case, is mainly due to non-ideal mirrors of the cavity or scattering losses in case of a semiconductor photonic bandgap cavity). The strong coupling regime implies that coherent exchange (called a Rabi flop) between the atom and the cavity is possible before decoherence takes over via spontaneous emission process or cavity loss. In particular, this regime is essential in a seminal proposal[6] for quantum state transfer between distant atom-cavity optical nodes, using a time dependent coupling $g(t)$.

The "traditional" cavity QED[1, 2] was further implemented in different solid-state systems, including semiconductor microcavity polariton (an exciton pair coupled to a photon)[8, 9], as well as in a cQED with a single semiconductor quantum dot (QD) that is playing the role of an atom while the photonic cavity is realized in a (2D) photonic bandgap structure[10, 11].

More recently, in the so-called circuit QED[12], the field has been revolutionized. Here the TLS is based on a Josephson junction (JJ) transmon qubit[12, 13], which is strongly coupled to a microwave (stripline) resonator. The strong coupling here comes from the large dipole moment of the JJ qubit (essentially a macroscopic quantum system) and also due to the (quasi) 1D-geometry of the microwave (MW) resonator, so that the confined cavity MW mode occupies a volume[13] $\mathcal{V}_{\mathrm{mode}} \approx 10^{-6}\lambda^3$. Combined with a low loss qubit/cavity elements (e.g. the qubit life time reaches a range from a microsecond to tens of microsecond in the case of a 3D-transmon[14]) the number of Rabi flops (coherent exchanges) reaches $\sim 40-100$ and also the regime of the so-called strong dispersive coupling[12, 13] is reached, which is suitable for schemes for quantum non-demolition qubit readout[13].

4. Phonons instead of photons

In an ideal crystal environment, phonons may play a role analogous to photons, though they propagate with the much slower speed of sound. In fact, the photon-phonon analogy is well supported in semiconductors at low energies: the conduction/valence bands play a role similar to the QED vacuum: the quasi-free electrons/holes propagate according to a quadratic dispersion, scatter off acoustic phonons that possess linear dispersion, and can be bound to charged impurity potentials to establish hydrogen-like atoms (donors or acceptors). Thus, the pattern of QED (matter and photon interactions) repeats itself in the behavior of electronic/hole quasi-particles in the solid (the "matter") and the phonon vibrations in a crystal (the "photons").

Consider first the phonons alone (both propagating and confined one). For small atomic displacements $\boldsymbol{u}(\boldsymbol{r})$ an expansion in normal modes gives

$$\boldsymbol{u}(\boldsymbol{r}) = \sum_{\boldsymbol{q},\sigma} \left(\boldsymbol{u}_{\boldsymbol{q}\sigma}(\boldsymbol{r})\, b_{\boldsymbol{q}\sigma} + \boldsymbol{u}^{*}_{\boldsymbol{q}\sigma}(\boldsymbol{r})\, b^{\dagger}_{\boldsymbol{q}\sigma} \right), \tag{2}$$

which approximately diagonalizes the phonon Hamiltonian: $H_{\rm ph} = \sum_{\boldsymbol{q},\sigma} \hbar\omega_{\boldsymbol{q}\sigma} \left(b^{\dagger}_{\boldsymbol{q}\sigma} b_{\boldsymbol{q}\sigma} + \frac{1}{2} \right) + H_{\rm anh}$. Higher terms correspond to small anharmonicity: $H_{\rm anh} = c\, b^{\dagger}_{\boldsymbol{q}\sigma} b_{\boldsymbol{q}'\sigma} b_{\boldsymbol{k}\sigma} + \cdots$, and are related to phonon self-interaction. The mode normalization in Eq. 2 is $\int d^3\boldsymbol{r}\, \boldsymbol{u}^{*}_{\boldsymbol{q}\sigma}(\boldsymbol{r}) \boldsymbol{u}_{\boldsymbol{q}\sigma}(\boldsymbol{r}) = \frac{\hbar}{2\rho\,\omega_{\boldsymbol{q}\sigma}}$, so that $b^{\dagger}_{\boldsymbol{q}\sigma}$ creates a phonon in the mode $\boldsymbol{q},\sigma$ with energy $\hbar\omega_{\boldsymbol{q}\sigma}$ in a material with mass density ρ. The vector $\boldsymbol{q}$ denotes a collective index of the discrete phonon mode defined via the phonon cavity boundary conditions and quantization volume $\mathcal{V}$. In particular, the plane wave expansion corresponds to rectangular periodic boundary conditions and $\boldsymbol{u}_{\boldsymbol{q}\sigma}(\boldsymbol{r}) = \left(\frac{\hbar}{2\rho\mathcal{V}\omega_{\boldsymbol{q}\sigma}} \right)^{1/2} \boldsymbol{\xi}_{\boldsymbol{q},\sigma} e^{-i\boldsymbol{q}\cdot\boldsymbol{r}}$ with wave vector $\boldsymbol{q}$, polarization $\boldsymbol{\xi}_{\boldsymbol{q},\sigma}$, and phonon branch σ.

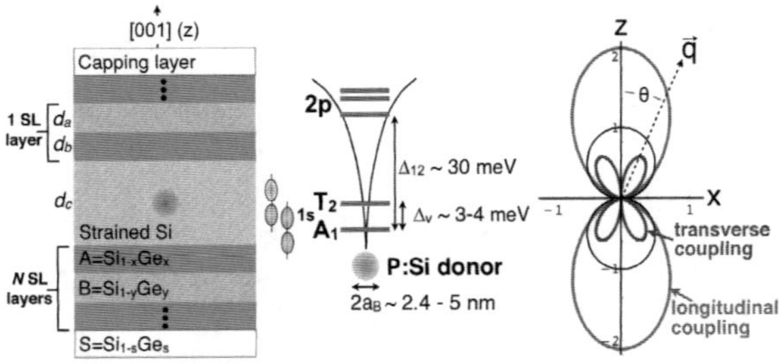

Figure 1. Ref.[15]:(a) A cavity-phoniton can be constructed in a Si/Ge heterostructure cavity as a hybridized state of a trapped single phonon mode and a donor TLS placed at a maxima of the phonon field. (b) The P:Si donor lowest $1s$ valley states, A_1, T_2, and upper levels ($1s$/ $2p$); their energy splittings can be controlled by the applied strain in the Si cavity. (c) Angular dependence of the coupling $g_q(\theta)$, Eq.(2) of [15], for the deformation potentials of Ref. [16] vs. dipole $\sim \cos\theta$-dependence (thin circles).

As an example of a phonon cavity, we have theoretically proposed[15] a (quasi) 1D-phononic bandgap micro-pillar silicon cavity (figure 1a). Here a Si cylindrical cavity is sandwiched between two acoustic distributed Bragg reflectors (DBR): the cavity mirrors. The boundary conditions are as for a free-standing cylinder, i.e. the stresses T_{zz} and T_{zr} are zero on the cylinder surface of the cavity. In the z (growth) direction of the Si cavity the displacement field is chosen in such a way, that (for a λ-cavity) one has nodes for the strain field on the Si-cavity z-boundaries. (This was our special choice for a donor:Si [15], we briefly comment below). In our prototype example, the DBR mirrors are made from SiGe layers of two different contents of Ge, thus forming a periodic (Bragg) structure (figure 1a). For an optimized DBR structure[15] we have estimated that the main phonon loss in the cavity is due to phonon leakage through the mirrors (extrinsic loss); however one can reach a high quality factors of $Q_{\rm ext} \approx 10^5 - 10^6$ using sufficient number of layers (e.g., for $N = 33$ gives $Q_{\rm ext} \simeq 10^6$ for the transition frequency of $\omega_{P:Si}/2\pi \simeq 730\,{\rm GHz}$, related to phosphorous donor in Si, see below). The corresponding cavity loss was estimated as $\kappa_{\rm ext} = \omega_{P:Si}/Q_{\rm ext} \simeq 2.8 \times 10^6\,{\rm s}^{-1}$. The intrinsic phonon cavity loss comes from phonon anharmonicity loss, $\Gamma_{\rm anh}$ and loss from scattering off isotopic (impurity) mass fluctuations, $\Gamma_{\rm imp}$ of which the latter loss rate is two orders of magnitude higher than anharmonicity loss rate (for low temperatures). This rate amounts to $\Gamma_{\rm imp} \simeq 7 \times 10^5\,{\rm s}^{-1}$, however it can be made one order of magnitude smaller for isotopically purified silicon. One should mention, that recent experiments on 1D and 2D phononic bandgap cavities shows possibility to achieve similar phononic cavity Q-factors of $10^5 - 10^6$ in silicon membrane nanostructures[17, 3], which will be discussed below.

5. Optomechanical cooling experiments

The fact that acoustic phonons can be quantum coherent (phonon coherence can be understood similar to photons, e.g. via the ability for interference and/or coherent quantum transfer) has been explored in a number of novel architectures, allowing seminal experiments in nanomechanical cooling via optomechanics [18, 19, 20, 3], trapping of phonons in artificial phononic bandgap cavities [21, 3], photon translation via phonons [22, 23], and indirect qubit-phonon coupling [24, 25].

We briefly discuss here the qualitative features of the optomechanical cooling[18, 19, 20, 3], that is closely related to the proposal of coherent photon-to-phonon conversion[4], we discuss later. For the cooling of the mechanical (phononic) cavity one couples it to an optical cavity via the photon optical pressure Hamiltonian[26, 27]: $H_{OM} = \hbar g_{OM} \hat{a}^\dagger \hat{a} (\hat{b}^\dagger + \hat{b})$; here the optical pressure force is proportional to the photon number operator $\hat{a}^\dagger \hat{a}$ in the cavity and H_{OM} can be simply related to the work to move a mechanical boundary[21] of an effective mass m_{eff}; notice that $\hat{b}^\dagger + \hat{b}$ is essentially the position operator $\hat{x}$ associated with the mechanical degree of freedom: a generic example of what we described could be a standard Fabry-Perot cavity with one of the mirrors being movable and coupled to a spring, representing a mechanical cavity of frequency ω_m. Then the optomechanical coupling g_{OM} can be related to the work performed by a single photon to move the mirror (mechanical boundary) at a distance given by zero-point mechanical fluctuations: $g_{OM} = \omega_c \frac{x_{ZPF}}{L} = (\omega_c/L)\sqrt{\frac{\hbar}{2m_{\text{eff}}\omega_m}}$.

The non-linear interaction, H_{OM}, leads to the presence of optical sidebands, most prominent being the closest one: the so-called blue and red sidebands at frequencies $\omega_c \pm \omega_m$, where ω_c is the cavity resonant optical frequency. It can be shown that in the sideband resolved limit $\omega_m \gg \kappa_{\text{opt}}$ higher sidebands are well suppressed. Thus, in a good approximation, if the optical input field is a red sideband photon of frequency $\omega_c - \omega_m$, it will scatter in the cavity to produce an output photon of higher frequency, ω_c, which "needs" a phonon from the mechanical cavity to "get out". The higher frequency optical channel is empty, which is plausible for, e.g., few hundred THz photons even at not very low temperatures, see below. Thus, this scattering process leads to cooling, and the effective phonon temperature will go down inversely proportional to the number of pumped red sideband photons in the optical cavity.

In the remarkable experiment of Ref.[3] the authors achieved an optomechanical cooling of a phononic bandgap nanostructure (similar to the 1D structure on a patterned Si membrane, shown on figure 2c). The "defect" in the 1D periodicity, as in the previous example of the micro-pillar DBR cavity, forms a phononic cavity, that can confine phonons (of frequency ≈ 4GHz). Since it is simultaneously a photonic bandgap structure, it also forms a photonic cavity of the same wavelength, $\lambda_{\text{opt}} = \lambda_m \approx 1500$ nm (corresponding to frequencies of $\omega_c \approx 200$ THz, i.e. telecom photons). Thus, it establishes a typical OM-structure, discussed above. For the base temperature of $T = 20$ K the OM-cooling process leads to an effective temperature of $T_{\text{eff}} = 0.2$ K with some $n_c \simeq 2000$ (red sideband) pumped photons in the cavity. Further cooling, by increasing n_c, is prevented by the noise entering the phononic cavity.

6. Phonon-to-photon translator proposal as a key element to probe phonon physics

The photon-to-phonon translator, proposed in Ref.[4], works in a similar fashion to the sideband cooling (a similar idea for quantum transduction was also discussed, e.g., in Ref.[28]). Just as for the cooling, the cavity confines two optical modes and a mechanical one, where the optical modes are designed to be sidebands to each other to resonantly increase the three-wave OM-coupling[4]: $H_{OM} = \hbar \tilde{g}_{OM} (\hat{a}_c^\dagger \hat{a}_p \hat{b} + \text{h.c.})$., Now, the higher frequency optical channel, $\hat{a}_c$ is not "empty" but is used to input/output an optical coherent signal, that is coherently translated to the phonon. The coherent nature of the translator is described by an effective beam-splitter type Hamiltonian[29]

$$H_{OM}^{\text{trans}} = \hbar G_{\text{eff}} (\hat{a}_c^\dagger \hat{b} + \hat{a}_c \hat{b}^\dagger) \tag{3}$$

The effective coupling is of the order of $G_{\rm eff} \sim \sqrt{n_c}\tilde{g}_{OM}$, i.e. it is enhanced by $\sqrt{n_c}$ (since the incoming photon or phonon "does not know" to which of the n_c red pump photons to scatter). Notice, that the translator works in both directions. The necessary conditions for this is the weak coupling regime[4], $G_{\rm eff} < \kappa_{opt}, \kappa_m$ (to avoid total internal reflection)

7. Non-linearity introduced by a single donor/acceptor impurity for Si nanostructures

Up to this point we demonstrated the coherent properties of phonons alone, and in particular, that optomechanical coupling allows probing of phononic structures by quantum optical means. To complete the analogy with photons one needs a non-linear element similar to an atom in cavity QED. Such an element has been proposed recently[15] where a single donor in silicon (P:Si or Li:Si) couples *directly* to confined phonons to form a hybrid state: a phoniton (direct analog of the polariton in cQED).

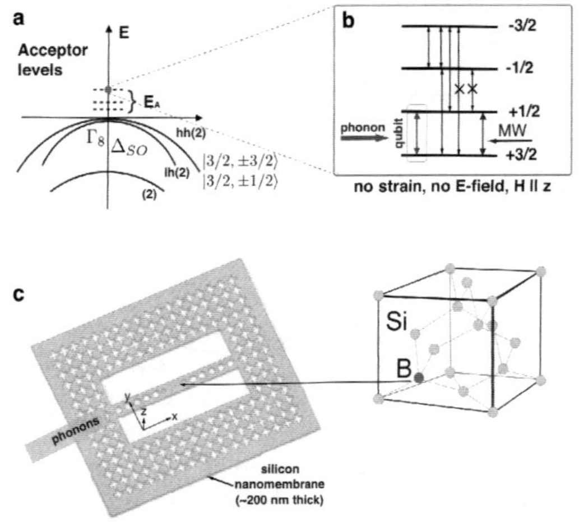

Figure 2. Ref.[30]: (a) Hole valence bands in Si; 4-fold degeneracy at the band top (and of lowest acceptor states) corresponds to particles of spin $J = 3/2$ (Γ_8 representation of cubic symmetry, see, e.g. Ref.[16]). (b) Ground state splitting is via external magnetic field along $[0,0,1]$ direction; allowed (forbidden) phonon transitions and qubit phonon driving are shown. Level rearrangement is via additional strain. System manipulation is via electric static/microwave fields. (c) Examples of a nanomechanical 1D phonon bandgap cavity reminiscent of already fabricated high-Q cavities in a patterned Si membrane [21, 3]. The acceptor is enclosed in the cavity and an on-chip phonon waveguide allows phonon coupling in/out of the system.

By considering low-energy acoustic phonons for electrons close to the band minimum the electron-phonon interaction can be written in a form similar to e.m. interactions in QED:

$$H^{\rm ac}_{\rm e.ph}(\boldsymbol{r}) = \sum_{ij} D_{ij}\varepsilon_{ij}(\boldsymbol{r}). \tag{4}$$

The operator $D_{ij} = -\hat{p}_i\hat{p}_j/m + V_{ij}(\boldsymbol{r})$ (with $\hat{\boldsymbol{p}} = -(i/\hbar)\nabla$ and a crystal model dependent $V_{ij}(\boldsymbol{r})$) coincides with the constant-strain deformation potential [16] and the strain $\varepsilon_{ij}(\boldsymbol{r}) = \frac{1}{2}\left(\frac{\partial u_i}{\partial r_j} + \frac{\partial u_j}{\partial r_i}\right)$ causes transitions between electronic states, just as e.m. fields do in QED. Despite the fact that this is a single impurity system, the impurity-phonon interaction may be large due to large deformation potential matrix elements:[16] $\langle\psi_{s'}|\hat{D}_{ij}|\psi_s\rangle \sim$ eV. In our proposal[15], the Si cavity was constantly strained, figure 1a, via suitable SiGe substrate in order to make the lowest $1s \to 1s$ transition in the P:Si ground state as small as possible (here it approaches 3 meV or 730 GHz for P:Si, and 0.586 meV or 146 GHz for Li:Si). This corresponds to a cavity length $d_c \sim \lambda_m \approx \{\text{few tens nm}\}$ in principle amenable to crystal growth engineering. We skip here calculational details[15] and mentioned two features differing our system from cQED. First, is the directionality of the coupling, figure 1c, related to the Si crystal symmetry under undirectional strain (along the growth z-direction). Also the impurity is "always there" vs. the finite traverse time for an atom in cQED. Estimations of the coupling strength shows that a strong coupling regime ($g > \{\Gamma_{\rm TLS}, \kappa_{\rm cav}\}$) (discussed above in the context of cQED) is well established; particularly the number of Rabi flops, $2g/(\kappa+\Gamma)$ can reach ≈ 100, comparable to circuit-QED[12]. This means that, in principle, experiments with single phonons could be as accessible as analogous experiments already performed with single MW photons, e.g. in Ref.[13].

The proposed donor system, however, requires very high frequencies[15] and can be difficult to integrate with other phonon components, which are now developed in the GHz range. While other impurities such as colour centers in diamond[25] or in III-V semiconductors are possible, a practical system in silicon would be highly desirable given recent demonstrations of high-Q cavities in silicon nanostructures [3], silicon's investment in materials quality, and compatibility with CMOS technology and silicon photonics.

In a recent paper [30] we propose a new quantum circuit element based on a single acceptor (such as B, Al, In, etc.) embedded in a patterned silicon nano-membrane, (figure 2), that would be compatible with established optomechanical components[21, 3]. The acceptor qubit (lowest two levels in the split acceptor ground state), figure 2b, is easily tunable in the range of $1-50$ GHz by external magnetic field, and also by additional electric field or strain. We show that the regime of strong resonant and also strong dispersive coupling of the acceptor qubit to a confined acoustic phonon is possible. Variation of the coupling and frequencies in time, as well as direct manipulation of the acceptor qubit via microwaves (figure 2b) would provide possibilities for state manipulation, readout, and quantum state transfer. We give explicit measurement protocols to observe the phonon vacuum Rabi splitting and readout of the phonon number state fine structure [30].

References

[1] Kimble H J 1998 *Physica Scripta* **T76** 127–137
[2] Raimond J M, Brune M and Haroche S 2001 *Rev Mod Phys* **73** 565–582
[3] Chan J, Alegre T P M, Safavi-Naeini A H, Hill J T, Krause A, Groeblacher S, Aspelmeyer M and Painter O 2011 *Nature* **478** 89–92
[4] Safavi-Naeini A H and Painter O 2011 *New Journal of Physics* **13** 013017
[5] Nielsen M A and Chuang I L *Quantum Computation and Quantum Information, Year = 2000* (Cambridge University Press, Cambridge, UK, 2000)
[6] Cirac J I, Zoller P, Kimble H J and Mabuchi H 1997 *Phys. Rev. Lett.* **78** 3221–3224
[7] Walls D F and Milburn G J 2008 *Quantum Optics* (Springer, Berlin 2008)
[8] Yamamoto Y, Tassone F and Cao H 2000 *Semiconductor Cavity Quantum Electrodynamics* (Springer, Berlin)
[9] Kavokin A and Malpuech G 2003 *Cavity polaritons, Vol.32* (Elsevier, Amsterdam)
[10] Pelton M, Vukovic J, Solomon G S, Scherer A and Yamamoto Y 2002 *Quantum Electronics, IEEE Journal of* **38** 170–177
[11] Bose R, Sridharan D, Kim H, Solomon G S and Waks E 2012 *Phys. Rev. Lett.* **108** 227402
[12] Blais A, Huang R S, Wallraff A, Girvin S M and Schoelkopf R J 2004 *Phys. Rev. A* **69** 62320
[13] Schuster D I, Houck A A, Schreier J A, Wallraff A, Gambetta J M, Blais A, Frunzio L, Majer J, Johnson B, Devoret M H, Girvin S M and Schoelkopf R J 2007 *Nature* **445** 515–518
[14] Paik H, Schuster D I, Bishop L S, Kirchmair G, Catelani G, Sears A P, Johnson B R, Reagor M J, Frunzio L, Glazman L, Girvin S M, Devoret M H and Schoelkopf R J 2011 *Phys. Rev. Lett.* **107** 240501
[15] Soykal Ö O, Ruskov R and Tahan C 2011 *Phys. Rev. Lett.* **107** 235502
[16] Bir G L and Pikus G E 1974 *Symmetry and Strain-induced Effects in Semiconductors* (Keter Publishing House, Jerusalem)
[17] Chan J, Safavi-Naeini A H, Hill J, Meenehan S and Painter O *arXiv* (*Preprint* 1206.2099v1)
[18] Kippenberg T J and Vahala K J 2008 *Science* **321** 1172–1176
[19] Rocheleau T, Ndukum R, Macklin C, Hertzberg J B, Clerk A A and Schwab K C 2010 *Nature (London)* **463** 72
[20] Teufel J D, Donner T, Li D L, Harlow J W, Allman M S, Cicak K, Sirois A J, Whittaker J D, Lehnert K W and Simmonds R W 2011 *Nature (London)* **475** 359–363
[21] Eichenfield M, Chan J, Camacho R M, Vahala K J and Painter O 2009 *Nature* **462** 78–82
[22] Dong C, Fiore V, Kuzyk M C, Tian L and Wang H *arXiv* (*Preprint* 1205.2360v1)
[23] Hill J T, Safavi-Naeini A H, Chan J and Painter O *arXiv* (*Preprint* 1206.0704v1)
[24] O'Connell A D, Hofheinz M, Ansmann M, Bialczak R C, Lenander M, Lucero E, Neeley M, Sank D, Wang H, Weides M, Wenner J, Martinis J M and Cleland A N 2010 *Nature* **464** 697
[25] Kolkowitz S, Jayich A C B, Unterreithmeier Q P, Bennett S D, Rabl P, Harris J G E and Lukin M D 2012 *Science* **335** 1603–1606
[26] Marquardt F, Chen J P, Clerk A A and Girvin S M 2007 *Phys. Rev. Lett.* **99** 093902

[27] Wilson-Rae I, Nooshi N, Zwerger W and Kippenberg T 2007 *Phys. Rev. Lett.* **99** 093901
[28] Rabl P, Cappellaro P, Dutt M V G, Jiang L, Maze J R and Lukin M D 2009 *Phys. Rev. B* **79** 041302
[29] Zhang J, Peng K and Braunstein S L 2003 *Phys. Rev. A* **68** 013808
[30] Ruskov R and Tahan C 2012 *arXiv* (*Preprint* 1208.1776v1)

Influence of the shielding currents lengthscale and anisotropy effects on the magnetic flux profiles of high-temperature superconductors

P Vanderbemden[1] and V Lovchinov[2]

[1] University of Liège, SUPRATECS research group, Department of Electrical Engineering & Computer science (B28), Sart-Tilman, B-4000 Liège, Belgium
[2] Institute of Solid State Physics, Bulgarian Academy of Sciences, 72 Tzarigradsko Chaussee Blvd., 1784 Sofia, Bulgaria

E-mail : Philippe.Vanderbemden@ulg.ac.be

Abstract. The so-called "magnetic flux profile" AC inductive technique is a powerful method for determining the critical current density J_c of bulk superconductors. In this work we aim at reporting analytical expressions for magnetic flux profiles of superconducting rectangular samples exhibiting a critical current density anisotropy. The results are used for examining the error resulting from approximating a rectangular cross-section by an "infinite cylinder" or "infinite slab" geometry. It is found that such approximations can lead to an artificial curvature of the flux profiles and errors of 10%-20% in the determination of J_c. Next, the effects of how planar defects (cracks, platelet boundaries,...) affect the magnetic flux profile signal are discussed. It is found that the magnetic flux profiles are much sensitive to the lengthscale of shielding currents, thereby providing means of investigation of the typical size of induced current loops in bulk superconductors. Finally some illustrative flux profile data measured on a bulk, large grain melt-processed YBCO single domain exhibiting J_c anisotropy are presented and discussed in relation with theoretical predictions.

1. Introduction

Thanks to their zero electrical resistance at cryogenic temperature, superconducting materials represent a promising way of designing attractive electrical engineering applications [1-3]. In particular, when $YBa_2Cu_3O_{7-\delta}$ (YBCO) superconductors are cooled at the liquid nitrogen temperature ($T = 77$ K), they can carry current densities larger than 10 kA/cm² [4-5] and trap magnetic flux densities in excess of 2 teslas. The key physical parameter determining the practical usefulness of superconductors is the maximum current density that can flow without dissipation, i.e. the critical current density J_c. Due the usually high values of J_c, the reliable and accurate determination of the critical current density is sometimes a challenge. Basically, either *transport* methods or *magnetic* methods can be used. In transport measurements, the current is directly injected through the specimen. Magnetic measurements consist in applying an external magnetic field which causes lossless shielding currents to flow mostly along the perimeter of the sample. In the case of strong magnetic flux pinning, the model proposed in the 60's by C.P. Bean in infinite geometry [6] assumes that the density of the screening currents is equal to the critical current density J_c. Therefore an appropriate mathematical treatment of magnetic measurements allows the critical current density J_c to be determined indirectly.

There are several methods for magnetic J_c measurements, which can be classified in AC or DC techniques [7]. The magnetic flux profile method, as proposed first by Campbell [8] and Rollins [9], is a convenient and straightforward AC measurement technique for determining J_c. It requires a primary coil generating an AC magnetic field and a secondary "pick-up" coil measuring the AC magnetic flux. Its advantage over other techniques is that it can probe volume properties of bulk superconducting samples of any size, in contrast to those relying on classical AC susceptometers and DC magnetometers which usually accommodate small samples (a few mm^3), with some exceptions [10-13]. Basically, the flux profile technique involves applying an AC magnetic field of amplitude H_0, and recording the flux amplitude $\Phi(H_0)$ in the sample for several increasing values of H_0, as illustrated schematically in figure 1a. If the critical current density is assumed to be independent of H_0, the penetration depth p of magnetic flux inside the sample is directly related to the derivative $d\Phi(H_0)/dH_0$ [8,14]. The plot of H_0 vs. p is called *flux profile*, whose slope is proportional to the local critical current density of the sample. A flux profile measurement represents therefore a convenient way of indirectly probing the local critical current density variations in a high-temperature superconductor [15]. Moreover, in the case of granular superconducting ceramics, the flux profiles have been shown to exhibit a pronounced kink, as shown in figure 1b; the two slopes of the flux profile are related to the *inter-* and *intra-* granular critical current densities [14,16]. A careful examination of the flux profiles measured at several frequencies and magnetic field configurations allows also more fundamental physical parameters to be determined, such as the thermal activation energy of flux lines [17], or pinning lengthscales and flux line lattice constants [18,19]. It can also be used to assess the quality and homogeneity of the bulk material, e.g. in melt-textured YBCO [20] as well as in MgB_2 [21].

The extraction of the critical current density from the flux profile also depends on the sample geometry. Flux profile calculations which have been published in the literature concern only isotropic samples having either an infinite slab or infinite cylinder geometry, as sketched in figure 2a. The aim of the present work is to extend calculations of flux profiles in the case of *parallelipipedic* samples of rectangular cross-section $a \times b$ (figure 2b), which are relevant to bulk single domain superconductors and single crystals. In addition, we will investigate how the anisotropy of J_c and / or the shielding current lengthscales affect the measured signal. Finally, flux profile measurements carried out on bulk melt-processed $YBa_2Cu_3O_{7-\delta}$ samples will be presented, and the results will be compared to the theoretical predictions.

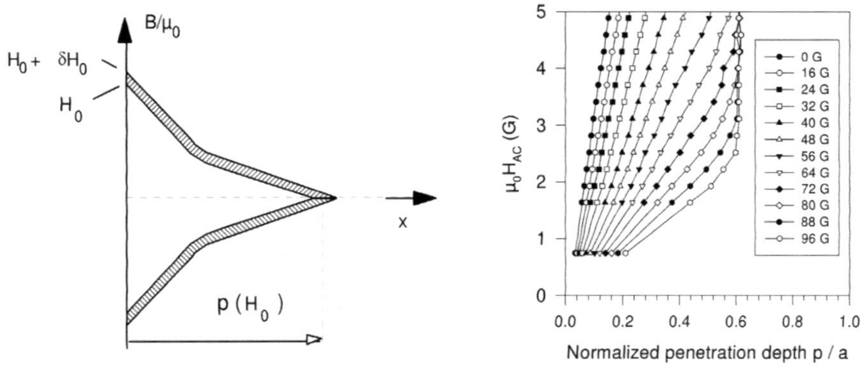

Figure 1. (a) Schematic representation of the magnetic flux density B at the edge of the sample as a function of distance x for two applied AC magnetic fields H_0 and $H_0 + \delta H_0$. (b) Flux profiles measured on a YBCO ceramic sample at 77K for different superimposed DC fields (see legend), showing clearly the appearance of a "kink" for the largest DC fields.

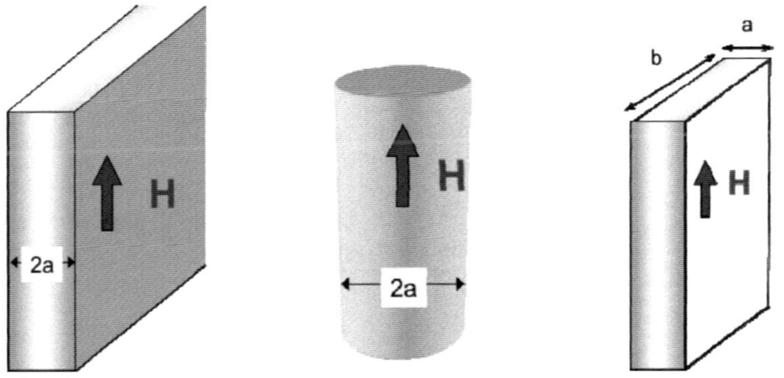

(a) (b)

Figure 2. (a) Two geometries for which analytical calculations of magnetic flux profiles are available in the literature (infinite slab and infinite cylinder). (b) Geometry investigated in the present work (infinite prism of rectangular cross-section).

2. Theory

2.1. General considerations

As recalled above, the key point of the Campbell method is that the derivative of the magnetic flux Φ with respect to the magnetic field amplitude H_0 gives access to the magnetic flux penetration depth $p(H_0)$. For a sample approximated either by an infinite slab of thickness $2a$ or by an infinite cylinder of diameter $2a$ (cf. figure 2a), the penetration depths are given respectively by [8,9,14]

$$p(H_0) = a\left(\frac{\dfrac{d\Phi}{dH_0}}{\dfrac{d\Phi_N}{dH_0}}\right) \quad \text{or} \quad p(H_0) = a\left[1 - \sqrt{1 - \left(\frac{\dfrac{d\Phi}{dH_0}}{\dfrac{d\Phi_N}{dH_0}}\right)}\right], \tag{1}$$

where Φ_N denotes the magnetic flux in the normal state (i.e. above the critical temperature T_c). The quantity $(d\Phi/dH_0)$ / $(d\Phi_N/dH_0)$ is involved in both above formulas and represents a normalized flux-derivative which can be determined easily by two experiments (one in the superconducting state below T_c and one in the normal state above T_c) , it will be noted hereafter by the letter F, i.e.

$$F = \frac{\left(\dfrac{d\Phi}{dH_0}\right)}{\left(\dfrac{d\Phi_N}{dH_0}\right)}. \tag{2}$$

The purpose of the present section is to find the analytical expressions for F as a function of the sample parameters (J_c and geometry) in the case of a bulk sample with a rectangular cross-section ($b \geq a$). The sample is assumed to be infinite in the direction parallel to the magnetic field H; in order to keep a finite magnetic moment however, we will consider a layer of thickness z. In the following calculations, we also assume that J_c is independant of H_0.

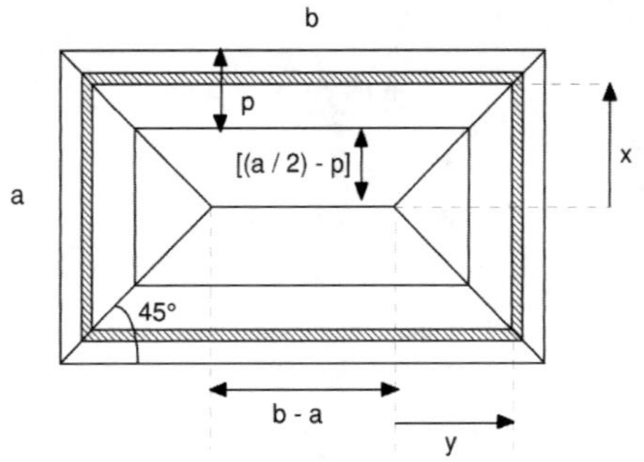

Figure 3. Cross-section of a rectangular the sample with dimensions *a* and *b* in the isotropic case. The supercurrents are assumed to flow along rectangular paths (shaded area). The solid interior lines show the field meeting lines when the sample is fully penetrated.

The principle of the flux profile determination involves calculating the magnetic moment *m* caused by rectangular shielding currents loops in the sample, as depicted in figure 3. Hence the magnetization *M* is obtained by dividing the magnetic moment *m* by the sample volume. By replacing $\Phi = ab\,\mu_0(H_0 + M)$ and $\Phi_N = ab\mu_0 H_0$ in Eq. (2) the normalized derivative *F* is given by

$$F = 1 + \frac{dM}{dH_0}. \tag{3}$$

2.2. Rectangular sample with isotropic critical current density $J_{cx} = J_{cy} = J_c$

Figure 3 shows a sketch of both sample geometry and current loops for an applied magnetic field of amplitude H_0. The current density is assumed to be the same for both directions, i.e. $J_{cx} = J_{cy} = J_c$. The penetration depth of magnetic flux, *p*, is equal to H_0 / J_c. The area delimited by the shaded current loop, situated at a distance *x* from the flux corner, is given by

$$A(x) = (2x)[2x + (b-a)]. \tag{4}$$

The total magnetic moment m is given by

$$m = \int_{a/2-p}^{a/2} A di = \int_{a/2-p}^{a/2} A(x) J_c t dx$$
$$= 2J_c zp\left[\frac{2p^2}{3} - \frac{p}{2}(a+b) + \frac{ab}{2}\right]. \tag{5}$$

In the particular case $p = a/2$ (full penetration condition), the above formula becomes

$$m\left(p = \frac{a}{2}\right) = \frac{J_c t a^2}{4}\left(b - \frac{a}{3}\right), \tag{6}$$

in accordance with the results published some time ago by Gyorgy et al. [22]. The resulting global magnetization *M* is negative and equal to the magnetic moment divided by the sample volume, i.e.

$$M = -|M| = -\frac{m}{V} = -\frac{m}{abz} = -\frac{2J_c}{ab}\left[\frac{2}{3}p^3 - \frac{(a+b)}{2}p^2 + \frac{ab}{2}p\right]. \tag{7}$$

Finally, the normalized flux-derivative F is given by

$$\begin{aligned} F &= 1 + \frac{dM}{dH_0} = 1 + \frac{1}{J_c}\left(\frac{dM}{dp}\right) \\ &= \frac{2(a+b)}{ab}p - \frac{4}{ab}p^2. \end{aligned} \tag{8}$$

This equation is only valid when the magnetic flux penetration depth, $p = H_0 / J_c$, is smaller than the half of the smallest side of the rectangular cross-section ($a/2$), even at the highest applied AC magnetic field amplitude H_0. Such a condition is realistic in practice since bulk (RE)BCO large grain superconductors (where "RE" denotes a rare-earth ion) are characterized by large J_c values [5].

2.3. Rectangular sample with anisotropic critical current density $J_{cx} \neq J_{cy}$

When the current density is anisotropic, the cases ($J_{cx} \geq J_{cy}$) and ($J_{cx} \leq J_{cy}$) need to be examined separately. Figure 4 shows a sketch of current loops within a rectangular sample ($b \geq a$) for which the critical current density along the x-direction J_{cx} is larger than along the y-direction J_{cy}. The anisotropy ratio J_{cx} / J_{cy} is denoted by r (in the case of figure 4, one has $r \geq 1$). Taking into account the conservation of the total current flow and defining p_y as the penetration depth of magnetic flux along the y-direction, is equal to ($p_y = H_0 / J_{cy}$), the magnetic moment m is given by

$$m = \frac{2J_{cy}p_y z a^2}{r}\left[\frac{2}{3}\left(\frac{p_y}{a}\right)^2 - \frac{1}{2}\left(\frac{p_y}{a}\right) + \frac{r}{2}\frac{b}{a}\left(1 - \frac{p_y}{a}\right)\right]. \tag{9}$$

Note that in the particular case $p_y = a/2$ (full penetration condition), the above formula becomes

$$m\left(p_y = \frac{a}{2}\right) = \frac{J_{cy}ta^2}{4}\left(b - \frac{a}{3r}\right), \tag{10}$$

in accordance with the calculations published in the literature [22]. The normalized flux-derivative F as defined by Eq. (2) is given by

$$F = \frac{2(a+br)}{abr}p_y - \frac{4}{abr}p_y^2. \tag{11}$$

The above calculations assumed that the critical current density along the x-direction J_{cx} was larger than along the y-direction J_{cy}, which led to an anisotropy ratio $r = J_{cx} / J_{cy}$ larger than unity. In the opposite case ($r \leq 1$), two different situations arise. If the ratio ($a / 2r$) is smaller than $b/2$, the current loops illustrated in Figure 2a are valid and the above calculations are still correct. If $a / 2r$ is larger than $b/2$ - which arises when $a > br$ - the current loops are similar to those shown in figure 2b. The corresponding formulations for magnetic properties can be easily obtained through interchanging x and y axes. This procedure leads to

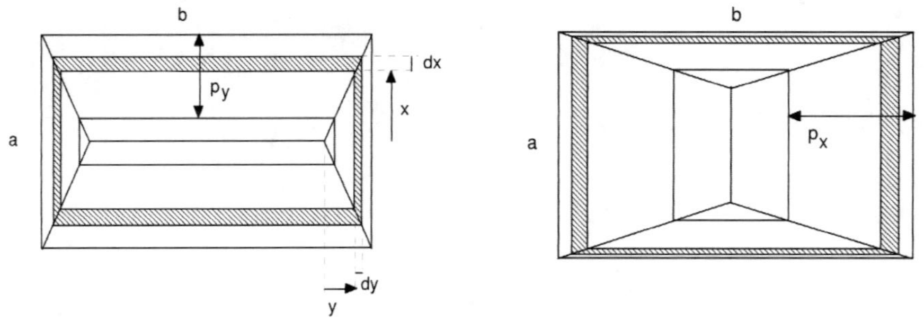

Figure 4. Supercurrent paths within a rectangular sample which exhibits a critical current density anisotropy, in the case $J_{cx} > J_{cy}$ (left) or $J_{cx} < J_{cy}$ (right).

$$m = 2J_{cx} p_x z r b^2 \left[\frac{2}{3}\left(\frac{p_x}{b}\right)^2 - \frac{1}{2}\left(\frac{p_x}{b}\right) + \frac{1}{2r}\left(\frac{a}{b}\right)\left(1 - \frac{p_x}{b}\right) \right]. \tag{12}$$

In the full-penetration limit $p_x = b/2$, the magnetic moment is given by

$$m = \frac{J_x t b^2}{4}\left(a - \frac{br}{3}\right). \tag{13}$$

Finally, the flux-derivative F is given by

$$F = \frac{2(a + br)}{ab} p_x - \frac{4r}{ab} p_x^2. \tag{14}$$

3. Discussion

3.1. The influence of the sample geometry

As can be seen in Eq. (8), (11) and (14), the normalized flux derivative F is a *parabolic* function of the penetration depth p, although J_c is assumed to be a constant. The non-linearity is thus a geometric effect which arises from the penetration of flux through *both* (orthogonal) faces of the sample. As a consequence, using a slab or cylinder approximation for a rectangular geometry always leads to the following errors : (i) an underestimation of J_c; the error being often in the range 10-20 % and (ii) a non-physical curvature of flux profiles.

3.2. The influence of lengthscale of shielding currents

The determination of J_c from the slope of the flux profiles assumes that shielding (critical) currents flow on a macroscopic lengthscale, fixed by the dimensions of the sample cross-section. Any defect in the microstructure (macro crack, grain boundary,...) may result in an alteration of this lengthscale, i.e. smaller shielding current loops. In the present section we want to investigate how a reduction of the lengthscale affects the flux profiles determined by the Campbell method. This has been done here for a superconducting sample with rectangular cross-section $a = s$, $b = 2s$, where s denotes a fixed and known length (e.g. $s = 1$mm). The magnetic moment m_1 generated by shielding currents in the sample shown in in the inset of figure 5a is given by

$$m_1 = m(s,2s) = 2J_c tp\left[\frac{2p^2}{3} - \frac{p}{2}(s+2s) + \frac{s(2s)}{2}\right]. \tag{15}$$

Suppose that a very weak link ($J_c \approx 0$) parallel to a divides the sample into two identical squared parts of side s; the lengthscale of shielding currents is now reduced, as shown in the inset of figure 5a The magnetic moment m_2 of this sample is given by

$$m_2 = 2m(s,s) = 4J_c tp\left[\frac{2p^2}{3} - \frac{p}{2}(s+s) + \frac{s^2}{2}\right] < m_1. \tag{16}$$

Similarly, a sample split into 4 identical parts is characterized by a magnetic moment m_4 given by

$$m_4 = 4m\left(\frac{s}{2}, s\right) = 8J_c tp\left[\frac{2p^2}{3} - \frac{p}{2}\left(s + \frac{s}{2}\right) + \frac{s}{2}\left(\frac{s}{2}\right)\right] < m_2 < m_1. \tag{17}$$

Suppose now that the Campbell flux profile method is applied on each of the 3 above samples, and that the flux penetration depth is calculated from the measured flux derivatives F by inserting the *whole* sample dimensions, $a = s$ and $b = 2s$, in Eq. (15), (16), (17). This procedure is equivalent to ignoring the weak links and assuming that shielding currents flow on a scale on the order of the sample size. The results are shown in figure 5a. The presence of one or two weak links located in the middle of the sample is clearly seen to affect the *slope* of the flux profiles, whatever the formula used for its determination. Such a situation is quite different from the case of granular samples for which a weak-link network results in a pronounced *kink* in the flux profile [14,16]. In ceramics, the kink is visible because of a non-negligible proportion of weak links in the sample. In the case of isolated weak links, the intergrain volume is small and all flux profiles are linear. As a consequence, the above simulation shows that isolated weak links may strongly affect the flux profiles determined by the Campbell method. However their effects cannot be distinguished from an intrinsically small critical current density.

3.3. Anisotropy effects

In this section we want to examine the effect of an anisotropic J_c on the problem of shielding current lengthscales discussed in Sect. 3.2 above. As an example, we compare the *slope* of the flux derivative, dF / dH_0, for a "single grain" (SG) rectangular sample with dimensions $a \leq b$, and for a "double grain" (DG) sample consisting in a juxtaposition of two rectangular samples of dimensions $a \leq (b/2)$. The magnetic moment m_{SG} of the SG sample and the corresponding normalized flux derivative, F_{SG}, are given by Eq. (9) and (11). From them we can calculate the initial slope (denoted S_{SG}) of the magnetic flux profile (H_0,F_{SG}) plot; this slope is given by

$$S_{SG} = \left\{\frac{d}{dH_0}F_{SG}\right\}_{H_0=0} = \frac{1}{J_{cy}}\left(\frac{2}{abr}\right)(a+br). \tag{18}$$

A similar procedure for the double grain (DG) sample, yields an initial slope (S_{DG}) given by

$$S_{DG} = \left\{\frac{d}{dH_0}F_{DG}\right\}_{H_0=0} = \frac{1}{J_{cy}}\left(\frac{2}{abr}\right)(2a+br). \tag{19}$$

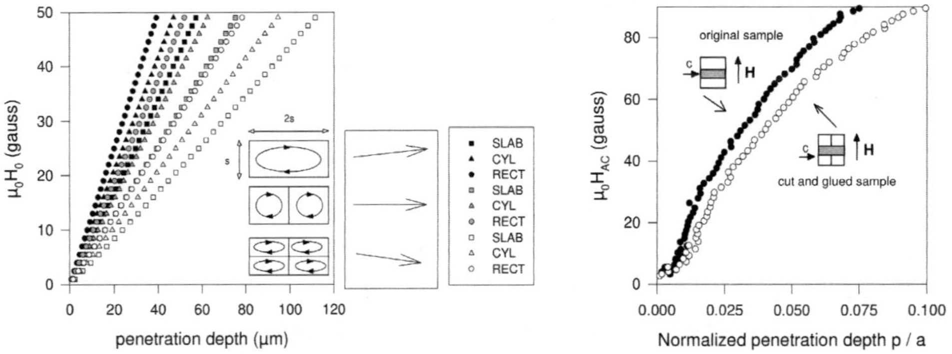

(a) (b)

Figure 5. (a) Comparison of the calculated flux profiles for the different "lengthscaled" samples 1, 2 and 4 schematically illustrated in the inset. The critical current density J_c is assumed to be equal to 10^4 A/cm². The flux profiles are calculated by approximating the rectangular cross-section by either a slab (squares) of thickness s or a cylinder (triangles) of equivalent radius $R = (2/\pi)^{1/2}\, s$. The circles refer to the flux profiles calculated using the formulas of the present paper, with no assumption on the cross-section. (b) Comparison of the flux profiles measured for two superconducting bulk YBCO samples, (i) the original single grain (SG) sample (black symbols) and (ii) a split double grain (DG) sample. The measurement is carried out at $T = 77$ K. The applied DC magnetic induction is 1500 gauss (0.15 T).

From Eq. (18) and (19), the ratio of the slopes of the magnetic flux profiles in the two cases, S_{DG} / S_{SG}, is given by

$$\frac{S_{DG}}{S_{SG}} = \frac{2a+br}{a+br} \quad (\geq 1) \tag{20}$$

Several observations can be drawn from the above equation. First, the slope $S = \mathrm{d}F / \mathrm{d}H_0$, which is inversely proportional to the "apparent" critical current density J_c, is clearly higher for the split sample than for the original one. However, in the limit case $r >> 1$, Eq. (20) shows that both slopes are the same. Indeed, in such a case, the equivalent "thickness" of current flow parallel to a, is negligible with respect to the sample dimensions, and currents in the vicinity of the cutting plane cancel out each other. This theoretical result will be compared to the experimental result in the next section.

4. Experiment

Flux profile measurements on a superconducting $YBa_2Cu_3O_{7-\delta}$ single domain are carried out in order to illustrate how the theoretical formulas developed above can be applied in practice. The single domain material is synthesised by a method described elsewhere [23,24]. The samples are characterized by the following superconducting parameters : critical current density $J_c = 10^4$ A/cm² and $T_c = 89$ K. Measurements of the critical current density for both directions gives an anisotropy ratio (J_c^{ab} / J_c^{c}) close to 3 at low magnetic inductions (< 0.5 T) [24,25].

Since the theoretical results clearly show (cf. Eq. (20)) that flux profile measurements are sensitive to the lengthscale over which shielding currents flow, the question we want to address is whether the shielding current loops when the magnetic field is applied parallel to the *ab* planes (and hence shielding currents do have a component parallel to the *c*-axis) are *macroscopic*, i.e. flow along the perimeter of the sample, or *microscopic*, i.e. flow along small circular loops because of the numerous microcracks and platelet boundaries in the single domain. To answer this question, magnetic flux profiles are first measured on a single grain sample with the applied field parallel to *ab* ("original" sample). Then the specimen is cleaved along the *ab* planes, both identical parts are repositioned side by side, glued with insulating GE varnish, and the same measurement is repeated ("cut and glued" sample). The sample geometries and results are summarized in figure 5b. When comparing both curves, it turns out that for a given magnetic field value, the AC penetration depth is higher in the case of the cleaved sample than for the original large specimen. The average slopes are found to differ by 12 %. From the results worked out above (Eq. 20), the ratio of slopes measured for single grain (SG) and double grain (DG) can be calculated analytically, knowing the dimensions of the original specimen cross section ($b \geq a$), and the anisotropy ratio r (= J_c^{ab} / J_c^{c}). In the present case, one has $b = 2a$ and $r \approx 3$. By assuming macroscopic shielding current loops, the theoretical ratio is thus given by $S_{DG} / S_{SG} = 1.14$. This result is in fairly good agreement with the measured data (1.12). In contrast, if the current loops were microscopic, i.e. much smaller than the sample dimensions, the flux profiles would not be influenced by the cutting / rejoining process. Consequently, the measurement result gives experimental evidence that shielding currents globally flow over a lengthscale comparable to the sample size, despite the presence of cracks and platelet boundaries. As discussed in ref. [24], the examination of *M*(*H*) loops on the same sample leads the same conclusion, but the sensitivity of the experiment is much more pronounced in the present case.

From this result we can conclude that the magnetic flux profiles are not only helpful in bringing out current density inhomogeneities but also in examining the influence of shielding currents lengthscales in bulk, large grain superconducting samples.

5. Conclusion

In this paper we developed analytical expressions relevant to the magnetic flux profile methods. Generally speaking, these methods consist in applying an external AC magnetic field and recording the magnetic flux Φ threading the sample cross-section, in order to deduce the penetration depth p as a function of the AC field amplitude H_0. If the AC field is superimposed to a bias DC field, the critical current density J_c may be assumed to be field-independent. In such a case, the slope of the flux profile gives the local critical current density. In the present work, flux profiles were calculated for samples of rectangular cross-section characterized by a field-independent but anisotropic critical current density. Such a model is quite appropriate for a crystal-like superconducting sample, e.g. a single crystal or a bulk, large grain YBCO superconductor. The effects resulting from approximating the parallelipipedic geometry by either a cylinder or a slab geometry were discussed. It was also shown that a few isolated weak links in the material affect the *slope* of the flux profile. Such a situation differs from a ceramic containing a weak link network, which results in a *kink* in the flux profile. Finally, we showed how the slope of flux profiles is influenced by the lengthscale of shielding currents. In the case of bulk melt-textured sample with applied field parallel to *ab* plane, the theoretical results were found to agree nicely with experiments.

Acknowledgements

We thank the University of Liège (ULg) and the Ministry of Higher Education of *Communauté Française de Belgique* for a research grant *Action de Recherches Concertées* (ARC 11/16-03). We thank Prof. M. Ausloos, A.M. Campbell, D.A. Cardwell for many fruitful discussions. The help of W. Lo and R. Cloots for supplying the superconducting sample is also acknowledged. This work is part of a collaboration programme financed by Wallonie Bruxelles International (WBI, Belgium) and the Bulgarian Academy of Sciences (BAS, Bulgaria).

Journal of Physics: Conference Series **398** (2012) 012012 doi:10.1088/1742-6596/398/1/012012

References

[1] Masson P J, Breschi M, Tixador P and Luongo C A 2007 *IEEE Trans. Appl. Supercond.* **17** 1533
[2] Ma G T, Lin Q X, Wang J S, Wang S Y, Deng Z G, Lu Y Y, Liu M X and Zheng J 2008 *Supercond. Sci. Technol.* **21** 065020
[3] Hsu C H, Yan Y, Hadeler O, Vertruyen B, Granados X and Coombs T A 2012 *IEEE Trans. Appl. Supercond.* **22** 7800404
[4] Cardwell D A *et al.* 2005 *Supercond. Sci. Technol.* **18** S173
[5] Devendra Kumar N, Rajasekharan T, Muraleedharan K, Banerjee A and Seshubai V 2010 *Supercond. Sci. Technol.* **23** 105020
[6] Bean C P 1964 *Rev. Mod. Phys.* **1** 31
[7] Frischherz M C, Sauerzopf F M, Weber H W, Murakami M and Emel'chenko G A 1995 *Supercond. Sci. Technol.* **8** 485
[8] Campbell A M 1969 *J. Phys. C* **2** 1492
[9] Rollins R W, Küpfer H and Gey W 1974 *J. Appl. Phys.* **45** 5392
[10] Vanderbemden P 1998 *Cryogenics* **38** 839
[11] Laurent P, Fagnard J F, Vanderheyden B, Hari Babu N, Cardwell D A, Ausloos M and Vanderbemden P 2008 *Meas. Sci. Technol.* **19** 085705
[12] Chen D X 2004 *Meas. Sci. Technol.* **15** 1195
[13] Trojanowski S and Ciszek M 2007 *Prz. Elektrotech.* **83** 26
[14] Küpfer H et al. 1988 *Cryogenics* **28** 650
[15] Eckert D and Handstein A 1976 *Phys. Status Solidi A* **37** 171
[16] Dang A, Godelaine P A, Vanderbemden P, Cloots R and Ausloos M 1995 *J. Appl. Phys.* **77** 3560
[17] Vanderbemden P, Destombes C, Cloots R and Ausloos M 1998 *Supercond. Sci. Technol.* **11** 94
[18] Otabe E S, Ohtani N, Matsushita T, Ishikawa Y and Yoshizawa S 1994 *Jpn. J. Appl. Phys.* 33 996
[19] Matsushita T, Otabe E S and Ni B 1991 *Physica C* **182** 95
[20] Hari Babu N, Rajasekharan T and Seshu Bai V 2000 *Physica C* **330** 203
[21] Ni B, Morita Y, Liu Z, Liu C, Himeki K, Otabe E S, Kiuchi M and Matsushita T 2009 *IEEE Trans. Appl. Supercond.* **19** 3529
[22] Gyorgy E M, van Dover R B, Jackson K A, Schneemeyer L F and Waszczak J V 1989 *Appl. Phys. Lett.* **55** 283
[23] Lo W, Dewhurst C D, Cardwell D A, Vanderbemden P, Doyle R A and Astill D M 1996 *Appl. Supercond.* **4** 507
[24] Vanderbemden P, Bradley A D, Doyle R A, Lo W, Astill D M, Cardwell D A and Campbell A M 1998 *Physica C* **302** 257
[25] Vanderbemden P, Cloots R, Ausloos M, Doyle R A, Bradley A D, Lo W, Cardwell D A and Campbell A M 1999 *IEEE Trans. Appl. Supercond.* **9** 2308

Atomic and electronic structure of N-terminated GaN$(000\bar{1})$ (1×1) surface

O Romanyuk[1], **P Jiříček**[1] **and T Paskova**[2]

[1] Department of Optical Materials, Institute of Physics, ASCR, Prague, Czech Republic
[2] Department of Materials Science and Engineering, North Carolina State University, Raleigh, NC 27617 USA

E-mail: `romanyuk@fzu.cz`

Abstract. The atomic and electronic structures of the polar GaN$(000\bar{1})$ surface were investigated by x-ray photoelectron diffraction (XPD) and *ab initio* density-functional theory (DFT) calculations. A well-ordered GaN$(000\bar{1})$ surface of a free-standing GaN wafer was prepared by sample annealing in a NH_3 atmosphere. A N-terminated bare (1×1) structure was confirmed. An electronic band structure was investigated by DFT calculations. The surface state at the top of the valence band was localized for a relaxed (1×1) structure. The energy loss function (ELF) was computed in the framework of the many-body perturbation theory. Plasmon loss peak positions were found to be in good agreement with reflection-electron energy loss spectroscopy (REELS) measurements.

1. Introduction

In recent years, many efforts have been focused on developing technology allowing semiconductor surface interfacing with biological macro objects for biomedical applications. The direct integration of the majority of semiconductor materials with biological objects is not possible, however, because of their toxicity [1]. A process of semiconductor surface bio-functionalization, i. e. surface passivation by hydrocarbons or large organic molecules, is required [2]. Knowledge about bare semiconductor substrate surfaces is necessary, therefore, for understanding the chemical reaction pathways and the absorption mechanisms on a surface during functionalization.

GaN is a promising candidate for semiconductor surface integration with biological objects: the functionalized GaN(0001) surface is stable in an aquatic environment and leaches a negligible amount of toxic Ga into the biological objects [2]. In the present work, a GaN free-standing substrate surface was investigated.

The conditions for preparing the GaN(0001) surface (Ga polarity) have attracted significant attention in the past [3]. Sharp (1×1) diffraction patterns were observed after surface etching and a three-step annealing procedure. The atomic structure of the GaN(0001) (1×1) reconstruction was investigated by diffraction techniques [4], photoemission spectroscopy (PES) [5] and *ab initio* calculations based on density functional theory (DFT) [6]. Under the most Ga-rich conditions, a pseudo-GaN(0001) (1×1) metallic surface was confirmed consisting of a Ga contracted bilayer [6]. Such a structure can be prepared by using a Ga flux and a low substrate temperature. On the other hand, the GaN(0001) (1×1) structure can also be prepared by sample annealing without Ga/N fluxes, i.e. out of equilibrium conditions [3].

The GaN(000$\bar{1}$) surface (N polarity) shows a number of reconstructions depending on atomic flux and substrate temperature variations [7]. The bare GaN(000$\bar{1}$) (1×1) structure was found to be thermodynamically unstable under N-rich conditions [8]. However, a stoichiometric (1×1) structure can be prepared by surface annealing without Ga/N fluxes [9]. Such a structure consists of a lower amount of Ga on the surface in comparison with GaN(0001) (1×1) phases and could be potentially suitable for biological applications.

Photoelectron spectroscopy measurements were carried out on the GaN(000$\bar{1}$) (1×1) reconstruction [10, 11]. Differences in surface preparation conditions, i.e. annealing temperatures and Ar or N_2 ion sputtering procedures, have led to discrepancies in the reported atomic and electronic band structures of the GaN(000$\bar{1}$) (1×1) surface [11, 12]. The N-terminated and Ga-adlayer (1×1) structure models have been suggested to explain the experimental data.

Recently, the atomic structure of the GaN(000$\bar{1}$) (1×1) reconstruction was investigated by quantitative low-energy electron diffraction (LEED) [9]. The Ga-adlayer model of the GaN(000$\bar{1}$) (1×1) structure was ruled out, whereas the N-terminated bilayer model was confirmed. Atomic relaxations were found to be in good agreement with DFT calculations.

Here, the atomic and electronic band structures of the GaN(000$\bar{1}$) (1×1) structure were investigated by XPD and DFT calculations, respectively. The electronic band structure of the relaxed N-terminated GaN(000$\bar{1}$) (1×1) surface was analysed. Based on electronic structure calculations, the dielectric properties of GaN were computed in the framework of a many-body perturbation theory (MBPT). An *ab initio* approach for dielectric function calculation was compared with the semi-classical theory for optical constant calculation from REELS spectra.

2. Experimental part

A GaN(000$\bar{1}$) sample [13] was grown by hydride-vapor phase epitaxy (HVPE) on a sapphire substrate along the [0001] direction in a standard vertical reactor. The as-grown GaN substrate had a thickness of about 1.5 mm with a dislocation density of about 10^5 cm^{-3}. Fe doping was used in order to compensate for the residual oxygen and silicon impurities, resulting in a free carrier concentration in the range of 10^{13} cm^{-3}. The GaN bulk wafer was self-separated from the sapphire substrate during the cooling down from the growth temperature. The N-side of the GaN substrate was mechanically polished and cleaned ex-situ in a HCl:DIH_2O (1:1) solution at 75°C for 10 min prior to loading into a ADES-400 angular-resolved photoelectron spectrometer. The GaN sample was degassed by flashing at 1000°C until the pressure inside the chamber was in the 10^{-9} mbar range. The GaN sample was heated indirectly through a sample holder consisting of a thin molybdenum plate; an electron beam was used to heat the plate, and the substrate temperature was monitored by an optical pyrometer. In order to obtain sharp LEED patterns, the sample was repeatedly annealed a few times for approximately 40 minutes at 1000°C in a NH_3 atmosphere of 2x10^{-6} Torr.

Standard mode angular resolved photoelectron diffraction intensity measurements were performed for a photoelectron kinetic energy of 825 eV (MgKα excitation from the N 1s core level) in two higher symmetry azimuths, namely the [1000] and [10$\bar{1}$0] directions (figure 1). The polar angle θ of the electron emission (analyzer) was varied over a range of 0° - 69° in 3° steps.

3. Computational details

Photoelectron diffraction from a large atomic cluster (2000 atoms, parabolically shaped) with a single atomic photoemitting layer placed at different depths under the surface was evaluated. The angle resolved intensities from the N 1s level were calculated by the code EDAC [14] taking multiple electron scatterings into account. The order of elastic scatterings was increased so as to reach sufficiently good convergence. An inelastic electron mean-free-path of 25 Å was used for electrons emitted with 825 eV kinetic energy, eight phase shifts were used in the muffin-tin type description of the atoms and an inner potential of 11 eV was used.

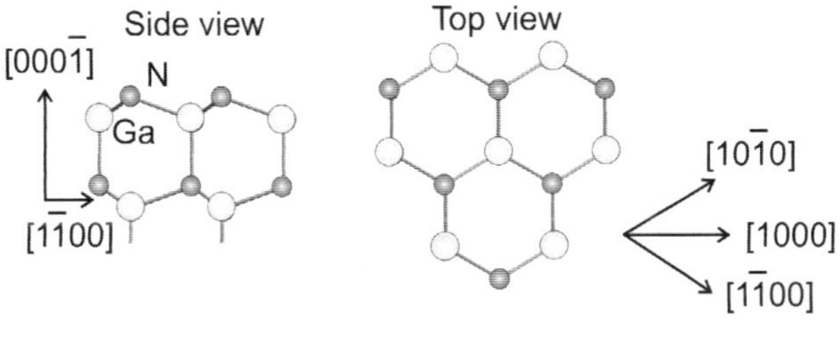

Figure 1. Ball-and-stick model of the GaN(000$\bar{1}$) (1×1) surface. The surface is N-terminated [9]. Ga and N atoms are marked by grey and red colors, respectively.

Ab initio DFT calculations of the GaN electronic band structure were carried out using the ABINIT computer code [15, 16]. The generalized gradient approximation (GGA) was used for the exchange-correlation energy functional, and norm-conserving pseudopotentials [17] of the Troullier-Martins type [18] were used to describe the atomic species. The Ga 3d electrons were treated as core electrons. The electronic wave functions were expanded in a plane wave basis. The kinetic energy cutoff was set to 30 Ha, and a k point set corresponding to $6 \times 6 \times 3$ for a bulk and $6 \times 6 \times 1$ per (1×1) surface Brillouin zone (BZ) [19] was used. The converged lattice constants a=3.197 Å and c=5.195 Å were used. A (1×1) surface structure was constructed using the repeated supercell approach with a GaN slab thickness of 8 Ga-N bilayers. The vacuum gap thickness was 13 Å. The bottom Ga layer of the slab was passivated by pseudohydrogens with 1.25 electronic charges. The atomic coordinates were adjusted until the interatomic forces became smaller than 10^{-4} Ha/Bohr, whereby only the three bottom layers (N-Ga-pseudo-H) were kept fixed.

The optical properties of the GaN were computed within a many-body perturbation theory implemented in Yambo computer code [20]. The relaxed ground state geometrical structure from ABINIT was used for the quasiparticle GW band structure calculations and the screening electron-hole interaction calculations, as well as for solving the Bethe-Salpeter equation by the Haydock recursion method [21, 22]. The screening function was calculated by using 40 k points in the irreducible BZ (in the convergence study, up to 112 k points were tested). Sixteen valence and sixteen conduction bands were found to be appropriate for the bulk dielectric function calculations. An energy onset of 0.5 eV for the energy loss function (ELF) curve was used to correct the volume peak plasmon position. An ELF resolution of 0.4 eV was used in the theoretical calculations, similar to the energy step used in the REELS measurements.

4. Results and discussion

In figure 1, a ball-and-stick model of the N-terminated GaN(000$\bar{1}$) (1×1) structure is shown. The direction to the neighboring atom within one atomic plane corresponds to the [1000] azimuth. The [10$\bar{1}$0] direction corresponds to the N-Ga atomic pair direction within an atomic bilayer.

Atomic structure of the GaN(000$\bar{1}$) (1×1) reconstruction is analyzed by XPD. In figure 2 a), the experimental and computed polar angle distributions of normalized XPD intensities are shown for [1000] ($\phi = 0°$) and [10$\bar{1}$0] ($\phi = 30°$) directions. The [1000] and [10$\bar{1}$0] atomic planes (perpendicular to the surface plane) of the GaN(000$\bar{1}$) (1×1) structure are shown in figure 2 b). The polar angles between the N emitters and the neighboring atoms are indicated. Since the GaN(000$\bar{1}$) (1×1) structure consists of two domains [9], the photoelectron intensities from the atomic planes rotated by 60° were averaged.

For the $\phi = 0°$ direction, the broad peak measured between 22°-32°, centred at 27°, could be interpreted as a convolution of the forward focusing peak intensity along the $\theta = 24°$ and $\theta = 32°$ directions (double layer). The peaks at 39° are close to the calculated peak position values at 42°. The experimental and theoretical peaks do agree at 51° and correspond to the N-N atomic pair direction. For the $\phi = 30°$ (90°) direction, the experimental peak positions at 26° are close to the N-N atomic pair direction at 28° [figure 2 b)]. The peak at 43° is a

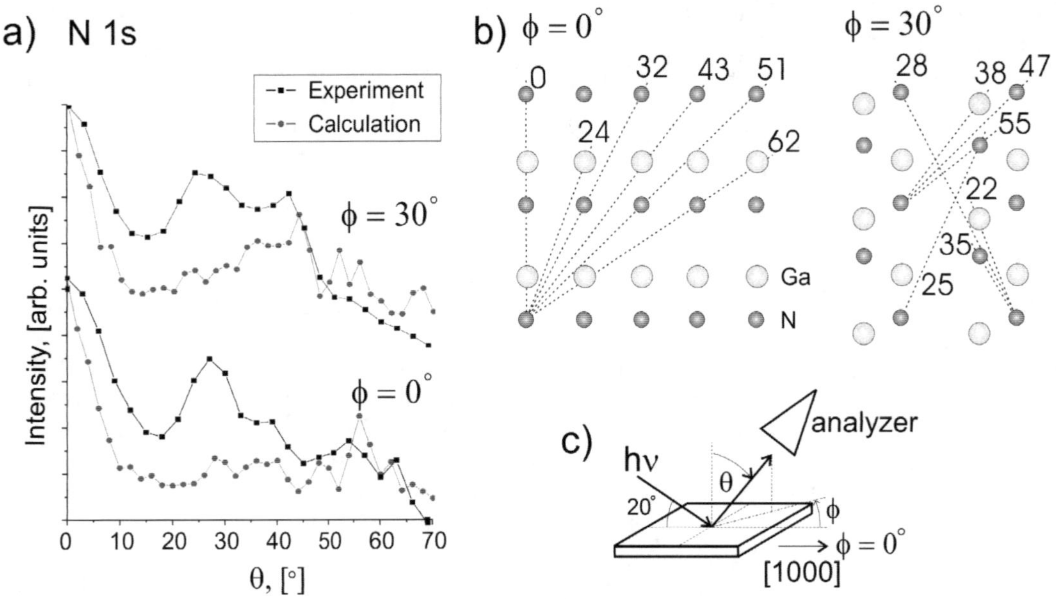

Figure 2. a) Polar dependence of the XPD intensities measured on the GaN(000$\bar{1}$) (1 × 1) surface along the [1000] ($\phi = 0°$) and [10$\bar{1}$0] ($\phi = 30°$) directions. Two symmetrical GaN planes are shown in b) with the corresponding polar angle θ values. The peak maxima in a) correspond to the forward focusing directions in b). An N 1s photoelectron kinetic energy of 825 eV (MgKα line) was used. The experimental and theoretical intensities were normalized. The experimental setup geometry is shown in c). The incidence angle of the photons was fixed to 20° from the surface along the [1000] direction.

convolution of the forward focusing peak intensity from the double layer along 38° and 47°. The experimental peak at 52° is close to the N-N pair direction at 55°. Thus, the experimental XPD curves agree qualitatively with the computed XPD intensities for the relaxed GaN(000$\bar{1}$) (1 × 1) surface: the N-polarity of the GaN crystal was confirmed. The N-terminated bare structure is in agreement with our previous study [9]. The structure consists of an N-terminated, relaxed bilayer on the top of the surface. A Ga adlayer on the top of the N layer [11] was ruled out [9].

In figure 3 a) and b), the electronic band structures of the bulk GaN and of the relaxed GaN(000$\bar{1}$) (1 × 1) surface are shown. There is a surface state s1 on the top of the valence band, which originates from the N 2p dangling bond orbital. The state is almost dispersiveless because of the weak interactions between the localized N dangling bonds [11]. This is a distinct feature of the the bare N-terminated GaN(000$\bar{1}$) (1 × 1) surface. This state is missing in case of the Ga-adlayer GaN(000$\bar{1}$) (1 × 1) structure [11, 12] since there is no N dangling bonds on the surface.

The total densities of the states (DOS) of the GaN bulk crystal and the relaxed N-terminated GaN(000$\bar{1}$) (1 × 1) surface are shown in figure 3 c) by red and black lines, respectively. In the case of surface DOS, the s1 peak is present on top of the valence band, while it is absent from the bulk DOS. A local DOS of the topmost N atom was computed to clarify the origin of the surface state. The N 2p orbital electron density is shown by the green line. There are a strong s1 peak and weak s2 and s3 peaks in the LDOS curve. The s2 and s3 peaks do overlap with the bulk Ga electronic states (**B** and **C**). Nevertheless, the surface state s1 is strong and can be used as a distinct feature of the clean N-terminated GaN(000$\bar{1}$) (1 × 1) surface.

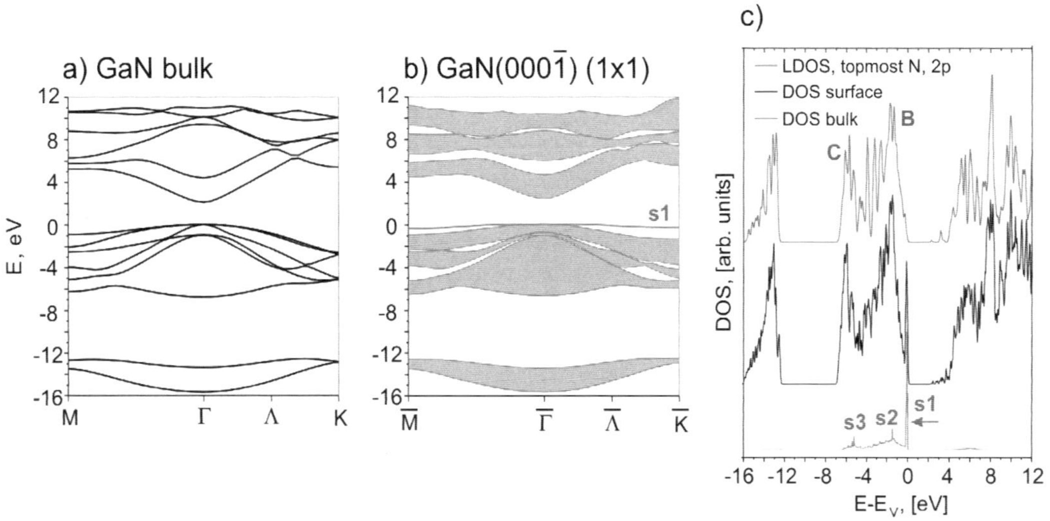

Figure 3. Electronic band structure of a) bulk GaN and b) GaN(000$\bar{1}$) (1×1) surface. There is a surface state on the top of the valence band in b) marked by a green line. c) The total density of states of the bulk GaN (red line) and (1×1) surface (black line). The local DOS of the N 2p states is shown by a green line. These states are formed by N dangling bonds on top of the GaN(000$\bar{1}$) (1×1) bare surface.

Molecular absorbtion during GaN surface functionalization alters the electronic structure and changes dielectric response of the clean GaN surface. Knowledge about the ELF of the clean N-terminated GaN(000$\bar{1}$) substrate is, however, important for surface functionalization study. The ELF of GaN(000$\bar{1}$) was recently investigated by REELS [23]. The ELF, Im($-1/\epsilon$), where ϵ is a dielectric function, was derived from REELS effective electron inelastic scattering cross-sections. Calculations were carried out within the framework of a semiclassical approach by using Drude-Lindhard type oscillator fitting [24]. Here, the ELF was computed from the first principles, thus avoiding the fitting procedures.

In figure 4, ELF curves derived from REELS spectra and computed by MBPT are shown. The plasmon loss peak positions at 9, 18.2, 23.5 and 28.5 eV agree well between the REELS- and DFT-derived curves. The peak intensities at 23.5 eV, however, are different. The peak corresponds to Ga 3d $\rightarrow$ 4p band transitions [25]. Agreement could not be improved by using modified pseudopotentials with the Ga 3d electrons as valence electrons. Such a difference can be explained by the presence of defects in the free-standing GaN substrate. Since N is preferentially desorbed from a surface during sputtering [3], an excess of Ga atoms (N defects) can be present on the surface. The appropriate annealing procedures in a NH_3 atmosphere could be required to recover the stoichiometry on the surface. In this sense, the ELF measured by REELS should approximate to the ELF derived by MBPT.

5. Conclusions

The atomic and electronic structure of a clean, ordered GaN(000$\bar{1}$) (1×1) surface were investigated by XPD. The polar dependence of the photoelectron intensities was computed taking multiple electron scatterings into account. The theoretical peak position maxima agree with the measured maxima for the N-terminated GaN(000$\bar{1}$) (1×1) reconstruction. The electronic band structure was computed in the framework of DFT. A surface state on the top of the valence

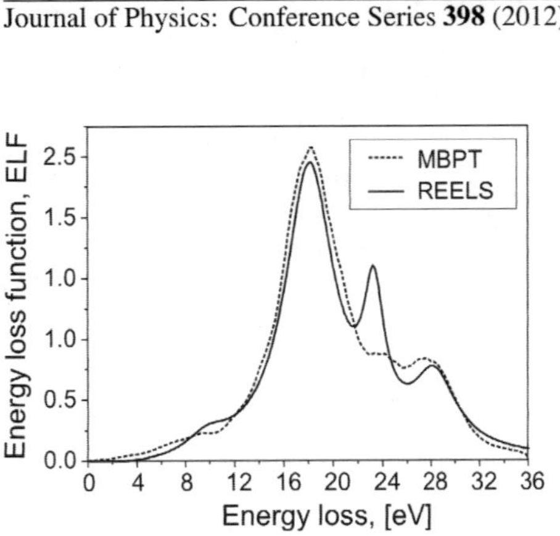

Figure 4. Energy loss functions derived from REELS spectra by a semiclassical approach (solid line) and computed within the MBPT (dashed line). The plasmon peak positions agree well, whereas the peak intensities at 23.5 eV are different. Atomic disorder on the surface can explain this intensity difference.

band was confirmed. The optical properties of GaN were derived by the MBPT. The loss peak positions derived by the MBPT were found to be in agreement with the REELS measurements.

Acknowledgments

Support by the Czech Academy of Sciences (No. M100101201) and by the Grant Agency of the Czech Republic (No. P204/10/P028) is gratefully acknowledged. The access to the MetaCentrum computing facilities provided under the programme "Projects of Large Infrastructure for Research, Development, and Innovations" LM2010005 funded by the Ministry of Education, Youth, and Sports of the Czech Republic is highly appreciated.

References

[1] Kotov N A, et.al. 2009 *Adv. Mater.* **21** 3970
[2] Jewett S A, Makowski M S, Andrews B, Manfra J and Ivanisevic A 2011 *Acta Biomaterialia* **8** 728
[3] Hattori A N, Endo K, Hattori K and Daimon H 2010 *Appl. Surf. Sci.* **256** 4745
[4] Yu Z X, Tong S Y, Xu S, Ma S and Wu H 2003 *Surf. Rev. Lett.* **10** 831
[5] Widstrand S M, Magnusson K O, Johansson L S O and Oshima M 2005 *Surf. Sci.* **584** 169
[6] Rosa A L and Neugebauer J 2006 *Phys. Rev.* B **73** 205346
[7] Ramachandran V, Lee C D, Feenstra R M, Smith A R, Northrup J E and Greve D W 2000 *J. Cr. Growth* **209** 355
[8] Smith A R, Feenstra R M, Greve D W, Neugebauer J. and Northrup J E 1997 *Phys. Rev. Lett.* **79** 3934
[9] Romanyuk O, Jiricek P and Paskova T 2012 *Surf. Sci.* **606** 740
[10] Kowalski B J, Iwanowski R J, Sadowski J, Kowalik I A, Kanski J, Grzegory I and Porowski S 2004 *Surf. Sci.* **548** 220
[11] Wang F H, Kruger P and Pollmann J 2001 *Phys. Rev.* B **64** 035305
[12] Chao Y C, Stagarescu C B, Downes J E, Ryan P and Smith K E 1999 *Phys. Rev.* B **59** R15586
[13] http://www.kymatech.com
[14] Abajo F J G, Van Hove M A and Fadley C S 2001 *Phys. Rev.* B **63** 075404
[15] Gonze X, et. al. 2002 *Comput. Mater. Sci.* **25** 478
[16] Gonze X, et. al. 2005 *Z. Kristallogr.* **220** 558
[17] Fuchs M and Scheffler M 1999 *Comput. Phys. Commun.* **119** 67
[18] Troullier N and Martins J L 1991 *Phys. Rev.* B **43** 1993
[19] Monkhorst H J and Pack J D 1976 *Phys. Rev.* **13** 5188
[20] Marini A, Hogan C, Gruning M and Varsano D 2009 *Comp. Phys. Comm.* **180** 1392
[21] Palummo M, Pulci O, Marini A, Reining L and Del Sole R 2006 *Phys. Rev.* B **74** 235431
[22] Benedict L X, Shirley E L and Bohn R B 1998 *Phys. Rev. Lett.* **80** 4514
[23] Romanyuk O, Jiricek P, Zemek J, Tougaard S and Paskova T 2011 *J. Appl. Phys.* **110** 043507
[24] Yubero F, Sanz J, Ramskov B and Tougaard S 1996 *Phys. Rev.* B **53** 9719
[25] Bermudez V M 1996 *J. Appl. Phys.* **80** 1190

Polarization dependence of III-nitride superlattice transmission coefficient

I S Asenova and E P Valcheva

Faculty of Physics, Sofia University, 5 J. Bourchier Blvd, Sofia, Bulgaria

E-mail: iasenova@phys.uni-sofia.bg

Abstract. We study the tunnelling process and behaviour of the transmission coefficient of an AlN/GaN superlattice as the internal polarizations in the composing layers change. Our calculations are based on the transfer matrix formalism and the exact solution of the Schrödinger's equation provided by the Airy functions. We also examine the dependence between the minizones width and the magnitude of the internal polarizations both for unbiased and biased structure.

1. Introduction

According to the recent researches [1], the III-nitride layered heterostructures have shown promising features for the manufacturing of devices for the optical communication region, high electron mobility transistors, ultra-violet lasers and resonant tunneling based structures for THz applications [1,2,3]. It has also bee shown that the internal piezo- and pyroelectric polarizations of the wurtzite modification have a significant effect upon the electronic properties of the barrier structures [4].

The calculations and the results we present here are inspired by the sample provided in [5] and are based on a previously presented model [6]. We use the effective mass approximation and the transfer matrix formalism. Our purpose is to demonstrate the behaviour of the transmission coefficient and the minizones in the conduction band for a ten-period wurtzite AlN/GaN superlattice in and out of the presence of external homogeneous electric field, while theoretically modifying the actual magnitudes of the polarizations in the layers. To assure a desired accuracy, we use the Airy function formalism when solving the one-dimensional Schrödinger's equation for the superlattice potential [7].

2. Theoretical framework

The superlattice we are interested in is composed of 11 layers wide-gap semiconductor material, i.e. AlN, which alternates with 10 layers of narrow-gap semiconductor material, i.e. GaN, and we examine the one-dimensional potential along the z-axis. The discontinuity in the conduction band of the contact materials is represented by U_0. On account of the internal, i.e. both piezoelectric and pyroelectric, polarization of the layers, the profiles of the wells and the barriers will not be rectangular even in the bias-free case. Instead, they assume trapezoidal form due to the internal electric fields, whose magnitudes are calculated as follows:

$$E_{\mathrm{w,b}} = \frac{\left(P_{\mathrm{b,w}} - P_{\mathrm{w,b}}\right)L_{\mathrm{b,w}}}{\left(\varepsilon_{\mathrm{b,w}}L_{\mathrm{w,b}} - \varepsilon_{\mathrm{w,b}}L_{\mathrm{b,w}}\right)\varepsilon_0} \tag{2.1}$$

in the wells and barriers regions, respectively. Here $P_{\mathrm{w,b}}$ represents the total polarization in the layer, $L_{\mathrm{w,b}}$ – its thickness, $\varepsilon_{\mathrm{w,b}}$ – the relative permittivities of the alternating well and barrier materials respectively, and ε_0 is the vacuum permittivity. Thus, the actual height of the potential barrier is defined to be:

$$U = U_0 + eE_{\mathrm{b}}L_{\mathrm{b}}, \tag{2.2}$$

where e is the elementary charge.

The bias, when applied, will further influence the potential profile of the structure. According to the direction we have chosen, the magnitude of the local field will increase the well regions and decrease in the barrier regions, as follows:

$$U_{\mathrm{w}n} = neE_{\mathrm{w}}\left(L_{\mathrm{w}} + L_{\mathrm{b}}\right) - e\left(E_{\mathrm{w}} + E_{\mathrm{ext}}\right)z \tag{2.3}$$

$$U_{\mathrm{b}n} = U_0 - (n-1)eE_b\left(L_{\mathrm{w}} + L_{\mathrm{b}}\right) + e\left(E_{\mathrm{b}} - E_{\mathrm{ext}}\right)z \tag{2.4}$$

in the n-th well and the n-th barrier, respectively. Here E_{ext} represents the magnitude of the external field. In our model, the bias is applied between the first and the last barrier, i.e. from $z=0$ to $z = L_{\mathrm{tot}} = 10\left(L_{\mathrm{w}} + L_{\mathrm{b}}\right) + L_{\mathrm{b}}$. A not to scale graphic, depicting the potential profile in the conduction band of biased and nonbiased structure, is shown on figure 1.

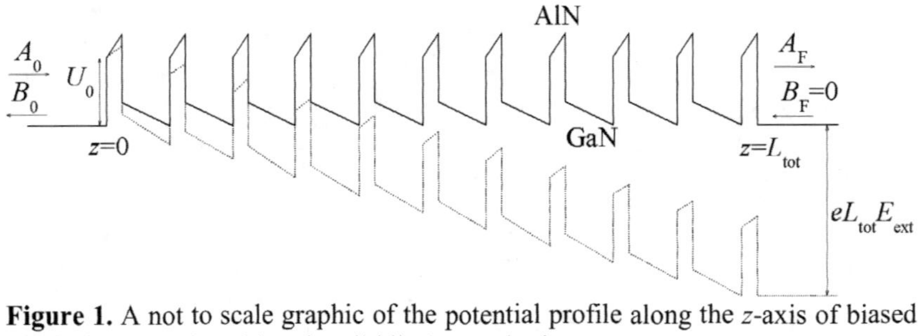

Figure 1. A not to scale graphic of the potential profile along the z-axis of biased (dashed line) and nonbiased (solid line) superlattice.

The solutions of the Schrödinger's equation, i.e. the wave functions for the one-dimensional potential of the superlattice considered, are given by the Airy functions and are presented in a previous study [6]. Outside the superlattice, the wave functions are represented by plane waves. We consider the incident and the reflected component of the wave functions at each interface and impose the effective mass dependent boundary conditions [8]. Thus we define 2×2 transfer matrices to correlate the amplitudes of the wave function on the right and on the left of each barrier and we are able to construct a transfer matrix for the multi-barrier system as follows:

$$\begin{pmatrix} A_{\mathrm{F}} \\ B_{\mathrm{F}} \end{pmatrix} = [R]\prod_{n=2}^{10}\left[T^{(12-n)}\right][L]\begin{pmatrix} A_0 \\ B_0 \end{pmatrix} = [\Omega]\begin{pmatrix} A_0 \\ B_0 \end{pmatrix}. \tag{2.5}$$

Here A_0 and B_0 are the amplitudes of the wave function in the region $z<0$, A_{F} and B_{F} are the amplitudes for $z > L_{\mathrm{tot}}$, $[R]$ and $[L]$ are the transfer matrices for the rightmost and the leftmost barriers, respectively, $\left[T^{(n)}\right]$ are the transfer matrices for each of the other barriers and $[\Omega]$ represents the 2×2 transfer matrix of the overall system. The particular components of each matrix are fully presented in [6].

In the case examined, the outgoing wave has no reflection component, i.e. $B_F = 0$ and the transmission coefficient is given by [9]:

$$\tau = \frac{|A_F|^2}{|A_0|^2} = \left|\frac{\det[\Omega]}{\Omega_{22}}\right|^2 = \frac{1}{|\Omega_{22}|^2}. \tag{2.6}$$

3. Numerical analysis

We calculate the bellow-barrier transmission coefficient on the basis of (2.6) for an AlN/GaN superlattice in range of internal polarizations and biases. We assume the conduction band discontinuity $U_0 = 1.8\,\text{eV}$ [5]; the internal polarizations $P_w = -0.032\,\text{C/m}^2$ and $P_b = -0.100\,\text{C/m}^2$ [10] in the well and the barrier, i.e. GaN and AlN layers, respectively; the relative permittivities $\varepsilon_w = 9.5$ and $\varepsilon_b = 8.5$ [1]; and the electron effective masses $m_w^* = 0.22m_0$ and $m_b^* = 0.30m_0$ [10], where m_0 is the electron mass. The electron effective masses outside the barrier structure, i.e. for $z < 0$ and $z > A_{tot}$ are chosen to equal m_0. The lattice constants of the well and barrier materials are $c_{GaN} = 5{,}19\,\text{Å}$ and $c_{AlN} = 4{,}98\,\text{Å}$ [1] respectively. The widths L_w and L_b of the wells and the barriers are defined by taking the number of unit cells equal to 9 and 3 respectively.

To examine the impact of internal polarizations on the transmission coefficient, we calculate the latter for ten, otherwise identical, superlattices that differ only in their internal polarizations. The magnitudes of the latter are modified by multiplying with suitable coefficients. The polarizations of the layers and the barrier heights of thereby generated structures are presented in table 1. The heights of the barriers are calculated after (2.1) and (2.2).

Table 1. Internal polarizations and barrier heights.

Coefficient of multiplication	P_w (C m^{-2})	P_b (C m^{-2})	U (eV)
1.0	-0.0320	-0.10	2.75
0.9	-0.0288	-0.09	2.66
0.8	-0.0256	-0.08	2.56
0.7	-0.0224	-0.07	2.47
0.6	-0.0192	-0.06	2.37
0.5	-0.0160	-0.05	2.28
0.4	-0.0128	-0.04	2.18
0.3	-0.0096	-0.03	2.09
0.2	-0.0064	-0.02	1.99
0.1	-0.0032	-0.01	1.89

For the study of the transmission coefficient in presence of uniform external electric field, the values of the applied voltage are chosen to be 50 mV and 150 mV.

4. Results and discussion

To study the effect of the magnitude of the internal polarization in the layers upon the transmittance of the barrier structures, on figure 2 we present the transmission coefficients of the all ten nonbiased superlattices, whose features are listed above.

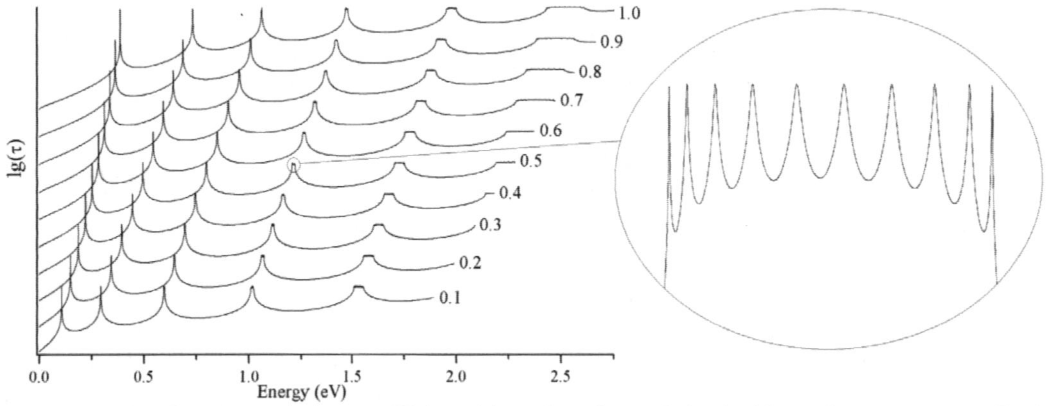

Figure 2. Bias free transmission coefficient (τ) as functions of the incident electron energy. Each curve is labelled with the factor that multiplies the polarizations in the layers and is shifted along the vertical axis.

In all cases the transmission coefficient shows resonant peaks up to $\tau = 1$ and the resonant energies are grouped in five to six minizones. A tendency of lowering the position of each minizone as the magnitude of the polarization field lowers is clearly observed. The number of the quasi-levels in each minizone equals the number of the wells, i.e. ten, in the superlattice, which can be clearly observed in figure 2.

To study the effect of external electric field upon the transmittance for the superlattices considered, on figure 3 we present our calculations of the transmission coefficient for applied voltages of 50 mV and 150 mV.

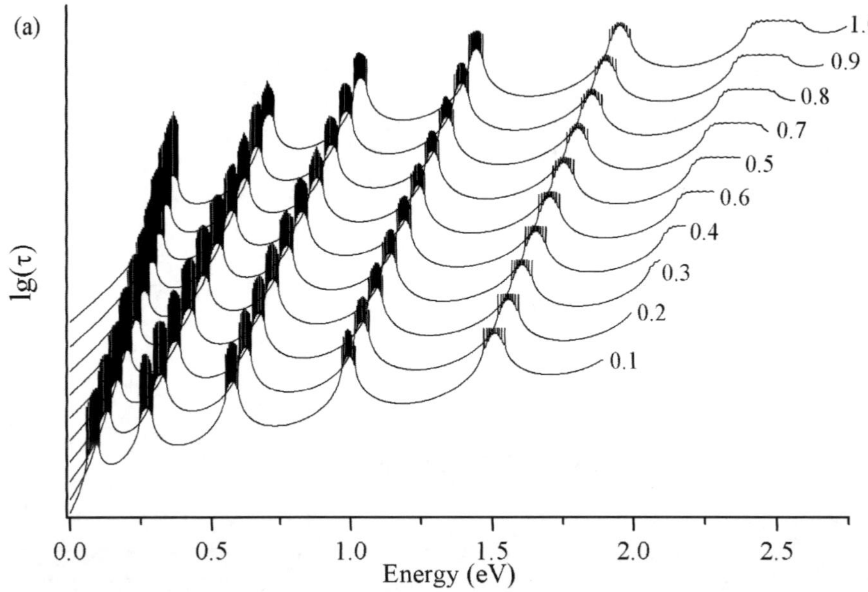

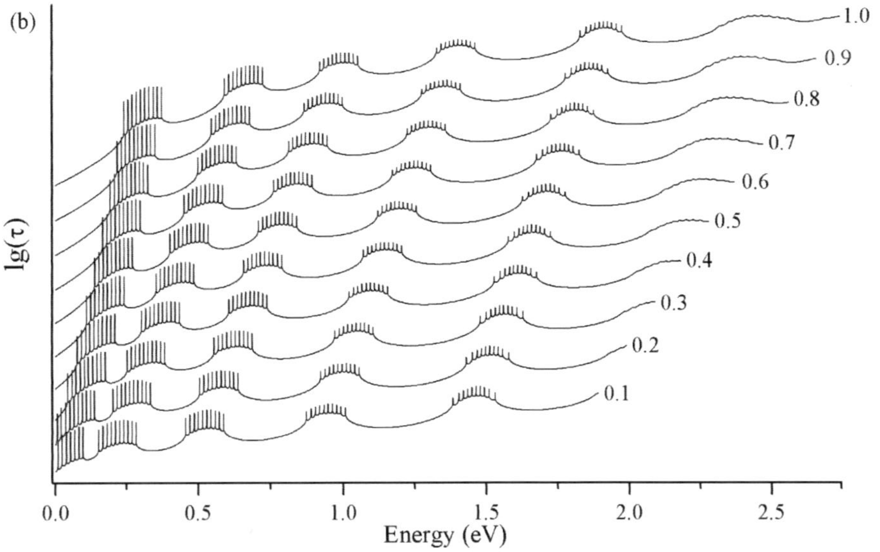

Figure 3. Transmission coefficients (τ) of biased superlattices as functions of the incident electron energy. Each curve is labelled with the factor that multiplies the polarizations in the layers and is shifted along the vertical axis. (a) For applied voltage of 50 mV. (b) for applied voltage of 150 mV.

As the bias grows the minibands are Stark shifted toward lower energies and tend to split. The transmission is suppressed in a greater degree for the miniband edge states than is for the middle of the minizones. It is also more suppressed for the superlattices, whose layer polarizations are multiplied by a higher coefficient. The Stark shift exhibited by the resonant energies is greater for the lower band edge states than for the upper band edge states.

The change of the minizones width versus the magnitudes of the internal polarizations is presented on figure 4. The measurements are made for the three consecutive lowest minizones in case of nonbiased superlattice and are taken from the lowest to the highest quasi-level in each minizone.

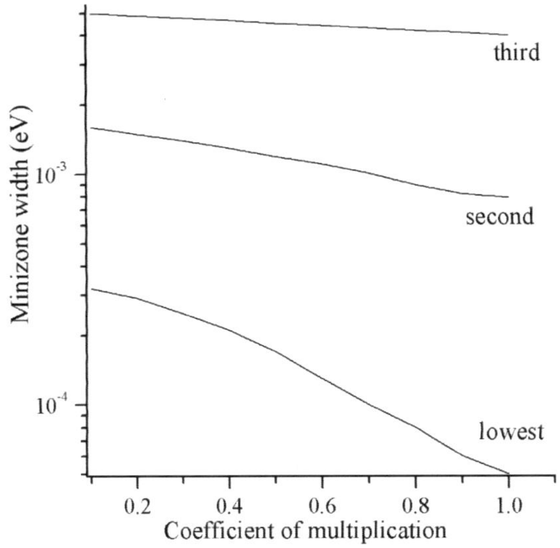

Figure 4. Width of the lowest consecutive three minizones of nonbiased superlattice as a function of the coefficient that multiplies the polarizations in the layers. Each curve is labelled according to the minizone, whose width it represents.

It is shown that the lowest minizone exhibits the greatest narrowing as the magnitudes of the polarizations increase. While applying external electric field, it is observed that all minizones exhibit broadening, which is notably greater for the lower minizones than for the upper ones.

5. Summary

In this work the effect of the magnitude of the polarization in the layers of a barrier structure upon the transmittance is studied. The calculations are based on the transfer matrix approach and the Airy functions formalism. The transmission coefficients for ten identical, but in the layer polarizations, superlattices are presented, in and out of external homogeneous electric field. It is shown that the transmission coefficient is suppressed to a greater degree for the superlattices, whose layer polarizations are multiplied by higher coefficients. The rearrangement in the positions of the minizones is shown and the dependence between their width and the internal polarization magnitudes is demonstrated in the nonbiased case.

Acknowledgements

This work was partially supported by the National Science Fund of Bulgaria and by the French-Bulgarian Bilateral Program EGGIDE (PAI-RILA). We would also like to thank to the Faculty of Physics, Sofia University, Bulgaria for the financial support.

References

[1] Monemar B 1999 *Journal of Materials Science: Materials in Electronics* **10** 227-254
[2] Cho A Y, Sivco D L, Hg H M, Gmachl C, Tredicucci A, Hutchinson A L, Chu S N G and Capasso F 2001 *J. Cryst. Growth* **227-228** 1-7
[3] Camacho A S, Gutiérrez R M and Ardilla G 2005 *Microelectronics Journal* **36** 922-927
[4] Ambacher O, Majewski J, Miskys C, Link A, M, Eickhoff M, Stutzmann M, Bernardini F, Fiorentini V, Tilak V, Schaff B and Eastman L F 2002 *J. Phys.: Condens. Matter* **14** 3399-3434
[5] Valcheva E, Kirilov K, Monemar B, Amano H and Akasaki I 2009 *Phys. Status Solidi* C **6** S751-S754
[6] Asenova I, Valcheva E and Arnaudov D (in progress)
[7] Gundlach K H 1966 *Solid-State Electronics* **9** 949-957
[8] Bastard G 1981 *Phys. Rev.* B **24** 5693-5697
[9] Tsu R and Esaki L 1973 *Appl. Phys. Lett.* **22** 562-564
[10] Vurgaftman I and Meyer J 2003 *Journal of Applied Physics* **94** 3675-3696

Annealing induced changes in ternary nanostructured $Zn_xCd_{1-x}Se$ thin films: structure and morphology

I Bineva[1], D Nesheva[1], B Pejova[2], M Mineva[1], Z Levi[1] and Z Aneva[1]
[1] Institute of Solid State Physics, Bulgarian Academy of Sciences, 72 Tzarigradsko chaussee blvd, 1784 Sofia, Bulgaria
[2] Institute of Chemistry, Faculty of Science, Sts. Cyril and Methodius University, Skopje, Macedonia

E-mail: irka@issp.bas.bg

Abstract. Single layers of $Zn_xCd_{1-x}Se$ with various compositions (x = 0.39, 0.59 and 0.8) were prepared by thermal vacuum evaporation at room substrate temperature. Consecutive deposition of small portions of ZnSe and CdSe with equivalent thickness of 0.12 or 0.37 nm was applied. X-ray diffraction and atomic force microscopy measurements were applied to explore the evolution of the crystal structure, microstructure, composition and surface morphology upon furnace annealing at 200 °C and 400 °C in an inert atmosphere. It has been found that as-deposited films were nanocrystalline with a grain size of around 10 nm and cubic structure. Upon annealing the size increased approximately three times and the cubic structure was preserved; no appearance of wurtzite phase was observed. It has been also ascertained that annealing caused significant reduction of the film surface roughness. Atomic force microscopy phase images revealed existence of a second phase on the surface of as-deposited films which disappeared after annealing. The effect of the preparation conditions on the film properties and annealing induced changes is discussed.

1. Introduction

The detection of blue-UV light has traditionally been the province of photomultiplier tubes, which are sensitive and fast but not well-suited for space and field applications as they are fragile and require high voltages. Enhanced Si detectors are thus often used, but their relatively low band gap results in high reflectivity, and dark currents that require cooling to improve sensitivity. This has led to interest in the prospects for II-VI [1,2] or III-V materials in short-wavelength detectors. The $Zn_xCd_{1-x}Se$ II-VI ternary alloys are direct-gap semiconductors in which the optical bandgap variation spans the entire visible range (from 1.75 to 2.7 eV) and the lattice parameter gradually decreases as the Zn content, x, is varied between 0 and 1. The bandgap tunability of the $Zn_xCd_{1-x}Se(S)$ systems means that they are candidates for use in tandem solar cells [3,4]. A more recent potential application of semiconductor thin films is in the generation of Terahertz radiation from photoconductive antennae.

For the preparation of ternary II-VI semiconductor layers mainly molecular beam epitaxy [5,6] and electrodeposition [7,8] techniques are used. On the other hand thermal evaporation in vacuum allows deposition of films with a low contamination level at a relative low cost. In previous papers [9,10] we have reported an original approach for preparation of $Zn_xCd_{1-x}Se$ thin films of various compositions in the range x = 0.39 - 0.8. The films were produced by alloying of ultra thin ZnSe and

CdSe films and the compositional variation were achieved by varying the equivalent thickness of the films.

In this study $Zn_xCd_{1-x}Se$ thin films with three different compositions are prepared by thermal vacuum evaporation and the effect of furnace annealing at 200 °C and 400 °C on the crystal structure, microstructure and surface morphology of the films is explored.

2. Experimental details

Simultaneous thermal evaporation of ZnSe and CdSe powders (Merk, Suprapure) from two independent tantalum crucibles was applied for the preparation of $Zn_xCd_{1-x}Se$ layers with $x = 0.39$, 0.59 and 0.8 and thickness of 400 nm. The layers were deposited on Corning 7059 glass substrates kept at room temperature. During the film deposition the substrates were rotated at a rate of 20 turns/min. During each turn pass they spent over each crucible 1/12 of the turn time and the time between the consecutive deposition of CdSe and ZnSe is 5/12 of the turn time. The film composition was set by using appropriate deposition rate (0.5 or 1.5 nm/sec) for each material. The rates were controlled by two previously calibrated quartz microbalance systems MIKI FFV. Thus, consecutively deposited small portions of ZnSe/CdSe with equivalent thickness of around 0.12/0.37, 0.37/0.37 and 0.37/0.12 nm made $Zn_xCd_{1-x}Se$ alloys with $x = 0.39$, 0.59 and 0.8, respectively. More details on the deposition procedure can be found in [9,10]. For each composition a part of the films were furnace annealed for 60 min in an argon atmosphere at annealing temperature $T_a = 200$ °C or 400 °C.

XRD experiments were performed on a Rigaku Ultima IV diffractometer with Cu K_α radiation, $\lambda = 0.154$ nm. The XRD patterns were registered via the step method applying 0.02° steps (2θ) and a counting time of 0.6 sec/step. The scattered X-ray radiation was detected with DteX detector. All the XRD patterns were taken within an angular interval 2θ from 10° to 80° where the main diffraction peaks of ZnSe and CdSe are located.

A Veeco Multimode V scanning probe microscope operating in tapping mode was used to investigate the surface morphology and phase composition of the $Zn_xCd_{1-x}Se$ films. Height, amplitude error and phase images were recorded in 500 nm, 1 µm and 5 µm scale. The scan rate was in the range 1 - 2 Hz and the image resolution was 512 lines per scan direction for the 1 µm and 5 µm images and 1024 lines for the 500 nm ones. At least three different points on the sample surface were explored. Aluminum coated silicon cantilevers TAP150-Al-G for soft tapping mode with a nominal resonance frequency of 150 kHz and a typical force constant of 5 N/m were used; the tip radius is less than 10 nm. All images were just flattened and no further processing was carried out. Roughness measurements were performed using the Nanoscope 7.30 programme. The grain size analysis was done on 500 nm images with SPIP^{tm} image processing software version 6.0.4 (Image metrology A/S) – evaluation copy, using Watershed method.

3. Results and Discussion

3.1. X-ray diffraction

In figure 1 (a) XRD patterns of as-deposited $Zn_xCd_{1-x}Se$ layers with $x = 0.39$, 0.59 and 0.8 are shown. The shape of all the patterns is typical for materials with polycrystalline structure. The diffraction peak at 2θ in the range 26-27° is the most intensive, narrow and well pronounced. Two additional very weak peaks at about 44° and 52° are also seen. The three peaks correspond to the {111}, {220} and {311} families of crystallographic planes in cubic $Zn_xCd_{1-x}Se$ [10]. The absence in the experimental patterns of peaks, which should correspond to {100}, {101} and {102} families (at around 24°, 29° and 37°, respectively) in the crystallographic planes of wurtzite $Zn_xCd_{1-x}Se$, clearly indicates that the films with all compositions have cubic symmetry. Based on the big difference in the peaks' intensity in figure 1 one can suggest that the films have a preferential {111} orientation.

The peak {111} is presented in an enlarged scale in figure 1 (b), which reveals that the decrease in the Zn content causes a proportional shift of the peak to lower angles. This shift is due to the substitution of Zn with Cd atoms that leads to an increase in the lattice parameter a (the cubic modifications of ZnSe and CdSe are isomorphous and completely soluble in the solid state). In order to get quantitative information about the {111} diffraction peaks all XRD spectra were background corrected and then least-square fitted for the estimation of their angular position, full width at half maximum (FWHM), area and height. Two analytical functions, Gaussian and Lorentzian, were initially tried for fitting, with better results obtained by applying the Lorentzian one. The results obtained from the Lorentzian fitting of the spectra of as-deposited films are summarised in table 1 and the peak positions are shown in figure 1 (b), inset. The dashed line in the inset represents the standard relationship between the {111} peak position and the Zn content x obtained using Bragg's equation $d_{\{hkl\}} = \lambda/2\sin\theta$ and taking into account that the lattice constant in cubic crystals $a = d_{\{hkl\}}(h^2+k^2+l^2)^{1/2}$ (h, k, l are the lattice planes; $d_{\{hkl\}}$ is the corresponding interplanar distance; for the {111} peak $a = d_{\{111\}}\sqrt{3}$) obeys Vegard's rule $a = x\,a_{ZnSe} + (1 - x)\,a_{CdSe}$ [11]. It is seen from the inset that the experimental points are fairly close to this line, which confirms that the suggested technique allows fabrication of $Zn_xCd_{1-x}Se$ films with a good composition control.

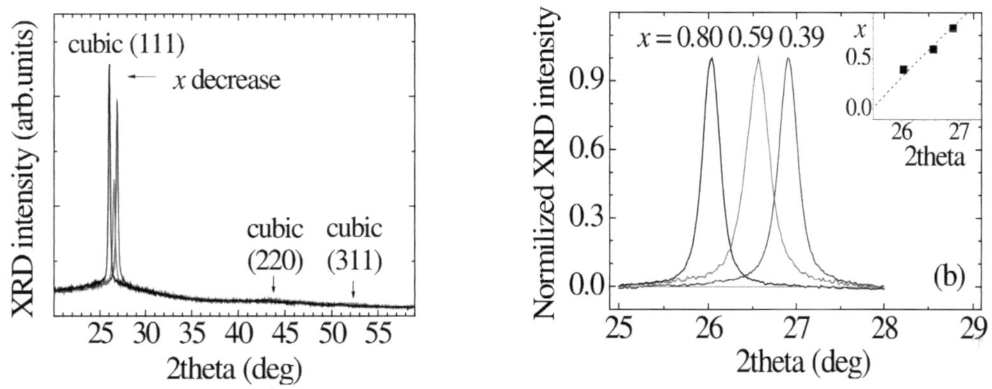

Figure 1. (a) XRD spectra of three as-deposited $Zn_xCd_{1-x}Se$ films with different compositions; (b) a part of the pattern showing the {111} peak. Inset: the dashed line represents the relationship between the {111} peak position and the Zn content x, the symbols correspond to the experimental positions.

Table 1. Angular position, lattice constant, full width at half maximum, area and height of as-deposited and annealed $Zn_xCd_{1-x}Se$ thin films determined by Lorentzian fit of the {111} peak in the XRD spectra.

$Zn_xCd_{1-x}Se$	Annealing	Center (deg)	Lattice constant (nm)	FWHM (deg)	Area	Height
$x = 0.8$	as-deposited	26.91	0.5734	0.245	13253	34391
	$T_a = 200$ °C	27.02	-	0.254	14235	35700
	$T_a = 400$ °C	27.01	0.5713	0.259	14371	35300
$x = 0.59$	as-deposited	26.55	0.5808	0.324	10008	19650
	$T_a = 200$ °C	26.56	-	0.296	10261	22043
	$T_a = 400$ °C	26.63	0.5791	0.269	11465	27068
$x = 0.39$	as-deposited	26.03	0.5922	0.211	13956	42017
	$T_a = 200$ °C	26.10	0.5906	0.201	14366	45515
	$T_a = 400$ °C	26.07	-	0.196	19555	63630

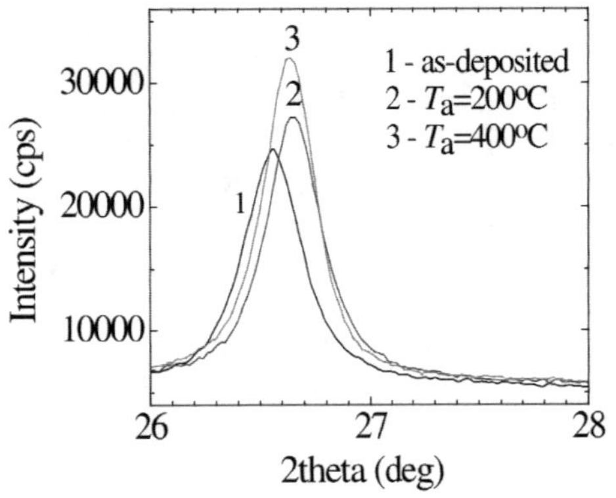

Figure 2. XRD spectra of three $Zn_xCd_{1-x}Se$ films with $x = 0.59$: 1 - as-deposited, 2 - annealed at 200 °C, 3 - annealed at 400 °C.

Figure 2 depicts the {111} peaks of an as-deposited $Zn_xCd_{1-x}Se$ film with $x = 0.59$ and two annealed films. It is seen that upon annealing the {111} peak slightly shifts to the larger angles and its intensity increases. Similar changes have been observed for the other two compositions. The results from the Lorentzian fitting (table 1) show that upon annealing the {111} peak displays a shift of 0.07° to 0.1° (in 2θ scale), which corresponds to a lattice constant decrease of 0.016-0.021 nm and in addition the annealing causes an increase of the peak area. These observations indicate an improvement of the film crystallinity, which is expected [12], as well as a slight densification. Since CdSe and ZnSe single crystals are usually characterized by the hexagonal wurtzite structure, normally the annealing induces a transition from cubic to hexagonal structure and also an increase of the grain size [12]. However, in our case even the films annealed at 400 °C show a cubic structure. In a previous study on the structure of self assembled CdSe nanocrystals embedded in a SiO_x matrix we revealed [13] that three quarters of nanocrystals in as-deposited films (grown at room substrate temperature) are of hexagonal structure and one quarter were cubic or both lattice types. For the samples annealed at 400 °C it was determined one half of the particles with cubic structure and another half with hexagonal or both lattice types [13]. These quite unusual observations give us basis to adopt the assumption made in [12] that the nanoparticle surface properties affect significantly their structure. Perhaps, the preparation technique we applied determines certain nanoparticle surface properties which impede cubic to hexagonal transformation. Annealing at higher T_a and/or longer annealing times will be carried out to verify this assumption. As for the grain size, it is frequently estimated from the FWHM of XRD bands. As seen from table 1 the annealing of the films with $x = 0.39$ and 0.8 causes a gradual FWHM decrease but an increase is observed for the films with $x = 0.59$. The obtained FWHM alterations are quite small and they are normally affected by both the grain size and fixed residual stress in the films. Therefore the annealing induced changes in the film microstructure, surface roughness and phase composition were explored by AFM measurements.

3.2. Atomic force microscopy

Two-dimensional surface images of two films with $x = 0.39$, as-deposited and annealed at 400 °C, are depicted in figure 3. They reveal that the layers are nanocrystalline and show an increase of the grain size upon annealing. Size increase has been observed for the films with all compositions. The obtained grain size of as-deposited films is around 10 nm while in the annealed samples the size is three times greater (~ 30 nm). The small grain size in the as-deposited films implies a high density of embryonic grains on the substrate surface which may be due to the applied deposition technique. Perhaps, the embryo density is high because of the combination of the very low average deposition rate and the low substrate temperature, resulting in low mobility of ad-atoms on the substrate surface.

Another effect observed upon the film annealing is a decrease of the surface roughness. It is illustrated in figure 4 for the films with $x = 0.8$. In as-deposited films a growth of quite high spikes is observed whose density is highest in the films with $x = 0.8$ (figure 4 (a)). Perhaps the low substrate

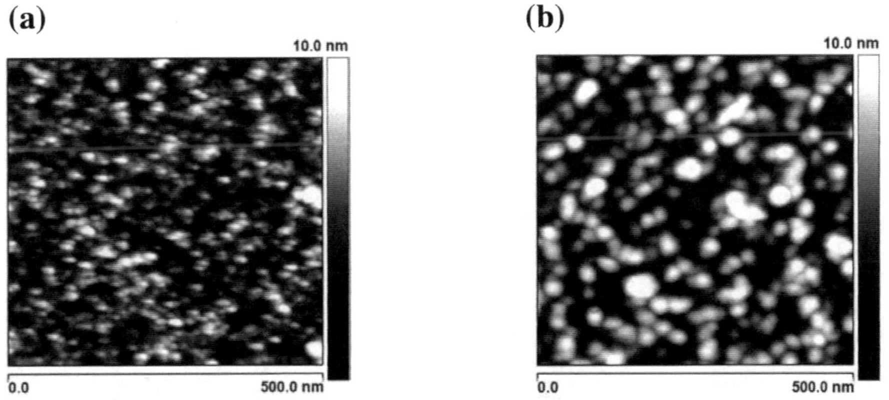

Figure 3. Two-dimensional surface images of two $Zn_xCd_{1-x}Se$ thin films with $x = 0.39$: (a) as-deposited and (b) annealed at 400 °C. The scale of both pictures is 500 nm in lateral direction and 10 nm in height.

temperatures, i.e. the low ad-atoms mobility, facilitate the spikes growth The annealing at 200 °C significantly improves the surface smoothness (figure 4 (b)); a decrease of the root mean square roughness from 4.5, 6.9 and 6.0 nm to 1.3, 2.1 and 1.6 nm for films with $x = 0.8$, 0.59 and 0.39, respectively, has been observed on 5×5 μm^2 surface area. The annealing at 400 °C results in some roughness increase (figure 4 (c)), which may be due to some re-evaporation of Se from the film surface.

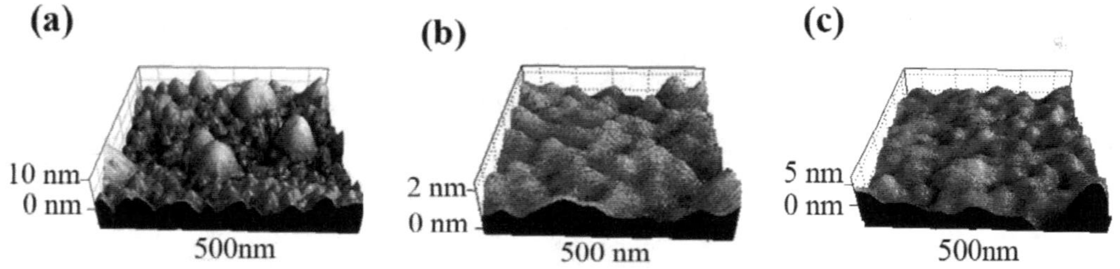

Figure 4. Three-dimensional surface images of three $Zn_xCd_{1-x}Se$ films with $x = 0.8$: (a) as-deposited, (b) annealed at 200 °C, (c) annealed at 400 °C.

Finally, we would like to consider the compositional homogeneity of the films as seen from the surface phase images. Figure 5 depicts the phase images of as-deposited and annealed $Zn_xCd_{1-x}Se$ films with $x = 0.59$. The phase signal is sensitive to variations in composition, adhesion, friction, viscoelasticity as well as other factors. Two different phases have been observed in all as-deposited films. The amount of the second phase (white colour) is highest in the films with $x = 0.59$ while in the films with the other two compositions it is relative small. This observation can be understood keeping in mind that the films with $x = 0.59$ were formed by allowing of ZnSe/CdSe layers with an equivalent thickness of 0.37 nm while in the films with $x = 0.39$ and 0.8 the equivalent thickness of one material is 0.12 nm. The registration of nanosized second phase on the film surface is in agreement with the conclusion made for the volume from Raman scattering experiments for existence of nanosized Cd-enriched regions [10]. Based on figure 5 (b) one can think that the applied annealing at 200 °C is not enough for full composition homogenization in the films with $x = 0.59$; it occurs at $T_a = 400$ °C. For the other compositions the second phase has not been seen upon annealing at $T_a = 200$ °C.

The observed small grain size makes as-deposited films interesting for chemical sensor applications because the integral interface area in such films is very large. Previous investigations on CdSe films annealed at 400 °C have revealed [14] dark conductivity increase and decrease in the

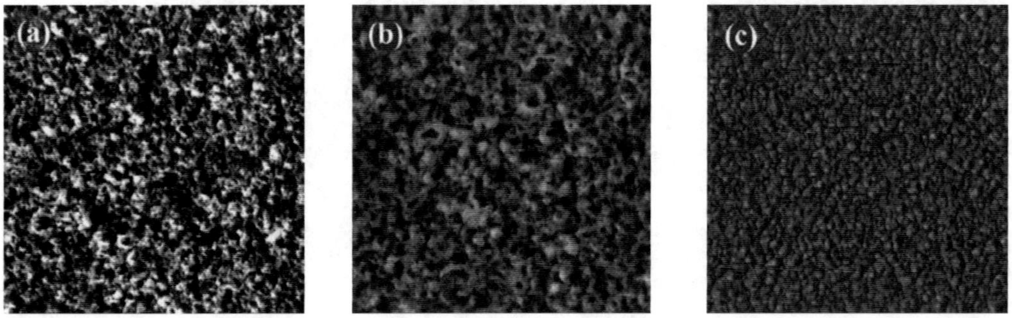

Figure 5. Phase surface images of three $Zn_xCd_{1-x}Se$ films with $x = 0.59$: (a) as-deposited, (b) annealed at 200 °C and (c) annealed at 400 °C. The scale of all pictures is 1μm in lateral direction.

density of 'slow' recombination centres. It may be expected that the above described annealing induced size increase in $Zn_xCd_{1-x}Se$ films will cause similar dark and photoconductivity changes and its combination with the observed effects of composition homogenization and surface smoothening increases the potential of the $Zn_xCd_{1-x}Se$ layers for application in short-wavelength detectors and solar cells.

4. Conclusions

Single layers of $Zn_xCd_{1-x}Se$ with $x = 0.39$, 0.59 and 0.8 were prepared by thermal evaporation of ZnSe and CdSe in vacuum. X-ray diffraction and AFM measurements revealed that as-deposited films are nanocrystalline with cubic structure and a small grain size (~ 10 nm). These observations have been related to the low substrate temperature and low average deposition rate. The annealing at 400 °C for 60 min resulted in size increase (approximately three times) but wurtzite phase was not detected. It has been assumed that the specific preparation approach determines certain nanoparticle surface properties which impede cubic to hexagonal transformation. Atomic force microscopy height images have shown spikes on the surface of as-deposited films whose density depends on the composition. The spikes growth has been assigned to low ad-atoms mobility during the film deposition. An anticipated roughness reduction has been observed upon annealing at 200 °C while the annealing at 400 °C caused some roughness increase related to re-evaporation of Se atoms. AFM phase images have revealed existence of a second phase on the surface of as-deposited films which disappeared after annealing. The largest amount of the second phase has been observed in the films with $x = 0.59$. This result has been related to the relative large equivalent thickness (0.37 nm) of both ZnSe and CdSe sublayers.

Acknowledgements This work has been financially supported by the Bulgarian Ministry of Education, Youth and Science under grant DMU 03-91.

References

[1] Hsiao C H, Changn S J, Hung S C, Cheng Y C, Huang B R, Wang S B and Lan B W 2010 *J. Cryst. Growth* **312** 1670
[2] Yoon Y-J, Park K-S, Heo J-H, Park J-G, Nahm S and Choi K J 2010 *J. Mater. Chem.* **20** 2386
[3] Tamargo M C (ed) 2002 *II–VI Semiconductor Materials and their Applications* (New York: Taylor and Francis) ch 2
[4] Chandramohan R, Mahalingam T, Chu J P and Sebastian P J 2004 *Sol. Energy Mater. Sol. Cells* **81** 371

[5] Gupta P, Maiti B, Maity A B, Chaudhuri S and Pal A K 1995 *Thin Solid Films* **260** 75
[6] Hankare P P, Chate P A, Asabe M R, Deleka S D, Mulla I S and Garadkar K M J 2006 *J. Mater. Sci.: Mater. Electron.* **17** 1055
[7] Katayama K *et al* 2000 *J. Cryst. Growth* **214–215** 1064
[8] Chen W R and Huang Ch J 2004 *IEEE Photon. Technol. Lett.* **16** 1259
[9] Nesheva D, Aneva Z, Šćepanović M J, Bineva I, Levi Z, Popović Z V and Pejova B 2012 *J. Phys.: Conf. Series* **253** 012035
[10] Nesheva D, Aneva Z, Šćepanović M J, Levi Z, Iordanova I and Popović Z V 2011 *J. Phys. D* **44** 415305
[11] Bis R F and Dixon J R 1969 *J. Appl. Phys.* **40** 1918
[12] R. J. Bandaranayake, G. W. Wen, J. Y. Lin, H. X. Jiang and C. M. Sorensen 1995 *Appl. Phys. Lett.* **67** 7
[13] Hofmeister H, Nesheva D, Levi Z, Hopfe S and Matthias S 2000 *Proc. EUREM 12*, Brno, Eds. Frank C L and Ciampor F (Brno: Czechoslovak Society for Electron Microscopy) p.365
[14] Nesheva D, Reynolds S, Aneva Z, Main C and Levi Z 2005 *J. Optoel. Adv. Mat.* **7** 517

Study of electrochromic APCVD WO_3-V_2O_5 films

G Bodurov[1], T Ivanova[1], M Abrashev[2] and K A Gesheva[1]
[1] Central Laboratory of Solar Energy and New Energy Sources, Bulgarian Academy of Sciences, Blvd Tzarigradsko Chaussee 72, 1784 Sofia, Bulgaria
[2] Faculty of Physics at Sofia University

E-mail: joro.bodurov@yahoo.com

Abstract. WO_3-V_2O5 thin films were deposited by atmospheric pressure chemical vapour deposition (APCVD). WO_3-V_2O_5 thin films are investigated related to their potential use as primary electrochromic layers (working electrodes) in Electrochromic Devices. A typical EC Device is a sandwich like structure with two conductive glasses and an electrolyte with working electrodes that possess electrochromic properties. APCVD has the advantages of scalability to large areas with uniform thickness and potentially low cost.

1. Introduction

An electrochromic device (ECD) is an optical system, consisting of two conductive glasses, over one of which the functional layer, the electrochromic (EC) material, is deposited. The second conductive glass could be either bare, or an "ion storage" film could be deposited on it, forming together the counter electrode. An electrolyte is deposited between the two electrodes. In the present work, WO_3-V_2O_5 thin films are investigated related to their use as working electrodes for such an EC Device.

If a small voltage pulse is applied to the device (contacts are made at the conductive glasses, on bare places left), electrons from the conductive glass get injected into the functional layer. For an electro neutrality of the circuit, ions from the electrolyte intercalate into the film, and in result the newly made film structure becomes absorptive, absorbs part of the solar light, and what we see is the complementary color. If WO_3-V_2O_5 is the functional layer, the color we see is dark blue, since the concentration of WO_3 dominates and this material absorbs in the near IR. In the area of the EC materials and devices, numerous publications are made, following the first publications of the discovering of the effect in that material by Satyen Deb [1]. Many publications related to the electrochromic effect in different thin film materials let to a handbook and many books [2-4]. The electrochromic materials gain great scientific interest as they find applications in automobiles, sunroof of buildings, display devices, light regulation and architectural glazing [2].

The advantages of the chemical vapor deposition at atmospheric pressure is related to the fact, that in a large-scale production a flow-through process is possible when the deposition goes at atmospheric pressure. CVD MoO_3, WO_3 and mixed films based on them are studied in details [5]. Among those transition metal oxides, tungsten trioxide appears to be the best electrochromic compound [2]. Its advantages are high coloration efficiency, good stability, and relatively low cost.. However a major disadvantage of the APCVD process for WO_3 deposition is the very low growth rate. Mixing of tungsten oxide with other transition metal oxide led to considerable increasing of the growth rate, improvement of layer properties especially in optical and electrochromic aspects [5]. We relate the role of vanadium oxide in the growth-rate increase with its lower-temperature CVD behavior. It vaporizes easier and forms growth centers, so the vapors could easily start growing. Films are much thicker at the same technological conditions as for the single WO_3 oxide films.

In this work, we present preparation and investigation of mixed WO_3-V_2O_5 films, deposited on glass and conductive glass substrates by APCVD method. Structural properties are studied by Raman and FTIR spectroscopies. Optical characterization is performed.

2. Experimental details

WO_3-V_2O_5 thin films were obtained by atmospheric pressure CVD in oxygen ambient. The precursor used was a physical mixture of $W(CO)_6$ and $V(CO)_6$ in ratio $W(CO)_6$:$V(CO)_6$=10:1. The substrate temperature was kept at 200°C. The sublimator temperature was chosen to be 70°C. The other important technological parameter is the ratio of the flow-rates of the carrying carbonyl vapors, namely the Ar flow, and the flow-rate of a reactive gas, oxygen (O_2). In this study, the gas flow-rates ratio was $Ar_{carbonyl}/O_2$ 1/32. Depositions were performed by using substrates of bare glass and substrates covered with conductive films of ITO (Donelly type). The observed low value of the thickness of the films on the bare glasses is related to the absence of "initially formed centers of growth". However, the thickness of the films deposited on the substrates covered with conductive layers of ITO, is relevantly high, due to the "centers of growth" of the ITO films themselves.

FTIR measurements were performed in the spectral range of 350- 4000 cm^{-1} by Shimadzu FTIR Spectrophotometer IRPrestige-21. The studied samples were also deposited on Si substrates and bare Si wafer (p-Si, with orientation <100>) was used as background. Raman spectra were also investigated. The Raman spectra were obtained using LabRAM HR Visible micro-Raman spectrometer. The 633 nm line of a He-Ne laser were used for excitation. An X100 objective was used both to focus the incident beam and to collect the scattered light in backscattering configuration. To prevent the samples by local laser overheating a low laser power (0.70 mW) was used. The range was 100-1280 cm^{-1}, with accumulation time – 60s X 5. UV-VIS spectrophotometric study was made by UV 3600 Shimadzu spectrometer in the spectral range of 300-1000 nm.

The surface morphology of the studied metal oxide films has been investigated by AFM (Atomic Force Microscope) DiMultimode V model (Veeco) in tapping mode. Scan rate was 1.5-2 Hz, the images resolution was 512samples/line. The samples were scanned at least two places. Silicon cantilevers (1-10Ohm-cm Phosphorus (n) doped Si) with a resonance frequency of 250-308 kHz, force constant 20-80 N/m, tip radius < 10 nm,50+/-10nm Al coating on the back side (MPP-11120-10, Veecoprobes) were used in the dynamic AFM experiments. Image processing was performed by means of Nanoscope 7.30.

3. Results and Discussions

Two of the films deposited on conductive glasses were measured in transmittance and reflectance modes in the visible and infrared regions. To study the structural transformation of as deposited films, one of them was annealed at 500 °C and the other at 400 °C.

APCVD WO_3-V_2O_5 films obtained on conductive glass show (figure 1) transparency above 65% in as deposited state and after thermal treatment at higher temperature (400 – 500 °C), the transmittance reaches up to 75 - 80% at 600 -700 nm. As it is seen from the figure, the reflectance after annealing decreases (around 10%), thus the overall optical behavior is improved. Usually the as deposited APCVD mixed oxide films possess amorphous like structure [2]. After annealing at elevated temperatures, the films undergo recrystallization, leading to higher degree of optical transparency.

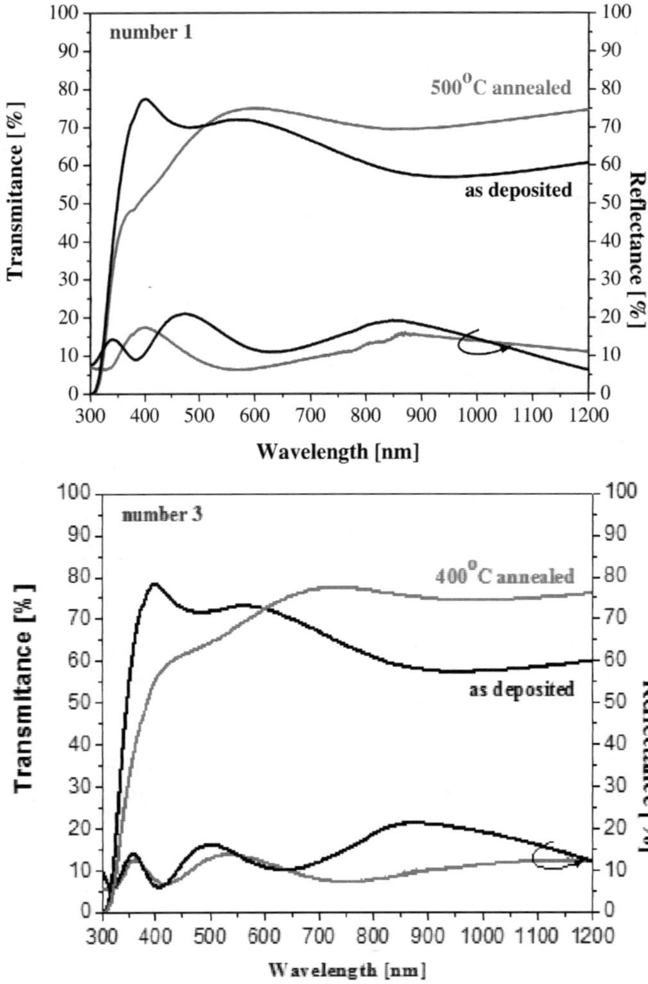

Figure 1. Transmittance and reflectance spectra of CVD WO_3-V_2O_5 films deposited on conductive glass substrates.

WO_3-V_2O_5 films were also investigated by FTIR Spectroscopy in as deposited state (figure 2, top) and after annealing at 400 and 500 °C in air for 1 hour (figure 2, bottom).

FTIR study shows that the spectra of WO_3 – V_2O_5 films are very similar in shapes to the corresponding spectra of pure tungsten oxide films [4], but the revealed absorption bands are with lower intensity. Several weak absorption bands have been detected that can be attributed to the vanadium oxide component.

In the as deposited films the characteristic absorption band for the single WO_3 is centred at 670 cm^{-1}, the band covers the spectral range 620 – 730 cm^{-1}. There can be found a contribution from several deformations, and stretching modes of WO_3 [5]. After thermal treatment at 400 °C and 500 °C, the broad band is not observed. Instead, the weak bands due to the stretching and bending vibrations of V-O bonding appeared in spectral range 400 – 550 cm^{-1} [6].

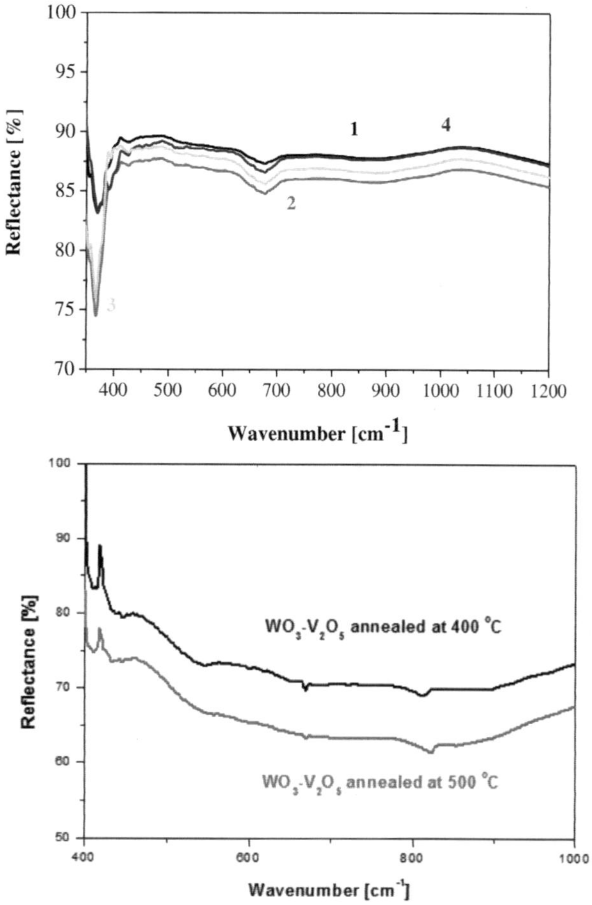

Figure 2. FTIR spectra of APCVD WO_3-V_2O_5 films as deposited (top) and annealed at 400 - 500°C (bottom).

In the range around 830 cm^{-1} some IR band assigned to asymmetrical stretching mode of V-O-V vibrations can be seen. The FTIR study clearly indicates that although the small amount of vanadium carbonyl in the precursor mixture, the vanadium oxide is present into the films structure and affects their properties.

The APCVD WO_3 – V_2O_5 films grown at these technological conditions also remain amorphous - like even after high temperature annealing at 500°C.

The Raman band at 248 cm^{-1} is due to the deformation modes of the two metal oxides. The band at 335 cm^{-1} is attributed to the bending vibrations of crystalline WO_3, which appears in the spectra of the as-deposited and annealed films. The bands at 682 and 840 cm^{-1} are related to the two different types of bridging oxygen of ν(O-W-O) vibrations[7]. The line at 985 cm^{-1} is assigned to terminal double bonds W=O or/and V=O bonds. The vanadium oxide component is represented by two weak Raman lines at 396 and 483 cm^{-1}, which can be assigned to the bending vibrations of V-O-V and to stretching V-O bonds of orthorhombic V_2O_5 [8]. The Raman peak at 483 cm^{-1} is observed also in the spectrum of the annealed film.

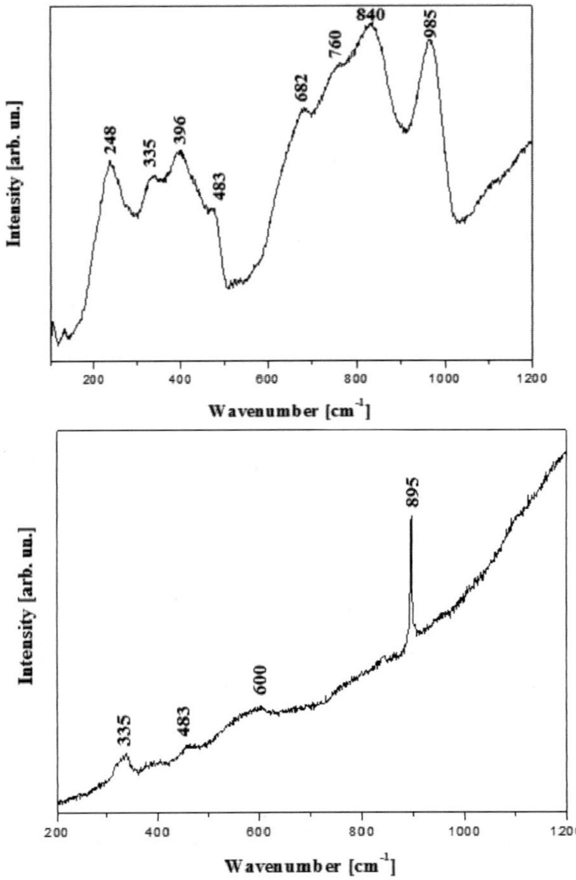

Figure 3. Raman spectra of APCVD WO_3-V_2O_5 films in as-deposited state (left) and after annealing at 500 °C (right).

As it is seen from figure 3 (bottom), a new larger Raman band appears centred at 600 cm^{-1}, which we consider as evidence for vanadium oxygen bonds. Studding thermochromic properties of vanadium oxide doped with tungsten, the authors of [9] observed a Raman peak around 600 cm^{-1}. The strong Raman peak at 895 cm^{-1} could be related to vanadium oxide crystalline phase. Surface observations during the Raman spectra measurements show existence of separate crystals immersed in a homogenous surface material, not exhibiting Raman features. Raman study of APCVD WO_3-V_2O_5 films reveals that the film structure is predominantly amorphous as the Raman bands are broad and with low intensity.

Having in mind the Raman results the separate crystals found in the AFM micrographs (figure 4, left), here we relate with the vanadium oxide phase, since our previous study of single WO_3 have shown that this material stays predominantly amorphous even after 500 °C [10]. When heated at 500 °C the films have the typical for the mixed transition metal oxides surface, namely cluster formation (figure 4, right). This would result in improved electrochromic performance, because of the increased area of the grain boundaries where ions can easier intercalate. The separate crystals are seen above the typical clustered surface.

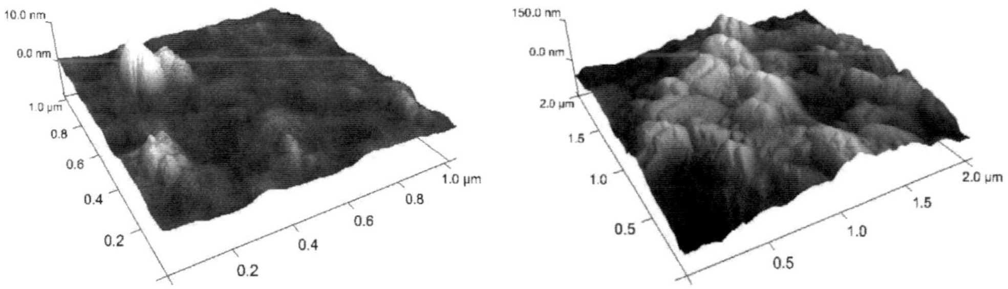

Figure 4. AFM Micrographs of as-deposited (left), and annealed at 500°C W-V-O film (right).

WO_3 – V_2O_5 thin films of the same type as the presented here samples, but obtained at higher sublimation temperature (80 °C) have shown well expressed electrochromic behaviour demonstrated in a three electrode configuration. The W-V-O film on conductive glass served as functional electrode, electrochemically derived NiO on conductive glass was used as a counter electrode, and a Calomel Electrode, as reference electrode. The electrolyte was a 0.6 M $LiClO_4$ in Propylene carbonate (PC), placed in a quartz cuvette. Film coloration was observed under voltage applied in the range -1 - +4 volts.

Further study will be related to detailed voltametric measurements to determine the colour efficiency and optical modulation.

4. Conclusions

Atmospheric pressure chemical vapour deposition is employed to produce mixed WO_3 – V_2O_5 films. Their optical transmittance and reflectance are measured for the as-deposited and annealed states. Infrared and Raman vibrational spectra are also measured. Films are predominantly amorphous, not only in as-deposited state but also after annealing at elevated temperatures. The lower temperature vanadium oxide undergo crystallization after annealing, forming separate crystals seen as immersed in amorphous – like homogenous film surface. Films have shown electrochromic effect when a small voltage of couple of volts is applied across them.

5. References

[1] Deb S 1969 *Appl. Opt. Suppl.* **1** 3192
[2] Granqvist C G 1995 *Handbook of Inorganic Electrochromic Materials* (Amsterdam: Elsevier Science)
[3] Monk P, Mortimer R J and Rosseinski D R 2007 *Electrochromism and Electrochromic Devices* (Cambridge, University Press)
[4] Gesheva K A 2007 *Thin Film Optical Coatings for Effective Solar Energy Utilization: Spectrally Selective Surfaces and Energy Control Coatings* (NY: Novapublishers)
[5] Reddy A M, Reddy A S and Reddy P S 2011 *Vacuum* B **85** 949
[6] Zhang Y Q, Xia X H, Tu J P, Mai Y Y, Shi S Y, Wang X L and Gu C D 2012 *J. Power Sources* **199** 413
[7] Songara S, Kupta V, Patra M K, Singh J, Saini L, Gowd G S, Vadra S R and Kumar N 2012 *J. Physics Chemistry solids* **73** 851
[8] Senthil K, Kwak G and Jong K *Applied Surface Science.* doi : 10.1016/j.apsusc(2012)04060
[9] Blackman C S, Piccirillo C, Binions R and Parkin I P 2009 *Thin Solid Films* **517** 4565–4570
[10] Gesheva K A, Szekeres A and Ivanova T 2003 *Journal: Solar Energy Materials and Solar Cells* A **76** 563-576

Journal of Physics: Conference Series **398** (2012) 012017 doi:10.1088/1742-6596/398/1/012017

Optical properties of thin films with plasmonic effect for light scattering

K Lovchinov, O Angelov, D Dimova-Malinovska

Central Laboratory of Solar Energy and New Energy Sources, Bulgarian Academy of Sciences, Sofia

E-mail: lov4@abv.bg

Abstract. Thin films of Al_2O_3 with embedded Cu and Ag nanoparticles (NPs) are deposited by r.f. co-sputtering on glass substrates. The samples are annealed at 400^{0}C and 500^{0}C for 3 h in forming gas (N_2+5% H_2) to stimulate Cu and Ag nanoclusters' formation. Multilayer structures (stacks) such as Al_2O_3/Ag/Al_2O_3, ZnO:Al/Ag/ZnO:Al and Ag/ZnO:Al/Ag/ZnO:Al are deposited on glass substrate by consequently sputtering of the individual layers. Before deposition of the capping layers of Al_2O_3 and ZnO:Al, the bi-layer structures are annealed in vacuum for 50 min at 220^{0}C or 400^{0}C in the case of glass/Al_2O_3/Ag, and at 220^{0}C in the case of glass/ ZnO:Al/Ag and glass/ Ag/ZnO:Al/Ag. The diffuse and specular reflectance and transmittance spectra are measured in the spectral range of 300-1200 nm to study the plasma resonance of Cu and Ag nanoparticles. The optical properties of the films and the stack structures are discussed.

1. Introduction

One of the main goals of the researchers' community involved in solar cells technology nowadays is decreasing of the cost of the solar cells together with increasing of their efficiency [1]. In parallel with the advance of research in nanotechnology, many researchers are now utilising nanophotonics in order to increase the efficiency of thin film solar cells [2]. Surface plasmon (SP) resonance can be employed to enhance light absorption in thin film solar cells, thanks to the scattering caused by noble metal nanoparticles or islands deposited on the semiconductor surface.

One topic that attracts a lot of attention is the use of metallic nanoparticles (NPs) at the front or at the rear side of thin film solar cells [3,4]. The introduction of noble metal nanoparticles into or on solar cells is able to induce a significant increase of the photocurrent since two main phenomena can take place nearby those objects: light scattering and near-field concentration of light. This can be realized by deposition of metal nanoparticles on the top surface of the solar cell and back reflector with metal nanoparticles at the bottom of the solar cell to direct the unabsorbed incident light to the adjacent junction [5, 6]. The absorption in the solar cell thus can be improved for an individual part of the spectrum or for a wide wavelengths range. This requires development of different thin films and multilayer structures with appropriate optical properties based on embedded metal nanoparticles with optimal size, shape and distribution to establish advantages other than those achievable with the standard light trapping methods [7]. Recently plasmon structures for light-trapping concept on

Published under licence by IOP Publishing Ltd

localizes surface plasmon polariton induced light scattering at nanostructures Ag back contacts of thin-film silicon solar cells has been proposed, designed and theoretically simulated [4].

In this article the results of study of the optical properties of thin films of Al_2O_3 with embedded Cu and Ag nanoparticles and of multilayer stacks, namely Al_2O_3(13nm)/Ag (10nm)/Al_2O_3(47nm), ZnO:Al(90nm)/Ag(10nm)/ZnO:Al(43nm) and Ag(100nm)/ZnO:Al(180nm)/Ag(10nm)/ZnO:Al(43nm) deposited by r.f. magnetron co-sputtering and sputtering on glass substrates are reported. The numbers in the bracket indicate the thickness of the corresponding films. The co-sputtered films with embedded Cu and Ag nanoparticles and the intermediate structures glass/Al_2O_3(13nm)/Ag(10nm) and glass/ZnO:Al(90nm)/Ag(10nm) (before deposition of the Al_2O_3 and ZnO:Al capping layer) are thermally treated at different temperatures and atmospheres to stimulate the formation of Cu and Ag nanoparticles with plasma resonance properties.

2. Experimental

The method of r.f. magnetron sputtering (13.56 MHz) is used to deposit the studied films and multilayer structures. The substrates are glass plates ultrasonically cleaned before deposition. Al_2O_3 films with embedded Cu and Ag are deposited by co-sputtering of Al_2O_3 ceramic target with both Cu and Ag plates in the zone of maximum erosion in atmosphere of Ar (0.5 Pa). The applied r.f power is 160 W. Two different sets of samples of Al_2O_3 with embedded Cu and Ag are deposited, where the area of Cu (S_{Cu}) and Ag (S_{Ag}) plates are different: set 1 - S_{Cu} = 45 mm^2 and S_{Ag} = 15 mm^2, and set 2 - S_{Cu} = 55 mm^2 and S_{Ag} = 20 mm^2. The substrates have not been deliberately heated during the deposition. The thickness of the deposited films is about 70 nm. The obtained films of Al_2O_3 with embedded Cu and Ag are annealed in forming gas - N_2 + (5%) H_2, for 3h at 400^{0}C or 500^{0}C. The multilayer structures glass/Al_2O_3(13nm)/Ag(10 nm)/Al_2O_3(47 nm) are deposited in sequence by sputtering of Al_2O_3 ceramic target, Ag target and Al_2O_3 ceramic target on glass substrate without heating.

The ZnO:Al films in the multilayer structures glass/ZnO:Al(90nm)/Ag(10 nm)/ZnO:Al (43 nm) are obtained by sputtering of target (ZnO + (2 wt.%) Al) and the Ag film of Ag target. The ZnO:Al films are deposited at substrate temperature, T_s, of 100^{0}C, and the Ag film without heating of the substrate. The r.f. power is 160 W in the case of deposition of oxide films (Al_2O_3 and ZnO:Al) and 65 W in the case of Ag films. The thickness of the films in the multilayer structures is indicated in the brackets.
Before deposition of the capping layers of Al_2O_3 and ZnO:Al, the bi-layer structures are annealed in vacuum 50 min at 220^{0}C and 400^{0}C in the case of glass/Al_2O_3/Ag, and at 220^{0}C in the case of glass/ZnO:Al/Ag or glass/ Ag/ZnO:Al/Ag.

The spectra of transmittance and reflectance are measured in the range of 300-1200 nm to study the optical properties of the thin films and the multilayer structures before and after thermal treatment. The spectra of the individual layers after deposition and annealing at different stage of preparation of the multilayer structures are presented as well. The haze ratio in transmission and in reflection for the multilayer structures are calculated using the spectra of specular and diffuse transmittance or reflectance, respectively, which demonstrate the light scattering from the metal nanoparticles. The reflectance and transmittance spectra of the films are measured by Schimadzu 3100 spectrophotometer. The films thickness is measured by profilometer "Talystep" with accuracy of 3%.

3. Results and discussion.

The technological conditions of deposition of the obtained layers and multilayer structures are presented in table 1. The data obtained from the optical characterization of the films and the structures (the peak positions of the band due to the plasma resonance in the transmittance and reflectance spectra) are given in table 2 and table 3, respectively.

Figures 1a and 1b show the transmittance and the reflectance spectra, respectively, of thin films Al_2O_3 with embedded Cu and Ag nanoparticles for both sets of samples.

Table 1. Technological condition of deposition of the thin films and multilayer structures. The thickness and the substrate temperature, T_s, of the individual layers in the multilayer structures are given in a sequence of the deposition.

Sample	Thickness, [nm]	T_s, [^{0}C]	Annealing
Glass/Al_2O_3:Cu:Ag – set 1 S_{Cu}=45mm^2; S_{Ag}=15mm^2	70	without heating	400^{0}C, 3h, N_2+H_2 or 500^{0}C, 3h, N_2+H_2
Glass/Al_2O_3:Cu:Ag – set 2 S_{Cu}=55mm^2; S_{Ag}=20mm^2	70	without heating	400^{0}C, 3h, N_2+H_2 or 500^{0}C, 3h, N_2+H_2
Glass/Al_2O_3/Ag/ Al_2O_3	13/10/47	all layers without heating	Ag(10 nm)- vacuum, 50 min., 220^{0}C or 400^{0}C
ilass/Ag/ZnO:Al/Ag/ZnO:Al	100/180/10/43	without heating/100^{0}C/ without heating/100^{0}C	Ag(10 nm)-vacuum, 50 min., 220^{0}C
Glass/ZnO:Al/Ag/ZnO:Al	90/10/43	100^{0}C/without heating/ 100^{0}C	Ag(10 nm)- vacuum, 50 min., 220^{0}C

Table 2. Positions of the plasma resonance bands in the transmittance spectra.

Sample	as deposited	$T_{an.}$, 220^{0}C	$T_{an.}$, 400^{0}C	T_{ann}, 500^{0}C	After capping (Ag-220^{0}C)	After capping (Ag-400^{0}C)
Al_2O_3:Cu:Ag - set 1 S_{Cu}=45mm^2; S_{Ag}=15mm^2	-	-	406 nm	405 nm	-	-
Al_2O_3:Cu:Ag - set 2 S_{Cu}=55mm^2; S_{Ag}=20mm^2	387 nm	-	406 nm	400 nm	-	-
Al_2O_3/Ag/ Al_2O_3	493 nm	469 nm	432 nm	-	472 nm	457 nm
Ag/ZnO:Al/Ag/ZnO:Al						
ZnO:Al/Ag/ZnO:Al	502 nm	512 nm	-	-	-	-

Table 3. Positions of the plasma resonance bands in the reflectance spectra.

Sample	as deposited	$T_{an.}$, 220^{0}C	$T_{an.}$, 400^{0}C	T_{ann}, 500^{0}C	After capping (Ag-220^{0}C)	After capping (Ag-400^{0}C)
Al_2O_3:Cu:Ag – set 1 S_{Cu}=45mm^2; S_{Ag}=15mm^2	-	-	420 nm	415 nm	-	-
Al_2O_3:Cu:Ag – set 2 S_{Cu}=55mm^2; S_{Ag}=20mm^2	-	-	432 nm	450 nm	-	-
Al_2O_3/Ag/ Al_2O_3	496 nm	476 nm	433 nm	-	497 nm	469 nm
Ag/ZnO:Al/Ag/ZnO:Al		430 nm				
ZnO:Al/Ag/ZnO:Al	495 nm	518 nm	-	-	459 nm	-

In the transmittance spectra of the as-deposited films the plasma resonance due to the Ag nanoparticles at about 400 nm is observed only in the set -2 (with higher area of Cu and Ag plates). The transmittance spectra of the films with embedded Cu and Ag NPs (figure 1a) for the both sets of samples after annealing demonstrate band of absorption at about 400 nm due to the plasma resonance of Ag nanoparticles more pronounced for the set 2 and after annealing at higher temperature - 500^{0}C. A smeared plasma resonance band due to the Cu in the spectral range of 600-800 nm could be scarcely seen only after high increasing of the scale performance.

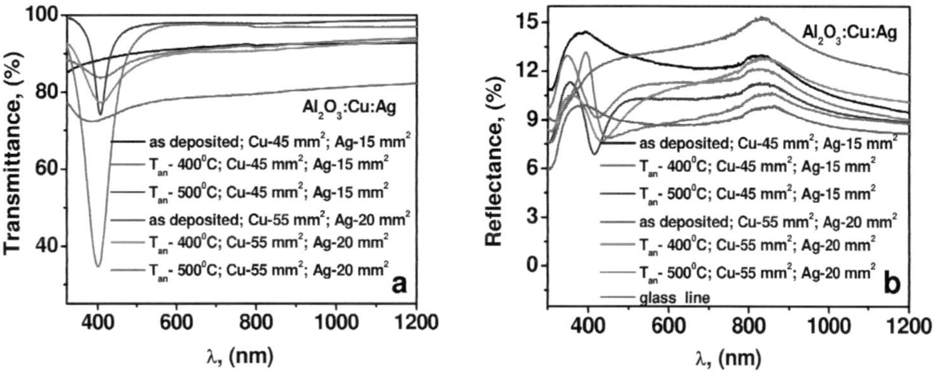

Figure 1. Transmittance (a) and reflectance (b) spectra of Al_2O_3 with embedded Cu and Ag for set 1 and set 2.

The reflectance spectra of the as-deposited films (figure 1b) for the both sets of samples demonstrate the absence of plasma resonance. The comparison with the spectrum of reflectance of the bare glass substrate in the same figure (figure 1b) confirms this. The band near 850 nm in the reflectance spectra is an artefact due to change of the detectors in the spectrophotometer. After annealing the band due to the plasma resonance of Ag NPs appears in the spectra of reflection at about 400 nm.

It is obvious that the higher annealing temperature provides better conditions for diffusion of metal nanoparticles and formation of larger NPs in the dielectric film. The formation of bigger metal nanoparticles with better size distribution results in more narrow plasma resonance band with higher intensity [8]. The observed plasma resonance in the samples based on dielectric Al_2O_3 thin film with embedded metal NPs demonstrate the possibility for their application in solar cell structures to enhance the light absorption.

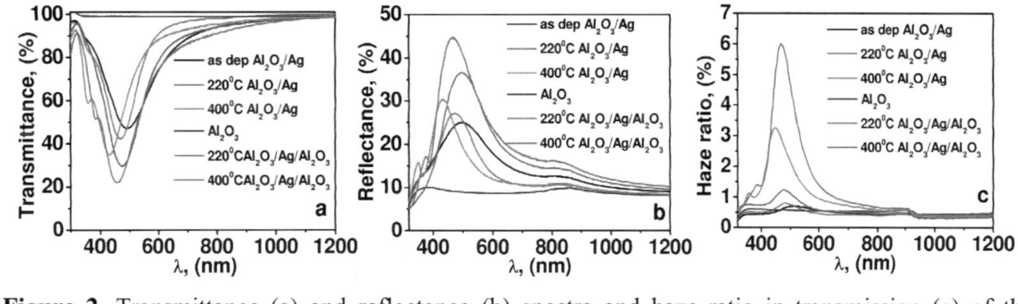

Figure 2. Transmittance (a) and reflectance (b) spectra and haze ratio in transmission (c) of the deposited films and structures at different stages of their preparation of the steak glass/Al_2O_3(13 nm)/Ag(10 nm)/Al_2O_3(47 nm) - as deposited and annealed in vacuum at 220^{0}C and 400^{0}C 50 min.

The spectra of transmittance, reflectance and the haze ratio in transmittance of multilayer structure glass/Al_2O_3(13nm)/Ag(10 nm)/Al_2O_3(47 nm), are presented in figures 2a, 2b and 2c, respectively The corresponding spectra on different stage of preparation of the multilayer structures are shown, as well. The spectra of pure Al_2O_3 thin film are given for comparison, as well.

The transmittance and reflectance spectra (figures 2a and 2b) after deposition of Ag (10 nm) layer in the stack glass/Al_2O_3(13 nm)/Ag(10 nm)) show broad band of absorption at about 493 nm due to the plasma resonance of Ag NPs. After annealing the plasma resonance band position shifts to the lower wavelengths (at T_{ann} = 400^{0}C it is observed at 432 nm) and its intensity increases. The shift of the plasmon band to the higher wave length higher than 400 nm indicates the presence of nanoparticles with shape different from the spherical [9].

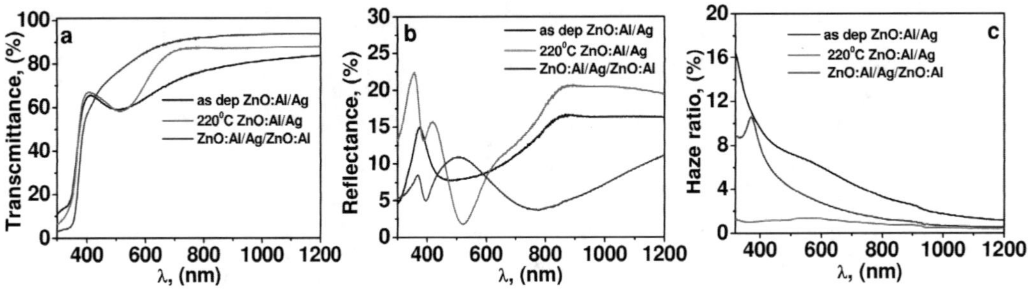

Figure 3. Transmittance (a) and reflectance (b) spectra and haze ratio in transmission (c) of thin films ZnO:Al(90 nm)/Ag(10 nm)/ZnO:Al(43 nm).

The haze ratio in transmission of glass/Al_2O_3(13 nm)/Ag(10nm) and glass/Al_2O_3(13 nm)/Ag(10 nm)/Al_2O_3(47 nm) multilayer structure has higher values and demonstrates higher pronounced scattering of light in the range of 400-600 nm after annealing of Ag (10 nm) layer at 400^{0}C. In the spectra of transmission, reflectance and of haze ratio in transmission of this stack before and after deposition of the capping Al_2O_3 layer, additional weak band of multiple plasma resonance bands at wave lengths lower than 400 nm are observed due to multiple absorption in the Ag nanoparticles [10].

The spectra of transmittance, reflectance and the haze ratio in transmission of multilayer structure glass/ZnO:Al(90 nm)/Ag(10 nm)/ZnO:Al(43 nm) are presented in figures 3a, 3b and 3c, respectively. The transmittance spectra of as-deposited and annealed structure glass/ZnO:Al(90 nm)/Ag(10 nm) demonstrate plasma resonance with broad band positioned at about 503 nm and 512 nm, respectively, more pronounced after annealing. The shift of the plasma resonance band to the higher wavelengths demonstrates that the shape of the Ag nanoparticles deviates from the spherical form and their size is different from the optimal one for Ag plasma resonance which is 200-250 nm [9]. After capping by ZnO:Al film of the annealed glass/ZnO:Al(90 nm)/Ag(10 nm) structures the band of plasma resonance in the transmission spectra smears.

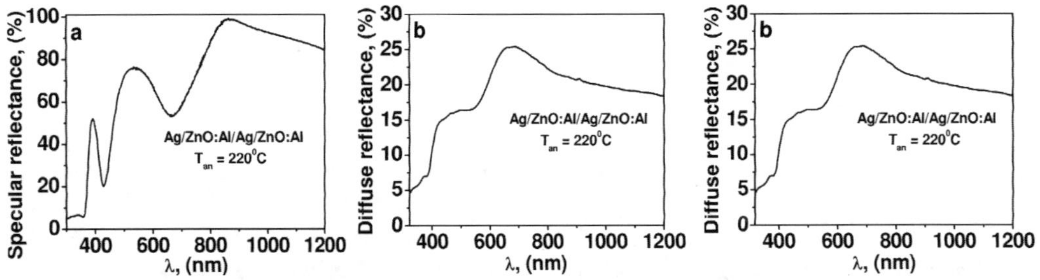

Figure 4. Specular reflectance (a) diffuse reflectance (b) spectra and haze ratio in reflection of multilayer structure Ag (100nm)/ZnO:Al (180nm)/Ag (10nm)/ZnO:Al (43 nm).

The haze ratio in transmission of the multilayer structure glass/ZnO:Al(90nm)/Ag(10 nm)/ZnO:Al(43 nm) has the highest value in the whole spectral region which demonstrate higher scattering of light. This kind of multilayer structure is deposited on glass coated with Ag reflector layer with thickness of 100 nm. The same consequence of technological processes and conditions of deposition and thermal treatment are applied. Figures 4a and 4b demonstrate the specular reflectance spectra and the diffuse reflectance spectra of the stack glass/Ag(100 nm)/ZnO:Al(90nm)/Ag(10nm) /ZnO:Al(43nm), respectively. The typical band of plasma resonance of Ag nanoparticles is observed at about 430 nm. The intensity of the band is less pronounced in the spectrum of diffuse reflectance (figure 4b). The values of reflectance of the stack deposited on the Ag reflector film are much higher in the whole spectral region under investigation than in the multilayer structures without Ag reflector. The haze ratio in reflectance (figure 4c) suggest for higher scattering of the light than in the multilayer structure without deposited Ag mirror on the glass substrate. This demonstrates that such multilayer structure deposited on Ag reflector layer can be applied on the back side of the thin film solar cells for increasing the back scattering of the unabsorbed light.

4. Conclusion.

The optical properties of thin films of Al_2O_3 with embedded Cu and Ag nanoparticles deposited by r.f. magnetron sputtering on glass are studied. The transmittance spectra of the films demonstrate absorption band of plasma resonance of Ag nanoparticles at about 400 nm more pronounced for the higher plates area of Ag and Cu and after annealing at higher temperature-500^{0}C. A very weak smeared band of plasma resonance at about 600-900 nm due to the Cu nanoparticles can be hardly observed. The plasma resonance peak intensity in transmittance spectra increases after annealing at 500^{0}C. The reflectance spectra of the as-deposited films demonstrate the absence of plasma resonance. After annealing the band of Ag plasma resonance in the reflectance spectra appears at about 400 nm. The haze ratio in transmission demonstrates scattering of light in a broad wavelengths range. The study of the optical properties of the multilayer structures reveal the presence of the plasma resonance band(s) in the spectra of transmission, reflection and haze ratio of transmission or reflection due to the presence of thin Ag nanostructured film. After deposition of the capping ZnO:Al layer the intensity of the plasma absorption band decreased and this need further investigation to establish the reason. The stack structure with ZnO:Al layers deposited on Ag back reflector demonstrates high values of specular and diffuse reflection. The obtained results show that such multilayer structure can be applied on the back side of the thin film solar cells for increasing the back scattering of the unabsorbed light. The observed plasma resonance in the films and the multilayer structures under investigation can be enhanced by optimal size, shape and distribution of the nanoparticles designed through new optimal technology conditions at future studies.

Acknowledgements: The work has been funded by the 7 European FP - project NanoPV № 246331

Reference

[1] Poortmans J 2011 *26th EUPVSEC (Hamburg, Germany)*
[2] Daif O, Drouard E, Gomard G, Kaminski A, Fave A, Lemiti M, Ahn S, Kim S, Roca P, Cabarrocas I, Jeon H and Seassal C 2010 *Optics Expres* **18** A293
[3] Pillai S and Green M 2010 *Plasmonics for photovoltaic applications, Solar Energy Materials and Solar Cells* **94** 1481
[4] Paetzold U, Moulin E, Pieters B, Carius R and Rau U 2011 *Optic express* **19** A 1219
[5] Fan Y, Burns M and Naughton M 2011 *Proc. SPIE* **8111** 811103-1 – 811103-7
[6] Ouyang Z, Beck F, Kunz O, Campbell P, Catchpole K, Varlamov S, Pillai S and Green M 2009 *Proc.24th EPVSEC (Hambourg, Germany)* 2608-2610
[7] Akimov Y, Ostrikov K and Li E 2009 *Plasmonics* **4** 107–113
[8] De G, Tapfer L, Catalano M, Battaglin G, Caccavale F, Gonella F, Mazzoldi P and Haglund R 1996 *Appl. Phys. Lett.* **68** 3820-3822

[9] Tan H, Santbergen R, Smets A and Zeman M 2012 *Nano Lett.* in press
[10] Link S and El-Sayed M 1999 *J. Phys. Chem. B* **103** **(21)** 4212–4217

Journal of Physics: Conference Series **398** (2012) 012018 doi:10.1088/1742-6596/398/1/012018

Structural and optical properties of electrochemically deposited ZnO films in electrolyte containing $Al_2(SO_4)_3$

K Lovchinov[1], M Ganchev[1], A Rachkova[1], H Nichev[1], O Angelov[1], V Mikli[2], D Dimova-Malinovska[1]

[1]Central Laboratory of Solar Energy and New Energy Sources, Bulgarian Academy of Sciences, Sofia

[2]Materials research centre, Tallinn University of Technology,19086 Ehitajate tee 5, Tallinn,Estonia

E-mail: lov4@abv.bg

Abstract. The present work concerns the electrochemical deposition of aluminium doped ZnO nanostructured thin films on SnO_2:F covered glass substrates. Doped with Al nanostructured ZnO (ZnO:Al) films are obtained by an electrochemical process using a three-electrode potentiostatic system with a saturated calomel electrode as reference electrode, in aqueous solution containing $ZnCl_2$, KCl and $Al_2(SO_4)_3$. The influence of the deposition parameters on the structural properties of the obtained ZnO:Al layers is investigated by Scanning Electron Microscopy (SEM), X-ray Diffraction (XRD) and Atomic Force Microscopy (AFM). Energy Dispersive X-ray Analysis (EDAX) is applied for measurement of the Al content in the films. The SEM micrographs and AFM pictures show that the ZnO:Al films consist of nanograins with a shape of walls. The XRD spectra demonstrate (100), (002), (101), (110) and (103) the characteristic reflexes of the ZnO. Influence of the Al concentration on the IR reflectance spectra and haze ratio of ZnO:Al thin films are presented and discussed.

1. Introduction

Quasi-one-dimensional nanoscale ZnO like nanowires, nanorods, nanowiskers, nanowalls etc. got up increasing attention over the past few years for their potential application in solar cells [1] as an n-type electrode [2], near UV-leasers [3], field effect transistors [4], chemical sensors [5], etc. ZnO nanostructured layers can be prepared by different methods, like RF magnetron sputtering [6], sol-gel method [7], electrochemical deposition [8], etc. Electrodeposition presents a simple, inexpensive upscalable method for covering of large surface areas al-lowing the controlled and patterned growth of ZnO nanostructures [9].

In this contribution we report the deposition of ZnO nanostructured thin films on the SnO_2:F covered glass substrates at temperature 80^0C from aqueous electrolyte containing $ZnCl_2$, KCl and $Al_2(SO_4)_3$. The concentration of doped Al, the surface morphology and the thickness of the ZnO layers are controlled by deposition potentials, temperature, plating time and $Al_2(SO_4)_3$ concentration.

2. Experimental

Thin ZnO nanostructured films are deposited by an electrochemical process from slightly acid aqueous solution of $ZnCl_2$ (5. 10^{-3} M) and KCl (0.5 M) with pH 4.0 at 80^0C and -1000mV (vs SCE) using a three-electrode electrochemical cell [9]. Different quantity of stock solution of $Al_2(SO_4)_3$ has been added to the electrolyte for varying of concentration of Al in the deposits. SnO_2:F coated glass substrates are used as working electrode. Spectrally pure graphite rod electrode is used as an anode. The electrolyte is agitated by magnetic stirrer. The deposition is carried out controlling the redox potential of the system. Duration of the ZnO deposition is between 30 and 120 min. The thicknesses of the prepared ZnO films are in the range of 1 and 2 μm. Different concentrations of the doping compound $Al_2(SO_4)_3$ in the electrolyte are explored: 1.10^{-5} M, 2.10^{-5} M, 3.10^{-5} M and 4.10^{-5} M $Al_2(SO_4)_3$. Samples without doping are deposited at the same conditions of the electrochemical process. Some of the samples are annealed at 400^0C in forming gas for 1 hour.

Quartz crystal resonators are used as microbalance to reveal the speed of electrodeposition of the ZnO nano-layer. The oscillators are HC-49 type package, 8 mm diameter polished quartz plates at AT-cut, 16 MHz base resonant frequency, with 600 nm thick gold electrodes with average surface area 0.1256 cm^2. The total oxygen content in solution is controlled in addition by DO&T meter Hanna Instruments 9146. The film structure is studied by X-Rays Diffraction spectrometry (XRD) using a Brucker D8 Advance spectrometer with Cu K_α radiation: λ_{Cu} $K_{\alpha1}$ =1.540560 Å and λ_{Cu} $K_{\alpha2}$ = 1.544426 Å (intensity half of that of λ_{Cu} $K_{\alpha1}$). The instrumental broadening in 2θ geometry is 0.04^0. The Al content in the films is determined by Energy Dispersive X-ray Analysis (EDAX) using Link AN10000 system analyser. The surface morphology and the thickness of the deposited films are studied by a Scanning Electron Microscope (SEM) Philips 515. Surface topography of the layers is studied by AFM using the NT-MDT Solver 47 Pro system operated in a "semi-contact" (tapping) mode. Spectra of reflectance in IR region are measured by a spectrophotometer Shimadzu IR Prestige – 21 in the range of 500 - 5000 cm^{-1}. The total and diffused transmittance spectra are measured by a spectrophotometer Shimadzu UV-3600 in the range of 320 - 2400 nm employing a 60 mm integrating sphere.

3. Results and discussion.

For determination of the speed of film growth several experiments with deposition of ZnO on quartz resonator surface are performed. The ZnO is deposited at 80^0C and potential of -1000 mV, in aqueous electrolyte of 0.5 mM $ZnCl_2$ and 0.5 M KCl.

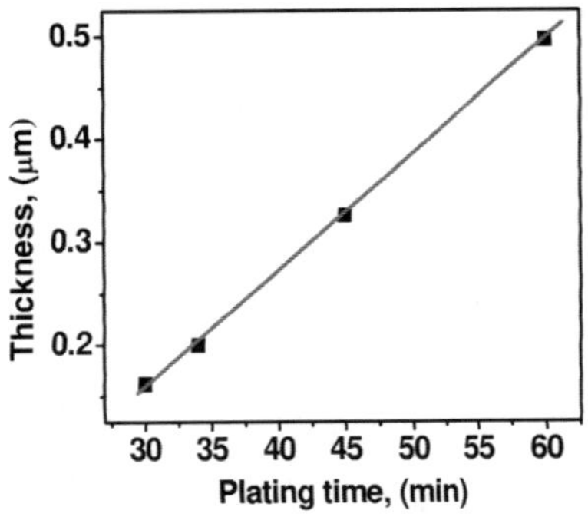

Figure 1. Dependence of the film thickness of ZnO deposited on quartz resonators on plating time.

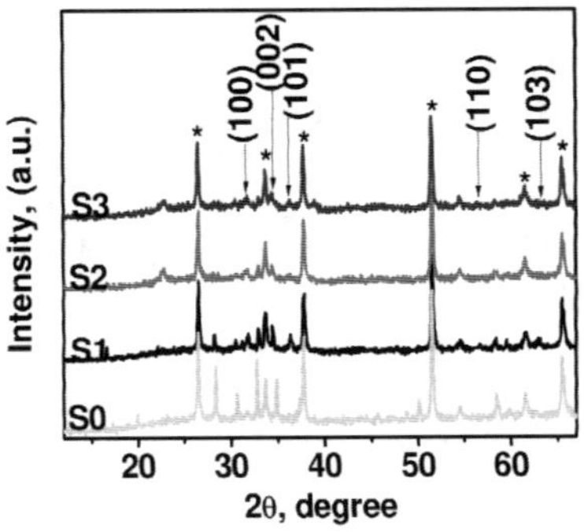

Figure 2. XRD spectra of the nanostructured thin films of ZnO, deposited with different concentrations of $Al_2(SO_4)_3$ in the electrolyte. XRD peaks of SnO_2:F are marked by asterisks. The Al concentration in the samples S0, S1, S2 and S3 are given in table 1.

Figure 1 shows the dependence of thin film ZnO thick-ness on deposition time. The thickness of the layers is calculated by change of resonant frequency caused of deposited ZnO film by the Sauerbrey equation (1) [10].The points could be fit by a line described by the equation:

$$d = 11.23\ t - 178.37, \qquad (1)$$

where d is the Sauerbrey thickness of the layer (in nm) and t is a the plating time (in minutes).
The slope of the line presents the value of the speed of the film growth equal to 11 nm/min. The negative part of the equation (1) indicates a delay of the start of the film growth (nucleation time). From the equation (1), it follows that the nucleation time (at d = 0) is about 953 sec, which coincides quite right with the observation, that there is no deposited film till the 15 minutes plating time.

Table 1. Data of the ZnO films: the $(Al_2SO_4)_3$ concentration in the electrolyte, the Al concentration in the ZnO films, C_{Al}, (at.%), the thickness, (μm), and the average roughness, (nm). For comparison the thickness and the average roughness of the SnO_2:F layer is given, as well.

Sample	$Al_2(SO4)_3$, *(M)*	C_{Al}, *(at.%)*	Thickness, *(μm)*	Average roughness, *(nm)*
SnO_2	-	-	0.9	30
S0	0	0.0	2.8	340
S1	2.10^{-5}	1.40	2.2	300
S2	3.10^{-5}	3.20	2.4	320
S3	4.10^{-5}	5.70	1.4	460

The value of the $Al_2(SO_4)_3$ concentration in the electrolyte used, the Al concentration in the deposited films, C_{Al}, the thickness and the average roughness of the ZnO films obtained from the AFM images are given in table 1. The corresponding data for the SnO_2:F and undoped ZnO film are given as well. The Al concentration obtained from EDAX analysis changes from 1.40 at.% till 5.70 at.% with increasing of the $Al_2(SO_4)_3$ concentration in the electrolyte.

Figure 2 presents the typical X-Ray Diffraction patterns of thin ZnO films, electrodeposited for 90 minutes without and with different concentration of $Al_2(SO_4)_3$ in the electrolyte (see Table 1). XRD diffraction pattern reveals the existence of ZnO with hexagonal wurtzite structure. The reflexes from the SnO_2:F are marked by asterisks. The reflexes from (100), (002), (101), (110) and (103) planes of ZnO are observed [11-13]. The intensity of the XRD peaks of ZnO decreases with increasing of $Al_2(SO_4)_3$ concentration in the electrolyte which demonstrates deterioration of the crystalline structure with Al concentration.

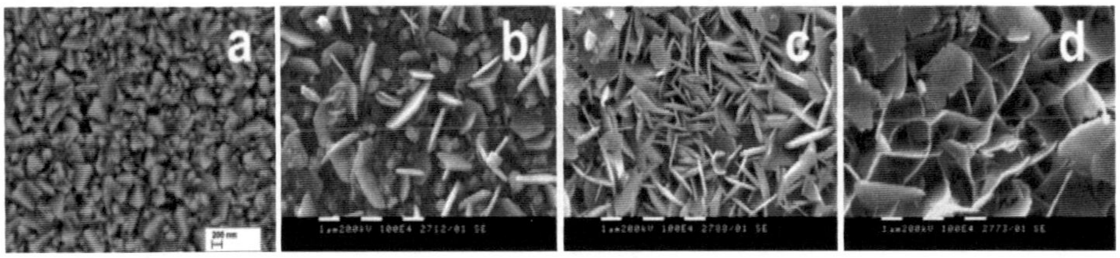

Figure 3. SEM pictures of the surface of SnO_2:F film (a), nanostructured thin films of ZnO - undoped (b) and doped with Al with different concentrations of $Al_2(SO_4)_3$ in electrolyte: (c) – 2.10^{-5} M (sample 1) and (d) – 4.10^{-5} M (sample 3). The markers correspond to 200 nm (a) and 1 μm (b, c and d).

Figure 3 shows SEM micrographs of undoped and doped with Al ZnO nanostructured thin films, deposited at constant charge density. For comparison the SEM of SnO_2:F film is given as well (figure 3a). Figure 3b displays the surface morphology of electrodeposited undoped ZnO layer. The picture exhibits the view of the film surface which consists of planar hexagonal ZnO walls with different length in the range of 1- 4 μm and 100-200 nm thick, stacked with the narrow side to the substrate surface. SEM image of ZnO film deposited with 2.10^{-5} M $Al_2(SO_4)_3$ is shown in figure 3c. The Al doped ZnO film has a similar morphology, as undoped one – the ZnO:Al nanoparticles represent hexagonal walls, almost vertically oriented to the substrate surface with smaller length (1-2 μm) than the particles in the undoped samples. With increasing of the Al concentration in the films up to 5.7 at.% the number of the nanowalls per unite area decreases.

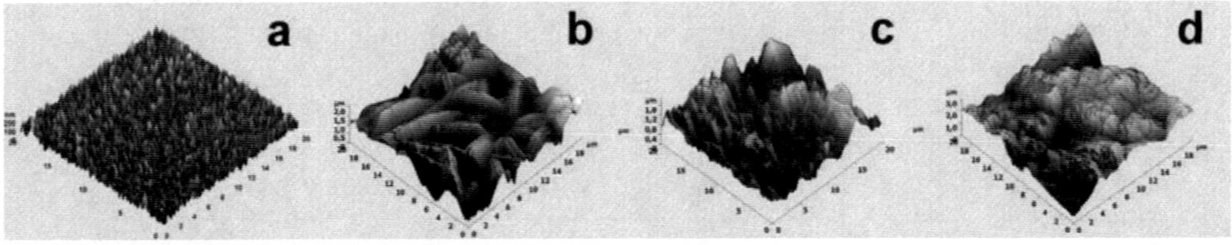

Figure 4. AFM pictures of the SnO_2 substrate (a) and nanostructured of ZnO thin films doped with Al: sample S0 (a), sam-ple S1 (c), and sample S2 (d)

The surface roughness of the deposited ZnO layers is determinate by AFM. An example of AFM pictures of un-doped and doped with Al ZnO film (samples S0, S1 and S3) are presented in figure 4. The average surface rough-ness increases with increasing of Al content in the grown ZnO:Al layers from 300 nm till 460 nm when Al increase from 1.40 at.% till 5.70 at.%. It has to be noted that the thickness of the doped ZnO films is lower than in undoped ones (table 1), thus the roughness depends on the conditions of growing in the presence of $Al_2(SO_4)_3$ in the electrolyte and not on the thickness of the layers.

The influence of the thermal annealing in forming gas for 1 hour on the surface morphology of deposited ZnO:Al is demonstrated in SEM images in figure 5. As can be seen that the annealing results in agglomeration of the nanowalls and increasing of their surface density with certain decrease in their size; additionally the small sized grains appear on their surface (figure 5b).

The optical properties of the films are studied by spectra of reflectance and haze ratio in the visible and IR region [13].

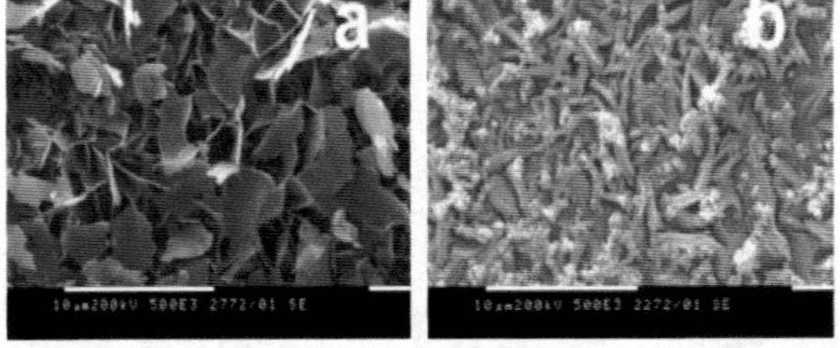

Figure 5. SEM pictures of as deposited ZnO:Al with 4.10-5M Al: as deposited - (a) and after annealing at 4000C in forming gas for 1 hour (b). The markers correspond to 10 μm.

Figure 6 shows the spectra of reflectance in the visible and near IR region of undoped ZnO and doped ZnO:Al nanostructured films deposited in electrolyte with a different $Al_2(SO_4)_3$ concentration before (figure 6a) and after annealing (figure 6b). The corresponding spectrum of SnO_2 substrate is given as well. It is seen that the electrochemically deposited ZnO films have lower value of the reflectance in the visible part of the spectra which make them suitable for application as a transparent antireflective coating in the solar cells.

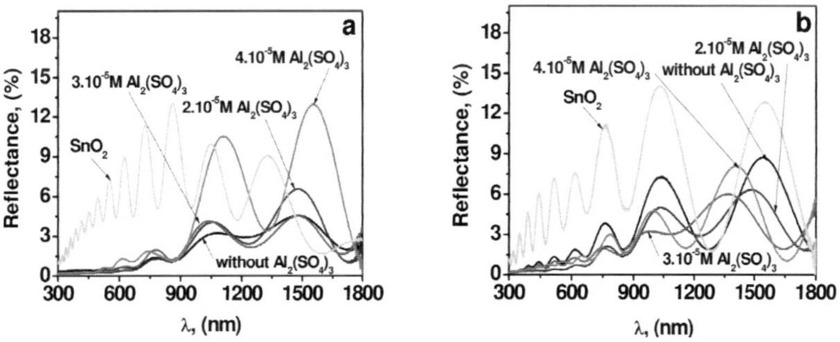

Figure 6. Spectra of reflectance of ZnO:Al films, deposited at different electrolyte concentrations of $Al_2(SO_4)_3$ before (a) and after (b) annealing

In figure 7 the reflectance spectra in the infrared region of undoped ZnO and doped ZnO:Al nanostructured films deposited in electrolyte with 4.10^{-5} M $Al_2(SO_4)_3$, as deposited and annealed at 400^0C for 1 hour are demonstrated.

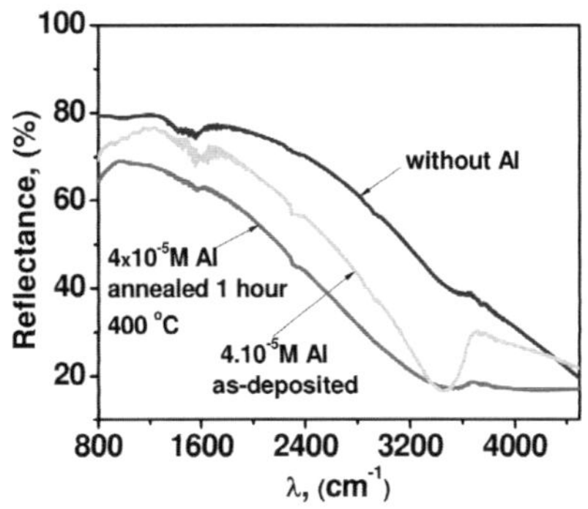

Figure 7. Infrared reflectance spectra of undoped and Al-doped ZnO films before and after anneaing.

The value of the reflectance decreases for Al-doped ZnO and this is more pronounced for the annealed sample. This behaviour could be explained by plasmon absorption by free electrons in the Al doped ZnO films. Spectra of haze ratio (H) of the undoped and doped with Al ZnO films, as deposited and after annealing in forming gas at 400^0C, are shown in figure 8(a) and (b), respectively. The value of haze ratio of the sample with C_{Al} = 1.4 at.% is higher than H of undoped ZnO film. The un-doped ZnO and doped one with 1.4 at.%.Al have values of the haze ratio in the visible region in the range of 60-90% which is much higher than that of SnO_2:F layer. With in-creasing of the Al concentration the haze ratio decreases in the all spectral region under investigation, however it is still higher than H of SnO_2:F. The annealing of the sample results in slight decrease of the haze (figure 8b). Probably different combinations of the size of the nanofeatures and roughness of the layers are a reason for differences in the haze ratio.

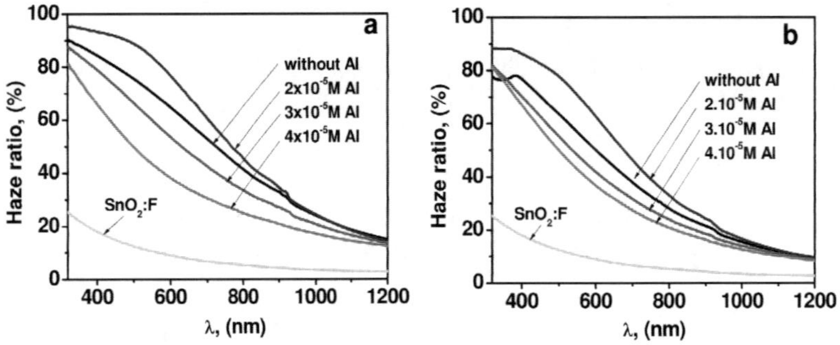

Figure 8. Spectra of haze ratio of ZnO samples prepared by electrochemical deposition undoped and doped with different concentration of Al: (a) – as deposited; (b) – annealed. The corresponding spectrum of SnO_2:F is presented as well.

4. Conclusion.

Detailed study of the structural and optical properties of nanostructured ZnO:Al layers deposited by electrochemical methods is performed. The speed of electrodeposition in the technological conditions used is calculated to be about 11 nm/min and follows linear dependence. It is observed that nucleation time does not depend on the process conditions. The concentration of the Al in the deposited ZnO layers increases with the concentration of the $Al_2(SO_4)_3$ in the electrolyte. The obtained films consist of nanosized particles with a shape of walls. It is observed that the size of the nanowalls increases with increasing of the Al concentration. The deposited films demonstrate low reflectance in the visible which makes them suitable of application as antireflective coating. The doped with Al ZnO layers have lower values of the reflectance in the IR spectrum than undoped ones due to the plasmon absorption by the free electrons. The electrodeposited ZnO layers on SnO_2:F coated glass substrate have higher value of the haze ratio than the SnO_2:F. AFM investigation shows an increase in the roughness with increase of Al concentration in the layers which is confirmed by the SEM study. The observed behaviour of the haze ratio, probably, is a consequence of alteration of nanowalls size and average roughness with increasing the doping concentration.

Acknowledgements: The work has been funded by the 7 European FP - project NanoPV № 246331

Reference

[1] Gregg B 2009 *MRS bull.* **30** 20
[2] Gonzales-Valls I and Lira-Cantu M 2009 *Energy&Environ. Sci.* **2** 12
[3] Service R 1997 *Science* **276** 895
[4] Park W, Kim J, Yi G, Bae M and Lee H 2004 *Appl. Phys. Lett.* **85** 505
[5] Ahn M, Park K, Heo J, Kim D, Choi K and Park J 2009 *Sensors&Actuators B-Chem.* **138** 168
[6] Dimova-Malinovska D, Tzenov N, Tzolov M and Vassilev L 1998 *Matter.Sci.Engineering* **B52** 59
[7] Lee J, Ko K and Park B 2003 *J.Crystal Growth* **247** 119
[8] Könenkamp R, Boedecker K, Lux-Striner M and Poschenrider M 2000 *Appl. Phys. Lett.* **77** 16
[9] Dimova-Malinovska D, Andreev P, Sendova-Vassileva M, Nichev H and Mikli V 2009 *Proc.24th EPVSEC (Hambourg, Germany)* 391-394
[10] Sauerbrey G 1959 *Zeitschrift für Physik* p 155 p 206
[11] JCPDS 1999 5-0664 ZnO
[12] JCPDS 1999 41-1445 SnO_2
[13] Minami T, Hirano T, Miyata T and Nomoto J 2012 *Thin Solid Films* **520** 3803

Al doped ZnO thin films – microstructure, physical and sensor properties

N Starbov[1], S Balabanov[1], I Bineva[1], A Rachkova[2], E Krumov[3] and K Starbova[1]
[1] Institute of Solid State Physics, Bulgarian Academy of Sciences
[2] Central Lab of Solar Energy and New Energy Sources, Bulgarian Academy of Sciences
[3] Institute of Optical Materials and Technologies, Bulgarian Academy of Sciences

E-mail: nstarbov@yahoo.com

Abstract. Thin ZnO films doped with Al are deposited by spray pyrolysis onto glass substrates using starting solution of Zn-acetate + n.AlCl (where 0.1 < n < 30 at.%). The ZnO phase composition and surface morphology are revealed via X-ray diffraction or atomic force and scanning electron microscopy respectively. UV/VIS transmittance/reflectance, as well as DC-conductivity measurements are applied in order to reveal the influence of the Al doping on the optical and electrical transport properties of the films studied. The sensing efficiency of the pure as well as of doped ZnO films for detection of noxious gases is checked via resistivity measurements under saturated vapours of ethanol, acetone, ammonia, dimethylamine and formalin at room temperature. Finally the results obtained are discussed concerning the application of the ZnO:Al films studied in the field of sensor technique.

1. Introduction

It was demonstrated earlier that the electrical resistance of spray-pyrolysis deposited ZnO thin films is very sensitive to the surface adsorption of several noxious gases or vapours [1]. The films were prepared from different precursors - Zn-nitrate or Zn-acetate, and thermal post-deposition methods were applied for improvement of their microstructure and related physical properties [2]. It was shown that the ZnO films obtained via pyrolysis of Zn-acetate were more sensitive to the studied noxious gas ambient than those deposited from Zn-nitrate precursor. For further improvement of the gas sensing efficiency a doping of ZnO-films with Al, In etc. could be applied [3-5]. According to Vasanelly et al. [3] the incorporation of Al^{+3} ions in the crystal lattice of ZnO is accompanied with formation of strong local electrostatic field. Thus, an increased surface concentration of the adsorption sites number could be obtained. Therefore the present paper aims to check the sensing properties of Al-doped ZnO thin films deposited via spray-pyrolysis of Zn-acetate.

2. Experimental

The undoped ZnO thin films were prepared by spray pyrolysis of $Zn(CH_3COOH)_2.2H_2O$ ethanol solution onto pre-cleaned soda-lime glass substrates. The same precursor modified by adding of $AlCl_3$ to the starting solution was used to deposit the doped samples. Thus, ZnO films with different Al concentration as 0.1, 0.6, 3, 10 and 30 % were obtained. Following the procedure established earlier [1, 2] the films with thickness about 200 nm were prepared by multiple spraying on the substrates initially hold at 350°C since at elevated temperature a substrates fracture occurs due to the glass thermal stress during the spraying. Thereafter, the samples were non-isothermally annealed for 16 hours at 560°C for completely finishing of the pyrolysis process [1].

The surface morphology of the ZnO:Al films was investigated via AFM (Atomic Force Microscopy) measurements using a Multimode V Veeco. Imaging was performed in tapping mode and height, amplitude and phase images were recorded. Scan rate was 2 Hz, the images resolution is 512 lines per scan direction. At least two different points on the sample surface were explored. Silicon probes with cantilever length of 125 μm and Al reflective coating on the backside (RTESPA, Veeco probes) were used in these experiments. RTESPA probes have a nominal resonance frequency of ca. 300 kHz and a typical force constant of 40 N/m. The tip nominal radius for these probes is less than 10 nm. For comparison the surface microstructure of the films was visualized under scanning electron microscope Philips SEM 515, the samples preparation being described in details earlier [2]. The phase composition of the aluminium doped ZnO films was identified via WAXRD (Wide Angle X-Ray Diffraction) using Philips PW 1050 diffractometer.

The measurements of the optical transmittance and reflectance spectra were performed using double-beam spectrometer Shimadzu UV-190 in spectral range 250-650 nm. For calculation of the total reflectance the diffuse reflectance spectrum was measured with an attachment consisting of an integrating sphere with a photomultiplier R 446.

The samples resistance was measured parallel to the substrate plane with narrow gold strip electrodes at an accuracy of ± 2%. The DC-conductivity was obtained in the dark ambient conditions using special designed vacuum thermostat, the sample temperature being varied between 20° and 150°C with an accuracy of ± 0.5°C. More details related to the resistance evaluation are described elsewhere [2]. The relative sensitivity of the samples for detection of different gases or vapours was evaluated according to the equation [6, 7] as $S = [(R_i - R_r)/R_i] \times 100\%$, where R_i and R_r are the ZnO film resistivity in inert atmosphere – dry pure nitrogen, or reagent gas respectively. For this purpose, saturated vapours of ethanol, ammonia, acetone, formalin or dimethylammine - DMA, were leaked in the evacuated thermostat vessel.

3. Results and Discussion

The surface morphology of the films studied is illustrated on figure 1. As seen, the undoped ZnO samples have granular microstructure, the mean grain size being between 50 and 150 nm. However, the doping with 3% Al is accompanied with distinct surface smooting. These peculiarities were confirmed also at atomic scale resolution on AFM 3D images (figure 2).

Figure 1. SEM micrographs (a-c) of spray pyrolysis deposited thin films of ZnO: undoped (a) or doped with 3% Al (b) and (c) - 30% Al

The crystal structure of the films studied was identified from the WAXRD spectra presented on figure 3 for virgin samples, as well as for doped with 3% or 30% Al. The spectra are typical for textured ZnO films with wurtzite structure the most intensive peak being at 2θ = 34.5° which corresponds to [002] crystallographic axis. This peak slightly increases with Al concentration up to 3 %. This indicates a better crystallinity as compared to the undoped samples. However, the [002] peak reduces twice by further 30 % rise of Al content in the films, but new phases were not identified.

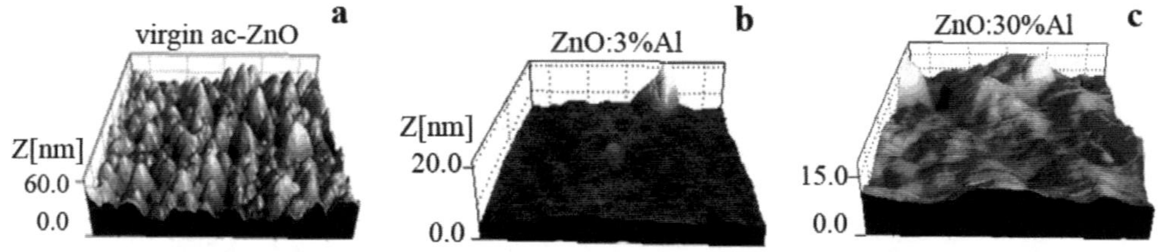

Figure 2. AFM maps of the samples studied. The symbols same as in figure1

Figure 4 illustrates the DC-conductivity of ZnO films measured at room temperature for samples doped with different Al amount. The observed maximum conductivity at or above 1% Al content corresponds well to the results obtained by J. de Marchant and M. Cocivera [8], the maximal σ in

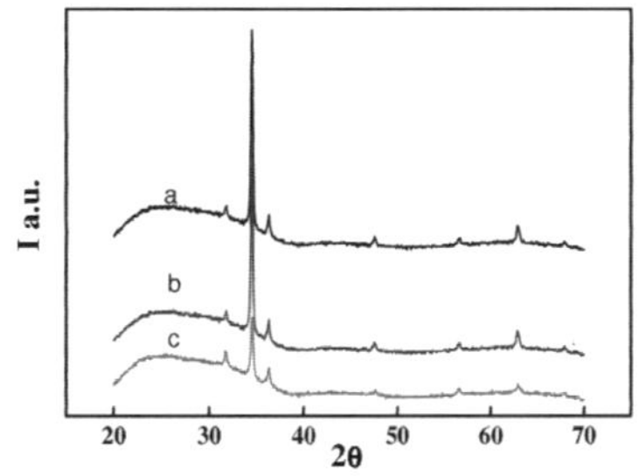

Figure 3. XRD spectra of annealed ZnO films. The symbols same as in figure 1

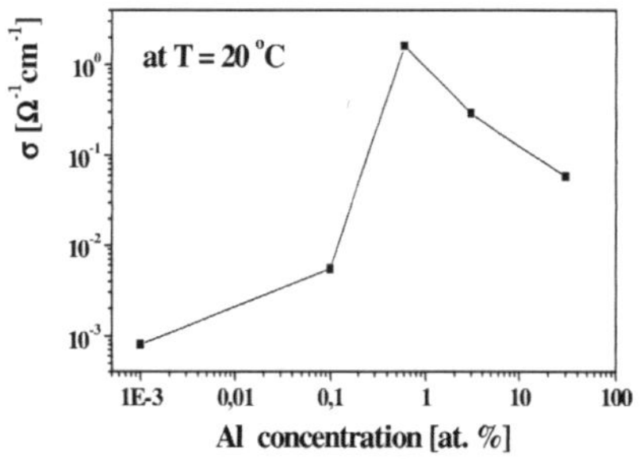

Figure 4. DC-conductivity of ZnO:Al films as dependent on the doping concentration

these experiments being measured at Al concentration of 2%. Similar dependence was observed in implanted with C or Si materials [9, 10]. The optical properties of ZnO:Al samples are presented on figure 5. The transmittance decreases from 97% for undoped ZnO films down to 85% at 30% Al concentration in accordance with the results of Sato et al. [11]. The total reflectance changes more weakly in conformity with the surface modifications when Al content in the ZnO thin films increases.

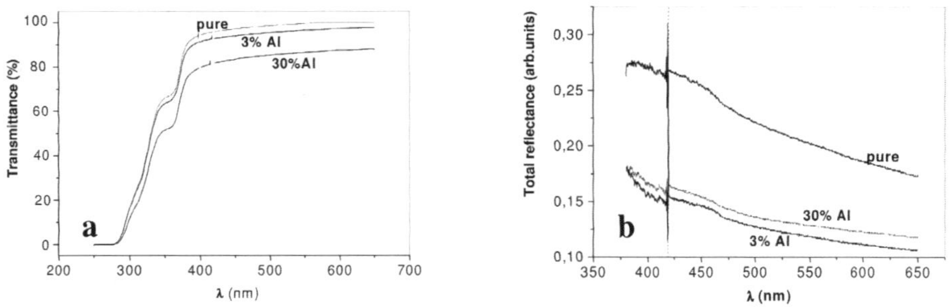

Figure 5. Optical transmittance – a) and total reflectance - b) of doped ZnO thin films

The sensing efficiency of the Al doped samples studied could be evaluated from figure 6, which presents S vs. doping concentration for the different noxious agents. The data for the virgin samples are not included in this graph for simplicity since in this case S is below 20% with an exception of the measured S in DMA vapours ambient - 1400% [1]. Thus, the ZnO:Al thin films have an increased sensing efficiency as compared to the sensitivity of undoped samples. However, there is not any general dependence of S on the Al content in the ZnO thin films studied.

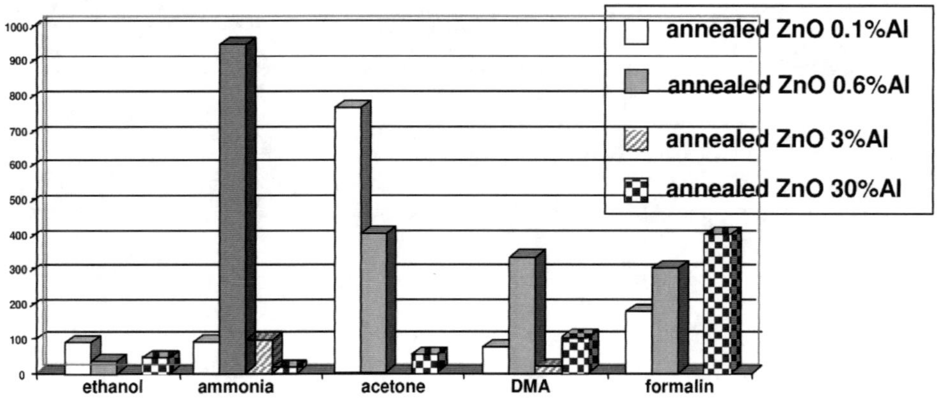

Figure 6. Relative sensitivity S of ZnO:Al films as measured for a number of hazardous agents

4. Conclusion

The results obtained in the present paper demonstrate that the Al doping has a noticeable effect on the sensing efficiency of ZnO thin films deposited via spray pyrolysis from acetate precursor. At the same time a definite general dependence of S from Al content in the samples was not observed. This means that the proposed by Vasanelli et al. [3] increased surface concentration of the adsorption sites created from the incorporated Al atoms in the ZnO crystal lattice has not decisive role for detection of the noxious vapours studied. However, the observed specific sensing response - maximal sensitivity at content 0.6% Al for ammonia, 0.1 % for acetone, 30% Al for formalin or the mentioned above for DMA in virgin samples, reveals a good prospective for preparation of selective chemical sensors based on ZnO thin films spray pyrolysis deposited from acetate precursor.

Acknowledgements This study was partial financed by Bulgarian Fund "Scientific Investigations" under grants VUF 1505 and DMU 03-91.

References

[1] Starbov N, Rachkova A, Karashanova D, Krumov E and Starbova K 2009 *JOAM* **11** 1375
[2] Starbov N, Krumov E, Kjosev B, Rachkova A and Starbova K 2007 *JOAM* **9** 245
[3] Penza M, Martucci C, Anismkin V I and Vasanelli L *Mater. Sci. Forum* 1996 **203** 137
[4] Nunes P, Fernandes B, Fortunato E, Vilarinho P and Martins R 1999 *Thin Solid Films* **337** 176
[5] Paraguay F D, Miki-Yoshida M, Morales J, Solis J and Estrada W L 2000 *Thin Solid Films* **373** 140
[6] Roy S and Basu S 2002 *Bull. Mater. Sci.* **25** 513
[7] Sahay P P, Tewari S, Iha S and Shamsuddin M 2005 *J. Mater. Sci.* **40** 1791
[8] de Marchant J and Cocivera M 1995 *Chem. Mater.* **7** 1742
[9] Tsvetkova T and Balabanov S 2008 *IAEA-TECDOC-1607* Viena p 9
[10] Balabanov S, Tsvetkova T, Borisova E, Avramov L and Bischoff L 2008 *J.Phys. Conf. Series* **113** 012038
[11] Sato H, Minami T, Takata S, Miyata T and Ishii M 1993 *Thin Solid Films* **236** 14

Effect of the gas composition on the structural and electrical properties of ZnO nanostructures obtained by oxidation of Zn at atmospheric pressure

S Kitova, R Kazakov, I Kalaglarski and R Stoimenov

Institute of Optical Materials and Technology "Acad. J .Malinowski", Bulgarian Academy of Sciences, Acad. G. Bonchev str., bl.109, 1113 Sofia, Bulgaria

e-mail: skitova@iomt.bas.bg

Abstract. In this work we report the synthesis of ZnO nanorods on glass substrates by a simple thermal evaporation of high purity Zn powder without the use of any catalyst. The product was obtained at atmospheric pressure by the oxidation of Zn vapour in a flow of Ar and O_2 gas mixture at low temperature of glass substrates (390 °C). The morphology, phase composition and crystallinity of the nanostructures were investigated by scanning electron microscopy and X-ray diffraction analysis. Their electrical properties were studied via current-voltage (I-V) measurements. XRD demonstrates that the nanorods deposited consist predominantly of the wurtzite phase of ZnO with preferred c-axis orientation. The results show that the oxygen content in the deposition atmosphere influences strongly the morphology and electrical resistance of the produced ZnO nanostructures.

1. Introduction

In the area of nanomaterial research, one-dimensional nanoscale semiconductors have attracted lots of interest due to their promising applications in electronics, photonics, sensors, transducers, energy generation, and so on [1-5]. The properties and performances of these new nanomaterials are mainly controlled by their nature, size and morphology, which in turn determine their electrical conductivity, optical absorption, UV and visible light emission or chemical reactivity [4]. The tendency of ZnO to grow in a variety of exotic morphologies has induced great interest to explore simple methodologies to grow such nanoarchitectures. After publication by Yang and co-workers of their seminal paper on synthesis of ZnO nanowires [6], many methods have been developed to produce ZnO nanostructures such as vapor phase methods (thermal evaporation and condensation, metalorganic chemical vapor deposition), solution phase method, etc. (for reviews see [1, 7, 8] and reverences cited there in). Different types of ZnO nano- and microstructures were reported in the literature, synthesized by the thermal evaporation processes using metallic zinc powder as source materials for zinc. ZnO micro and nanorods were synthesized on silicon substrates in a horizontal quartz furnace by the thermal evaporation of metal zinc powder under oxygen ambient [9, 10]. Vertically aligned ZnO nanowires were prepared through vaporization of metal zinc powder at low temperature using the Au-coated silicon substrates [11]. In all of the results presented previously, the scientists used either metal catalysts or crystalline substrates for the growth of ZnO nanostructures.

In this paper, we report the synthesis of ZnO nanorods on glass substrates at relatively low temperature (390 °C) without use of any metal catalyst or seeds, via a simple thermal evaporation of zinc powder at atmospheric pressure. The influence of the oxygen content in deposition atmosphere on the texture, crystallinity, grain size, morphology and electrical resistance of the nanostructured layers obtained has been studied.

2. Experimental

ZnO nanostructures were grown on glass substrates using a metal vapour deposition method. The schematic diagram of the experimental setup for the fabrication of ZnO nanostructures is shown in Figure 1. Zinc powder (99.9999 %) was placed in a quartz boat as the Zn source in the center of a quartz tube in a furnace. The quartz tube was kept at atmospheric pressure by flowing high purity Ar (99.99%) with a flow rate of 300 sccm and heated up to 710 °C. Ar carrier gas with different O_2 content varying from 2 to 20 vol. % flows through the quartz tube with the rate of 500 sccm at reaching the synthesis temperature. All gas flows were controlled by mass flow meters controllers. The growth time was set to 30 min before terminating oxygen flow and cooling down to room temperature.

The layers were deposited on glass substrates, carefully cleaned. The cleaning procedure first involved washing with a 1% detergent solution and then with deionized water in an ultrasonic cleaner. The substrates were placed horizontally at the downstream side of the Zn source at a distance of about 25 cm. The temperature of the substrate, controlled by a thermocouple, was 390 °C. The thickness of the layers of about 500 nm was measured by a profilometer type Talystep.

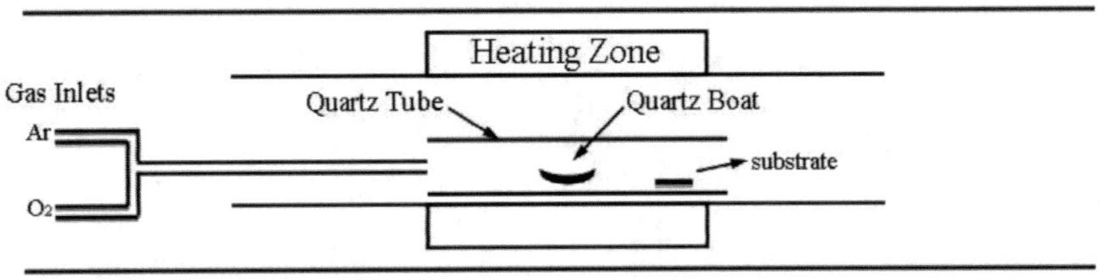

Figure 1. Schematic diagram of the experimental setup for the fabrication of different shape ZnO nanostructures.

The crystal structure of the layers was characterized by X-ray diffraction (XRD), using a Philips (PW 1710) apparatus with Cu-Kα radiation separated by a graphite focusing monochromator. Smoothing processes of 5 point Fast Fourier transform followed by an appropriate baseline correction were applied to all diffraction spectra. The intensity and full width at half maximum (FWHM) of an (hkl) plane were determined by profile fitting procedure, using an Gaussian or pseudo Voigt peak shape after Kα2 stripping procedure.

The surface morphology of the layers was examined by a scanning electron microscope (SEM Philips 515). The resistivity measurements were provided with 2 mm planar narrow strip Al electrodes, 200 nm thick, with a distance of 2.9 mm between each other. They were vacuum deposited through a precise stainless steel mask on the free ZnO surface. Current – voltage (I-V) measurements were performed in dark at 25 °C with Keithley 230 voltage source and Keithley 617 electrometer.

3. Results and Discussion

The layers deposited were white in color. The phase of the layers and their crystallographic orientation were determined by XRD spectra. XRD pattern of the layers deposited on glass plates at indicated O_2 content in Ar/O_2 gas mixture during their deposition are given in Figure 2. The XRD results show that the ZnO layers are polycrystalline. Five peaks observed at 31.55, 34.2, 36.0, 47.30 and 56.4° match the values given in powder diffraction file 05-0664 [12] while the relative intensities of these peaks are distinct from those of ZnO powder. These peaks can be indexed to (100), (002), (101), (102) and (110) planes of the hexagonal wurtzite structure of ZnO crystal.

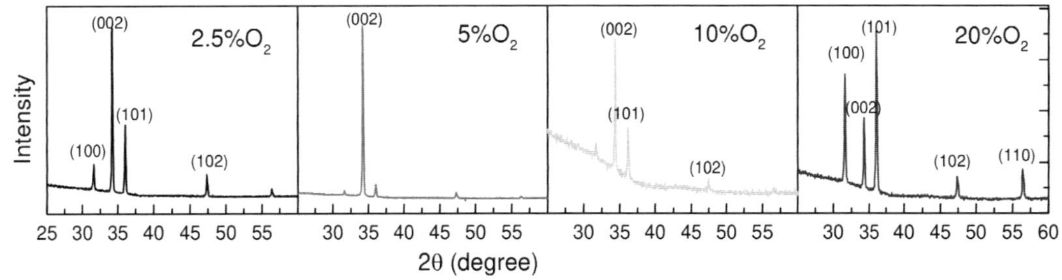

Figure 2. XRD patterns of the layers deposited on glass plates at indicated O_2 content in Ar/O_2 gas mixture.

In order to quantify the preferred orientation (PO), often called texture, in the crystalline films, the texture coefficient (T_c) was defined as [13]

$$T_c(hkl) = \frac{I(hkl)/I_o(hkl)}{N^{-1}\sum_n I(hkl)/I_o(hkl)} \tag{1}$$

where *I(hkl)* is the measured intensities of a (hkl) plane converted into the corresponding intensity of an infinitely thick sample. $I_o(hkl)$ is the intensity of a random oriented ZnO powder standard measured at the same conditions, N is the reflection number and n is the number of the diffraction peaks. A sample with randomly oriented crystallites presents T_c(hkl) = 1, while the larger the value, the larger the abundance of crystallites oriented at the (hkl) direction.

The average grain size (D) of crystallites was calculated using well-known Scherrer's formula [14]:

$$D = \frac{0.9\lambda}{\beta\cos(\theta)} \tag{2}$$

where λ is the wavelength of X-rays used (=1.54059 Å), β - the broadening of diffraction line measured at half its maximum intensity in radians and θ - the angle of diffraction in degree.

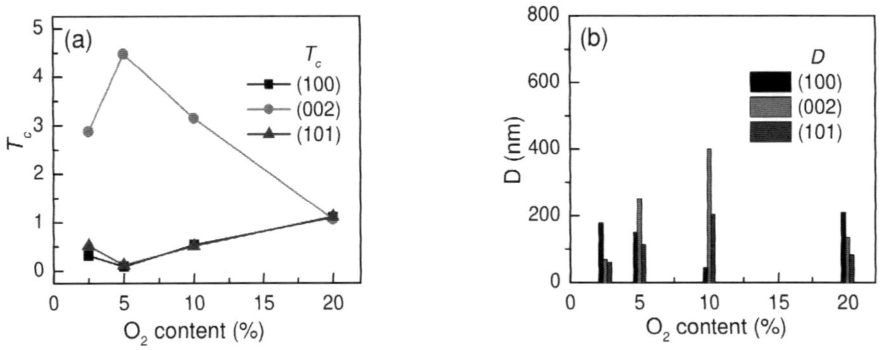

Figure 3. Texture coefficient T_c (a) and grain size D (b) calculated from XRD patterns of the (100), (002) and (101) peaks in dependence of O_2 content in the gas mixture.

Figure 3 presents the influence of the oxygen content in deposition atmosphere on T_c and average grain size calculated by from XRD patterns of the (100), (002) and (101) peaks. The higher value of T_c obtained for (002) peak than the other ones is an indication of preferable c-axis phase with good crystalline quality formed on pure glass substrates. As a whole, the increasing O_2 content above 10% degrades c-axes PO and leads finally to more random orientation with lower crystalline quality. It is seen that the average grain size of c-phase also decreases since the growth of the grain along substrate surface is enhanced.

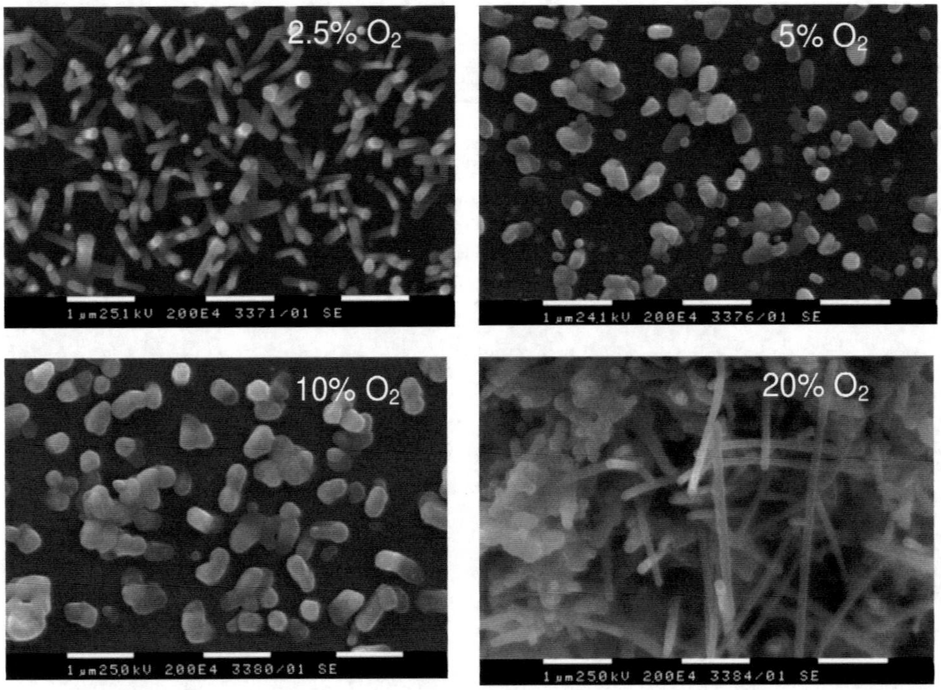

Figure 4. Top-view SEM images of the layers grown at indicated O_2 content in the gas mixture.

Figure 4 shows SEM images of the layers grown on pure glass substrates at different oxygen content in carrier gas. It is seen that the layers deposited at O_2 content up to 10 % in the gas mixture consist of ZnO nanorods grown perpendicular to the substrate surface. ZnO nanorods were produced regularly on a whole substrate (30 mm x 20 mm). All the observed nanorods exhibit hexagonal surfaces throughout their length with hexagonal facets. The average diameter of grown nanorods increases with the O_2 content, which is consistent with the data by XRD analysis. A further increase in the amount of oxidizing agent in the carrier gas destroys the uniformity of the layer. The substrate is covered with small drops consisting of randomly distributed ZnO nanowires shown in the figure.

The XRD and SEM analyses shed some light on the growth mechanism. The formation of ZnO nanostructures with different morphologies is obviously due to the different O vapor pressure since the morphologies of ZnO layers are mainly influenced by the concentration of oxygen gas. Most probably the growth of ZnO nanorods and nanowires followed a self-seeding vapor–solid (VS) growth mechanism [7]. In the initial growth stage, the self-seeding occurred on the glass substrate and then the ZnO nanorods grew on the seeds by feeding with the evaporated zinc and oxygen elements. In addition, with a sufficient supply of oxygen sources, some ZnO seeds grow laterally on the substrate with formation of nanowires. Further study however, is needed for understanding the exact nuclei and growth mechanism.

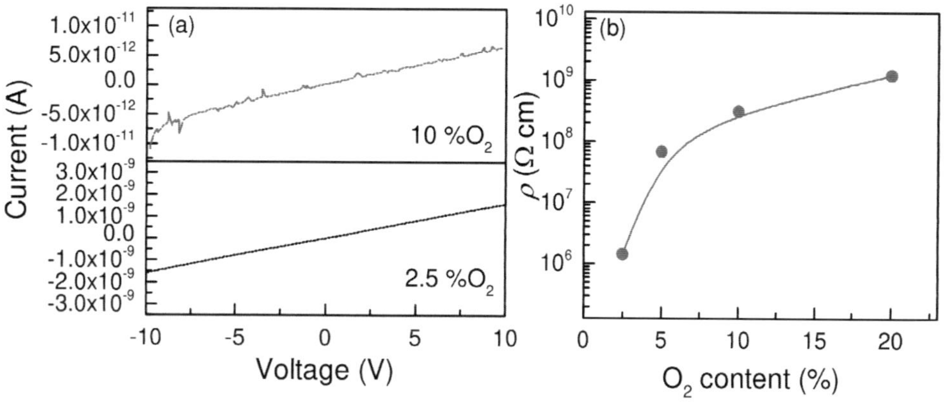

Figure 5. (a) I-V characteristics of ZnO layers obtained at indicated O_2 content; (b) Resistivity (ρ) of the ZnOlayers in function of O_2 content in the gas mixture.

Figure 5a depicts the current–voltage (I–V) characteristics measured between pairs of the fabricated ohmic contacts. They clearly show an ohmic behaviour which implies good quality ohmic contacts and that the ZnO layer is conducting. Figure 5b plots the resistivity of layers in dependence of the O_2 content in the gas mixture. It is seen that the resistivity of ZnO nanostructures increases from $7.2x10^6$ Ω cm to $1.2x10^9$ Ω cm as the oxygen content in carrier gas increases from 2.5 to 20 %. This result is in correspondence with the data of other authors published in the literature [15].

As it was well known undoped ZnO thin films have n-type conduction which is caused by a deviation from stoichiometry due to native defects. Various kind of native point defects are present in ZnO layers [15]. They include oxygen vacancies (V_O), zinc vacancies (V_{Zn}), Zn interstitials (Zn_i), oxygen interstitials (O_i), Zn atoms at oxygen anti-sites (ZnO), and oxygen atoms at Zn anti-sites (O_{Zn}). Previously, (V_O) and Zn_i have been assumed to be the main sites of donors in ZnO crystals. However, recent first principles calculations verified that V_O is a deep donor and it cannot be the origin of n-type conduction [16]. In contrast, Zn_i can act as a shallow donor. The high value of ρ of about $1.2x10^9$ Ω cm obtained for ZnO layers deposited at high O_2 content in gas mixture indicates that the number of defects in ZnO layers have diminished. More probably in our case with the increase in oxygen content in the gas mixture during the deposition some oxygen vacancies and especially interstitial Zn disappear thus causing the increase in film resistivity.

4. Conclusion

The polycrystalline nanostructured ZnO layers consisting of nanorods and nanowires were obtained at low temperature of glass substrates without the use of any catalyst or seeds at atmospheric pressure. The XRD data indicate that the nanorods exhibit the hexagonal wurtzite structure with preferred c-axis orientation. The crystallinity, texture and the grain size of the deposited nanostructures depend on the oxygen content in the carrier gas. The high O_2 content degrades c-axes phases and leads finally to phases with lower crystal quality and more random orientation. The results by I-V measurements show that the resistivity of the nanostructures increases with increasing amount of oxidizing agent in the gas.

It can be expected that due to their characteristics these nanostructured layers may be used as building blocks for preparing chemical and biosensors as well as photocatalysts where inherently large surface to volume ratio of structured materials are important prerequisite for enhanced sensitivity.

Acknowledgements
The authors would like to thank E. Milanov for XRD study and G. Maleshkova for SEM observations. This work was supported by Grant # DO02-207 of the Bulgarian Science Fund.

References

[1] Djurisic A, Ng A and Chen X 2010 *Progress in Quantum Electrinics* **34** 191
[2] Ozgur U, Alivov Y, Liu C, Teke A, Reshchikov M, Dogan S, Avrutin V, Cho S and Morkoc H 2005 *J. Appl. Phys.* **98** 041301-1-103
[3] Duan X, Huang Y, Agarwal R and Lieber C 2003 *Nature* **421** 241
[4] Xia Y, Yang P, Sun Y, Wu Y, Mayers B, Gates B, Yin Y, Kim F and Yan Y 2003 *Adv. Mater.* **15** 353
[5] Wang Z 2004 *J. Phys.:Condens. Matter,* **16** R829
[6] Huang M, Mao S, Feick H, Yan H, Wu Y, Kind H, Weber E, Russo R and Yang P 2001 *Science* **292** 1897
[7] Wang Z 2009 *Materials Science and Engineering* R, **64** 33
[8] Heo Y, Norton D, Tien L, Kwon Y, Kang B, Ren F, Pearton S and LaRoche 2004 *J Materials Science and Engineering* R **47** 1
[9] Zhao D, Liu Y, Shen D, Lu Y, Zhang L and Fan X 2003 *J. Appl. Phys.* **94** 5605
[10] Umar A, Im Y and Hahn Y 2006 *Journal of Electronic Materials* **35** 758
[11] Zhang Y, Jia H, Wang R, Chen C, Luo X, Yu D and Lee C 2003 *Appl. Phys. Lett.* **83** 4631
[12] Powder Diffraction File 05-0664 for hexagonal Zinc oxide (2001 JCPDS-International center for Diffraction data)
[13] Barret C and Massalski T 1980 *Structure of Metals* (Oxford: Pergamon Press) p 204
[14] Cullity B and Stock S 2001 *Elements of X-Ray Diffraction* (Prentice Hall, 3nd ed) p161
[15] Ma Y, Du G, Yanga T, Qiu D, Zhang X, Yang H, Zhang Y, Zhao B, Yang X and Liu D 2003 *Journal of Crystal Growth* **255** 303
[16] Janotti A and Van de Walle C 2007 *Phys. Rev. B* **76** 165202

Journal of Physics: Conference Series **398** (2012) 012021 doi:10.1088/1742-6596/398/1/012021

Influence of titanium precursor on the photocatalytic properties of TiO_2 sprayed films under visible light

V Blaskov[1], M Shipochka[1], I Stambolova[1,*], S Vassilev[2], A Eliyas[3], P Stefanov[1] and A Loukanov[4]

[1] Institute of General and Inorganic Chemistry, BAS, Acad. G. Bonchev St, bl. 11, 1113, Sofia, Bulgaria
[2] Institute of Electrochemistry and Energy Systems, BAS, Acad. G. Bonchev bl. 10, 1113, Sofia, Bulgaria
[3] Institute of Catalysis, BAS, Acad. G. Bonchev, bl. 11, 1113, Sofia, Bulgaria
[4] Dept. of Engineering Geoecology, University of Mining and Geology, St. Edrev St., 1700 Sofia, Bulgaria

E-mail: stambolova@yahoo.com

Abstract. Thin nano-sized TiO_2 films were deposited on aluminum foil substrates by the spray pyrolysis method, using $Ti(i\text{-}OPr)_4$ (TIP) and $TiCl_4$ (TC) as precursors. The films were characterized by means of X-ray Diffraction (XRD), Scanning Electron Microscopy (SEM) and X-ray Photoelectron Spectroscopy (XPS). According to the XRD patterns the as-deposited films appear to be amorphous. The thermal treatment at 400°C leads to the formation of anatase nano-crystallites. The XPS analyses showed that the Ti2p broad photoelectron peak of as-deposited TC films indicated a mixture of Ti^{3+} and Ti^{4+} oxidation states. After treatment at 400°C the Ti2p peak displays only Ti^{4+} oxidation state for both TIP and TC films. The number of hydroxyl groups on the surface is decreased after the thermal treatment.
The photocatalytic activity of the films was studied towards degradation of azo dye Reactive Black 5 (RB5) as model wastewater pollutant under visible light illumination. It was found out that the films obtained from TC were more active than those obtained from TIP films. The thermally treated samples are better photocatalysts than those non-treated, because they posses anatase crystalline phase and stoichiometric TiO_2. The TOC measurements showed minimal concentration of total organic carbon in the dye solution after 180 min of visible light irradiation.

1. Introduction

Water pollution due to colored effluents from textile- and paper-industries has attracted much attention among scientists from all over the world. The classical methods for purification of the wastewaters after colorization such as separation, membrane filtration, oxidation etc. are unable to decompose effectively the dyes and can produce some toxic metabolites [1]. In general, photocatalysis is used as a pre-treating step to degrade non-biodegradable organic pollutants to biodegradable compounds. This requires the development of effective photocatalysts for their degradation. Titanium dioxide (TiO_2) is a promising photocatalyst for the treatment of polluted air and water [2]. Recently, TiO_2 films have been widely studied in photocatalytic degradation of the various textile-colouring dyes, because it overcomes the main disadvantage – the necessity of post-treatment separation in a system and allows longer term operation. The spray pyrolysis method has been applied for the synthesis of TiO_2 films with promising photocatalytic activity [3-6]. So far the papers deal mainly with the influence of deposition temperature and type of the substrate on the photocatalytic activity. There are no available data on the influence of the type of titanium precursor on the properties of TiO_2 sprayed films.

The aim of the present study was to investigate both the effects of the type of the titanium precursor and of the thermal treatment on the photocatalytic activity of TiO_2 sprayed films.

2. Experiment

2.1. Films deposition

Two types of titanium precursor solutions were prepared for the spray pyrolysis procedure. Titanium tetraisopropoxide (TIP) was hydrolyzed in a water-ethanol mixture at molar ratio TIP:H_2O:C_2H_5OH=1:1:30. Acetyl acetone (AcAc) was added as a complexating agent in such amount that one obtains the ratio TIP:AcAc =1:1 (solution A). Titanium chloride ($TiCl_4$) was used as inorganic titanium precursor and it was diluted in ethanol (solution B). Each one of the solutions A and B was diluted with a mixture of isopropanol and butyl carbitol ($C_4H_9OC_2H_4OC_2H_4OH$) and then stirred intensively for 1h at room temperature. The as-prepared mixtures were used for the spray procedure. The aluminum foil sheets (76mm x 25mm) were cleaned successively in hot ethanol and then in acetone. The aerosol of the precursor solution was generated by pneumatic glass nebulizer and transported to the heated substrate. The spray coating process was repeated 5 times for each film at 10 sec intervals. Afterwards, one part of the deposits was treated in air raising progressively the temperature from 20 to 400°C for 1 or 2 hours, another part remained untreated. The theoretically estimated TiO_2 content in the films was 2.2 mg/cm^2.

2.2 Characterization

The phase composition of the samples was studied by X-ray diffraction (XRD) with CuK_α-radiation (Philips PW 1050 apparatus). The crystallite size was estimated based on the XRD spectra. The composition and electronic properties of the films were investigated by X-ray photoelectron spectroscopy (XPS). The measurements were performed in VG ESCALAB II electron spectrometer using AlK_α radiation with energy of 1486.6 eV. The binding energies were determined with an accuracy of ±0.1 eV. The chemical composition of the films was investigated on the basis of peak areas and binding energies of O1s, Ti2p and Al2p photoelectron peaks (after linear subtraction of the background and Scofield's photoionization cross-sections). A scanning electron microscope (SEM) JSM-5510 of JEOL was used for morphology observations of the films.

2.3. Photocatalytic tests

The photocatalytic experiments were conducted with powerful visible-light irradiation source (TUNGSRAM 500 W lamp, 9700 Lm). The photocatalytic reactor contained aqueous solution of Reactive Black 5 dye (20 ppm concentration), which was being stirred continuously. The TiO_2 coated substrate was placed into the vessel and covered completely with 150 ml dye solution. The photocatalytic degradation was evaluated by taking aliquot of the solution and measuring the residual concentration using spectrophotometer UV-1600PC in the wavelength range from 200 to 800 nm at regular time intervals.

2.4 Total organic carbon measurements

Total Organic Carbon (TOC) analyzer, made by Shimadzu Co. (Japan), model VCSH, is based on total combustion of the organics, contained in the aqueous solution and the result is expressed in milligrams of carbon per liter of water. The use of TOC analyzer allows us to follow the course of the reaction by evaluating the concentration (mg C/L) based on the decreasing absorbance of the dye solution with the time. The analysis of a solution of unknown concentration was done by 5 injections and taking the mean value – in most cases 5 injections were sufficient to obtain a value of the coefficient of variation below 2% (after the microprocessor of the TOC excludes automatically the outliers). Only for the initial solution the deviation was higher, which necessitates some additional manual injections.

3. Results and discussions

X-ray diffraction patterns revealed that the sprayed thin films without any additional treatment are amorphous. The thermal treatment leads to a formation of well crystallized anatase phase with average crystallites size about 20-23 nm.

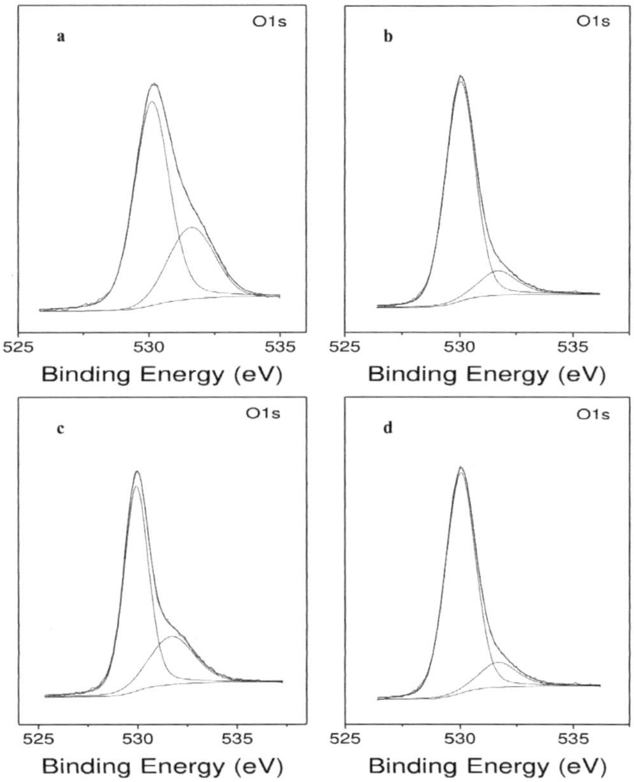

Figure 1. Deconvolution of O1s peaks of TC films not treated (a), TC films, treated at 400°C, (b) TIP films, not treated (c), TIP films treated at 400°C (d).

The surface composition and chemical state of the TiO_2 films were investigated by XPS analyses. The O1s peaks of TC and TIP films are wide and asymmetric and could be divided. They are deconvoluted by Lorentzian–Gaussian curve fitting into two components at 530.0 eV and 531.6 eV, respectively (figure 1). The first component at 530.0 eV is related to O^{2-} ions in the TiO_2 crystal lattice. The second component at 531.6 eV is attributed to oxygen atoms in hydroxyl groups. A decrease in the number of the hydroxyl groups on the films surface after treatment is also seen in the figure. Figure 2 shows the Ti2p photoelectron spectra of the films obtained from both titanium precursors. The Ti $2p_{3/2}$ peaks of TiO_2 films, obtained from $TiCl_4$, are located at 458.6 eV for the as-deposited film and 458.8 eV for the film treated thermally at 400°C. The as-deposited film exhibited broad Ti2p photoelectron peak, corresponding to a mixture of Ti^{3+} and Ti^{4+} oxidation states. After treatment at 400°C, the binding energy and the shape of the Ti2p peak became characteristic of Ti^{4+} oxidation state. The position of Ti $2p_{3/2}$ peaks of TiO_2 films, prepared from titanium isopropoxide are at 458.8 eV, which is typical for TiO_2.

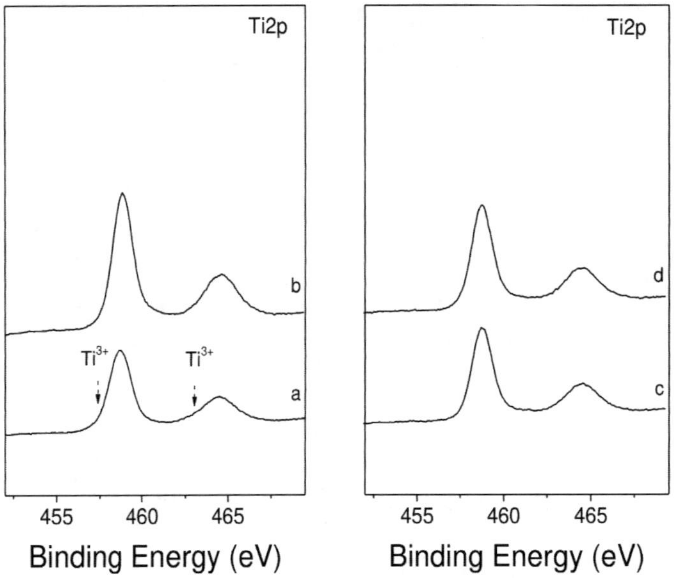

Figure 2. Ti2p core level spectra of the surface of the non-treated TC films (a), TC films treated at 400°C (b), non-treated TIP films (c), TIP films treated at 400°C (d).

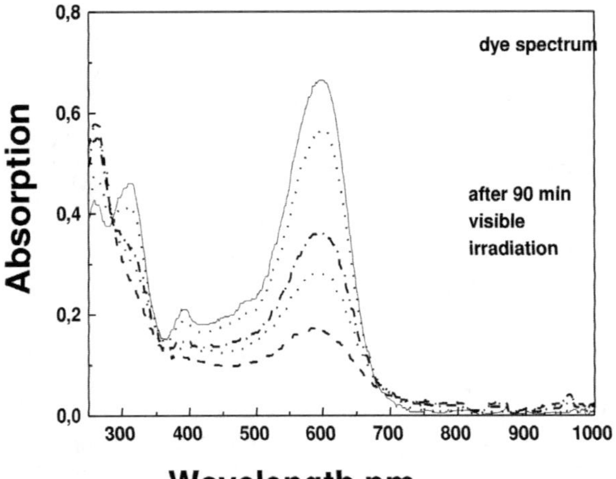

Figure 3. Decolorization of the dye after visible light illumination for 90 minute period. The photocatalyst is TC film, treated at 400°C.

Figure 3 illustrates the typical spectrum of the dye, decolorized by TiO_2 sprayed film, obtained from $TiCl_4$. It is seen in the figure that the process of its degradation can be observed by the changes in the position and intensity of the peaks, corresponding to aromatic ring at 313 nm and azo bond at 599 nm.

Photocatalytic activity tests with the films showed strong influence of the type of the precursor solution (figure 4 and figure 5). The films, obtained from inorganic precursor (TC) are more efficient photocatalysts than those obtained from titanium isopropoxide. The films, obtained from organic precursor probably have greater extent of carbon-compound contaminations (originated from TIP and AcAc) than TC films. Duminica et al. [6] have proved that the carbon residues in the titania films are detrimental for the ptotocatalytic activity. The thermal treatment after spray pyrolysis affects also the films properties. As a rule the photocatalyst samples treated thermally at 400°C are the better ones than those non-treated, because they possess anatase crystalline phase with stoichiometric composition.

An exception appear to be the non-treated TC films (figure 4, curve 5), which exhibit behavior similar to the treated TC films, regardless of their amorphous nature. The films probably have higher surface area and contain mixtures of Ti^{4+} and Ti^{3+} ions (figure 2), which could be the reason for their activity.

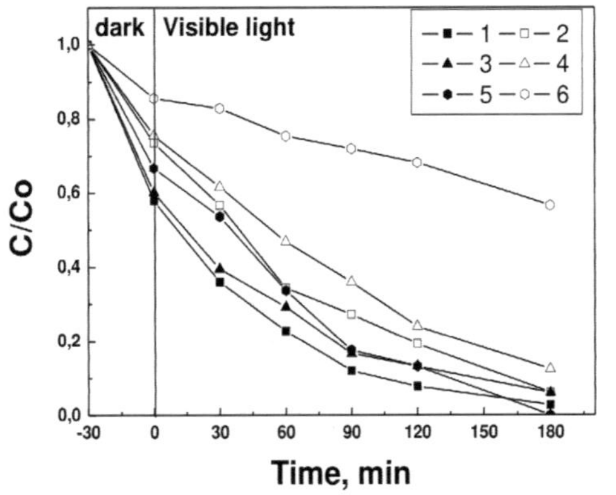

Figure 4. Degradation of RB5 with the time on TC and TIP films, estimated by the change of absorption peak at 599 nm (corresponding to azo bond) ; TC, 400°C-2h (1); TIP, 400°C-2h (2); TC, 400°C-1h (3) TIP, 400°C-1h (4); TC, non-treated (5) and TIP, non-treated (6.)

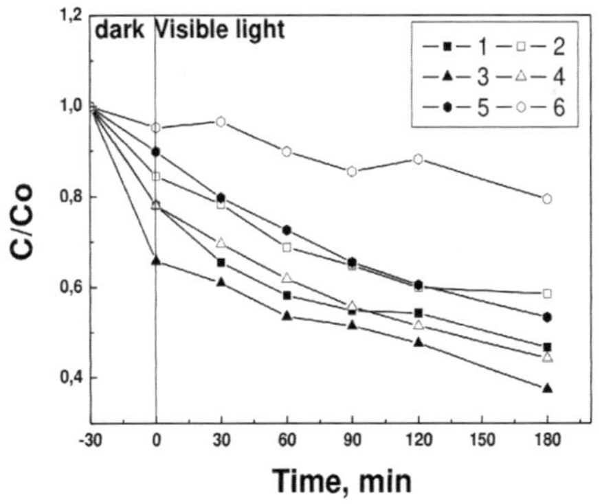

Figure 5. Degradation of RB5 with the time on TC and TIP films, estimated by the change of absorption peak at 313 nm (corresponding to naphthalene ring), TC, 400°C-2h (1); TIP, 400°C-2h (2); TC, 400°C-1h (3) TIP, 400°C-1h (4); TC, non-treated (5) and TIP, non-treated (6)

The TOC measurements showed that the concentration of total organic carbon content in the sample after 180 min of photocatalytic activity testing is 2.26 mg/L and 23 mg/L for TC and TIP films, respectively. This result revealed that the inorganic titanium precursor is more suitable for preparation of efficient photocatalyst films for azo dye degradation in aqueous solutions.

4. Conclusions

Nanosized TiO_2 films were prepared by spray pyrolysis method. The films treated thermally possess crystalline anatase phase, while those non-treated are amorphous. According to XPS analysis the peak, corresponding to Ti2p state indicates only Ti^{4+} oxidation state after thermal treatment of the films at 400°C. The photocatalytic properties towards degradation of textile dye in aqueous solution depend both on the type of the titanium precursor and on the thermal treatment. The samples obtained from $TiCl_4$ precursors showed higher photocatalytic activity than the films, prepared from $Ti(i\text{-}OPr)_4$. As a rule the films treated thermally are more active photocatalysts, than those non-treated. This result can be explained with formation of anatase crystalline phase with stoichiometric composition of the treated TiO_2 films. TOC measurements showed minimal concentration of total organic carbon in the dye solution after 180 minutes visible light irradiation.

Acknowledgement
The authors acknowledge the financial support by project DVU 02-36 (2010) of the National Science Fund of Bulgaria.

References
[1] Saquib M Muneer M 2003 *Dyes Pigments* **56** 37
[2] Hoffmann H Martin S Choi W and Bahnemann D 1995 *Chemical Reviews* **95** 69
[3] Lopez A Acosta D Martinez A I and Santiago J 2010 *Powder Technol* ***202*** 111
[4] Abou-Helal M O and Seeber W T, 2002 *Applied Surface Science* **195** 53
[5] Uzunova-Bujnova M Kralchevska R Milanova M Todorovska R Hristov D and Todorovsky D 2010 *Catalysis Today* **151** 14
[6] Duminica F D Maury and F Abisset S 2007 *Thin Solid Films* **515** 7732

Ultrasonically sprayed ZnO films: optical, electrical and gas sensing properties

O Dimitrov[1], D Nesheva[2], Z Levi[2], I Miloushev[2], T Tenev[2], V Blaskov[3] and I Stambolova[3*]

[1] Institute of Electrochemistry and Energy Systems, BAS, Acad. G. Bonchev bl. 10, 1113, Sofia, Bulgaria
[2] Georgi Nadjakov Institute of Solid State Physics, Tzarigradsko Chaussee Blvd. 72, 1784, Sofia, Bulgaria
[3] Institute of General and Inorganic Chemistry, BAS, Acad. G. Bonchev St, bl. 11, 1113, Sofia, Bulgaria

E-mail: stambolova@yahoo.com

Abstract. Highly transparent, polycrystalline ZnO thin films were deposited by ultrasonic spray pyrolysis. The phase and chemical composition of the films were characterized by X-ray analysis (XRD) and X-ray photoelectron spectroscopy (XPS). According to the XRD analysis the films possess wurtzite structure with preferential (002) orientation. The morphology of the ZnO films was investigated by scanning electron microscopy (SEM). Optical transmittance higher than 80% is found in the visible region. The influence of both the substrate and postdeposition annealing temperatures on the dark conductivity and its activation energy were investigated. It has been found that the conductivity increases with increasing of the substrate and annealing temperatures while the activation energy decreases. These effects have been related to an increase of the density of oxygen vacancies. The sensing properties of the samples were tested in the presence of vapours of water, ammonia, ethanol and acetone. It has been revealed that the substrate temperature influences the gas sensitivity. The films deposited at 200°C have shown the highest sensitivity with the best results registered in the presence of ammonia vapours.

1. Introduction

ZnO thin films received much attention in the past decades due to their semiconducting and piezoelectric properties. With a direct wide band gap of 3.37 eV and a large exciton binding energy of 60 meV at room temperature [1,2], nanostructured ZnO films have a wide range of applications in the modern technologies. They are suitable material for electronic, optoelectronic and electromechanical devices [3], as well as for photocatalysis [4]. The physicochemical properties depend strongly on the preparation methods of the ZnO thin layers. The most commonly used physical methods include sputtering [5], pulsed laser deposition [6], molecular beam epitaxy [7] or electron beam evaporation [8]. Other techniques such as chemical vapour deposition [9], sol-gel [10] and spray pyrolysis [11] are also applied. The ultrasonic spray pyrolysis as a variant of classical spray pyrolysis has an advantage of activating the precursor solution and thus promoting the formation of nuclei [12]. The type of zinc precursor (organic or inorganic) affects strongly the morphology of ZnO layers [13-16]. Zinc acetate [13, 14], zinc chloride [13] and zinc nitrate [15] are the most commonly used precursors for deposition of ZnO films by spray pyrolysis. The zinc-ammonia complex solution is applied by Li et. al. [16] for the preparation of ZnO powders with high surface area. Wang et. al. [12] have successfully obtained column-like, flower-like and tube-like arrays of ZnO crystals from zinc nitrate-ammonia complex solution. The zinc-ammonia complex easily undergoes decomposition even at low temperatures - below 100°C [17]. In this study a solution of zinc-ammonia complex was selected as a suitable precursor and was ultrasonically sprayed on heated glass substrates with various deposition parameters. There are two aims to their work: (i) to obtain ZnO thin nanosize films by ultrasonic spray pyrolysis and (ii) to investigate their optical, electrical and gas sensing properties.

2. Experiment

High purity ZnO powder was used as zinc source, which was dissolved in ammonia aqueous solution to obtain 0,025M zinc-ammonia complex. The acidity of the solution was adjusted with acetic acid to pH 12. Glass microscopic slides (76 x 26 mm, Waldemar Knittel Glasbearbeitungs - GmbH) were used as substrates and were ultrasonically cleaned in ethanol for three minutes then thoroughly washed with de-ionised water. The aerosol of the precursor was generated ultrasonically and was transported to the heated substrate by means of an air flow. The substrates were placed on a stainless steel holder. They were heated to temperatures in the range of 200-350°C, chosen on the basis of preliminary experiments. The temperature was monitored using fine thermocouple fixed on the substrate. The postdeposition annealing temperatures (T_a) were 300 and 400°C. The phase composition of the samples was studied by X-ray diffraction (XRD) with CuK$_\alpha$-radiation (Philips PW 1050). The crystallite size was estimated from the XRD spectra. A scanning electron microscope (SEM) JSM-5510 of JEOL was used for the morphology investigation of the films. The films thickness was determined with Talystep (Rank Taylor Hobson) profilograph, using needle with thickness of 0.12 µm. The optical transmittance of the samples was measured with Lambda 1050 (Perkin Elmer) spectrophotometer with a standard InGaAr photodiode.

Dark current and steady state photocurrent (under white light illumination) dc measurements were carried out by means of a Keithley 6487 Picoammeter using sputtered gold co-planar contacts on the top surface of the layers; the contacts produce a gap cell with an active area of ~ (1×10) mm^2. Current-voltage (*I-V*) characteristics, measured at applied electric fields in the range 1 - 10^2 V cm^{-1}, showed Ohmic-like behaviour of the contacts. Temperature dependences of the dark current were measured at a heating rate of 0.1 °C.sec^{-1}. Prior to each measurement, the samples were heated in dark up to 423 K in order to achieve charged carrier equilibrium and to desorb water molecules captured on the film surface. No changes in the dark current and the slope of its temperature dependence have been observed when the measurement was repeated. A preliminary test of the films gas sensitivity has been performed by measuring the dc dark current or photocurrent at room temperature before (in a vacuum cryostat evacuated to ~ 1 Pa), $I_o^{d,p}$, and after exposure $I_g^{d,p}$ to ammonia, water, ethanol, acetone and isopropanol vapours in air as the carrier gas. Vapour content in the air entering the cryostat was determined by the vapour drawn into the cryostat from a soaked piece of cloth fixed to the valve entrance. Normally, the evacuation of the cryostat for 3 - 5 minutes was enough to completely return the samples in the initial state (i.e. to recover the initial I_d values).

3. Results and discussions

Table 1. Thickness of the ZnO layers in nanometers, T_a = 300°C

Deposition Time	Substrate Temperature		
	200°C	250°C	300°C
30 min	100	110	170
60 min	90	145	240
90 min	50	75	340

The films thickness was in the range of 50-340 ± 10 nm. As seen in table 1 it is affected by the films deposition time and substrate temperature (T_{sub}). The observed decrease in thickness with deposition time at lower T_{sub} can be explained with dissolving of the amorphous phase. It is known that amorphous substances are markedly more soluble than their crystalline counterparts due to the free energy difference between amorphous and crystalline forms. The films transition from amorphous phase to crystalline state starts at T_{sub} 250°C and by 300°C the decomposition reactions are completed and ZnO is well crystallized. That is why at higher T_{sub} the increase of deposition time leads also to thickness increase.

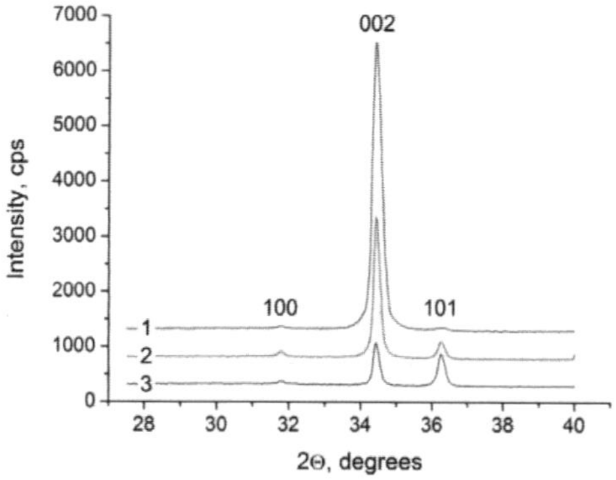

Figure 1. XRD spectra of ZnO films obtained at: 1 - T_{sub} = 250°C, T_a = 300°C; 2 - T_{sub} = 300°C, T_a = 300°C; 3 - T_{sub} = 300°C, T_a = 400°C.

The phase composition was estimated by XRD measurements (figure 1). They revealed formation of wurtzite hexagonal ZnO phase. The films are preferentially oriented in direction (002). The intensity of the (002) peak decreases with the enhancement of the deposition (T_{sub}) and annealing temperatures (T_a), which is in good agreement with the literature [18,19].

Figure 2 shows SEM images of two samples obtained at different substrate temperatures. They are dense with smooth surface morphology, without cracks. Secondary agglomerate particles are seen on the film surface; their size increases with the increase of substrate temperature. The formation of particles on the surface is also reported by Znaidi L. et.al. for ZnO sol gel films, obtained from diluted solution [20].

The transmittance spectra of the films are presented in figure 3; the best samples possess transparency of over 80%. Figure 3 (a) shows a transmittance decrease with the enhancement of substrate temperature. This result correlates with the data of the thin film thickness, which grows with the temperature. The annealing of the samples leads to more transparent layers, which is in accordance with the literature [21]. This effect is explained with the full decomposition of the zinc precursor and the lack of organic residues. A characteristic minimum is seen which is centered in the range 365-370 nm (3.4-3.35 eV) i.e. at the optical band gap energy of ZnO (3.37 eV). No dependence of the minimum position on the preparation parameters has been found. Such feature has been observed by other authors in both room temperatures transmittance [22] and absorption [23] spectra of ZnO films and has been related to exciton absorpton [23].

Typical temperature dependence of the dark current, I_d, is shown in figure 4 (a) in semi-log coordinates. The observed linearity, which was seen for all sample, indicates that:

$$I_d \sim \exp(-E_d/kT) \qquad (1)$$

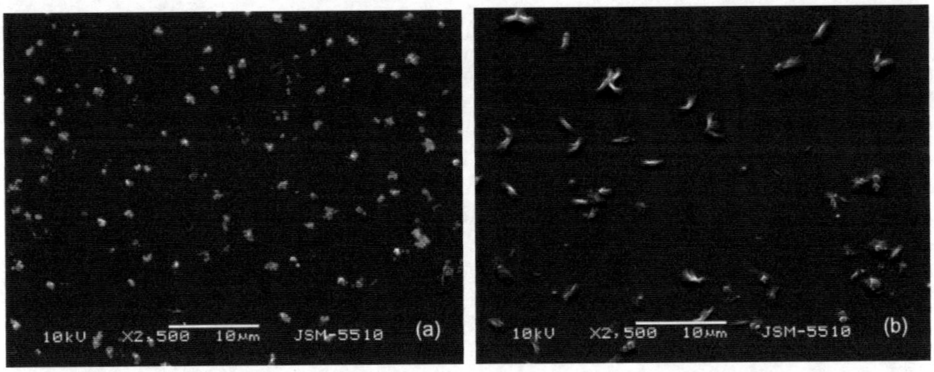

Figure 2. SEM images of ZnO films obtained at: (a) T_{sub} = 250°C; (b) T_{sub} = 300°C

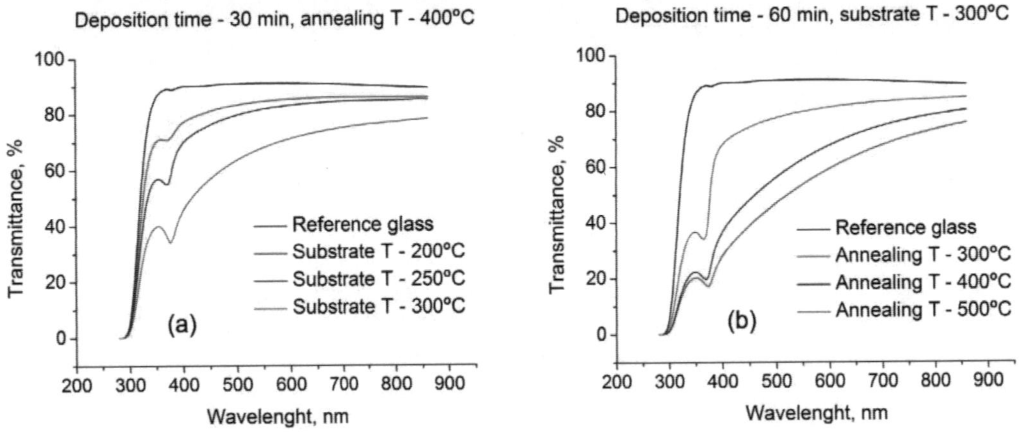

Figure 3. Transmittance spectra of ZnO thin films. The substrate and annealing temperatures are denoted in the figures.

which corresponds to a thermally activated process. In eq.(1) E_d is the dark current activation energy which is sum of the Fermi level position, E_F, and the height of intergrain potential barriers, E_b ($E_d = E_F + E_b$), k is the Boltzmann constant and T is the measuring temperature. The values of the activation energy E_d for a series of samples obtained at different substrate and annealing temperatures are summarized in figure 4 (b). It is seen that the activation energy decreases from 0.8-0.9 eV down to ~ 0.3 eV with increasing T_{sub}. Assuming electron transport in the conduction band of these undoped films, the observed substantial E_d change with increasing T_{sub} can be explained with a Fermi level shift to the conduction band bottom and lowering of E_b. Hence one can assume a donor concentration enhancement with the temperature but the donor nature is unclear.

A comparison of the temperature dependences of the photocurrent (I_p) and dark current of two samples is shown in figure 5. It is seen that the samples exhibit high photoconductivity ($I_p/I_d = 10^3$-10^4 at room temperature) which is interesting for photodetector applications. The photocurrent activation energy may be used to determine E_b ($I_p = I_{p0} \exp(-E_p/kT) \sim \mu_D \tau$, where $\mu_D = \mu_{D0} \exp(-E_b/kT)$ is the carrier drift mobility and τ is the photoexcited carrier lifetime. It has been ascertained that it is 0.16-0.20 eV for the films deposited at T_{sub}=200°C and decreases to ~ 0.1 eV for the films prepared at T_{sub}=300°C. This observation supports the above assumption for E_b lowering with increasing T_{sub}.

The detected sensitivity of the films, estimated by the change of the dark current and photocurrent caused by the adsorption of gas molecules, is summarized in table 2. A dark current increase was registered after exposure to the tested vapours which indicates that the absorbed gas molecules

Journal of Physics: Conference Series **398** (2012) 012022 doi:10.1088/1742-6596/398/1/012022

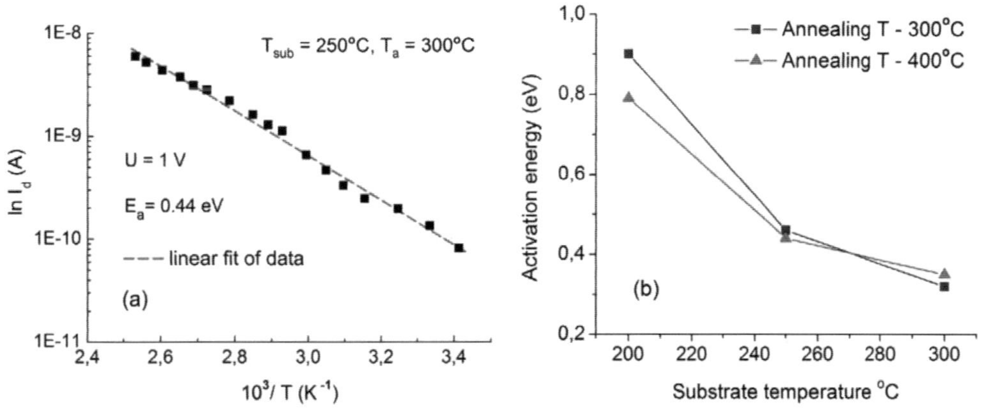

Figure 4. (a) I_d *vs* $1/T$ plot for a ZnO specimen; (b) dark current activation energy *vs* T_{sub} for films, treated at two different temperatures denoted in the figure.

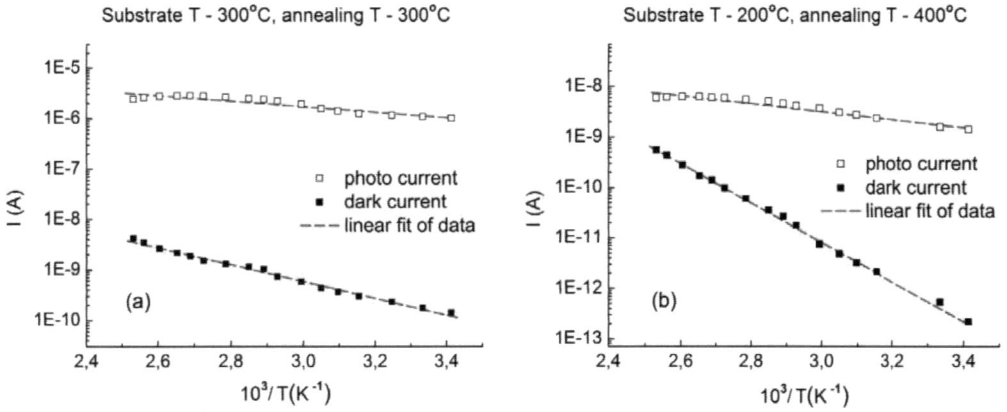

Figure 5. Temperature dependence of the dark current and photocurrent of two ZnO films.

may decrease the height of the intergrain potential barriers. On the contrary, upon analyte exposure the photocurrent decreases, which is probably due to τ decrease but the changes are relatively small. The films response time, i.e. the time for achieving the maximum dark current (photocurrent) alteration, was quite short, between 5 and 30 sec. As seen from the table the I^d changes are strong for the samples obtained at the lowest substrate temperature (200°C) and the ratio $I_g^d/I_o^d = 3\times10^5$ under exposure to ammonia vapour is significantly higher than $I_g^d/I_o^d = 5\times10^4$ obtained for the water vapour. The sensitivity of these films to the other tested vapors (ethanol, acetone) is much lower, which indicates a good selectivity of the films. With increasing substrate temperature the film dark conductivity increases by 3-4 orders of magnitude. This observation supports the above assumption for a donor concentration increase. The conductivity increase reduces the intergrain potential barriers and this probably cause the strong gas sensitivity decreases.

Table 2. Sensitivity of ZnO films to water, ethanol, acetone and ammonia vapours in carrier gas air

Substrate / annealing temperatures (T_{sub} / T_a)	Current in vacuum (A)	Water	Ethanol	Acetone	Ammonia
	I_o^d (dark)	sensitivity (I_g^d/I_o^d)			
200°C / 300°C	2.10^{-12}	5.10^4	15	15	3.10^5
200°C / 400°C	5.10^{-12}	4.10^3	6.10^3	6	9.10^3
300°C / 300°C	1.10^{-8}	2.3	1	1	15
300°C / 400°C	2.10^{-9}	1	1	1	1.4
	I_o^p (photo)	sensitivity (I_o^p/I_g^p)			
200°C / 300°C	5.10^{-9}	2.5	2	2.4	4
200°C / 400°C	8.10^{-9}	3.5	5	2	5
300°C / 300°C	6.10^{-6}	4	1	1.5	4
300°C / 400°C	1.10^{-6}	4	1.4	1.3	4.5

4. Conclusions

Highly oriented (002) ZnO thin films are obtained by ultrasonic spray pyrolysis. In the visible range the films show over 80% transmittance. It is found that the film dark conductivity increases and its activation energy decreases with increasing substrate and annealing temperatures. Based on these observations an increase of the donor density with the temperature increase is assumed. The substrate deposition temperature also influences the gas sensitivity of the films. Within this study the films deposited at 200°C show sensitivity to the tested vapours; it is highest to ammonia vapours.

Acknowledgements

The authors thank to the Bulgarian Ministry of Science and Education for the financial support by project DVU 02/36 (2010).

References

[1] Ozgur U, Alivov Y I, Liu C, Teke A, Reshchikov M A, Dogan S, Avrutin V, Cho S J and Morkoc H 2005 *J Appl Phys* **98** 041301
[2] Look D C 2001 *Mater Sci Eng B* **80** 383
[3] Liu K W, Ma J G, Zhang J Y, Lu Y M, Jiang D Y, Li B H, Zhao D X, Zhang Z Z, Yao B and Shen D Z 2007 *Solid-State Electronics* **51** 757
[4] Stambolova I, Blaskov V, Shipochka M, Vassilev S, Dushkin C and Dimitriev Y 2012 *Mater Chem Phys* **121** 447
[5] Chen K, Chiang K S, Chan H P and Chu P L 2008 *Optical Materials* **30** 1244
[6] Claeyssens F, Freeman C L, Allan N L, Sun Y, Ashfold M N R and Harding J H 2005 *J Mater Chem* **15** 139
[7] Heo Y W, Ip K, Pearton S J, Norton D P and Budai J D 2006 *Appl Surf Sci* **252** 7442
[8] Varnamkhasti M G, Fallah H R and Zadsar M 2012 *Vacuum* **86** 871
[9] Nam K H, Kim H, Lee H Y, Han D H and Lee J J 2008 *Surf Coat Tecnol* **202** 5463
[10] Kaur R, Singh A V and Mehra R M 2005 *J Mater Sci: Mater in Electronics* **16** 649
[11] Allah F K, Abé S Y, Núñez C M, Khelil A, Cattin L, Morsli M, Bernéde J C, Bougrine A, Del M A Valle and Díaz F R 2007 *Appl Surf Sci* **253** 9241
[12] Wang Z, Qian X, Yin J and Zhu Z 2004 *Langmuir* **20** 3441
[13] Romero R, Leinen D, Dalchiel E A, Ramos-Barrado J R and Martin F 2006 *Thin Solid Films* **515** 942
[14] Messaoudi C, Sayah D and Abd-Lefdil M 1995 *Phys. Stat. Sol.(a)*, **151** 96

[15] Puetz J, Gasparro G and Aegerter A 2003 *Thin Solid Films* **422** 40
[16] Li D and Haneda H 2003 *J Photochem Photobiol A: Chemistry* **155** 171
[17] Gao X D, Li XM, Yu WD, Li L, Qiu JJ and Peng F 2007 *Solar En Mater Solar Cells* **91** 467
[18] van Heerden J L and Swanepoel R 1997 *Thin Solid Films* **299** 72
[19] Kitova S and Danev G 2012 *J Phys: Conference Series* **253** 012031
[20] Znaidi L 2010 *Mater Sci Eng B* **174** 18
[21] Singh P, Kumar A, Deepak and Kaur D 2007 *J Crystal Growth* **306** 303
[22] Xu L, Li X and Yuan J 2008 *Superlatt Microstuctr* **44** 276
[23] Yano M, Koike K, Sasa Sh and Inoue M 2006 *Zinc Oxide Bulk, Thin Films and Nanostructures* ed C. Jagadish and S. Pearton, (Elsevier) chapter 11

Optical properties of porous chalcogenide films for sensor application

A Lalova, R Todorov

Institute of Optical Materials and Technologies "Acad. J. Malinowski", Bulgarian Academy of Sciences, Acad. G. Bonchev Str., bl. 109, 1113 Sofia, Bulgaria.

E-mail: rossen@iomt.bas.bg

Abstract. The object of the present work is investigation of the optical properties of obliquely deposited thin films from As – S – Ge system. Aiming to obtain high porous coatings the deposition rate was varied in the range of 0.05-10 nm/s. The conditions for deposition of thin As – S - Ge films with columnar structure and high porosity were established. The role of the actual deposition conditions on the optical properties is examined. The optical constants (refractive index, *n* and absorption coefficient, α) and thickness, *d* as well as the optical band gap, E_g, and slope parameter *B* in dependence of the deposition angle and rate are determined from specrophotometric measurements in the spectral range 400-2000 nm applying the Swanepoel's envelope method and Tauc's procedure. Increasing of the value of *n* from 2.40 to 1.83 for thin film with composition $As_{10}Ge_{30}S_{60}$ with increasing deposition angle from 0° to 75° is observed. The possibility of using the thin films for optical sensing of SO_2 and H_2S was examined. Reversible changes of the refractive index, Δn = 0.015 were observed as a consequence of treatment virgin - exposure to H_2SO_4 vapors– annealing at 120 °C.

1. Introduction

Fabrication of materials with high porosity is one approach for development of novel nanomaterials. The porous materials possess unique physical properties due to the extremely large surface-to-volume ratio and quantum confinement effects [1, 2]. Progress in synthesis, characterization and modelling of such materials would enable technological innovations in various applications ranging from microelectroncs to optoelectronics and specifically biosensing [3].

Changes of the electrical and optical properties of porous materials due to filling of the pores with different gases make them attractive for application in gas sensing [4, 5]. The propylamine and nitrogen dioxide sensing properties of thin porous chalcogenide films from As_4S_3 and As-Ge-Te were investigated from Tsiulyanu et al. [6]. It was reported in [7] that thin films from As-S-Ge system could be applied as NH_3 gas sensors for concentrations higher than 250 ppm.

The object of investigation of the present work is the influence of the deposition conditions on the optical properties of chalcogenide films from $As_{40-x}Ge_xS_{60}$ system. The possibility for deposition of thin chalcogenide films with high porosity and their potential application as optical sensors of sulphur containing compounds such as SO_2 and H_2S was examined.

2. Experimental details

Bulk glasses from the system As - S - Ge were synthesized in a quartz ampoule from elements of 99,999 % purity by the method of melt quenching [8]. Thin films from previously synthesized bulk glasses from $As_{40-x}Ge_xS_{60}$ system (for x = 0, 10, 30, 35) were deposited by thermal evaporation on rotated substrates or by the well known oblique deposition technique at angle of deposition 75°.The bulk materials were evaporated from a Ta boat ("Knudsen type"). The deposition rate was varied in the range of 0.05 - 1 nm/s. In the case of substrates continuously rotated during the evaporation process the substrate holder is a dome-shaped calotte that can be considered as a segment of a sphere. The evaporation sources are located close to (approximately at) the geometric centre of this sphere and the vapours of the evaporated compound fall normally on the substrates.

The composition of the thin films obtained was determined by a scanning electron microscope Jeol Superprobe 733 (Japan) with an X-ray microanalyser. The experiments were performed at an accelerating voltage of 20 kV, current of 1.4 nA and scanning time of 200 s for each spectrum. The transmittance (T) and reflectance (R) spectra of thin films were measured by UV-VIS-NIR spectrophotometer Carry 5E (Australia) in the region 400 - 2000 nm to an accuracy of ± 0.1 % and ± 0.5 %, respectively.

3. Results and discussion

Thin films from $As_{40-x}Ge_xS_{60}$ system for x = 0, 10, 30, 35 were thermally evaporated. The composition of the layer was controlled by X-ray microanalysis and was determined to be close to the expected one. The SEM cross-section images of thin films from As-Ge-S system are presented in figure 1.

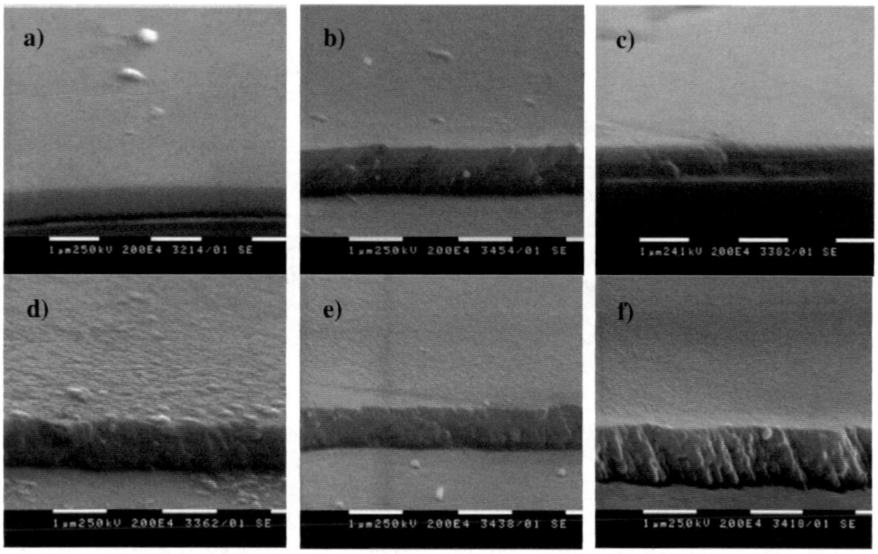

Figure 1. Cross-section images obtained by SEM of thin $As_{40-x}Ge_xS_{60}$ films deposited at different rates: $As_{40}S_{60}$ (0.4 nm/s) (a); $As_{40}S_{60}$ (0.1 nm/s) (b); $As_{40}S_{60}$ (0.06 nm/s) (c); $As_{30}Ge_{10}S_{60}$ (0.4 nm/s) (d); $As_{10}Ge_{30}S_{60}$ (0.1 nm/s) (e) and $As_{10}Ge_{30}S_{60}$ (0.06 nm/s) (e) at 75° angle of incidence of the vapours. (The marker corresponds to 1 μm.)

The germanium containing layers clearly demonstrate columnar structure even at deposition rate of 0.4 nm/s while for thin $As_{40}S_{60}$ films columns were observed at 0.1 nm/s. At very low deposition rates (0.06 nm/s) of the thin $As_{40}S_{60}$ film no columns were formed (figure 1c).

The refractive index, n and thickness, d of the thin film were calculated from transmission spectra using Swanepoel's method [9]. The program used to calculate n will determine it to an accuracy of ± 0.5 % for an error in the transmittance of ± 0.1 % [10]. In figure 2 a comparison is given of the dispersions of the refractive index of thin films deposited normally or obliquely at 75° angle of incidence of the vapors. Due to their microstructure the obliquely deposited films demonstrated lower refractive index for all compositions. The most drastically decreased values of *n* were found for the thin film with composition $As_{10}Ge_{30}S_{60}$. The results for *n* are summarized in table 1.

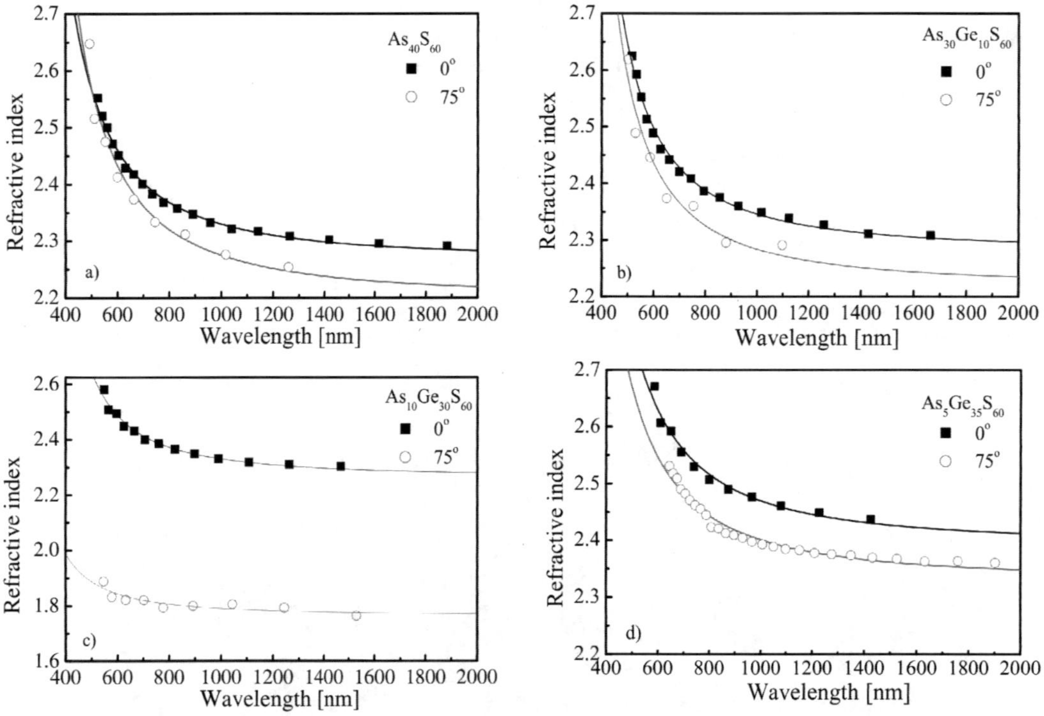

Figure 2. Dispersion of the refractive index of thin $As_{40-x}Ge_xS_{60}$ films for x = 0 (a); 10 (b); 30 (c); 35(d); evaporated on rotated substrates (0°) and oblique deposition (75°).

Table 1. Calculated values for the optical parameters (refractive index, *n*, optical band gap, E_g and structural parameter, *B*), thickness, *d*, average coordination number, Z and porosity, ϕ of thin As-Ge-S films.

Composition	Z	Conditions of deposition	d, □m	*n* (at λ = 1550 nm)	E_g [eV]	B [$cm^{-1/2}.eV^{-1/2}$]	ϕ
As_2S_3	2.40	normal deposition	1.23	2.29	2.35	770	0.001
		oblique deposition at 75°	0.81	2.23	2.33	738	
$As_{30}Ge_{10}S_{60}$	2.50	normal deposition	1.08	2.31	2.32	745	0.036
		oblique deposition at 75°	0.76	2.10	2.48	644	
$As_{10}Ge_{30}S_{60}$	2.70	normal deposition	0.87	2.28	2.15	591	0.277
		oblique deposition at 75°	1.3	1.78	2.63	378	
$As_5Ge_{35}S_{60}$	2.75	normal deposition	0.88	2.42	2.05	568	0.035
		oblique deposition at 75°	1.41	2.35	2.00	501	

The reflectance and transmittance spectra were used for calculation of the linear absorption coefficient, α and optical band gap, E_g through the equations:

$$T = (1-R)^2 \exp(-\alpha d) \tag{1}$$

where T is transmittance, R is reflectance and d is thin film's thickness. Analysis of the strong absorption region ($10^4 \leq \alpha \leq 10^5$ cm^{-1}) was carried out using the following well-known quadratic equation, often called *Tauc's* law [11]:

$$(\alpha h\nu)^{1/2} = B(h\nu - E_g^{opt}) \tag{2}$$

where B is a substance parameter, which depends on the electronic transition probability, ($h\nu$) is the photon energy and E_g^{opt} is the so-called *Tauc's* gap. The comparison for the refractive indices, n for wavelength λ = 1550 nm, optical band gap and slope parameter B for thin film prepared on rotated substrates and obliquely deposited films at angle 75° is given in table 1. Significant variation of the band gap when changing the angle of incidence of the vapours is observed for thin $As_{10}Ge_{30}S_{60}$ layer. The lower values of B for the thin films with columnar structure are due to larger structural disorder.

We used the effective medium theory for determination of the porosity of the thin films. According the Bruggeman model a porous semiconductor material can be considered as a mixture of two phases and its effective refractive index n_{eff} in the non-absorbing region follows the equation:

$$1-\phi = \frac{\left(\dfrac{n_{eff}^2}{n_c^2} - \dfrac{n_d^2}{n_c^2}\right)}{\left[\left(\dfrac{n_{eff}^2}{n_c^2}\right)^{1/3}\left(1-\dfrac{n_d^2}{n_c^2}\right)\right]} \tag{3}$$

where n_c and n_p are the refractive indices of the continuous media and pores, respectively, while ϕ is the porosity. We have considered as continuous media the thin films deposited at normal incidence and duly used the data obtained for their refractive index. The results for porosity calculated applying equation 3 are presented in table 1. We found that the values for ϕ varied in the range of 0.002 to 0.277. The maximal value for ϕ was obtained for the thin film with composition $As_{10}Ge_{30}S_{60}$ where the average coordination number is Z = 2.7. We suggest that the observed behaviour of the porous parameter, ϕ for thin As - Ge - S films can be related to the structure of the thin films. According to [12] the free volume and atomic compactness of the glasses from nonstoichiometric section As_2S_3-Ge_2S_3 reach maximal values at coordination numbers around 2.7.

Furthermore we probed the possibility of potential application of the obliquely deposited coatings for optical sensing of different gasses. The idea for application of porous materials as optical sensors is based on the changes of the effective refractive index due to filling of the pores with gas. We performed initial sensing experiments by exposing the thin films to the vapours of sulphur acid (H_2SO_4). The results from the testing of thin films with composition $As_{40}S_{60}$ and $As_{30}Ge_{10}S_{60}$ are presented in figure 3. Notable changes of the refractive index are observed for unexposed coatings, while the exposed ones remain unaffected by the gas exposure. Increase of the values of n was observed after exposure of thin film to the vapours of the acid. It is well known that the population of the As-As bonds is higher in as-deposited thin films [13]. These bonds interact with S-S bonds during the illumination and forming As-S bonds. We can expect that the vapours of H_2SO_4 consist of H_2S and SO_2 molecules and these sulphur containing species interact with As-As.

The variation of the refractive index with different treatments is shown in figure 3d. It was observed that after 20 min of gas exposure the refractive index changed with 0.015. This change is reversed after 3 min. annealing at 120 °C. We can suggest that the sulphur containing species in the H_2SO_4 vapours create weak bonds that are broken after short annealing.

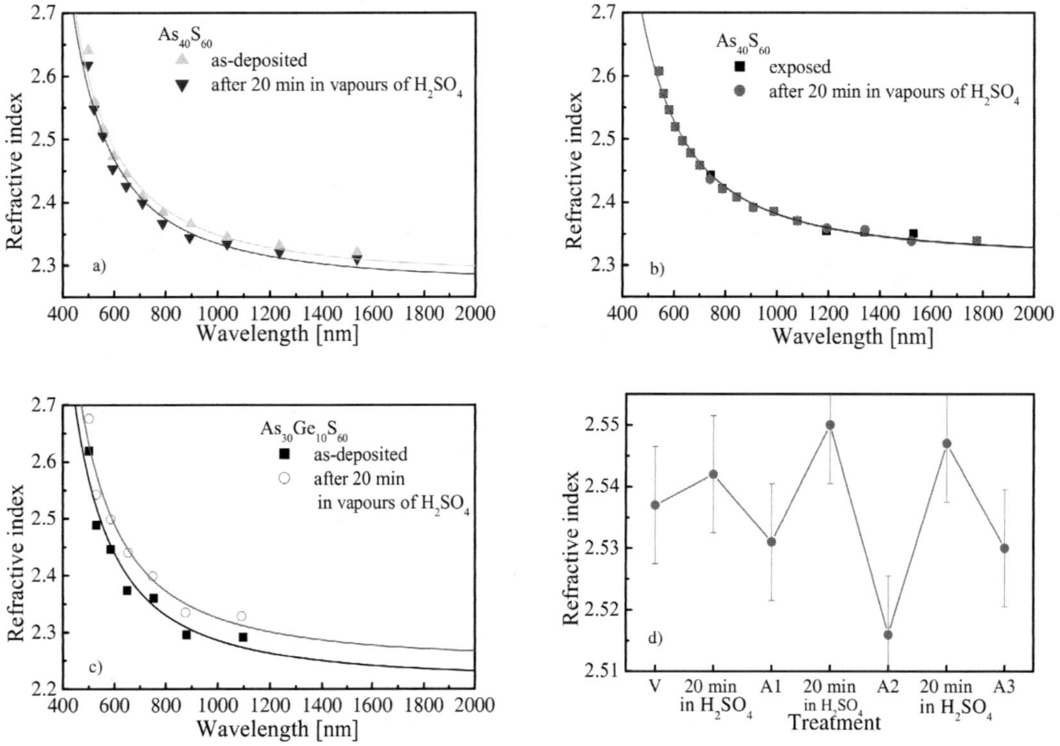

Figure 3. Refractive index of thin films with composition $As_{40}S_{60}$ unexposed (a); $As_{40}S_{60}$ exposed (b) $As_{30}Ge_{10}S_{60}$ unexposed (c) before and after treatment with sulphur acid vapours and periodic changes of the value of n for thin $As_{40}S_{60}$ layer in consequences of treatment virgin (V) -exposure to vapours for 20 minutes and annealing (A) (d)

In figure 3d 3 cycles of interaction with vapours and subsequent annealing for thin As_2S_3 film are demonstrated. This result demonstrates the possibility for application of thin As-Ge-S layers as optical sensors. For a complete understanding of this effect additional information for the binding energy of different kind of bonds and structural units and surface of thin films is required.

4. Conclusions

The conditions for preparation of thin As-Ge-S films with columnar structure were investigated. The SEM images showed that columnar structure is observed at deposition rates in the range of 0.1 – 0.4 nm/s. The refractive index of the films is varied in very wide range from 1.78 to 2.42 at $\lambda = 1550$ nm depending on layers' composition and their microstructure. It was found that maximal porosity possess the thin films with composition $As_{10}Ge_{30}S_{60}$ and average coordination number $Z = 2.7$. The possibility was demonstrated for the application of thin As-Ge-S films in optical sensing of sulphur compounds. The large variation in n and sensitivity to gasses provide an opportunity for the films from the studied system to be applied in multi-layered structures where the response to small changes in the optical parameters of thin films is greater.

Acknowledgement
The authors would like to acknowledge for the financial support from project BG05PO001-3.3-005/0001 Science & Business" branch, operational program "Human Resource Development".

References

[1] Braun M M, Pilon L 2006 *Thin Solid Films* **496** 505
[2] Zimin S P, Gorlachev E S, Naumov V V, Buchin E Yu, Zogg H 2011 *J. Phys: Conf. Ser.* **291** 012023
[3] Garahan A, Pilon L, Yin J, Saxena I 2007 *J. Appl. Phys.* **101** 014320
[4] Bi Z, Anderoglu O, Zhang X, MacManus-Driscoll J L, Yang H, Jia Q, Wang H 2010 *Nanotechnology* **21** art. no. 285606
[5] Barreca D, Gasparotto A, Maccato C, Maragno C, Tondello E, Comini E, Sberveglieri G 2007 *Nanotechnology* **18** art. no. 125502
[6] Tsiulyanu D, Martin S, Liess H-D, Eisele I 2003 *J Optoelectron Adv Mat* **5** 1349
[7] Georgieva V, Yordanov Tz, Pamukchieva V, Arsova D, Gadjanova V, Vergov L 2010 *AIP Conf. Proc.* **1203** 1079
[8] Todorov R, Petkov K 2001 *J. Optoelectron. Adv. Mater.* **3** 311
[9] Swanepoel R 1984 *J. Phys. E: Sci. Instrum.* **17** 896
[10] Todorov R, Tasseva J, Babeva Tz, Petkov K 2010 *J. Phys. D: Appl. Phys.* **43** 505103
[11] Tauc J 1974 *Amorphious and liquid semiconductors* (New York: Plenum Press)
[12] Kovalskiy A 2001 *J. Optoelecron. Adv. Mat.* **3** 323
[13] Martin T P 1984 *J. Phys. Chem.* **80** 170

AES study of Pt-doped thin Sn-O-Te films

G Spasov[1,2], B Georgieva[2,3] and I Podolesheva[2]

[1] Technical University of Sofia, 8, Kliment Ohridski Blvd., 1000 Sofia, Bulgaria
[2] Institute of Optical Materials and Technologies "Acad. J. Malinowski", Bulgarian Academy of Sciences, Acad. G.Bonchev Str., Bl.109, 1113 Sofia, Bulgaria

E-mail: biliana@iomt.bas.bg

Abstract. Thin SnO_2-based films with Sn/Te ratio ($R_{Sn/Te}$) ≈ 2.1, prepared by thermal vacuum co-evaporation of Sn and TeO_2 and doped with Pt before or after thermal annealing, are studied by Auger electron spectroscopy (AES). Thermal annealing leads to changes in the chemical composition and structure (crystallization) of the amorphous as-deposited films. The depth profiling has revealed that the film composition is homogeneous and Pt is dispersed throughout the whole volume of the samples irrespective of their structure and the sequence of doping and thermal annealing. Information is obtained about the oxidation processes taking place during the preparation and thermal annealing of the films, about the influence of the substrate material and the correctness of the evaporation process modeling. It is shown that the significantly higher ethanol sensitivity of the films doped with Pt before annealing is not related to the Pt-distribution.

1. Introduction

Resistivity gas sensors on the base of metal oxide semiconductors, such as SnO_2, have been widely investigated during the last decades mostly due to their high sensitivity, fast response, small size and low power consumption [1]. To enhance their selectivity, doping with different additives like noble metals or other oxides is used [2]. The precise choice of the doping element and working temperature gives the possibility to improve the gas sensing properties, mainly the sensitivity and selectivity towards reducing gases.

The method developed in our Institute for obtaining doped SnO_2 by vacuum co-deposition of TeO_2 and Sn is described in detail elsewhere [3]. It allows preparing thin films with desired $R_{Sn/Te}$ which determines to a great extent the gas sensing properties. As-deposited amorphous films with $R_{Sn/Te} \approx 0.8$ have very good characteristics as humidity sensors operating at room temperature [4]. At a ratio of about 2.1, after additional Pt-doping and crystallization by thermal annealing, high sensitivity to ethanol is observed [5].

From previous investigations of films with $0 < R_{Sn/Te} < 2$ it is known that the as-deposited films are homogeneous in the whole compositional range and that in films with $R_{Sn/Te} > 2$ no segregation processes could be expected upon annealing at temperatures up to 200 °C [3]. In order to check this at higher temperatures and to find out if the in-depth distribution of Pt depends on whether the crystallization is carried out in the presence of Pt or not, a quantitative depth profiling had to be used. It turned out that dealing with a composite system with a low content of one of the components represents an interesting topic for Auger analysis.

2. Experimental

Sn-O-Te films with thickness (d) 40 – 60 nm and $R_{Sn/Te} \approx 2.1$ were prepared by thermal vacuum co-deposition of Sn and TeO_2 using separate Knudsen type cells, under vacuum better than 10^{-4} Pa and on stationary substrates at room temperature. During the whole evaporation process the fluxes of both substances were controlled with quartz oscillators. The amount of each substance in the films was calculated on the base of the measured flux using computer programs [6] and was controlled by Energy Dispersed Spectroscopy in a scanning electron microscope (SEM/EDS). The thickness of the films was measured with a profilometer. The samples intended for SEM/EDS- and AES-analysis were deposited onto silica and carbon substrates. Doping with Pt was realized in a separate vacuum cycle using tungsten basket as a source. Before or after the Pt-doping the samples were crystallized by thermal annealing up to 360 °C with a step of 40 °C, for 15 min at each temperature.

To study the in-depth distribution of the main components (Te, Sn and O) and of the dopant (Pt) in as-deposited (amorphous) and thermally annealed (crystalline) films, Auger profiling was used. In the case of annealed films the influence of the doping/crystallization-sequence on the compositional distribution was investigated as well. The spectral Auger analysis was performed in the differential mode at primary electron energy E_p = 3 keV (I_p = 0.07 μA, V_{mod} = 4 V_{ptp}, ΔE/E = 0.3%). The most intense Auger peaks of Sn (430-437 eV), Te (483-491 eV), O (510 eV), Pt (64 eV), C (272 eV) and Si (92 eV) were monitored. The intensity of the doublets was measured from the top of the low-energy peak to the bottom of the high-energy peak, and the intensity of the rest of the peaks – peak-to-peak.

The depth profiling was carried out by sputtering with an Ar^+ ion beam with energy E_I = 3 keV and ion beam current I_I = 2 μA, followed by AES analysis at each etching step.

3. Results and Discussion

3.1. Auger analysis data

The spectrum of a SnO_2 standard sample with peaks of carbon, tin and oxygen, obtained in the energy window 205-555 eV, is shown in figure 1.

The data obtained during the depth profiling of an as-deposited Pt-doped Sn-O-Te film with $R_{Sn/Te} \approx 2.1$, presented as peak intensities vs. sputtering time, are displayed in figure 2.

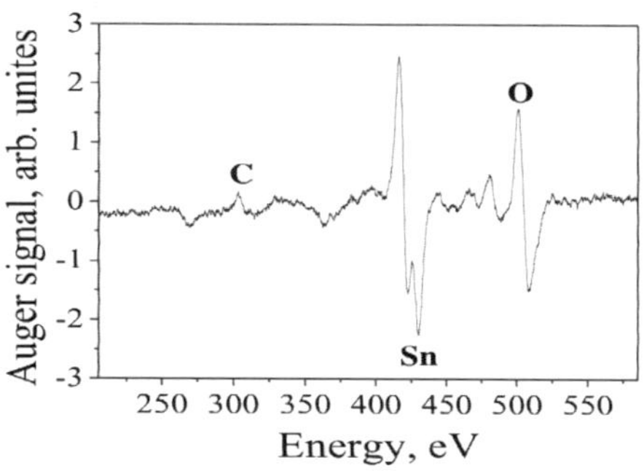

Figure 1. Spectrum from a SnO_2 standard with peaks of carbon, tin and oxygen.

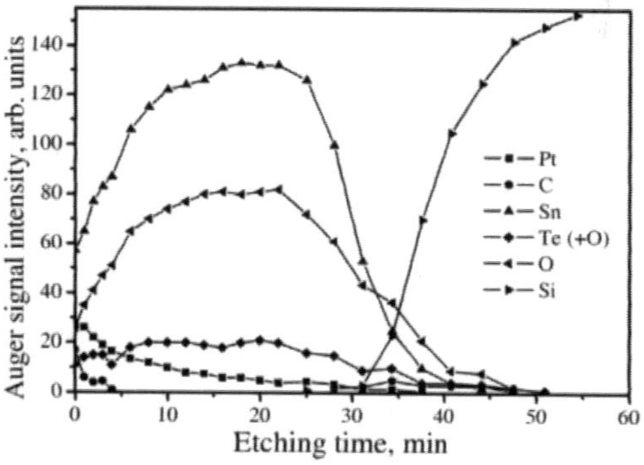

Figure 2. Profile "intensities/sputtering time" from an as-deposited Pt-doped Sn-O-Te sample.

3.2. Quantification of the sputtering profiles

3.2.1. Calibration of the depth scale (x-axis of the profile). As it will be shown later, the profiling reveals that (i) the composition of the films is homogeneous over the entire thickness of the films and (ii) the distribution of Pt shows no peculiarities at the interface film/substrate, so the interface region needs no special attention and the detailed profiling can be limited to its beginning. The latter we have defined as the depth where the Auger signals from the film drop by 10%. Both experimental results have given us the reason not to perform detailed calibration of the x-axis. We have assumed that the distance between the film surface and the film/substrate-interface is equal to the film thickness measured with the profilometer and that the sputtered depth is proportional to the sputtering time.

3.2.2. Calibration of the concentration scale (y-axis of the profile). Here the analytical difficulties in the quantification are arising from the interference between the constitutive peaks of the doublets, leading to hard-to-evaluate changes in the intensity. One of the reasons for this interference (broadening of the peaks due to changes in the modulation energy or the energy resolution of the spectrometer) we eliminate using own standards. Of a greater importance are, however, the so-called "chemical changes" in the peaks (in this case due to oxidation of the analyzed element). They alter the linear dependence of the doublet intensity on the element content, which underlies the AES quantification. The "chemical changes" of an Auger peak manifest themselves in: (i) energetic shift, (ii) peak shape changes (as a rule causing changes in the elemental sensitivity) and (iii) a possible interference between the peaks of the element (due to (i)), if in the analyzed sample volume the element is present in more than one chemical state. The energetic shift itself does not influence the Auger quantification, that is why we are not going to discuss this problem; the analysis distinguishes between the elemental and oxidized state, regardless of the oxidation level (SnO and SnO_2) [7]. In the case of Sn we have studied the second and third effect experimentally. With the aim to examine whether the elemental sensitivity factor (ESF) changes with the chemical state, we have compared the Sn-signals from Sn (metal) and SnO_2 (figure 3). To the oxidized Sn we have applied matrix corrections for the electron backscattering [8], as well as for the atomic densities and IMFP (nonelastic mean free electrons path) [9], with the coefficient for oxides 0.55 [10]. As the normalized intensities differ within 20% – i.e. within this accuracy - it could be assumed that the elemental sensitivity factors for Sn (metal) and SnO_2 are identical [7,11,12]. Superposing both signals in different ratios we have modeled their interference (nanosized mixture "Sn-SnO_2" with prevailing dioxide). It turned out that the total Sn signal – up to the important for our investigations 25% – weakly depends on the relative amount of Sn in the mixture. That is why the relation between this intensity and the intensity of O (figure 4) changes much slower than predicted by the relation: $(1 + x)/(1 - x)$, that is valid for the case of two elements with constant ESF (according to this relation at 25% the relative intensity is 2.58). In figure 4 the maximal relative nonlinearity of the curve in the interval of interest is 1.9%. It should be stressed that, as stated in the literature, the value of the ratio $I_{Sn/O}(SnO)/I_{Sn/O}(SnO_2)$ – 1.32 [7,12] or 1.41 [13] – differs only by 4 to 10% from the value 1.27, obtained by replacing SnO with the mixture having the same composition (0.75 SnO_2, 0.25 Sn).

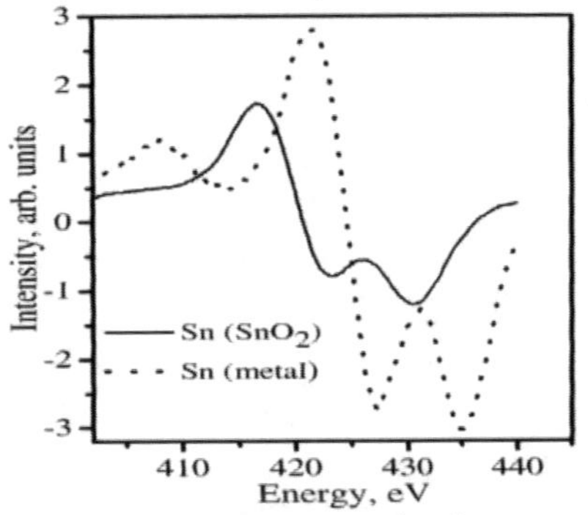

Figure 3. Auger spectra in the energy window 402-445 eV from the Ar^+ cleaned surface of SnO_2 and Sn standard (tenfold statistical accumulation).

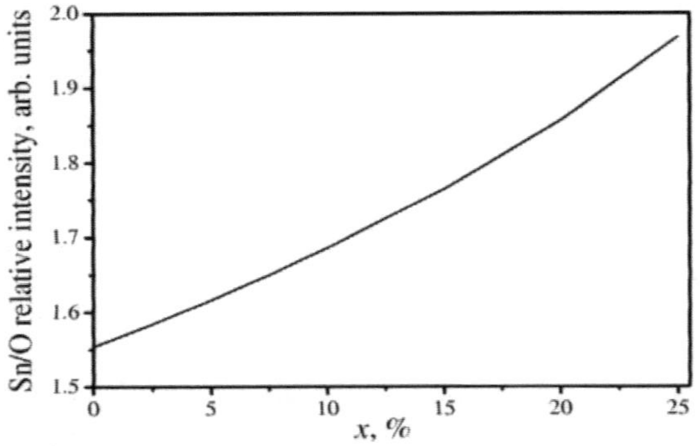

Figure 4. Dependence of the intensity of the superposed Sn signal to the O one $I_{Sn}[xSn_{metal}+(1-x)Sn_{SnO2}]$ $I_O[(1-x)O_2]^{-1}$ from the nanosized mixture $[xSn_{metal}+(1-x)SnO_2]$ as a function of the Sn_{metal} content x.

As for the Te-doublet ,(490 eV) an additional difficulty arises from its interference with the O-peak, so it is not possible to perform the classical Auger quantification. Thus we have calculated the local content of Te using the ratio $R_{Sn/Te}$, estimated by the SEM/EDS-analysis. The reason for applying this approach is based on the following two experimental results: (i) on the fact that $R_{Sn/Te}$ does not change upon heating up to 360 °C (2.1 before and 2.05 after the thermal annealing) and coincides well with the predetermined value (calculated by means of the computer programs [6]), and (ii) on the already mentioned in-depth film homogeneity (table 1). The latter gives us the reason to assume that Te follows the distribution of Sn, i.e. the integral SEM/EDS-result for $R_{Sn/Te}$ is valid for every point of the film. (In the quantification terminology, Te is determined using an external integral standard.).

Table 1. Characterization of the plateau in the Sn/O-dependence for the samples studied ("Film structure" – before Pt-doping, "Thermal annealing" – after Pt-doping)

Sample No	Substrate	Film structure	Thermal annealing	Sn/O intensity	Error, *ΔI/I*,%
1	Graphite	crystalline	No	1.085	0.5
2	Graphite	amorphous	Yes	1.122	1.6
3	Si	amorphous	No	1.638	1.3
4	Si	amorphous	Yes	1.252	1.5

The calibration of the concentration axis (y) for Sn, O, Pt, Si, and C occurs according to the ESF method, as the Te content represents 1/2.1 of the Sn content. The standards used and the experimental ESF (relative to Si) are given in table 2. The statistical error of their determination does not exceed 0.8%. For Sn the value of ESF in the table is correct only in the case of compositions close to SnO_2.

Table 2. Experimental relative (to Si) element sensitivity factors

Element	Si	O	Pt	Au	C	Sn
Standard	Si	Thermal SiO_2	Metal	Metal	Graphite	SnO_2
Rel. ESF	1	1.03	0.87	0.70	0.31	3.23

3.2.3. Quantification and its verification. Knowing the mean relative content of Te, there are several ways of determining the homogenous Te profile. According to the procedure, chosen by us, first the mean ESF_{Te} for the film was calculated (using the averaged relative intensity Sn/Te for the film, ESF_{Sn} and $R_{Sn/Te}$ value obtained by SEM/EDS). Later on, at every point of the film the concentration of the components was determined using the standard ESF method. The profiles obtained are shown in figure 5. Table 3 summarizes the integral (averaged) concentrations of the elements in the film.

Table 3. Averaged concentrations of the film constituents (before and after excluding the carbon content) and concentration of Pt related to the total metal content (Pt/m,%)

Sample No	Concentration, at.%					Conc. (– carbon), at%				Pt/m,
	Sn	O	Te	Pt	C	Sn	O	Te	Pt	%
1	17.0	50.8	8.4	5.7	18.1	20.8	62.0	10.2	7.0	18.4
2	19.1	57.4	9.5	5.7	8.3	20.9	62.7	10.3	6.2	16.6
3	23.5	45.5	11.4	7.9	11.5	26.6	51.5	12.9	9.0	18.6
4	22.1	56.8	10.6	5.9	4.6	23.2	59.6	11.1	6.2	15.2

Additionally, the measured by SEM/EDS content of Pt can be used to verify the quantification. It shows which (integral) part of the metal content represents Pt. The analogous parameter, calculated on the base of the Auger results (integrating along the depth of the film), reveals an average systematic increase of 39%. Taking into consideration the facts, that the Auger-signal of Pt is weak and that Pt is introduced into a "very thin" (for SEM-EDS analysis) film, the coincidence can be qualified as good.

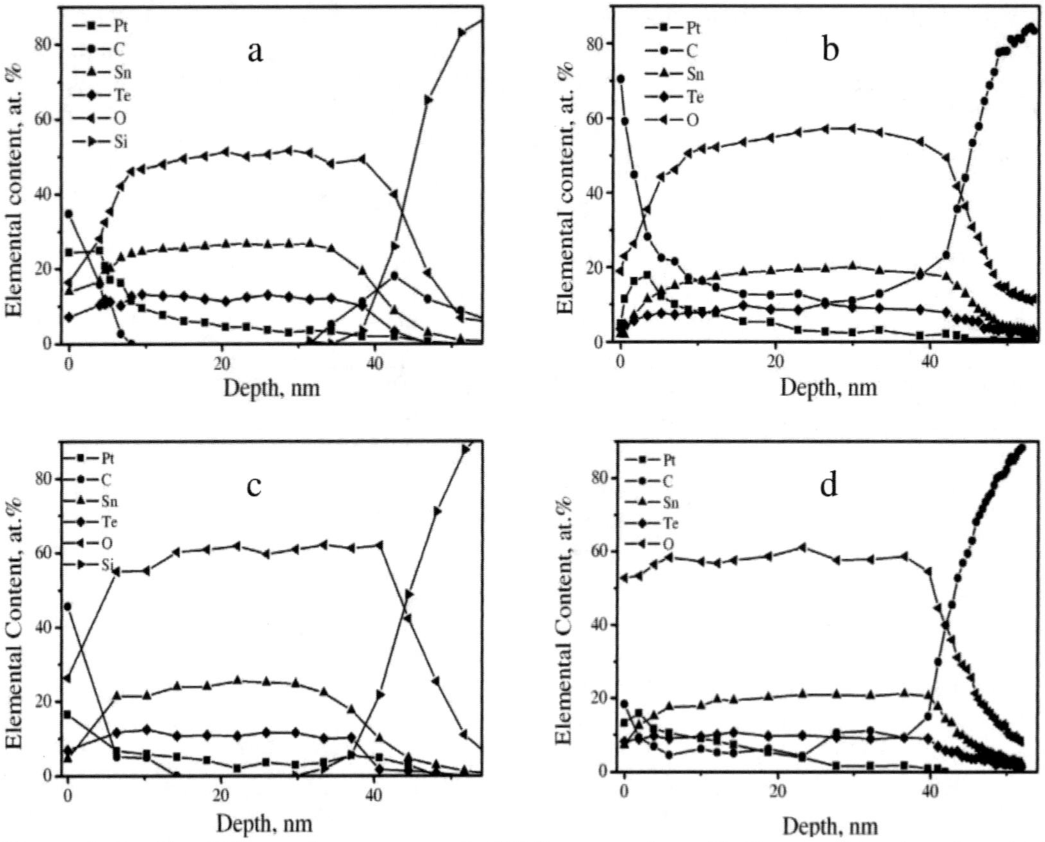

Figure 5. Profiles "Elemental content vs. depth" from samples with $R_{Sn/Te} \approx 2.1$: a) as-deposited amorphous Pt-doped film, Si substrate; b) film, doped with Pt after crystallization, graphite substrate; c) and d) films, doped with Pt before crystallization, Si and graphite substrates, respectively.

A second method was applied to evaluate the oxygen content relative to the content of the elements in the substrate. It makes use of the direct relation of the signal intensities of the element *X* from a sample and a standard with its corresponding concentration. Despite this relation reflects the essence of the Auger quantification, it is not used independently in the analytical practice, mainly because it does not take into account the inequality of the sample and the standard. If however two points of a profile are representing the sample and the standard, a big part of the problems is overcome. Additionally we have made the method more precise taking the corrections for the matrix differences. Relating the intensity of the substrate (C or Si) to that of oxygen the mean oxygen content (for the whole film) was determined. The results for the four samples differ from the corresponding values determined by means of the RESF-method on average by 10.6%. The small difference between the results, obtained using two independent methods, is indirect evidence that the quantification has been correctly carried out.

3.3. Conclusions from the Auger Analyses

The measured mean relative intensity Sn/O is approximately equal to that of SnO_2, i.e. the oxygen in the film is almost sufficient to oxidize Sn to SnO_2. This result could be expected due to the greater affinity of the oxygen towards Sn than towards Te [14]. So the amount of oxygen, which is not sufficient to oxidize the whole film, is predominantly bound to Sn. As Sn is present in the films mainly in the form of or close to SnO_2 it can be used as standard. On the other hand, the similarity between the standard and the analyzed composition allows us to substitute ESF for Sn in the film with that for SnO_2. Thus, the quantification of Sn was performed – for simplicity and with an acceptable error – and not by the curve in figure 4. The other results obtained can be summarized as follows:

I. All films with $R_{Sn/Te} \approx 2.1$ (amorphous and crystalline) show a uniform in-depth composition.

II. The level of film oxidation depends on the film preparation and increases upon thermal annealing in air. In the as-deposited amorphous films the oxygen content is sufficient to oxidize almost all Sn to SnO_2. The oxidation of Te begins (if the oxidation process continues) after all Sn is oxidized. The end result is the total oxidation of Sn and Te.

III. The Pt nanoparticles penetrate the full depth of the film during the film doping. The distribution of Pt is slightly inhomogeneous (its content decreases towards the film-substrate interface) but it is not affected by thermal annealing.

IV. The type of the substrate does not influence the composition and the doping of the film (provided that it does not react with the film); this has been shown for graphite and Si (as well as for Au).

V. The film analysis confirms the correctness of the evaporation process modeling [6].

VI. The important experimental result that thermally annealed Pt-doped thin Sn-O-Te films with $R_{Sn/Te} \approx 2.1\text{-}2.3$ possess significantly higher ethanol sensitivity if the doping precedes the annealing is not related to differences in the in-depth-distribution of Pt. Perhaps it is due to differences in the morphology of films crystallized in the absence or presence of Pt.

4. Summary

Auger quantification of the system Sn-Te-O and Pt is proposed. The elemental sensitivity of Sn in the metal and oxidized state is studied and for the analysis the non-stoichiometric compound SnO_x is modeled as a mixture of SnO_2 and Sn. The distribution of Te is supposed using the data from the SEM/EDS-analysis and the in-depth profiling.

Using Auger electron spectroscopy technique the distribution of the elements in as-deposited and thermally annealed films of one and the same composition ($R_{Sn/Te} \approx 2.1$) was studied. The depth profiling has revealed that the film is homogeneous and Pt is dispersed throughout the whole volume of the samples irrespective of their structure (amorphous or crystalline). Moreover, the in-depth distribution of Pt does not depend on the sequence of doping and thermal annealing.

Information is obtained about the oxidation processes taking place during the preparation and thermal annealing of the films, about the influence of the substrate material and the correctness of the evaporation process modeling. It has been shown that the significantly higher ethanol sensitivity of the films doped with Pt before the annealing is not related to the Pt-distribution.

References

[1] Yamazoe N, Sakai G, Shimanoe K 2003 *Catalyst Surveys from Asia* **7** 63
[2] Rumyanceva M, Safonova O, Boulova M, Ryabova L and Gashkov A 2003 *Russian Chem. Bulletin, International Edition* **52** 1217
[3] Podolesheva I, Dimov V, Tarassov M, Platikanova V and Konstantinov I 1998 *J. Vac. Sci. Technol.* **A 16**(4) 2608
[4] Georgieva B, Podolesheva I, Pirov J and Platikanova V 2005 *.J. Optoel. Adv. Mater.* **7** 2595
[5] Georgieva B, Podolesheva I, to be published
[6] Podolesheva I, Gushterova P, Platikanova and Konstantinov I 1998 *J. Vac. Sci. Technol.* **A 16**(2) 674
[7] de Kluizenaar E 1983 *J. Vac. Sci. Technol.* **A 1** 1480
[8] Ichimura S, Shimizu and Langeron J P 1983 *Surf. Sci.* **124** L49
[9] Seah M P and Dench W A 1979 *Surf. Interface Anal.* **1** 2
[10] Seah M P 1986 *Surf. Interface Anal.* **9** 85
[11] Frankenthal R P and Siconolf D J 1980 *J. Vac. Sci. Technol.* **17** 1315
[12] Powell R A 1979 *Appl. Surf. Sci.* **2** 397

[13] Draudy J, Hoflund G, Gardner S, Yngvadottir E and Schryer D 1990 *Surf. Interface Anal.* **16** 369
[14] *CRC Handbook of Chemistry and Physics*, 49th ed. (Chemical Rubber, Cleveland, 1969)

One dimensional photonic crystals from As_2S_3 and PMMA films for photonic and sensor applications

T Babeva[1], G Marinov[1], J Tasseva[1], A Lalova[1] and R Todorov[1]

[1]Institute of Optical Materials and Technologies "Academician Jordan Malinowski", Bulgarian Academy of Sciences, Acad. G. Bonchev Str., Bl. 109, 1113 Sofia, Bulgaria

E-mail: babeva@iomt.bas.bg

Abstract. One dimensional photonic crystal (1D-PhC) in form of 19 layered stack was prepared through layer-by-layer deposition of spin coated Poly (methyl methacrylate) (n = 1.49 at 1.55 μm) and vacuum deposited As_2S_3 (n = 2.27 at 1.55 μm) as low and high refractive index materials, respectively. The optical properties of the single layers and the multilayered stack are studied at normal and oblique incidence of linearly polarized light. The possibility of using the 1D-PhC as an omnidirectional reflector was exploited theoretically and confirmed experimentally. Besides, the prospect of using the studied structure for chemical sensing with optical read-out is demonstrated.

1. Introduction

One dimensional (1D) photonic crystals (PhC) are structures comprising two media with different refractive indices arranged in a 1D periodic manner. Among the numerous areas for their application the most attractive ones are in organic light emitting devices to enhance their white light emission [1], as omnidirectional reflectors in hollow optical fibers [2], dielectric reflectors and filters [3].

The widely accepted concept of 1D - PhC is a quarter-wave stack of alternating low- and high-refractive index layers (n_L and n_H) ($nd = \lambda/4$, n - refractive index, d - physical thickness, λ - operating wavelength) [4]. For a wave propagating normally to the stack (zero angle of incidence) 1D photonic band gap (i.e high reflectance band) exists that is shifted towards shorter wavelengths with increasing of the angle of the light incidence (oblique incidence). For application of 1D PhC as reflector in optical fibres an omnidirectional reflectance (ODR) is required that means reflectance of the stack to be close to 100 % for all incident angles and polarizations of light. It has been shown that if the optical contrast Δn ($=n_H - n_L$) and number of the layers are sufficiently high, an omnidirectional reflectance band would be open [5]. Additional requirements towards n_L and n_H values are placed for ODR band to be open: $n_H > 2.264$ [6], $\Delta n > 0.75$ [6] and a large value of the ratio n_L / n_0, (n_0 being the refractive index of the incident medium) [7]. Apart from the mentioned optical requirements for ODR band, it is very important the chemical and mechanical compatibility of the layers in the stack to be guaranteed in order to prevent multilayers from cracking and delaminating.

One good alternative to high temperature processed systems is the combination of chalcogenide glasses ($n_H > 2.2$ at 1550 nm) and polymer layer ($n_L \sim 1.5\text{-}1.7$) [8-10]. Quarter-wave stacks from

$As_{33}Se_{67}$ (n_H = 2.64), $Ge_{20}Se_{80}$ (n_H = 2.58) and $Ge_{25}Se_{75}$ (n_H = 2 .35) as high refractive index materials and polyamide–imide films (PAI) (n_L = 1.67), polystyrene (PS) (n_L = 1.53) and polyvinyl-butyral (PVB) (n_L = 1.47) as low refractive index materials were prepared by alternating vacuum deposition of chalcogenide thin films and spin coating of polymer films [8-10]. The stacks comprised between 15 and 17 layers and exhibited high reflection band in NIR (near infrared) region both for normally and obliquely incident s-polarized light.

It was demonstrated that broadening of ODR band could be achieved in $Ge_{33}As_{12}Se_{55}$ / polyamide-imide polymer stacks by photo- or thermo- doping of chalcogenide glass with silver resulting in increasing of the optical contrast due to higher refractive index of the doped chalcogenide glass [7].

It was shown that the deposition of a thin 50 nm Au film on the substrate under $Ge_{33}As_{12}Se_{55}$ (n_H= 2.55) / PAI (n_L= 1.65) stack provided higher angle-averaged reflectance for a low number of layers as compared to all-dielectric case [11]. Furthermore significantly increased omnidirectional bandwidth was observed compared to nonmetallized mirrors comprising the same materials [12].

Another possibility for achieving ODR band is combining two suitably chosen chalcogenide glasses, i.e fabrication of all-chalcogenide reflectors. Stacks from 15 alternating layers of Ge-S and Sb-Se [13] with optical contrast higher than 1 were fabricated by thermal evaporation. Further 17 layered stack from exposed As_2Se_3 / GeS_2 with optical contrast higher than 0.8 have been already realized [14].

Planar chalcogenide NIR filters (Fabry-Perot filters) were prepared as quarter wave stacks made of Ge–S / Sb–Se pairs surrounding low refractive index Ge-S spacer [15] and alternating As_2S_3 and $Ge_{15}As_{25}Se_{15}Te_{45}$ layers with As_2S_3 spacer layer [16, 17].

In this study we designed, fabricated and investigated one-dimensional photonic crystals from As_2S_3 and Poly (methyl methacrylate) (PMMA) film as high and low refractive index films deposited layer-by-layer using thermal evaporation and spin coating methods, respectively. The good optical contrast (~0.8), mutual compatibility and low optical losses are the main factors that have determined the particular choice of the materials. The possibility of using the 1D PhC as omnidirectional reflector was explored theoretically and confirmed experimentally. Besides, the prospect of using the studied structure for chemical sensing with optical read-out is discussed.

2. Experimental details

The stock solution of the polymer was prepared by dissolution at ambient temperature of one gram of PMMA (Poly (methyl methacrylate)) in 10 ml of Dichloroethane (Aldrich) using magnetic stirrer for accelerating the process. The polymer films with different thicknesses were obtained by the method of spin coating using the stock solution further diluted by adding dichloroethane. Polymer layers with thicknesses of 260 nm were obtained by pouring a drop of 0.5 ml of 2.45 w/v % polymer solution on the preliminarily cleaned substrate. The speed and duration of spinning were 2000 rpm and 30 s. To remove the extra solvent the samples were annealed for 30 minutes at temperature of 60 ^{o}C.

The bulk As_2S_3 glass was synthesized in a quartz ampoule by the method of melt quenching from elements of purity 99.999 % [18]. The layers were deposited by thermal evaporation at deposition rate of 0.5 – 0.7 nm/s in vacuum of 10^{-3} Pa. The X-ray microanalysis showed that the film composition is close to that of the bulk samples. The thickness was controlled in situ by quartz oscillator monitoring and ex situ by optical and profiler measurements.

Multilayered stacks comprising 19 layers were prepared by alternating vacuum evaporation of As_2S_3 with target thickness of 170 nm and spin coating of PMMA with target thickness of 260 nm. The target thicknesses of both layers were calculated aiming at their optical thicknesses (*nd*) to be a quarter of the operating wavelength (1.55 μm in our case).

Transmittance, *T* and reflectance, *R* spectra of single layers and multilayered stacks were measured at normal and oblique incidence of light with high precision UV-VIS-NIR spectrophotometer Cary 5E (errors in *T* and *R* at normal incidence are 0.3% and 0.5 %, respectively). For oblique measurements a high quality Glan-Taylor polarizer for obtaining linear polarization and an in-house made Al-mirror [19] as a reference mirror were used (error in *R* for oblique incidence is less than 1%).

3. Results and discussion

3.1. Characterization of a single layer

The knowledge of the optical parameters of the stack's building blocks is very important for a correct design of the stack. Figure 1 presents refractive index, *n* and extinction coefficient, *k* of As_2S_3 with thickness, *d* of 170 nm and PMMA layers with thickness of 260 nm. For determination of *n*, *k* and *d* the As_2S_3 layers were deposited simultaneously on two types of substrates - optical glass BK7 and Si-wafer. Optical constants and thickness were calculated from transmittance and reflectance measurements at normal light incidence using already developed calculating procedure [20]. PMMA polymer layers were deposited onto high refractive index substrate - flint glass with refractive index in the range 1.67 - 1.71 in order to guarantee high optical contrast between the substrate and the investigated layers. *n*, *k* and *d* of PMMA were determined through minimization of a function consisting of the discrepancies between measured and calculated *T* and *R* spectra using nonlinear curve fitting [21].

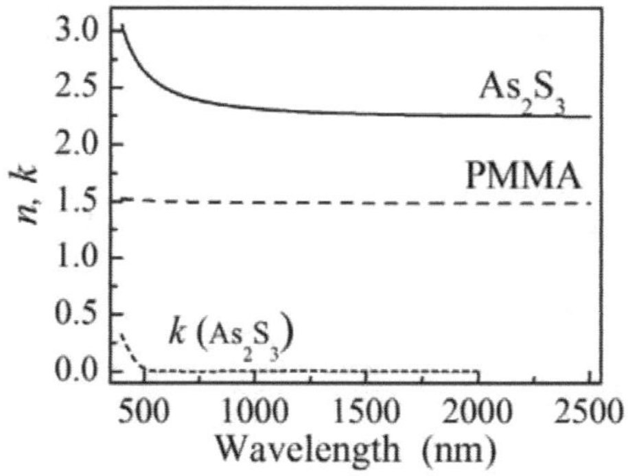

Figure 1. Dispersion curves of refractive index *n* and extinction coefficient *k* for As_2S_3 (170 nm) and PMMA (260 nm) films, calculated by the procedures described in the text.

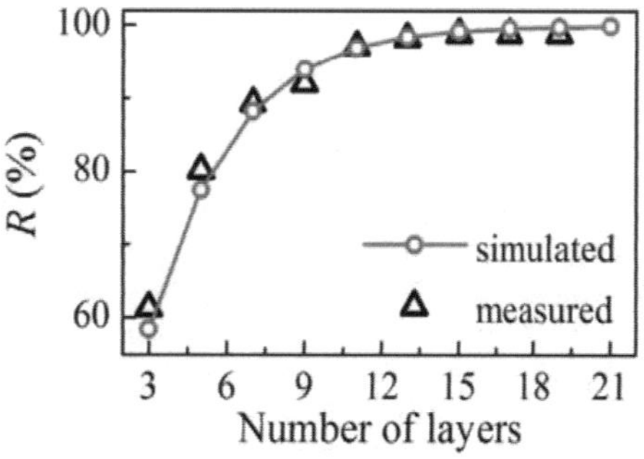

Figure 2. Maximum Reflectance (*R*) in the photonic band for normal light incidence as a function of the number of the layers in the quarter-wave stacks (n_H=2.27,n_L=1.49). Triangles present the experimental points

The refractive indices (figure 1) at the wavelength of 1550 nm are 2.27 and 1.49 for As_2S_3 and PMMA films, respectively (Δn = 0.78) that means that the two conditions for existing of ODR band are fulfilled. Both layers are transparent at 1550 nm - the calculated values for extinction coefficients of layers are insignificant and almost zero within the framework of the experimental errors.

3.2. Theoretical modelling of the optical performance of the stack

For design and optimization of the optical behavior of the stack we used the transfer matrix method. A very comprehensive description of the calculations can be found in [6, 22]. Figure 2 presents the simulated and measured values of maximum reflectance at normal light incidence in the stop band of the quarter wave stack with n_H = 2.27 and n_L = 1.49 as a function of the number of the layers in the stack. It is important to note that each stack in our calculations consists of odd number of layers: the first and last layers of the stack are from the high refractive index material. In this case the reflectance is higher as compared to the case when the stack is terminated with low refractive index material.

It is seen from figure 2 that the increase of R is mostly pronounced until the deposition of the 15th layer. The deposition of next layers leads to a slight rise. Very good agreement between the calculated and measured values was obtained.

3.3. Characterization of multilayered stack

Figure 3 (a) presents the positions (in nm) of the reflectance edges of the photonic bands measured at a level R = 80 % as a function of the incident angle for p- and s-polarization. The ODR band is defined between the short reflectance edge at 0° and long reflection edge at 90° for p-polarization (see for example [6]). It is seen that for angles smaller than or equal to 70° a quasi ODR band centered at wavelength of 1396 nm with width of 48 nm exists in our case. It is located between the short R edge of 0° and long R_p edge for 70°. This means that the quarter-wave stack of alternating As_2S_3 and PMMA layers exhibits R > 80 % for incident angle range 0 - 70° and all polarizations in the spectral range 1372-1420 nm. Theoretical calculations have shown that ODR band with width of 2 nm is opened at wavelength of 1369 nm.

Figure 3 (b), (c) and (d) present the measured spectra at 0°, 50° and 70° for both polarizations. The widening of s- and narrowing of p-reflection band as well as their shifts toward smaller wavelengths with the increasing of the incidence angle are clearly seen.

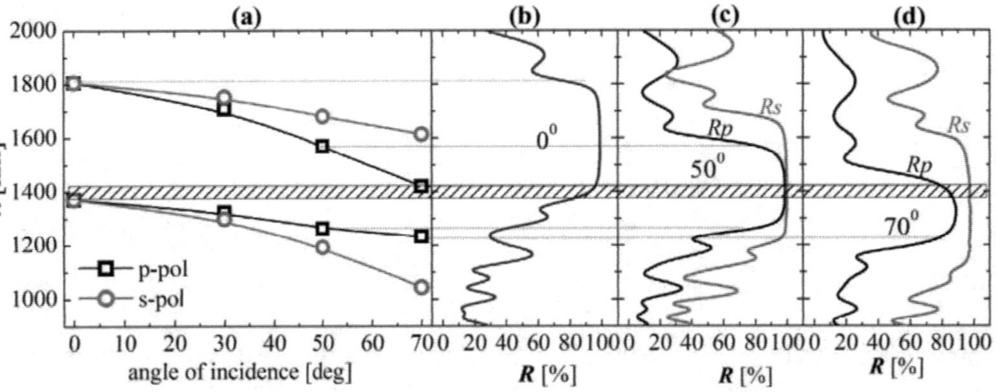

Figure 3. a) Measured reflection band edges at a level R = 80 % for p- and s-polarization as a function of the incident angle. Measured reflectance of stack comprising 19 alternating layers of As_2S_3 and PMMA as a function of wavelength for p- and s-polarization at incident angles of b) 0°, c) 50° and d) 70°. The shaded area represents the quasi-ODR band for the angular range 0° - 70°

It is worth noting that the position of ODR band can be tuned simply by changing the thicknesses of the layers building the stack. Widening of the band can be achieved by increasing the optical contrast between the layers through selecting chalcogenide glass with a higher refractive index (GeSbSe for example [23]) or using the well known photo- and thermo-induced phenomena in chalcogenide glassy materials [14, 15, 24]. For example the photodarkening after illumination by light or the photodoping with silver increases the refractive index of As_2S_3 layer with 0.12 [24] and 0.25 [25], respectively.

3.4. Sensing experiments

Considering that PMMA swells when exposed to organic vapors [26, 27] we expected that this would lead to a shift of the photonic band gap of the stack. For the sensing experiments we prepared stacks with a 'defect' layer in the middle illuminated for 1 h (luminescence lamp, 200 W) in order to guarantee stable layers insensitive to further illumination. The defect layer is a PMMA film with doubled thickness (520 nm) which results in a transmission peak in the stop band (figure 4). We performed initial sensing experiments by exposing the fabricated stack to the vapors of water, ammonia and chloroform. The sensing experiments are conducted by putting the sample above the analyte vapor originated from evaporation of the liquid of chloroform or ammonia analyte at ambient temperature and pressure or water at 100°C. The transmittance spectra are measured prior to and after the exposure to the analyte. Special attention is paid on conducting the measurements at one and the same spot.

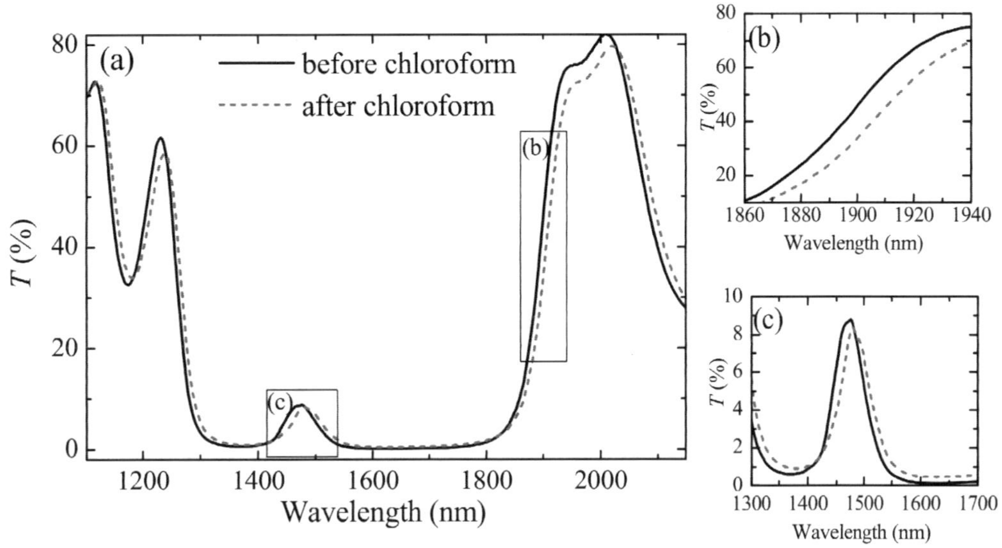

Figure 4. a) Transmittance spectra of As_2S_3 / PMMA stack before and after exposure to chloroform vapors; b) and c) selected magnified regions of the spectra denoted on the figure

Figure 4 presents the T spectra before and after chloroform exposure. It is seen that the chloroform vapors bring forth a shift of the band toward longer wavelength (about 10 nm) and consequent decrease of transmittance with value of 10 % at a wavelength of 1906 nm that is an easily detectable difference. Exposure to water and ammonia vapors provokes no changes in T spectra.

These sensing experiments aimed at showing a possible application of the fabricated stack as a chemical sensor with optical read-out. We are aware of the fact that although the initial results are promising, a lot of work is still to be done. The scientific efforts will be focused on enhancing the porosity of the layers enabling the analyte to penetrate easily into the multilayered structure. Besides the photonic band gap shift at different temperature and vapor pressure will be study.

4. Conclusions

The design, fabrication and characterization of one-dimensional photonic crystal (1D-PhC) are presented. The 1D-PhC consists of 19 quarter-wavelength layers of alternating As_2S_3 ($n_H = 2.27$ at 1.55 μm) and Poly (methyl methacrylate) ($n_L = 1.49$ at 1.55 μm) films as high and low refractive index layers deposited by thermal evaporation and spin-coating, respectively. The optical parameters derived from single layers characterization were used for theoretical modeling of the stacks performance. It was experimentally demonstrated that a 'quasi' ODR band was open for incident angles up to 70° with the central wavelength of 1396 nm and width of 48 nm.

A 19-layer quarter-wave stack from As_2S_3 and PMMA with a defect layer of PMMA with thickness of 520 nm in the middle was fabricated. The initial sensing experiments, conducted by exposure of the stack to the vapors of water, ammonia and chloroform showed that As_2S_3 / PMMA stack could be used for selective detection of chloroform. A shift of the transmittance band of about 10 nm toward longer wavelengths and consequent decrease of *T* with 10 % at wavelength of 1906 nm are observed due to exposure to chloroform. No changes are detected for water and ammonia vapors. Further experiments focused on enhancing the porosity of stack's building blocks are needed.

Acknowledgement
T. Babeva is grateful to Prof. S. Sainov from IOMT-BAS for supplying the PMMA material. The financial support of the Project DO2-723/2.08.2012 (under the project BG051PO001/3.3-05-001 "Science and Business") is highly appreciated.

References

[1] Park B, Kim Mi-Na, Kim S and Park J 2008 *Opt. Express* **16** 14524-31.
[2] Chiu H, Hsiao Fu-Li, Chan C H and Chen C C 2008 *Opt. Express* **16** 15069-73.
[3] Kohoutek T, Orava J, Prikryl J, Wagner T and Frumar M 2011 *J. Non-Cryst. Sol.* **357** 157–160.
[4] Joanopoulus J D, Meade R D and Winn J N 1995*Photonic crystals: Molding the Flow of Light* (Princeton, N.J.: Princeton University Press,).
[5] Fink Y, Winn J N, Fan S, Chen C, Michel J, Joannopoulos J D and Thomas E L 1998 *Science* **282** 1679-82.
[6] Kim S H and Hwangbo C K 2002 *Appl. Opt.* **41**, 3187-92
[7] Clement T J, Ponnampalam N, Nguyen H T and DeCorby R G 2006 *Opt. Express* **14**, 1789-96
[8] Kohoutek T, Wagner T, Orava J, Krbal M, Ilavsky J, Vlcek Mil and Frumar M 2008 *J. Non-Cryst. Sol.* **354,** 529–532.
[9] Kohoutek T, Orava J, Hrdlicka M, Wagner T, Vlcek Mil and Frumar M 2007 *J. Phys. Chem. Sol.* **68** 2376-80.
[10] Kohoutek T, Wagner T, Orava J, Krbal M, Ilavsky J, Vesely D and Frumar M 2007 *J. Phys. Chem. Sol.* **68** 1268-71.
[11] Ponnampalam N and DeCorby R G 2008 *Appl. Opt.* **47** 30-37.
[12] DeCorby R G, Nguyen H T, Dwivedi P K and Clement T J 2005 *Opt. Express* **13** 6228-33.
[13] Kohoutek T, Orava J, Wagner T, Hrdlicka M, Vlcek Mil, Frumar M 2008 *J. Phys. Chem. Sol.* **69** 2070–4.
[14] Todorov R, Tasseva J, Babeva T and Petkov K 2010 *J. Phys. D: Appl. Phys.* **43** art. 505103.
[15] Kohoutek T, Orava J, Prikryl J, Wagner T, Vlcek Mil, Knotek P and Frumar M 2009 *J. Non-Cryst. Sol.* **355** 1521-5.
[16] Kondakci H E, Yaman M, Koylu O, Dana A and Bayindir M 2009 *Appl. Phys. Lett.* **94**, 111110 - 111110 -3.
[17] Kondakci H E, Yaman M, Koylu O, Dana A and Bayindir M 2010 *Appl. Opt.* **49** 3596 -3600.
[18] Tasseva J, Todorov R, Tzankov D and Petkov K 2007 *J. Opt. Adv. Mat.* **9** 344-7.
[19] Babeva T, Kitova S, Mednikarov B and Konstantinov I 2002 *Appl. Opt.* **41** 3840-6.
[20] Konstantinov I, Babeva T and Kitova S 1998 *Appl. Opt.* **37** 4260-7.
[21] Babeva T, Todorov R, Mintova S, Yovcheva T, Naydenova I and Toal V 2009 *J. Opt. A: Pure Appl. Opt.* **11** 024015 (8pp).
[22] Todorov R, Tasseva J and Babeva T 2011 "Thin Chalcogenide Films For Photonic Applications" in: *Photonic Crystals / Book 2*, A. Massaro Ed., (Rijeka, Croatia: INTECH).
[23] Dikova J, Todorov R and Babeva T 2012 *J. Physics: Conf. Series* **356** art. no. 012023.
[24] Tasseva J, Todorov R and Petkov K 2009 *J. Opt. Adv. Mat.* **11** 1257-60.
[25] Tasseva J, Todorov R, Babeva T and Petkov K 2010 *J. Optics* **12**, 065601.
[26] Capan I, Tarimci C, Hassan A. K. and Tanrisever T 2009 *Mat. Sci. Eng. C* **29**, 140-3.
[27] Capan I, Tarimci C, Erdogan M and Hassan A. K. 2009 *Mat. Sci. Eng. C* **29**, 1114-17.

Tunable Bragg stacks from sol-gel derived Ta_2O_5 and MEL zeolite films

B. Gospodinov[1], J. Dikova[1], S. Mintova[2] and T. Babeva[1]

[1] Institute of Optical Materials and Technologies "Acad. J. Malinowski", Bulgarian Academy of Sciences, Acad. G. Bonchev str., bl. 109, 1113 Sofia, Bulgaria.
[2] Laboratoire Catalyse & Spectrochimie, Université de Caen, 6, boulevard du Maréchal Juin, 14050 Caen Cedex, France.

E-mail address: babeva@iomt.bas.bg

Abstract. In this paper we investigated sol-gel derived Ta_2O_5 and nanosized MEL zeolite films obtained by spin coating of Tantalum sol and colloidal zeolite solution, respectively. Refractive index and thickness of the films were determined using non-linear curve fitting of measured reflectance spectra. The influence of the post deposition annealing on the optical properties and thickness of the films was studied. Besides tunable Bragg stacks were designed and prepared by layer-by-layer deposition of Ta_2O_5 and MEL suspensions with quarter-wave thicknesses. The influence of water, acetone and methanol on the optical behavior of Bragg stacks was discussed.

1. Introduction

Bragg stacks are multilayered systems comprising high and low refractive index materials with quarter-wave optical thicknesses: $nd = \lambda_c/4$, n refractive index, d-physical thickness, λ_c - wavelength of incident light, arranged in periodic manner. When Bragg stack is illuminated with light, multiple reflected waves occur at each boundary that are in phase and interfere constructively, which is resulting in appearance of high reflectance band centered at λ_c. Because the position of the reflectance band is extremely sensitive to the changes in refractive index and thickness of the constituent layers it is possible the position of reflectance band to be controlled upon the influence of external stimuli. In this way tunable Bragg stacks, very attractive for sensing applications have been fabricated [1,2].

Ta_2O_5 is a typical high refractive index semiconductor with an advantage over TiO_2 of UV transparency. Among various deposition techniques used for production of thin films (layers) from Ta_2O_5, the sol-gel method attracts considerable scientific interest because of its versatility, low cost and low temperature processing. Besides, it allows control of the microstructure of the coating and produces durable and chemically stable films [3,4]. Additionally, thick films can be easy produced using sol-gel synthesis approach [5,6].

The most widely used starting material for sol-gel preparation of Ta_2O_5 is tantalum ethoxide ($Ta(OC_2H_5)_5$). Usually it is hydrolyzed in parent alcohol in the presence of catalyst, most often acid in order to form inorganic oxide network [7]. Because of the high sensitivity of $Ta(OC_2H_5)_5$ to moisture the process of obtaining clear sol is not trivial. It is well known that the rate of hydrolysis and condensation are very fast, causing complete gelation in few hours [8]. Therefore the sol stabilization is crucial for obtaining clear sols and good quality films. This was achieved by addition of a complex ligand - methacrilic acid, [9] or most often glacial acetic acid was used [3,10]. Both acids act as chelating agents and decrease the reactivity of the employed metal alkoxide by formation of hybrid complexes in which the number of reactive alkoxide groups decrease [9].

In general, the porous structure of sol-gel films has smaller refractive index values - *n*, in comparison to films obtained by other methods [11]. Besides, the heat treatment leads to an increase of the *n* value [3,10,12].

Nanosized zeolite films are appropriate for low refractive index counterpart of tunable Bragg stacks. Zeolites are a unique class of industrially significant materials with tunable physicochemical properties [13]. Their great utility in standard and advanced applications in diverse media is due to their large internal surface area, i.e., the presence of voids of controllable dimensions at the atomic, molecular, and nanometer scales [14]. The possibility to control their composition, pore shape and size, hydrophobicity, hydrophilicity and overall particle size and morphology are attractive advantages for using the zeolites in preparation of tunable Bragg stacks. Besides, the zeolites can be prepared with nanosized dimensions [15] and thus will act as optical transducers of physico-chemical processes that occur within the micropores and interparticle mesopores. The zeolite-analyte interactions such as sorption/desorption, ion exchange, diffusion etc. will cause changes of the refractive index and thickness of the zeolite films and subsequent shift of the reflectance band of Bragg stack is expected.

In this paper we investigate sol-gel derived Ta_2O_5 and nano-sized MEL zeolite films obtained by spin coating of tantalum sol and colloidal zeolite suspension, respectively. The water free sol-gel route for obtaining clear and stable sol of Ta_2O_5 is developed. The influence of the post deposition annealing on the optical properties of the films is studied. Besides tunable Bragg stacks were designed by layer-by-layer deposition of Ta_2O_5 and MEL. The influence of water, acetone and methanol on the optical behavior of the Bragg stacks is discussed.

2. Experimental details

The coating solution (tantalum sol) was prepared using tantalum ethoxide $Ta(OC_2H_5)_5$ (99.98 %, Sigma-Aldrich) as a chemical precursor, isopropyl alcohol (Merck) as solvent, glacial acetic acid CH3COOH (Sigma-Aldrich) as catalyst and chelating agent, and diethanolamine DEA (Merck) as a surfactant to prevent sediment impurities in the solution. Although the synthesis procedure for the Ta coating solution used in this paper is similar to that of Tepehan [10], three distinctive changes are made: (i) water free, (ii) isopropyl alcohol used instead of ethanol, and iii) small amount of DEA added.

The solution A was prepared by mixing of 35 ml of isopropyl alcohol with 1 ml of glacial acetic acid and then 1.5 ml $Ta(OC_2H_5)_5$ is slowly drop wise added. The solution B was prepared by mixing 2 ml glacial acetic acid with 15 ml isopropyl alcohol. After 30 min stirring both solutions were mixed and then 1 ml DEA was added. The final mixture was transparent and colourless with pH of about 5. The obtained solution is subjected to slow stirring for 18 h. The solution was very stable and can be kept at ambient temperature for extended time without precipitates. There was no evidence of gelation after 30 days.

The Ta_2O_5 films were deposited by dropping of 0.2 ml of the coating solution on pre-cleaned Si wafer as substrates and then spin at a rate of 2000 rpm. The Si wafers were cleaned in a detergent solution, washed with isopropanol and then dried at room temperature. Prior to the deposition of the coating solution it was filtered with 0.2 μm pore-sized filter. After deposition, the films were annealed at different temperatures in the range 90-650 °C. The Ta_2O_5 film thickness after annealing at 500° C for 30 min was 40 nm. The coating procedure was repeated several times in order to obtain thicker films. The intermediate films were dried at 60° C to increase their mechanical stability.

Nanosized MEL zeolite crystals were synthesized from a colloidal precursor suspension with the following chemical composition: SiO_2 : 0.3TBAOH (tetrabutylammonium hydroxide) : 4.0EtOH (ethanol) : $18.0H_2O$, at 90 °C for 68 h [16]. After purification the nanocrystals were redispersed in ethanol to obtain a suspension with a solid concentration of 5 wt. %. The zeolite ethanol suspensions were used for preparation of zeolite films by spin coating method: 0.3 ml of zeolite suspension with desired concentrations was dropped onto the Si substrate. The speed and duration of spin-rotation were varied between 1000 to 2000 min^{-1} for 30-90 s. The zeolite films were annealed at 500 °C for 30 minutes with a heating rate of 10°C / min. The zeolite films under investigation had a thickness in the range 100 - 300 nm.

Bragg stacks with 5 layers were prepared by alternative deposition of Ta_2O_5 and MEL zeolite films with target thicknesses of 72 and 110 nm, respectively on the Si substrate. The target thicknesses of both films were calculated aiming at their optical thicknesses - the product of physical thickness and refractive index, to be a quarter of the operating wavelength - 520 nm in our case. In order to achieve the correct thicknesses of the films, the concentration of the zeolite suspension was optimized by addition of appropriate amount of solvent - ethanol. The Ta-sol was used as prepared but two deposition steps were required with intermediate drying of layers with hot air. An annealing at 320°C for 15 min was applied after deposition of each film. Finally the stack was subjected to calcination at 500 °C for 45 min.

3. Results and discussions

Refractive index *n*, extinction coefficient *k*, and thickness *d*, of the films were determined from reflectance measurement at normal incidence in the spectral range 400 – 1000 nm using Wemple - DiDomenico dispersion equations for *n* [17] and exponential decay function for *k* according to the following:

$$n(E) = \sqrt{1 + \frac{E_0 E_d}{E_0^2 - E^2}}, \qquad k(E) = a_0 \exp(\frac{E}{A}), \tag{1}$$

where E is the light energy in the measured spectral range. The coefficients E_0, E_d, a_0 and A in eq. 1 along with d were determined through minimization of a goal function G (eq. 2) consisting of the discrepancies between measured R_{meas} and calculated R_{calc} reflectance spectra. A non-linear subspace trust region method combining the interior-reflective Newton method with a preconditioned conjugate gradients method [18] was used for minimization:

$$G(E_d, E_0, a_0, A, d) = \sum_i \{R_{meas} - R_{calc}\}^2, \tag{2}$$

where i is the number of the points in the used spectral range. For accurate and unambiguous minimization, proper initial values for the unknown parameters are required. Because such information is not available we used the approach explained in details elsewhere [19]. Briefly, the minimization procedure was run using a wide grid of initial values for the unknown parameters and the error function *Err* of the minimization was calculated as the residual value of the goal function at each solution:

$$Err = \sqrt{\sum_i |G(E_d, E_0, a_0, A, d)|^2}. \tag{3}$$

The values of E_0, E_d, a_0, A and d in the global minimum of *Err* were used as initial values in the next minimization step and the calculation algorithm is continuously repeated until the value of *Err* at the minimum did not change any longer. The *n*, *k* and *d* of film were obtained by substitution of the final values of E_0, E_d, a_0, A and d in eq.1.

Journal of Physics: Conference Series **398** (2012) 012026 doi:10.1088/1742-6596/398/1/012026

Figure 1 presents refractive index and thickness of films from MEL zeolites and sol-gel derived Ta_2O_5 as a function of temperature annealing. The values are averaged over 5 samples and the error bars present the deviations from the average values. The increase of *n* for Ta_2O_5 and decrease for MEL films with annealing are clearly seen. The removing of residual solvent and organic additives as acetic acid and DEA leads to densification of the Ta_2O_5 films and formation of inorganic oxide network manifesting itself in fast decrease of thickness and increase of the refractive index. Annealing at 500 ^oC for 30 min is sufficient to produce stable films. Further annealing at 650^oC does not lead to significant changes of *n* and *d*. The values of *n* in our case are lower as compared to those obtained in [12,20], which can be explained with different thickness and increased porosity of the films. In the case of MEL film the pores empty during calcination and thus the refractive index decreases. Simultaneously a small collapse of the film volume is observed as well.

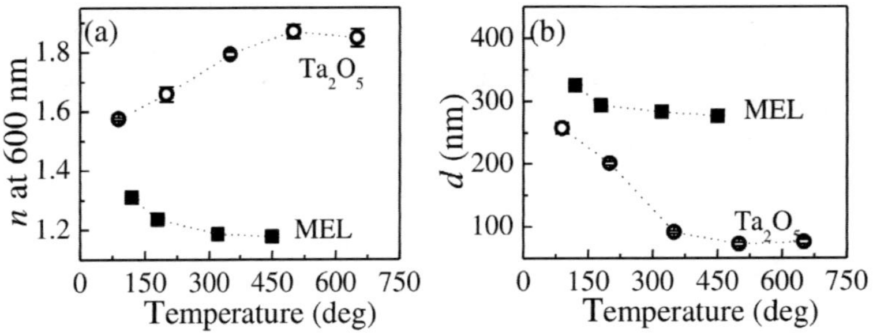

Figure 1. Dependence of refractive index (a) and thickness (b) of the films as a function of annealing temperatures.

The good optical contrast i.e the difference of refractive indices between the films - about 0.7, is sufficient for obtaining high reflectance Bragg stacks using comparatively low number of films. Simultaneously very good linear dependence of MEL thicknesses on the concentration is observed (not shown in the paper) that enable us to choose precisely the appropriate suspensions concentration to obtain the correct quarter wave thickness of the MEL films for the stack.

The sensing experiments were performed by soaking the samples into specific liquids - water, methanol and acetone, for 180 s and the reflectance spectra of the films prior and after soaking were measured. Figure 2 presents the reflectance spectra *R* of single Ta_2O_5 and MEL films and Bragg stack consisting of 5 alternating films of Ta_2O_5 - d = 72 nm and MEL - d = 102 nm, before and after exposure to acetone, methanol and water liquid. The highest changes of *R* obtained for the films are shown in figure 3. It is seen that for all liquids the changes of *R* for single films are smaller as compared to the changes of *R* for the Bragg stack. The change of reflectance is associated with increase of the effective optical thickness - *nd*, which could be mainly due to the increase of the effective refractive index. The liquid penetrates into the pores of the films and substitutes the air that occupies the pores. Considering that the liquids have refractive index higher than the air the increase of effective refractive index could be expected.

The measured reflectance spectra presented on figure 2 were used to calculate the optical thickness of the films. An increase of 1.25%, 2.4% and 3.06% was observed for Ta_2O_5 and 3.4%, 0.9% and 0.1% for MEL exposed to acetone, methanol and water respectively. The films of Ta_2O_5 are more sensitive to water and less sensitive to acetone while the opposite trend is observed for the MEL zeolite film.

Recently, we have shown that the strength of the optical response of MEL zeolites depends on the interplay between hydrophobicity, polarity, surface tension, refractive indices and molecular size of the liquids under investigation [21]. Thus, due to the high hydrophobicity the optical response of MEL zeolite in the case of water exposure is very weak, while the hydrophilic nature of Ta_2O_5 is the most probable reason for the increased optical response toward water. The smaller polarity of methanol and especially acetone leads to weaker response of the Ta_2O_5 film. Considering that acetone has the highest refractive index among the studied liquids, the strongest response of the MEL film towards acetone may be expected [21].

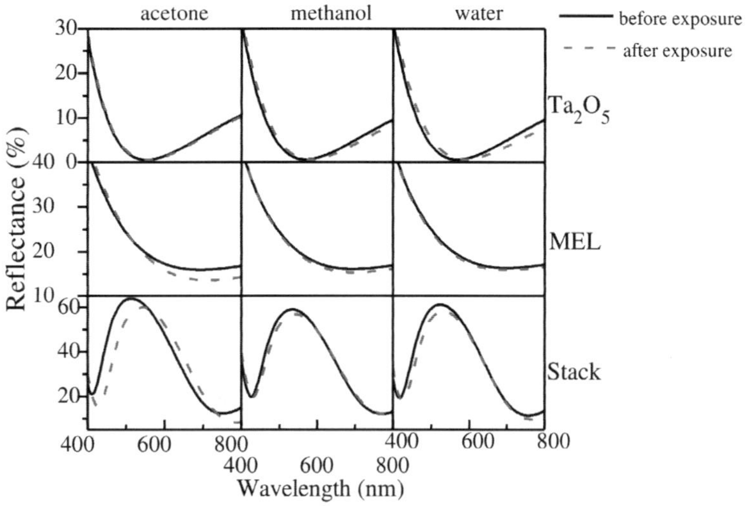

Figure 2. Reflectance of Ta_2O_5 and MEL single films and Bragg stack consisting of 5 alternating Ta_2O_5 and MEL films before (solid line) and after (dashed line) soaking in denoted liquid analytes for 180 s.

In the case of the Bragg stack, the acetone exposure leads to shift of the reflectance band with 30 nm towards longer wavelengths, while methanol and water result in considerably smaller shift of 6 and 8 nm, respectively. Due to the band shift a decrease of the stack reflectance with 20 % is observed in the case of acetone and 5 % and 8 % for methanol and water, respectively.

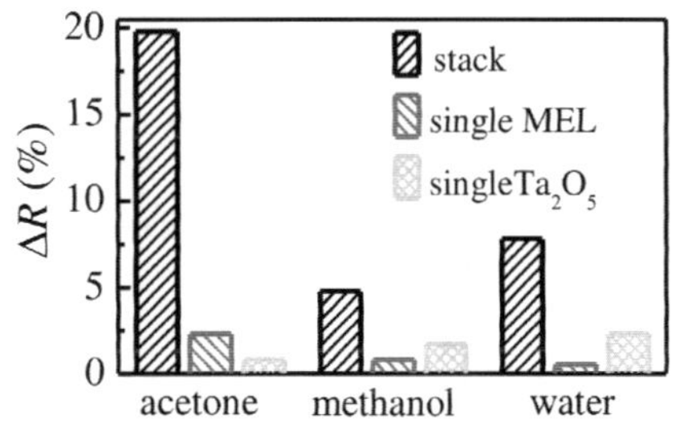

Figure 3. The changes of reflectance of Bragg stack and single films under exposure to acetone, methanol and water liquids.

The selective change in R of the Bragg stack when exposed to acetone is clearly seen in figure 3.

The contribution of Ta_2O_5 to the overall methanol and water response is stronger as compared to the pure MEL zeolite film. As already mentioned due to the more hydrophilic nature of Ta_2O_5 as compared to the MEL zeolite, methanol and especially water penetrates more easily through and thus changes the effective optical thickness which is resulting in the reflectance change.

4. Conclusions

In this paper we investigated sol-gel derived Ta_2O_5 and nanozeolite MEL films obtained by spin coating of sol solution and zeolite suspension, respectively. It is demonstrated that clear and stable Ta sol can be obtained by water free sol-gel synthesis approach with addition of diethanolamine as stabilizer and acetic acid as chelating agent. Optical characterization has shown that the annealing at 500 °C is sufficient to obtain stable and high refractive index Ta_2O_5 films.

Tunable Bragg stacks consisting of five sol-gel derived Ta_2O_5 films and nanosized MEL zeolite films are fabricated by alternative spin coating deposition with quarter-wave thickness at 520 nm. A selective shift of reflectance band with 30 nm towards longer wavelength is observed when the Bragg stack is exposed to liquid acetone. In the case of methanol and water the shifts are 6 and 8 nm, respectively. Due to acetone exposure a substantial change of reflectance of the Braag stack with almost 20 % is observed that is important prerequisite for application of the Ta_2O_5 / MEL stacks as sensors with optical encoding.

Acknowledgement

The authors are very grateful to Prof. V. Bluskov from the Institute of General and Inorganic Chemistry of Bulgarian Academy of Sciences and Prof. F. Z Tepehan from Istanbul Technical University, Turkey for fruitful discussion and advices in preparation of Tantalum sol.

References

[1] Colodrero S, Ocana M and Miguez H 2008 *Langmuir* **24**, 4430-39
[2] Choi S, Mamak M, Freymann G, Chopra N and Ozin G 2006 *Nano Letters* **6** 2456-61
[3] Ozer N and Lampert C 1997 *J of Sol-Gel Sci and Tech* **8** 703-9
[4] Sun Y, Vong M and Sermon P 1997 *J of Sol-Gel Sci and Tech* **8** 493–7
[5] Ndiege N, Wilhoite T, Subramanian V, Shannon M and Masel R 2007 *Chem. Mater.* **19** 3155-3161
[6] Subramanian V, Ndiege N, Seebauer E, Shannon M and Masel R 2008 *Thin Solid Films* **516** 4784–92
[7] Sun Y, Sermon P and Vong M 1996 *Thin Solid Films* **278** 135-9
[8] Ling H, Yan M and Rhodes W 1989 *Proceeding Euro-Ceramic '89* Vol. 1 p. 285.
[9] Oubaha M, Elmaghrum S, Copperwhite R, Corcoran B, McDonagh C and Gorin A 2012 *Optical Materials* **34** 1366–70
[10] Tepehan F, Ghodsi F, Ozer N and Tepehan G 1999 *Sol. Energy Mat.&Sol. Cells* **59** 265-75
[11] Ghodsi F, Tepehan F and Tepehan G 1997 *Thin Solid Films* **295** 11-15
[12] Rehg T, Ochoa-Tapia J, Knoesen A and Higgins B 1989 *Appl. Opt.* **28** 5215 -21
[13] Mintova S, Valtchev V, Onfroy T, Marichal C, et. al 2006, *Micropor. Mesopor. Mater* **90** 237.
[14] Martínez C and Corma A 2011**,** *Coord. Chem. Rev.* **255** 1558.
[15] Ng E.-P and Mintova S 2008 *Micropor. Mesopor. Mater* **114** 1.
[16] Mintova S, Petkov N, Karaghiosoff K and Bein T 2001 *Micropor. Mesopor. Mater* **50** 121-128
[17] Wemple S and DiDomenico M 1971, *Phys. Rev. B* **3** 1338-51.
[18] Coleman T and Li Y 1996 *SIAM Journal on Optimization* **6** 418
[19] Babeva T, Todorov R, Mintova S, Yovcheva T, Naydenova I and Toal V 2009 *J. Opt. A: Pure Appl. Opt.* **11** 024015.
[20] Ghodsi F E and Tepehan F Z 1999 *Sol. Energy Mat. & Sol. Cells* **59** 367-375
[21] Babeva T, Todorov R, Gospodinov B, Malinowski N, Fallah J El and Mintova S 2012 *J. Mater. Chem.* **22** 18136-8.

Erythrocyte dynamics in flow affects blood rheology

V Vitkova[1,4], A Farutin[2], B Polack[3], C Misbah[2] and T Podgorski[2]
[1] G. Nadjakov Institute of Solid State Physics, Bulgarian Academy of Sciences, Sofia, Bulgaria
[2] Laboratoire Interdisciplinaire de Physique – UMR 5588, CNRS – Université Joseph Fourier, Grenoble I, BP 87, 38402 Saint Martin d'Hères, France
[3] TIMC (Techniques de l'Ingénierie Médicale et de la Complexité) /TheREx/ CNRS-UJF UMR 5525, INSERM IFR, 130 CHU de Grenoble, France

E-mail: victoria@issp.bas.bg

Abstract. Normal blood consists of highly deformable particles (red blood cells, RBC, or erythrocytes) suspended in a Newtonian fluid (blood plasma). As a rough physical model of erythrocytes, giant unilamellar vesicles (GUV) are successfully used to probe their membrane properties. In shear flows vesicles and red blood cells show rich variety of dynamical behaviours influencing the rheological properties of their suspensions. Here, we focus on new experimental aspects of the problem in the case, when a combination of an oscillatory shear rate and a basic constant shear rate is applied to suspensions. Experimental examples with concentrated RBC suspensions are presented together with a discussion on the importance of the superposition of a constant shear flow to the pure oscillation, which is usually used to extract the viscoelastic properties of a complex fluid.

1. Introduction

Representing a suspension of formed elements (red blood cells (RBC) or erythrocytes, white blood cells (leukocytes), and platelets) in aqueous solution of proteins and salts (plasma), blood is a prominent example of a complex fluid, whose physical properties have been in the scope of interest for researchers from diverse scientific fields [1-4]. The RBCs consist of a thin elastic membrane filled with a hemoglobin solution that is a Newtonian fluid with viscosity about 10 mPa.s [5, 6]. Besides the transport of oxygen and carbon dioxide, another important property of intracellular hemoglobin is its contribution to internal viscosity, which determines the rheological properties of erythrocytes [7]. The other particulate constituents of blood, white blood cells and platelets usually occupy only less than 1% of the total volume of blood cells in normal human blood, thus, the RBCs govern the flow properties of blood [3, 8]. RBCs are readily deformable and it allows them to pass through the smallest capillaries [2, 9]. Their ability to deform in flow is an essential feature [10] determined by the viscosity of the carrier fluid, its osmolarity (affecting the viscosity of the hemoglobin solution), the shear rate of the flow as well as of biological factors such as in-vivo aging for example [11].

Erythrocyte membranes as all biomembranes are characterized by a “fluid mosaic” structure [12] unifying in a perfectly functioning dynamical ensemble its lipidic and proteinic constituents with functionally important lipid asymmetry between the two membrane leaflets [13, 14]. The presence of a cytoskeleton attached to the inner surface of RBC membranes makes them resistant to shear

[4] To whom any correspondence should be addressed.

deformations, thus inducing additional dynamical regimes of cells in flow [15, 16]. Thanks to the biphilic nature of lipid molecules the lipid matrix of biomembranes is easily reproducible in laboratory conditions and closed structures, giant vesicles, are readily formed via elaborated experimental protocols with controlled physicochemical parameters [17-19].

So far, it has been revealed both theoretically and experimentally that in shear flows, vesicles and red blood cells show a rich variety of dynamical behaviours having a signature on the rheological properties of their suspensions [20, 21]. Several types of motion in linear hydrodynamic fields have been predicted and observed for vesicles and RBC, namely *tank-treading* (*TT*), *tumbling* (*TB*) (or flipping), *vacillating-breathing* (*VB*) (sometimes called also "trembling" or "swinging") for vesicles and *swinging*, for RBC, and *spinning* or *kayaking*, where the main axis describes a cone about the perpendicular to the plane of the shear flow (see [22, 23] and the references therein).

The elasticity of the spectrin network leads to specific features of red cells' dynamics in flow. As it has been predicted [15] (under the assumption of fixed cell shape) and observed [16], in very viscous solvent and depending on the applied shear stress, erythrocytes can execute several types of motion. At low shear stresses (less than 0.1 Pa [16]) cells tumble, at intermediate shear rates they are in swinging mode, while at higher shear stresses erythrocytes perform, similarly to vesicles, pure tank-treading.

The question about the intrinsic viscosity $\tilde{\eta}$ of suspensions has intrigued the physicists for more than hundred years. Numerous theoretical works have been performed since the first study of Einstein, who derived the expression for the effective viscosity $\eta = \eta_0\left[1 + \tilde{\eta}\phi + O(\phi^2)\right]$ of dilute suspension of rigid spheres with volume fraction ϕ in suspending medium with viscosity η_0 [24, 25]. Recently, expressions have been provided for the intrinsic viscosity of dilute suspensions of tank-treading nearly spherical vesicles as well as for suspensions of tumbling vesicles [26, 27].

The experimental parameters relevant to our study are as follows: (*i*) ***excess area*** $\Delta = 4\pi\left(\nu^{-2/3} - 1\right)$ (or ***reduced volume*** $\nu = 3V / 4\pi(S/4\pi)^{3/2}$) of a vesicle (or blood cell) with surface S and volume V, quantifying its deflation from a sphere and thus determining its shape; (*ii*) ***viscosity ratio*** (or ***viscosity contrast***) $\lambda = \eta_{in}/\eta_0$ between the viscosity η_{in} of the internal solution (hemoglobin solution for RBC) and the viscosity η_0 of the suspending medium; (*iii*) ***relative viscosity*** $\eta_r = \eta/\eta_0$ of suspensions; and (*iv*) ***capillary number*** $Ca = \eta_0\dot{\gamma}(S/4\pi)^{3/2}/k_c$ accounting for the deformability of a vesicle with membrane bending modulus k_c in flow with shear rate $\dot{\gamma}$.

In the present work, experimental results about the effective viscosity and viscoelasticity of concentrated RBC suspensions are presented and discussed. The focus is put on new aspects of the problem in the case, when a combination of an oscillatory shear rate and a basic constant shear rate is applied to the suspension [28, 29].

2. Materials and methods

Fresh whole blood was supplied by Etablissement Français du Sang (EFS) and Centre Hospitalier Universitaire (CHU) in Grenoble, France from haematologically healthy donors. Phosphate buffer saline (PBS) with pH 7.4 and osmolarity 290 ± 10 mOsm/kg and dextran, a biocompatible polymer of anhydroglucose, with three different average molecular weights (15000-30000, ~70000 and ~500000) were purchased from Sigma-Aldrich Chemie (Saint-Quentin Fallavier, France). Due to the high hygroscopicity of the polymer, all dextran solutions in PBS were prepared from fresh, newly open flasks.

2.1. Preparation of RBC suspensions

Blood washing by gentle centrifugation in PBS buffer solution assured the removal of plasma proteins responsible for the cell-cell adhesion. The volume fractions of RBCs (hematocrits) in all experimental samples were measured by cell counting as described in detail in [20]. In order to obtain the desired viscosity ratio λ in each sample and taking into account the final hematocrit needed, erythrocytes were dispersed in an external solution, containing PBS and/or a given concentration of dextran. Dextran type and concentrations were adapted to cover all the studied range of viscosity ratios $\lambda \in [0.05,20]$ taking care to avoid any dextran-induced aggregation of cells in suspensions [30]. The cytoplasmic viscosity of RBCs was determined on the basis of the mean corpuscular hemoglobin concentration [5]. For each RBC suspension, a control sample with the same dextran concentration in the PBS was prepared separately for the independent measurement of the external viscosity η_0.

2.2. Rheological measurements

Viscosities of RBC suspensions were measured as a function of the viscosity ratio λ between the inner and outer fluids. Viscoelasticity measurements were made with two instruments: a stress-controlled Bohlin Gemini 150 rheometer (Malvern Instruments, Orsay, France) with a cone-plate geometry (60 mm diameter, 2° angle), and a stress- or shear rate-controlled Physica MCR 301 rheometer (Anton Paar GmbH, Graz, Austria) with a cone-plate (50 mm diameter, 0.5° angle) and Couette (2 mm gap, 40 mm length) geometries. All measurements were performed at constant temperature of 22°C and in conditions preventing the undesirable drying of the sample.

3. Results and discussion

The effective viscosity η of concentrated RBC suspensions was measured for a wide range of viscosity ratios λ (figure 1). Various models for the viscosity of concentrated suspensions of non-aggregated erythrocytes have been developed so far, based on the differential medium approach [31]. In figure 1, in order to avoid any model-dependent artefacts, instead of the intrinsic viscosity, the relative viscosity of RBC suspensions η_r is represented as a function of the viscosity contrast.

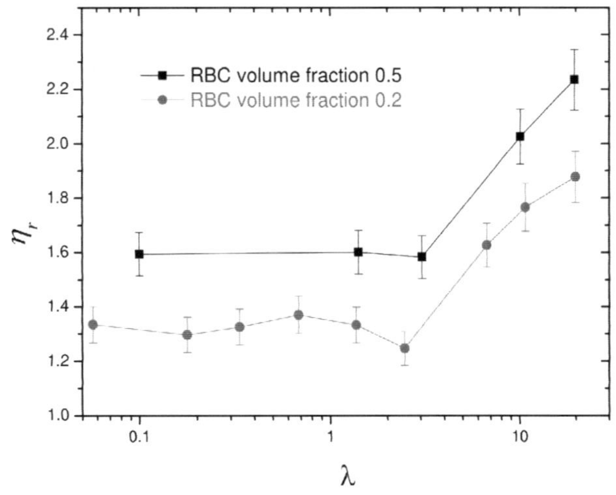

Figure 1. Relative viscosity of suspensions with RBC volume fractions $\phi = 0.2$ and $\phi = 0.5$ as a function of the viscosity contrast.

In our previous study [20], a clearly distinguishable transition between the *TT* and *TB* dynamical regimes with a minimum of the intrinsic viscosity around the transition point ($\lambda \approx 2$ [32, 33]) was obtained experimentally for dilute ($\phi = 0.05$) erythrocyte suspensions. For concentrated suspensions ($\phi > 0.1$), the hydrodynamic interactions between cells are considerable and alter the individual cell dynamics in flow. Consequently, pure tumbling motion is expected to be hindered unlike for erythrocytes in dilute suspensions above a critical value of the viscosity contrast. Nevertheless, our rheological measurements of suspensions with $\phi = 0.2$ and $\phi = 0.5$ showed λ-dependent behaviour of their relative viscosity, which remains practically constant for small λ, while a steep increase of η_r is obtained for higher values of λ. These results lead to the conclusion that rheology of concentrated suspensions is also sensitive to the cell dynamics similarly to the dilute limit. Future experimental and theoretical advances on hydrodynamic interactions of vesicles and red cells in flow could shed light on the origins of such increase in the relative viscosity of suspensions for small external viscosities (high values of λ).

Concentrated RBC suspensions were subjected to oscillatory shear stresses $\tau(t) = \eta^*(\omega)\dot{\gamma}(t)$ in order to extract their complex viscosity $\eta^* = \eta' + i\eta''$ and complex modulus $G^* = G' + iG'' = \eta''/\omega - i\eta'/\omega$, where ω is the angular velocity of oscillations. In the cases, when $G' > G''$, the elastic response of the suspension is higher than the viscous dissipation. Such an example is given in figure 2 for two volume fractions of erythrocytes $\phi = 0.2$ and $\phi = 0.33$ and for viscosity ratio $\lambda = 1.2$, which is still in the *TT*-branch of the curve $\eta_r(\lambda)$ but close to the transition point. This signature of a viscoelastic medium has been previously observed for RBC suspensions [34], with high volume fractions $\phi > 0.2$, and is believed to be a consequence of the elasticity of the cells, packed more densely at high concentrations.

Blood flow in aorta and large arteries is pulsatile, which motivates the investigation of RBC rheology in conditions similar to the real ones, i.e. when a combination of an oscillatory flow at constant frequency and a steady flow is applied to the suspension:

$$\tau(t) = \tau_0(1 + \alpha \cos \omega t) \tag{1}$$

where $\alpha\tau_0$ is small enough in order to expect linear response from the suspension.

Such experimental studies on whole blood have been previously conducted and some features of blood viscoelasticity were revealed and attributed to the RBC disaggregation and deformation in combined steady and oscillatory shear flows [29]. Very recently, it was shown theoretically that the complex viscosity of dilute vesicle suspensions subjected to superimposed steady and oscillatory shear stress reveals much more insightful properties of the suspension than its complex viscosity in the case when only oscillating shear stress is applied [28]. It has been obtained that in weak flows vesicles relax to *TT* motion monotonously, while for strong flows and high enough viscosity contrasts the relaxation is achieved after a transient of damped oscillations. It was shown that different relaxation regimes result in qualitatively distinct dependencies $\eta^*(\omega)$ and that the relaxation times of vesicle dynamics govern the quantitative behaviour of $\eta^*(\omega)$ [28].

Due to instrumental limitations we were able to perform reproducible measurements only with concentrated RBC suspensions with $\phi \geq 0.2$. Results for the complex viscosity of RBC suspensions with $\phi = 0.2$ and $\phi = 0.3$ and with $\lambda = 0.2$ are presented in figure 3. The applied shear stress according to Eq. (1) had constant part $\tau_0 = 0.1$ Pa (swinging motion expected [16]) with oscillation amplitude $\alpha\tau_0 = 0.01$ Pa. Fitting with three parameters (η_0, A, and ν) was performed on the experimental data points for the real part of $\eta^*(\omega)$ using the real part of the complex equation $\eta = \eta_0 + A/(\nu - i\omega)$, namely $\mathrm{Re}(\eta) = \eta' = \eta_0 + A\nu/(\nu^2 + \omega^2)$. The same values of parameters were used to trace the theoretical curve for the imaginary part of $\eta^*(\omega)$, $\mathrm{Im}(\eta) = \eta'' = A\omega/(\nu^2 + \omega^2)$. For all studied ranges of λ and ϕ, the values of η_0, obtained by the fit, are in excellent agreement with the values of the dynamic viscosity, experimentally determined via independent measurements in steady flow. In all studied cases, the values of the fitting parameter ν seem to be independent of the applied shear rate. The model [28] suggests that $1/\nu$ is a relaxation time of the RBC dynamics under constant shear flow. To probe the origins of the characteristic time related to ν, further experiments are needed in a wide range of shear rates and for various λ.

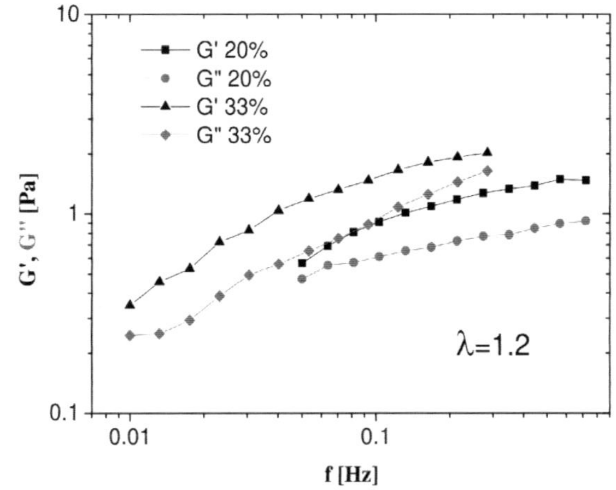

Figure 2. Complex modulus of suspensions with volume fractions of RBC $\phi = 0.2$ and $\phi = 0.33$ and viscosity ratio $\lambda = 1.2$.

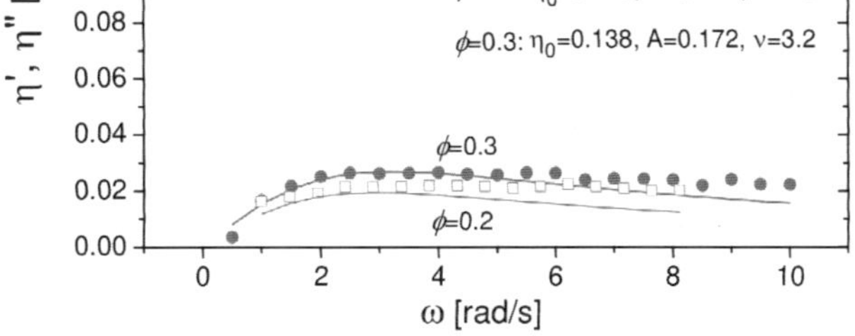

Figure 3. Complex viscosity of suspensions with RBC volume fractions $\phi = 0.2$ and $\phi = 0.3$ with viscosity ratio $\lambda = 0.2$. The applied shear stress is $\tau_0 = 0.1$ Pa with oscillation amplitudes $\alpha\tau_0 = 0.01$ Pa.

4. Conclusion

In the present work, the rheological properties of concentrated RBC suspensions were studied in order to reveal the correlation between the individual dynamics of cells and the suspension rheology. Blood disorders strongly affect the deformability of erythrocytes as well as their dynamics in the blood stream leading in some cases to dramatic physiological consequences for the organism [35]. Numerous examples with strong sociological impact can be given, such as sickle cell anemia, iron deficiency anemia, thalassemia, chronic alcoholism, etc, affecting the morphology and/or mechanical properties of erythrocytes. Since the rheological measurements presented here are sensitive to the physical properties of the cells, they can be used to capture pathological deviations in the shape, the reduced volume, and/or rigidity of cells by such macroscopic assay of suspension rheology. Future efforts must be focused on the relation between the rheological properties of suspensions and the contents of pathological cells in them.

Acknowledgments

Authors thank G. Coupier and C. Verdier for experimental advice and helpful discussions. Support by the National Science Fund of Bulgaria (Grant No. DMU03-80/12.12.2011) is gratefully acknowledged.

References

[1] Usami S 1982 *Biorheology* **19** 29
[2] Chien S, Usami S, Dellenback R J and Gregersen M I 1967 *Science* **157** 827
[3] Schmid-Schönbein H, Rieger H and Fischer T 1980 *Angiology* **31** 301
[4] Fedosov D A, Pan W, Caswell B, Gompper G and Karniadakis G E 2011 *Proc. Natl Acad. Sci.* **108** 11772
[5] Ross P D and Minton A P 1977 *Biochem. Biophys. Res. Commun.* **76** 971
[6] Snabre P and Mills P 1999 *Colloids and Surf. A: Physicochem. and Eng. Aspects* **152** 79
[7] Fischer T M, Stohr-Lissen M and Schmid-Schonbein H 1978 *Science* **202** 894
[8] Barshtein G, Ben-Ami R and Yedgar S 2007 *Expert. Rev. Cardiovasc. Ther.* **5** 743
[9] Chien S 1970 *Science* **168** 977
[10] Chien S 1987 *Annu. Rev. Physiol.* **49** 177
[11] Pfafferott C, G. B. Nash and Meiselman H J 1985 *Biophys. J.* **47** 695
[12] Singer S J, Nicolson, G. L. 1972 *Science* **175** 720
[13] Manno S, Takakuwa Y and Mohandas N 2002 *Proc Natl Acad Sci USA* **99**
[14] Bevers E, Comfurius P, Dekkers D, Harmsma M and Zwaal R 1998 *Biol Chem* **379** 973
[15] Skotheim J M and Secomb T W 2007 *Phys. Rev. Lett.* **98** 078301
[16] Abkarian M, M. Faivre and Viallat A 2007 *Phys. Rev. Lett.* **98** 188302
[17] Dimova R, Aranda S, Bezlyepkina N, Nikolov V, Riske K A and Lipowsky R 2006 *J. Phys. Condens. Matter* **18** S1151
[18] *Giant Vesicles* 2000, ed P L Luisi and P Walde (Chichester: John Wiley & Sons, Ltd.)
[19] Montes L-R, Alonso A, Goni F M and Bagatolli L A 2007 *Biophys. J.* **93**
[20] Vitkova V, Mader M, Polack B, Misbah C and Podgorski T 2008 *Biophys. J.* **95** L33
[21] Kantsler V, Segre E, and Steinberg V 2008 *Europhys. Lett.* **82** 58005
[22] Vlahovska P M, Podgorski T and Misbah C 2009 *C. R. Physique* **10** 775
[23] Vitkova V and Misbah C 2011 Dynamics of lipid vesicles – from thermal fluctuations to rheology, *Advances in Planar Lipid Bilayers and Liposomes*, Ed A Iglic (Burlington: Academic Press) chapter 9 pp 258-292
[24] Einstein A 1906 *Ann. Phys. Leipzig* **19**
[25] Einstein A 1911 *Ann. Phys. Leipzig* **34**
[26] Misbah C 2006 *Phys. Rev. Lett.* **96**
[27] Danker G and Misbah C 2007 *Phys. Rev. Lett.* **98**
[28] Farutin A and Misbah C 2012 *J. Fluid Mech.* **700** 362
[29] Vlastos G, Lerche D, Koch B, Samba O and Pohl M 1997 *Rheol. Acta* **36** 160
[30] Pribush A, D. Zilberman-Kravits and Meyerstein N 2007 *Eur. Biophys. J.* **36** 85
[31] Pal R 2003 *J. Biomech.* **36** 981

[32] Goldsmith H L and Marlow J 1972 *Proc. R. Soc. Lond. B.* **182** 351
[33] Fischer T M and Schmid-Schonbein H 1977 *Blood Cells* **3** 351
[34] Thurston G B 1972 *Biophys. J.* **12** 1205
[35] Horne M K 1981 *Am J Med* **70** 288

Experimental study of the bending elasticity of charged lipid bilayers in aqueous solutions with pH5

D Mitkova[1], **A Stoyanova-Ivanova**[1], **Yu A Ermakov**[2], **and V Vitkova**[1,3]
[1] G. Nadjakov Institute of Solid State Physics, Bulgarian Academy of Sciences, Sofia, Bulgaria
[2] The Frumkin Institute of Physical Chemistry and Electrochemistry, Russian Academy of Sciences, Moscow, Russia

E-mail: victoria@issp.bas.bg

Abstract. Exposure to high concentrations of contaminations due to air polluting gases, vapours and aerosols and possibly altering the normal pH in the body could lead to undesirable changes in the properties of biological cells. Here, we study experimentally the mechanical properties of synthetic phospholipid bilayers containing increasing molar fractions (up to 0.15) of charged lipid (synthetic phosphatidylserine) in aqueous solutions with controlled ionic strength and at pH 5, which is slightly lower than the physiological values of pH. Our observations in phase contrast and fluorescence testified to the coexistence of two phases in membranes for temperatures below 29°C. Micro-sized inhomogeneities in vesicle membranes were systematically observed at temperatures lower than 29°C and for molar fractions of phosphatidylserine in the bilayer higher than 0.1. For the quantitative determination of the membrane bending rigidity, we applied thermal fluctuation analysis of the shape of quasispherical lipid vesicles. As far as the liquid-crystalline state of the bilayer is a necessary condition for the application of the experimental method, only vesicles satisfying this requirement were processed for determination of their membrane bending rigidity. The value obtained for the bending modulus of bilayers with 0.15 molar content of charged lipid is about two times higher than the bending modulus of uncharged membranes in the same bathing solution. These findings are in qualitative agreement with our previous results for the bending rigidity of charged bilayers, measured by vesicle micromanipulation.

1. Introduction

Air polluting gases, vapors and aerosols are treated as potentially dangerous if their concentration for a certain period of time can become so high as to cause damage to health. Exposure to high concentrations of pollutants influencing the normal pH values (pH 5.5) in the body could lead to undesirable consequences for cells' properties and functioning [1-3]. Mechanical constants of biological membranes are somehow related to all vital processes in cells [4]. In this respect, the investigation of membrane mechanics and its alteration by external factors is of importance to our knowledge about the impact of emissions of various types on living organisms.

Lipid bilayers model some basic physical characteristics of biomembranes such as their permeability and resistance to bending and stretching. In water, lipid bilayers spontaneously form closed structures, which in the case of unilamellar quasispherical objects with diameters of the order of

[3] To whom any correspondence should be addressed.

tens of micrometers are called 'giant unilamellar vesicles' (or GUVs). During the last decades GUVs have been exploited as a rough physical model of simple biological cells and have been used to probe the membrane properties [5] and cell hydrodynamics in flows [6].

Thanks to elaborated experimental protocols, GUVs are easy to prepare in laboratory conditions with a good control of the chemical composition of membranes and bathing solutions. Due to the considerable progress achieved so far, the application of advanced experimental methods allows quantifying the membrane bending rigidity [7, 8]. Phosphatidylserine is one of the major phospholipid components in cellular membranes [9] and its contribution to the membrane bending rigidity has been in the scope of our interest so far [10].

The present work aims at the experimental determination of the bending elasticity modulus of phosphatidylcholine lipid membranes, containing various molar concentrations of synthetic phosphatidylserine, in aqueous solutions with pH 5.

2. Materials and methods

Lipid vesicles were prepared from 1–stearoyl–2–oleoyl–sn–glycerol–3–phosphocholine (SOPC, 788.14 g/mol, Avanti Polar Lipids Inc., USA) and molar fraction of 1,2-dioleoyl-sn-glycero-3-phospho-L-serine (DOPS (sodium salt) 810.03 g/mol, Avanti Polar Lipids Inc., USA) via electroformation [11, 12] in aqueous solutions of 0.19 M of sucrose ($C_{12}H_{22}O_{11}$, 342.3 g/mol, Sigma-Aldrich, Germany) and sodium salt (NaCl, 58.44 g/mol, INEOS, UK) with different pH. For the preparation of all aqueous solutions double distilled water, obtained by quartz distiller, was used. Hydrochloric acid (HCl, 36.46 g/mol, Sigma-Aldrich, Germany) with concentration of 0.1 M was added in aqueous solutions of sucrose and sodium salt in order to achieve the desired acidity. The electroformation protocols were adapted to the presence of 0.01 M NaCl in the aqueous phase [13]. All pH measurements were performed by a combined pH-meter with Ag/AgCl electrode Lovibond (Germany). Fluorescent label Lissamine™ Rhodamine B 1,2-Dihexadecenoyl-sn-Glycero-3-Phosphoethanolamine, Triethyl-ammonium Salt (Rh-DHPE, Molecular probes, USA) was used for visualisation of the membrane phase separation. For the quantitative determination of the membrane resistance to bending, analysis of the thermal fluctuations of shapes of quasispherical vesicles was applied [7, 14].

Membrane fluctuations of stroboscopically illuminated [15] quasispherical vesicles without defects were observed and recorded by means of an inverted light microscope by Zeiss, Axiovert 100 (Germany) in phase contrast with a thermo-controlled long-working distance oil-immersed objective (100x, NA 1.25). CCD camera (C3582, Hamamatsu Photonics, Japan) was mounted on the microscope and connected to the video input of a frame grabber board (DT3155, Datatranslation, USA) in a computer for proper digitization (768x576 8-bit pixels, pixel size: 0.106 μm/pix). Fluorescence observation regime of the microscope with image intensifier (M4314, Hamamatsu Photonics, Japan) was exploited to visualize phase separations in lipid membranes.

All measurements of the bending rigidity of charged membranes were performed at constant temperature of (30±0.1)°C using thermo-controlled oil-immersed objective (100x, NA 1.25) and microscope stage (Zeiss, Germany).

3. Results and discussion

Vesicles with inhomogeneous membranes are not appropriate for contour analysis and determination of the bending elasticity modulus due to the limitations of the applied theoretical model [16, 17] assuming liquid state of the vesicle membrane.

Flat non-fluctuating domains in vesicle membranes were systematically observed at temperatures lower than 29°C and for molar fractions of phosphatidylserine in the bilayer higher than 0.1. Recently, it has been shown theoretically that charged membranes show complex phase behavior at various salt concentrations in the aqueous solution [18]. Up to our knowledge the phase diagram of fully hydrated two-component lipid mixtures (SOPC lipid matrix with DOPS charged lipid) in presence of sucrose (0.19 M $C_{12}H_{22}O_{11}$) and 0.01 M NaCl at pH 5.0 (0.1 M HCl) has not been explored systematically, yet.

Our observations of DOPS/SOPC vesicles in the mentioned-above aqueous surroundings revealed the existence of micro-sized inhomogeneities in the membranes at temperatures below 29°C. This phase separation was detected by means of phase-contrast microscopy (see figure 1) as well as in fluorescence regime (figures 2 and 3). Figure 1a represents two subsequent images of the equatorial cross-section of a giant vesicle with membrane containing 20 mol% DOPS in SOPC matrix in aqueous solution of sucrose (0.19 M $C_{12}H_{22}O_{11}$) with 0.01 M NaCl at pH 5.0 (0.1 M HCl). Two adjacent contours of a vesicle, whose membrane is composed of 50 mol% DOPS in SOPC in the same aqueous environment, are shown in figure 1b. Stiffer non-fluctuating parts on the membrane contours of fluctuating vesicles were readily observable during experiment. White arrows on figure 1 point at such flat domains positioned at the equator of observed vesicles.

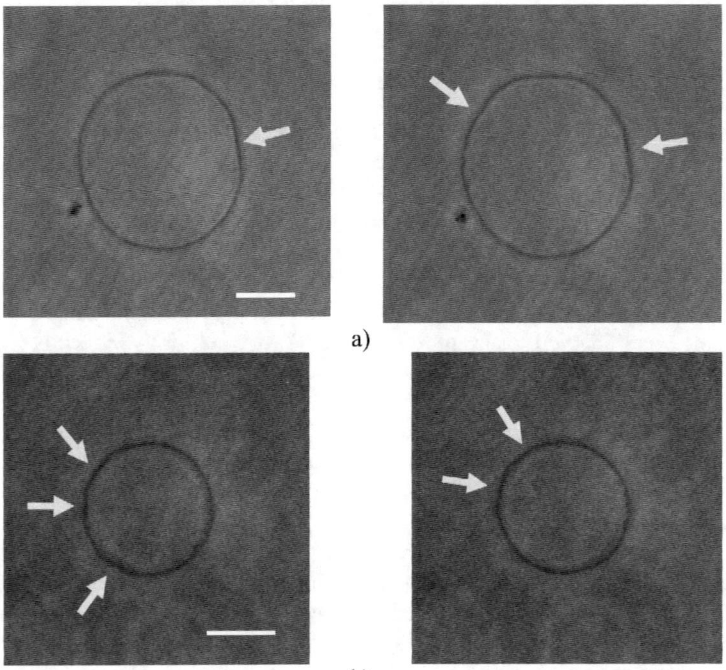

a)

b)

Figure 1. Phase contrast microscopy: a) Two-component lipid vesicles (SOPC lipid matrix with 20 mol% DOPS charged lipid) in aqueous solution of sucrose (0.19 M $C_{12}H_{22}O_{11}$) containing 0.01 M NaCl with pH 5.0, temperature 23°C; b) Two-component lipid vesicles (SOPC lipid matrix with 50 mol% DOPS charged lipid) in the same aqueous solution and at same temperature as in (a). The bars correspond to 10 µm. Arrows point at flat lipid domains (see text).

In order to better visualize the flat non-fluctuating domains in vesicle membranes, fluorescence microscopy was applied. Rhodamine dyes have been commonly used in membrane biophysics although some evidences exist for their contribution to the lipid peroxidation related to large-raft formation in membranes [19, 20]. Rh-DHPE was used in our study with 1.5% molar concentration in the membrane. Initially, observations were performed on vesicles with ternary membranes SOPC/DOPS/Rh-DHPE (0.785/0.2/0.015) in aqueous solutions of sucrose (0.19 M $C_{12}H_{22}O_{11}$) and sodium chloride (0.01 M NaCl) with pH 5.0 (0.1 M HCl) as presented in figure 3. Micrometric dark domains were systematically observed on vesicle membranes at $T<29^{\circ}C$, testifying to phase separation in bilayers.

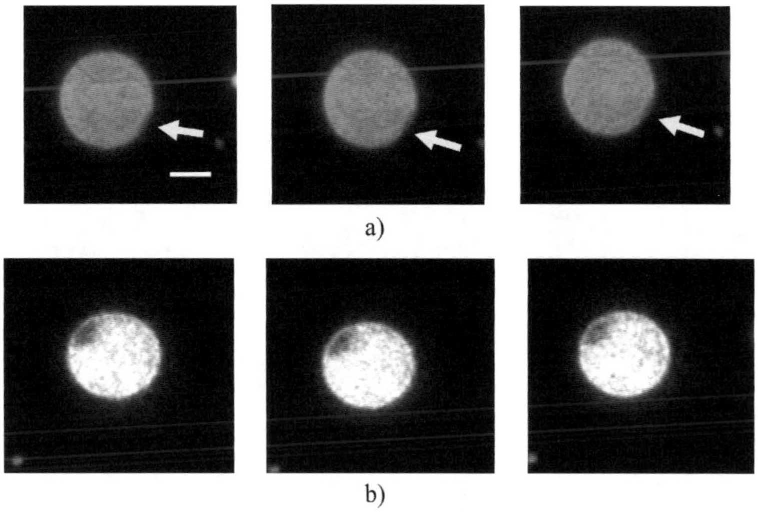

Figure 2. Fluorescence microscopy of a ternary lipid vesicle (SOPC lipid matrix with 20 mol% DOPS and 1.5 mol% Rh-DHPE) in aqueous solution of 0.19 M $C_{12}H_{22}O_{11}$ and 0.01 M NaCl with pH 5.0 at: a) temperature 23°C; b) temperature 27°C. The bar corresponds to 10 μm.

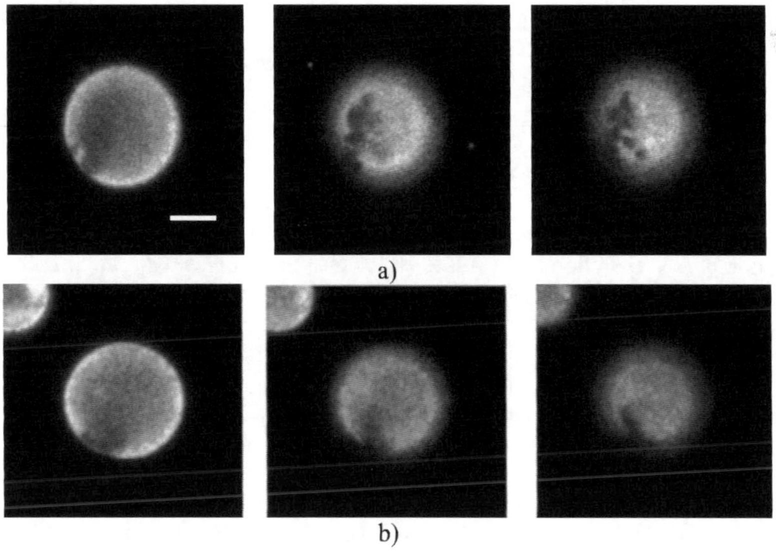

Figure 3. Fluorescence microscopy of lipid vesicles with three-component membranes (SOPC lipid matrix with 20 mol% DOPS and 1.5 mol% Rh-DHPE) in double distilled water at: a) temperature 23°C; b) temperature 27°C. The bar corresponds to 10 μm.

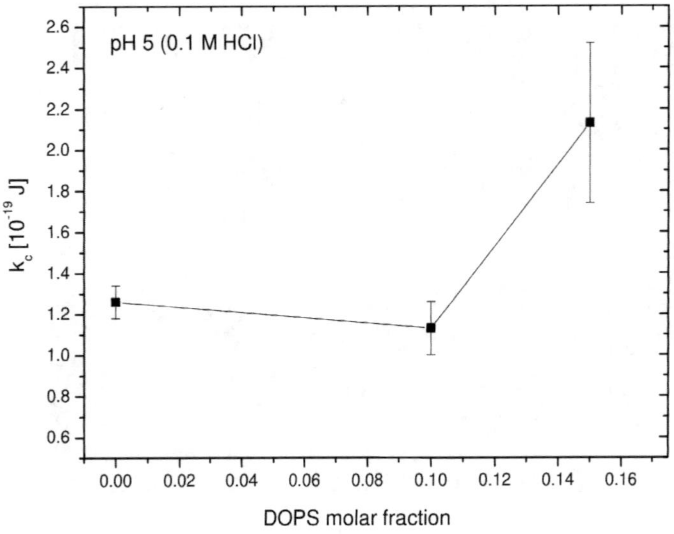

Figure 4. Bending elasticity modulus of two-component membranes (SOPC lipid matrix with DOPS charged lipid) in aqueous solutions containing 0.19 M $C_{12}H_{22}O_{11}$ and 0.01 M NaCl with pH5 (0.1 M HCl).

Control experiments in double distilled water were performed with the same ternary vesicles in attempts to reveal the contribution of the hydrophilic additives on the phase behavior. Images of such vesicle with ternary membrane in double-distilled water are shown in figure 3. Dark domains can be clearly distinguished on vesicle membranes. Phase separation has been observed in similar ternary membranes composed of natural lipids egg-PC/brain-PS/Rh-DOPE with 0.89/0.1/0.01 molar ratios at temperatures around 15°C (G. Staneva, unpublished).

The experimental fact that flat domains were observed even in two-component SOPC/DOPS membranes (figure 1), provides evidence against the hypothesis for a possible Rh-DHPE-related lipid peroxidation promoting phase separation [19]. Future effort must focus on the exploration of the phase diagram of the studied system and the revelation of the contribution of pH and solutes on the membrane phase state [18].

Our experimental observations led to the conclusion that the studied membranes are in their homogeneous liquid state at temperatures above 29°C and consequently, all measurements of the membrane bending elasticity were performed at T=30°C. At least twenty vesicles were recorded for each DOPS content in the membrane (0, 10 or 15 mol %). Less than a half of the registered vesicles, i.e. around ten vesicles for each DOPS molar fraction, satisfied all criteria for goodness during the contour analysis (Bivas et al., to be published) and were used for calculation of the membrane bending elasticity modulus. The experimental results from thermal fluctuation analysis showed that the presence of 15 mol% of phosphatidylserine in the bilayer lead to almost a double increase of the membrane bending rigidity at pH 5 as depicted in figure 4. Points with the error bars represent values of the membrane bending modulus, calculated by taking into account the error of single measurements. This experimental finding is in qualitative agreement with our previous results from micromanipulation experiments for the bending elasticity of SOPC bilayers containing up to 40 mol% brain-PS [10].

4. Conclusion

In the present work a systematic experimental study was conducted for the determination of the bending rigidity of phosphatidylcholine lipid (SOPC) membranes, containing increasing molar concentrations (up to 0.15) of synthetic phosphatidylserine (DOPS), in aqueous solutions with controlled ionic strength and at pH 5. Two-phase coexistence was observed by phase-contrast and fluorescence microscopy in membranes with molar fractions of phosphatidylserine in the bilayer higher than 0.1 and at temperatures lower than around 30°C. By means of thermal shape fluctuation analysis of nearly spherical vesicles, the bending elasticity was quantified for homogeneous liquid membranes. The value of the bending modulus of bilayers containing 0.15 molar fraction of charged lipid was obtained to be about two times higher than the bending modulus of uncharged membranes in the same surrounding medium.

Further experimental investigations with higher molar concentrations of DOPS in SOPC membranes and at lower pH values will shed light on how the acidity of the bathing solution alters the bending elasticity of charged membranes, thus helping to elucidate the relevance of the physicochemical parameters of the aqueous surroundings to the membrane mechanics.

Acknowledgments

Authors wish to thank Prof. Isak Bivas for sharing his expertise in thermal fluctuation analysis and Assoc. Prof. Galia Staneva for fruitful discussions. This study was supported by Grant No. DMU03-80/12.12.2011 from the National Science Fund (Bulgaria). ASI acknowledges financial contribution of ISSP-BAS (Grant BK-01-12/29.03.2012). This work was performed in the frames of the bilateral cooperation between the Bulgarian Academy of Sciences and the Russian Academy of Sciences.

References

[1] Mar T, Larson TV, Stier RA, Claiborn C, Koenig JQ 2004 *Inhal Toxicol* **1613** 809
[2] Linn W, Szlachcic Y, Gong H, Jr, Kinney P and Berhane K 2000 *Environmental Health Perspectives* **108** 427
[3] Barraza-Villarreal A *et al.* 2008 *Environmental Health Perspectives* **116** 832
[4] Alberts B, Johnson A, Lewis J, Raff M, Roberts K and Walter P 2002 *Molecular Biology of the Cell* (New York: Carland Publishing, Inc.)
[5] Dimova R, Aranda S, Bezlyepkina N, Nikolov V, Riske K A and Lipowsky R 2006 *J. Phys. Condens. Matter* **18** S1151
[6] Vlahovska P M, Podgorski T and Misbah C 2009 *C. R. Physique* **10** 775
[7] Mitov M D, Faucon J F, Meleard P and Bothorel P 1992 Thermal fluctuations of membranes, ed G W Gokel (Greenwich: JAI Press Inc.) pp 93-139
[8] Hess S T, Gudheti M V, Mlodzianoski M and Baumgart T 2007 Shape Analysis of Giant Vesicles With Fluid Phase Coexistence by Laser Scanning Microscopy to Determine Curvature, Bending Elasticity, and Line Tension *Methods in Membrane Lipids*, ed A M Dopico (Humana Press) chapter 25 pp 367-387
[9] Leventis P A and Grinstein S 2010 *Annu Rev Biophys* **39** 407
[10] Vitkova V, Genova J, Finogenova O, Ermakov Y, Mitov M D and Bivas I 2004 *C. R. Acad. Bulg. Sci.* **57** 25
[11] Angelova M and Dimitrov D 1986 *Faraday Discuss. Chem. Soc.* **81** 303
[12] Vitkova V, Antonova K, Popkirov G, Mitov M D, Ermakov Y A and Bivas I 2010 *J. Phys.: Conf. Ser.* **253** 012059
[13] Pott T, Bouvrais H and Meleard P 2008 *Chem. Phys. Lipids* **154** 115
[14] Faucon J F, Mitov M D, Meleard P, Bivas I and Bothorel P 1989 *J. Physique* **50** 2389
[15] Genova J, Vitkova V, Aladgem L and Mitov M D 2005 *J. Optoel. Adv. Mater.* **7** 257
[16] Helfrich W 1973 *Z. Naturforsch.* **28c** 693
[17] Schneider M B, Jenkins J T and Webb W W 1984 *J. Physique* **45** 1457

[18] Shimokawa N, Komura S and Andelman D 2011 *Phys. Rev. E* **84** 031919
[19] Ayuyan A G and Cohen F S 2006 *Biophys. J.* **91** 2172
[20] Bouvrais H, Pott T, Bagatolli L A, Ipsen J H and Méléard P 2010 *Biochimica et Biophysica Acta* **1798** 1333

Feasibility study of the proton yield from the reaction $D(^3He,p)^4He$ as a possible tool for radiotherapy treatment

N M Uzunov[1, 6], N Liguori[2], C L Fontana[2,6], Y Baneva[3], K Atroshchenko[2], M Bello[4], G Moschini[4], A Rosato[5], V Rigato[4], B Doyle[7] and P Rossi[2,6,7]

[1] Faculty of Natural Sciences, Shumen University "K. Preslavsky", 115 Universitetska str., 9712 Shumen, Bulgaria
[2] Dept. of Physics "Galileo Galilei", University of Padua, 8 Francesco Marzolo str., 35131 Padua, Italy
[3] Dept. of Physics and Biophysics, Medical University, 150 Tzar Liberator str., 9000 Varna, Bulgaria
[4] National Laboratories of Legnaro, INFN, Viale dell'Universita 2, 35020 Legnaro (PD), Italy
[5] Dept. of Oncology and Surgical Sciences, University of Padua, 64 Gattamelata str., 35128 Padua, Italy
[6] INFN section of Padua, 8 Francesco Marzolo str., 35131 Padua, Italy
[7] Sandia National Laboratories, USA

E-mail: nikolay.uzunov@lnl.infn.it

Abstract. Recent achievements in proton and carbon ions therapy have shown the importance of the hadron therapy methods. Aiming at radiotherapy applications such as dermatological and intra-operative procedures, where a short range treatment is needed, we have studied the use of nuclear reactions induced by low energy ions from small accelerators. A very suitable reaction is $D(^3He,p)^4He$, using $^3He^+$ ions with energies of about 800 keV. The resulting protons have energies above 17 MeV and could deliver significant radiation dose depending on the accelerator $^3He^+$ beam current and the irradiation time. The deuterium containing target was prepared by reactive magnetron sputtering of titanium in Ar and Ar+D_2 radiofrequency plasma on a substrate of Silicon. The Ti-D_x stoichiometry and deuterium content was determined by Ion Beam Analysis. The accelerated $^3He^+$ beam was provided by the 2.5MV Van de Graaff accelerator at the National Laboratories of Legnaro, INFN, Italy. Proton yield as a function of the beam current at different forward scattering angles has been studied for the energies of the incoming $^3He^+$ in the 700keV – 800keV energy interval. The irradiated volume and the radiation dose in biological tissues as a function of the proton energy and proton yield has been estimated. Possible applications in small animal treatment studies as well as potential clinical radiotherapy applications are discussed.

1. Introduction

Therapies based on the use of accelerated hadrons (hadron therapy) nowadays are a very important aspect of medical radiation therapy of cancer. Hadrons possess a very useful property: the strong increase of the linear energy transfer (LET) at the end of the charged particle trajectory in matter, the so-called Bragg Peak. Such a high energy release allows an in-depth "conformal" treatment of tumor tissues [1, 2]. The main radiobiological advantage of such a conformal therapy is the release of the

Journal of Physics: Conference Series **398** (2012) 012029 doi:10.1088/1742-6596/398/1/012029

highest possible dose to the tumor, while protecting the surrounding normal tissues as much as possible.

Using modern imaging technologies such as Computed Tomography (CT), Single Photon Emission Computed Tomography (SPECT), Positron Emission Tomography (PET), Magnetic Resonance Imaging (MRI), etc., it is possible to obtain a three-dimensional image of the tumor. Then the tumor irradiation can be conducted from diverse angles thus decreasing drastically the dose released in the surrounding tissues. This is the principle of three-dimensional conformal radiotherapy (3DCRT) [2,3]. One way to realize 3DCRT is to use a broad hadron beam passing through collimators designed to match the tumor contour shape at each angle of irradiation. The in-depth dose distribution is shaped by specific milled compensators to reduce the range of hadrons [4, 5]. Another way to deliver the dose in the tumor is to decrease the hadron beam size to a narrow mono-energetic "pencil" beam and to make it scan across the object to achieve the planned dose. The scanning is conducted in x,y plane perpendicular to the beam direction and the depth scan (in z direction) is realized applying appropriate energy of the hadrons' beam [5].

Proton beams, being the lightest hadron particle, features several advantages as for the radiation therapy. However, its mass is high enough to have a relatively low lateral side scatter in the tissue. As a consequence, the proton beam has a low divergence, maintains the preliminary shape and delivers the planned dose at the tumor. In addition the calculation of the stopping power in the tissues is simple; the computer simulations are working quite well and finally there is no significant tail after the Bragg Peak. This is important when vital or radiation fragile organs, such as the genital system, eyes, brain, the spinal column, etc., are situated behind the tumor tissue.

Typical energies to conduct effectively radiation therapy of tumors situated in-depth of human body are in the range 50 to 250MeV for protons and about 120 – 400MeV/nucleon for ions like carbon [6]. These energies can be supplied by charged particles accelerators such as linear accelerators (linacs), cyclotrons or synchrotrons. The choice of the accelerator depends on the particular requirements (type of hadrons, beam currents, etc.) but in general cyclotrons and synchrotrons are the accelerators mostly used for hadron radiotherapy. The equipment for the hadron therapy includes the accelerator, the beam guidance system, the gantry (in the case of patient irradiation from many directions), and it requires a lot of space, especially in case of high energies or heavy ions. However, application-oriented proton accelerators, such as the ones for industrial and medical purposes, are simple and less expensive. The development of the modern technologies and the application of superconducting materials allow substantially reducing the size of the cyclotron, which can easily mounted in the gantry. This also removes the need of a separate cyclotron control room and a proton transport beam line.

In this article we present the results from an experimental study about the use of the high Q-value (Q = 18.35MeV) exothermic nuclear reaction $D(^3He,p)^4He$ for possible proton radiotherapy applications. Protons with energies up to 17.4MeV are obtained using relatively low-energy helium beams (about 800keV), supplied by a small-size accelerator, adequate for hospital usage. The idea to exploit the reaction $D(^3He,p)^4He$ for proton therapy has been first proposed in [7] and a thorough analysis of possible applications has been conducted. The aim of the experiment described in the present paper is to measure experimentally the proton yield from the reaction for a particularly designed deuterium-containing target and to draw conclusions about possible radiotherapy applications.

2. Materials and methods

2.1. Target preparation

The deuterium containing target was prepared by reactive magnetron sputtering of Titanium in Ar and Ar+20%D_2 radio-frequency plasma at the total pressure of 3.5 10^{-3} mBar, on a substrate of Silicon (nominal thickness 85μm, Boron doped). The deposition was performed in two steps. First a thin layer of Titanium for adhesion purposes was deposited using pure (N60) Ar plasma. Then a second layer of Ti-D_x compound was formed using Ar+20%D_2 gas mixture.

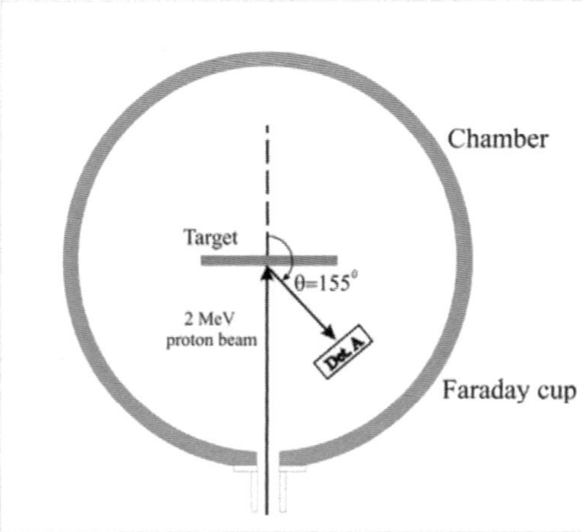

Figure 1. RBS setup for target characterization

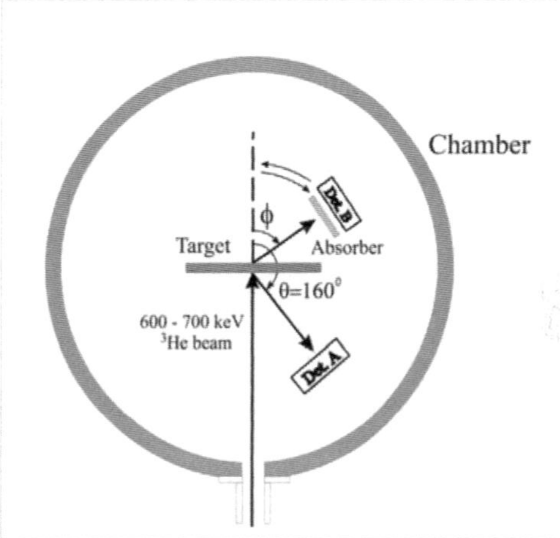

Figure 2. Setup for particles yield measurements from the reaction D(^{3}He,p)^{4}He.

2.2. Target characterization

To determine the target layer structure we performed a Rutherford Back-Scattering (RBS) analysis using a proton beam with energy of 2MeV (at the AN2000 accelerator of the National Laboratories of Legnaro, INFN, Italy). The target was placed with the "deuterated" layer facing the beam. The backward scattered protons were detected at 155^0 from beam direction using a ion-implanted-silicon charged-particle detector with minimum depletion depth of 100μm (detector A of figure 1). The whole chamber acted as a Faraday cup to measure the accumulated beam charge. A spectrum of the backscattered protons from the target at an angle $\theta = 155^0$ is shown in figure 3.

The amount of Deuterium was determined from the back-scattering spectrum (figure 3), which offers the ratio between the quantities of Titanium (N_{Ti}) and Deuterium (N_D): $N_D/N_{Ti} = 1.28$. We also measured the Deuterium areal density, $\rho_D = 1.7 \times 10^{18}$at/cm^2 [8]. The measured thickness of TiD_x was about 250nm.

3. Proton yield measurements

To study the reaction D(^{3}He,p)^{4}He we have used a ^{3}He beam delivered by the same AN2000 accelerator. As it has been shown in [9-11] the cross-section for this reaction has highest values for energies in the range 700keV – 800keV that we have picked up for our measurements. The Proton yield at different forward scattering angles has been studied.

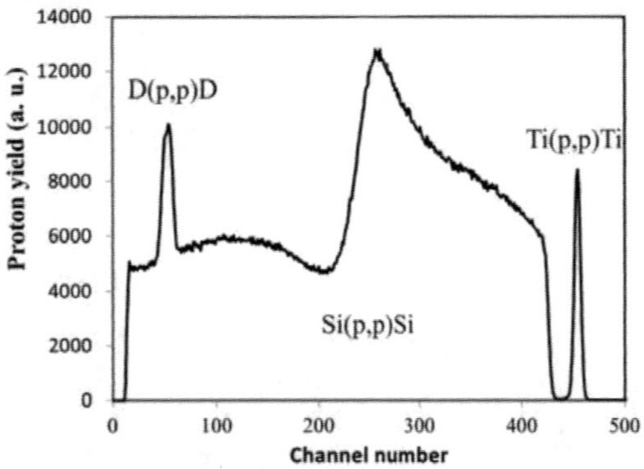

Figure 3. Elastic back-scattering spectrum ($\theta=155^0$) of 2MeV protons on the target

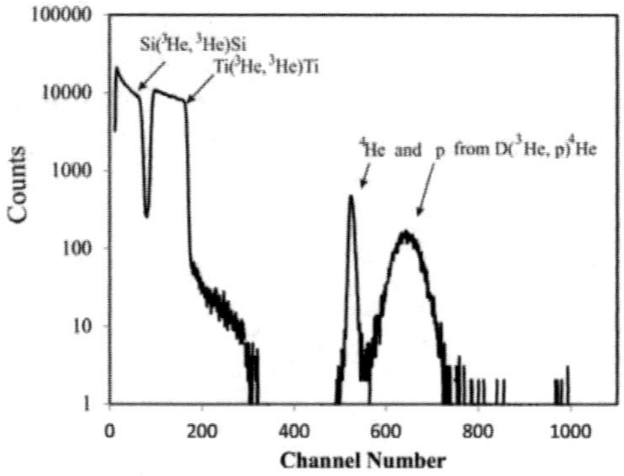

Figure 4. Back-scattering spectrum of 3He ions on the target and particles from the $D(^3He,p)^4He$ and $D(^3He, ^4He)p$ reactions, measured with detector A at $\theta=160^0$

The experimental setup including two detectors is shown in figure 2. A partially depleted Si detector (minimum depletion thickness 100μm) 100μm Si detector was placed at 160^0 from the beam direction for measurements in the backward direction and for the assessment of the beam charge for spectra normalization. Another Si detector (Detector B in figure 2; totally depleted 1018μm thick) was used to detect the particles in the forward direction. The detector angular position ϕ was varied from -70^0 to +70^0 with respect to the beam direction. A passive Silicon layer with a thickness of 1008μm was mounted in front of the Detector B in order to stop all the protons in the detector depleted volume.

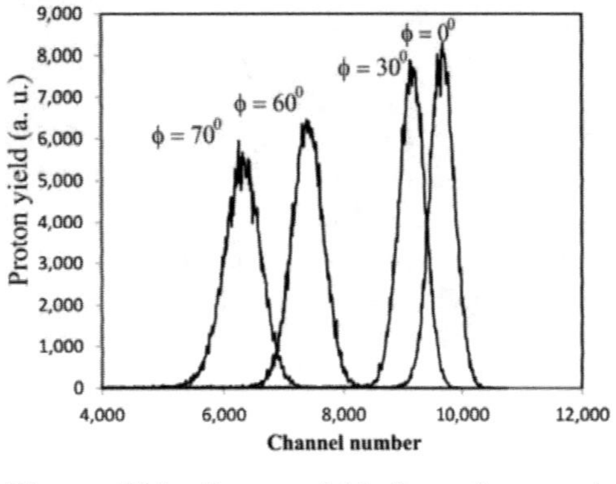

Figure 5(a). Proton yield from the reaction $D(^3He,p)^4He$ at beam energy $E(^3He)$ = 800keV, measured with detector B

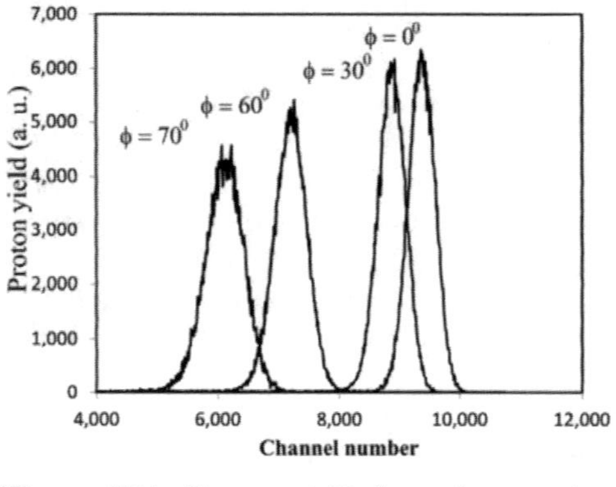

Figure 5(b). Proton yield from the reaction $D(^3He,p)^4He$ at beam energy $E(^3He)$ = 700keV, measured with detector B

A spectrum of the particles (elastically scattered He-3 and reaction products) in the backward direction is shown in figure 4. The low energy part reveals the target layer structure and was also used for spectrum normalization, while protons and ^4He particles originated from the $D(^3He,p)^4He$ and $D(^3He, ^4He)p$ reactions appear at higher energy values.

Proton spectra of the $D(^3He,p)^4He$ reaction have been collected at different forward angles in the range -70^0 to $+70^0$ using ^3He+ beams at 700 keV and 800 keV. Spectra of the detected particles in the forward direction, including the relative particle yield at different angles for both energies, are shown

Journal of Physics: Conference Series **398** (2012) 012029 doi:10.1088/1742-6596/398/1/012029

in figure 5(a) and 5(b). Proton yield as a function of the angle of detection in the forward direction is shown in figure 6(a) (for E(^{3}He) = 700keV) and in figure 6(b) (for E(^{3}He) = 800keV).

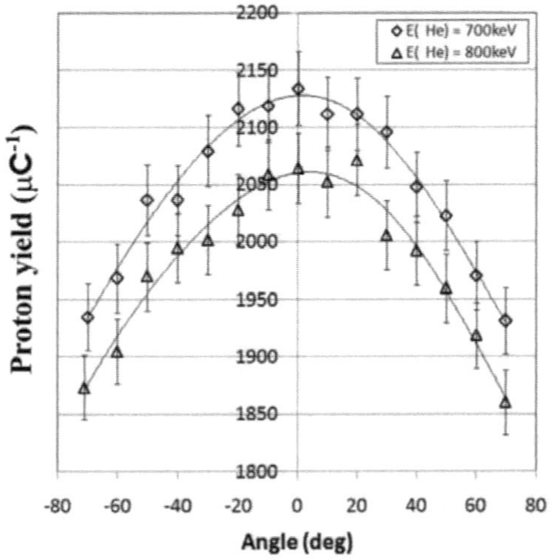

Figure 6(a). Measured proton yield from the reaction D(^{3}He,p)^{4}He at beam energies E(^{3}He) = 700keV and E(^{3}He) = 800keV

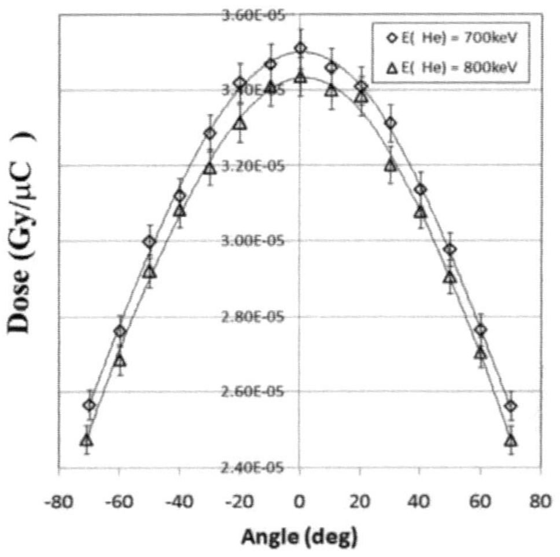

Figure 6(b). Calculated dose in Gy/μC for protons from the reaction D(^{3}He,p)^{4}He at beam energies E(^{3}He) = 700keV and E(^{3}He)= 800keV

4. Discussion and conclusion

From the kinematics of the reaction D(^{3}He,p)^{4}He it follows that maximum energy values of the protons in the forward directions, after passing through about 90μm thick Si substrate at angle of 0^0, is about 17MeV (at beam energy E(^{3}He) = 800keV) and 16.6MeV (at E(^{3}He) = 700keV). The range of these protons for in biological tissues is less than 3.2mm. For example, the range of protons in muscles for the two energies is 3.1mm and 2.9mm respectively (SRIM [12]). This fact limits possible applications to short range treatments only. A very thin ^{3}He beam, bombarding the target, would irradiate a hemispherical tissue volume of 0.125cm^3, if it is situated very close to the target [7]. Our measurements show that the integral delivered dose for such volume equals to 1.75×10^{-4} Gy/μC (at 700keV ^{3}He beam) and 1.705×10^{-4} Gy/μC (at 800keV ^{3}He beam). It is clear that to achieve radiation doses of 0.5 – 2 Gy, often used in proton therapy, one needs to go up with the beam current and the time of irradiation: for example to a beam current around 1mA for tens of seconds. However, in this case the main problem is not only the beam current limitations but also the sputtering of the layer caused by ^{3}He and the heat dissipated in the target, although the thermal conductivity of Ti hydride is of same order of that of pure Titanium [13] allowing for efficient cooling schemes.

When a broader and more uniform beam is needed, as it is desired in case of dermatological or intra-operative treatment, the distance from the deuterium target to the biological object has to be increased up to several centimeters and the size of the target can be enlarged. Following [7] such geometry is referred as "topical radiation field". Obviously, in this case, the dose will decrease with the square of the distance leading to larger irradiation times. An advantage however, is the possibility to enlarge the beam diameter to decrease the target heating.

To increase the number of emitted protons one could also increase the thickness of the deuterium-containing layer. A change of layer thickness of a factor of two to four would proportionally decrease the irradiation time or the beam current. Such a change can be obtained by sequential deposition of several Ti and TiD_x layers for mechanical stability. A further increase of the proton yield may be obtained by improving the deposition method to reach the TiD_2 stoichiometry.

Though the range 2.5 – 3mm of 17MeV protons in tissue is generally short, there are a number of possible applications such as dermatologic applications (skin cancer treatment), intraoperative treatment, tumor therapy of the vocal cords, nasal tumors, etc. [7]. Another application of this method is for R&D irradiation of laboratory small animal (mice), which requires a substantially lower range of the protons. Precise measurements of the radiation dose and therapeutic effect of protons in animals are scant today. One can believe that less-intensive proton beams can be successfully applied for experimenting radio-biological effects in small animals.

Acknowledgements
One of the authors (N Uzunov) would like to acknowledge the support received from the ICTP, Trieste, Italy and in particular the Program for Training and research in Italian Laboratories.

Sandia National Laboratories is a multi-program laboratory operated by Sandia Corporation, a wholly owned subsidiary of Lockheed Martin Company, for the U.S. Department of Energy's National Nuclear Security Administration under contract DE-AC04-94AL85000.

References
[1] Amaldi U (2005) *Nuclear Physics* **A 751** 409
[2] Paganetti H Proton Thearpy:History and Razionale *Proton Therapy Physics* (Series in Medical Physics and Biomedical Engineering) ed. H. Paganetti (CRC Press Taylor & Francis group)
[3] Khan F M *The Physics of Radiation Therapy (Fourth Eddition)* (Lippincott Williams & Wilkins a Wolters Kluwer business 2010) chapter III 413
[4] Slopsema R Beam Delivery Using Passive Scattering *Proton Therapy Physics* (Series in Medical Physics and Biomedical Engineering) ed. H. Paganetti (CRC Press Taylor & Francis group)
[5] Xu X G, Bednarz B and Paganetti H (2008) *Phys. Med. Biol.* ***53*** R193
[6] Regler M, Benedikt M and K Poljanc K *CERN Accelerator School (Intermediate Accelerator Physics)* Hephy-PUB-757/02
[7] Horn K M, Doyle B, Segal M N, Hamm R W, Adler R. J. and Glatstein E (1995) *Nucl. Instr. Meth. Phys. Res.* **B 106** 606
[8] Ding W, Shi L Q, Long X G, Luo S Z, Peng S M, Zhang B and Xia X J (2009) *Nucl. Instr. Meth.* **B267** 2341
[9] Möller W and Besenbacher F (1980) *Nucl. Instr. and Meth.* **168** Issues 1-3 111
[10] Kunz W E (1955) *Physical Review* Vol. 97 Issue.2 456
[11] Alimov V Kh and Mayer M (2005) *Nucl. Instr. Meth. Phys. Res.* **B234** Issue 3 169
[12] SRIM-2008 *Stopping Power and Range of Ions in Matter* NEA-0919/06
[13] Ito M, Setoyama D, Matsunaga J, Muta H, Kurosaki K, Uno M and Yamanaka S (2006) *Journal of Alloys and Compounds* **420** 25

Optical properties of plastic materials for medical vision applications

N G Sultanova[1], S N Kasarova[1] and I D Nikolov[2]

[1] Dept. of Mathematics and Physics, University "Assen Zlatarov"- Burgas,
1 Prof. Yakimov Str., Burgas-8010, Bulgaria,
[2] Dept. of Optics and Spectroscopy, Faculty of Physics, Sofia University,
5 J. Bourchier Blvd., Sofia-1164, Bulgaria

E-mail: sultanova@btu.bg

Abstract. Several types of optical polymer materials suitable for ophthalmic or medical vision applications have been studied. We have measured refractive indices of studied plastics at various wavelengths in the visible and near-infrared spectral regions. Important optical characteristics as Abbe numbers, dispersion coefficients and curves, principal and relative partial dispersion have been evaluated. Calculated refractometric data at many laser emission wavelengths used for medical surgery, therapy and diagnostics is included. As an example of a medical vision application of plastics, optical design of a micro-triplet for use in disposable endoscopes is presented.

1. Introduction

The variety of medical applications of optical plastic materials expands in consumer optics as well as in high precision vision devices. Over 50 percent of all eyeglass lenses sold nowadays are made of organic polymers. Except in ophthalmology, many surgical and diagnostic instruments as laparoscopes, arthroscopes, cystoscopes, endoscopes, etc. apply plastic materials [1]. The use of polymer optics is found in a wide range of not only clinical vision applications such as digital cameras, fiber photonic elements and systems, microscopes, surgical head-mounted video displays, LCD projectors, videoconferencing devices and mobile imaging. The rapid development of microelectronics and semiconductor industries places high demands on the performance of the applied miniature optical elements. Improvements in polymer and integrated hybrid glass-polymer components and systems offer efficient and compact solutions to meet these demands [2].

Optical polymers (OPs) have several key advantages over glasses as reduced weight and cost, high impact and shatter resistance and ability to integrate proper mechanical and optical features [1]. The injection-moulding process is very effective to reproduce not only spherical but sophisticated optical shapes such as aspheres, and other complex geometric surfaces or elements with noncircular apertures. Obviously, unique priority of polymers in medical applications is safety. The major drawbacks of plastic materials are that they have less scratch resistance and are more sensitive to environment changes such as temperature and humidity. Plastics lenses can be supplied with hard, anti-abrasion coatings with anti-static and hydrophobic properties, helping to maintain the transparency of the surfaces and making them easier to clean [3]. Broad thermal range functionality and better correction of aberrations of medical vision components and devices can be achieved by usage of hybrid glass-polymer optics. Incorporation of plastics and glass may result in low-cost, high-performance and high-volume optics.

Application of OPs in the design of optical elements and devices requires the knowledge of their refractometric and dispersive characteristics. We have studied optical properties of more than twenty types of transparent plastics, including principal and some new development polymers [4, 5]. A few of them, however, are suitable for ophthalmic or medical vision applications. Most popular for this purpose are materials as acrylic based polymers, polycarbonates, allyl diglycol carbonates, cyclo olefin polymers and copolymers, polystyrenes, alicyclic methacrylate copolymers. On the base of measured extensive refractometric data of such plastic samples at various wavelengths in the visible (VIS) and near-infrared (NIR) regions we have evaluated optical characteristics of studied materials as Abbe numbers, dispersive coefficients and curves, principal and relative partial dispersions. Influence of temperature on refraction of polymer media is experimentally determined and their thermo-optic coefficients are presented. Comparison to some glass types is carried out to confirm the optical quality of OPs and optimize material selection in accordance to consumer's preferences or vision system requirements.

2. Experimental

Several methods were applied to obtain refractometric data of OPs. In the VIS range of spectrum we have used the Carl Zeiss Jena Pulfrich-Refractometer PR2 with its V-type SF3 glass prism. Measurements are based on the deviation angle method. Samples were produced as injection moulded plates or cubes with two fairly well polished, mutually perpendicular surfaces to obtain good refractometric data [4,5]. Proper immersion emulsions with a suitable refractive index were used to ensure the optical contact between the plastic samples and the prism. Measuring temperature of 20°C was maintained and temperature regulation was possible with stability of 0.2°C. Refractive indices were measured at the emission wavelengths of the spectral lamps of the PR2 instrument in the VIS region: for the mercury source at the green e-line 546.07 nm and blue g-line 435.83 nm, for the helium source at the yellow d-line 587.56 nm, and for the hydrogen source at the blue F-line 486.13 nm, and red C-line 656.27 nm. Measurement accuracy of the Pulfrich-Refractometer is 2×10^{-5} [6].

The V-type prism with its thermostatic housing is suitable to obtain refractometric data at varying temperature. Some of the polymers were measured in the range between 10 and 50 °C using a MLW thermostat U4 with water bath, made in Germany. Thus influence of temperature on refraction of OPs was investigated.

Additional goniometric set-up with the same prism, a white lighting module (a 250 W halogen lamp and a condenser system) with interference filters (IFs) and a photo detector device was applied for measuring in the entire VIS and NIR regions. The transmission maxima of the applied IFs are as follows: 548, 589, 597, 659, 703 nm for the VIS region, and 752, 804, 833, 879 and 1052 nm for the NIR spectrum. A G5-LOMO goniometer with an accuracy of one arc second was used to measure the deviation angle. Comparison between the obtained results by the PR2 refractometer and the goniometric set-up in VIS light was possible. The metrological tests in [4] pointed out that our measurements of the OPs' indices guarantee accuracy better than ±0.001. A number of polymer specimens have been measured with this goniometric set-up and a He-Ne laser at 632.8 nm as a lighting module [7].

Transmission of studied OPs was measured in the spectral range from 400 to 2500 nm using a UV-VIS-NIR spectrophotometer Varian Carry 5E. For this purpose thin polymer films with varying thickness have been produced by dissolving some of the polymers in proper organic solvents [8].

3. Optical characteristics

3.1. Refraction

Among studied polymers with medical purpose applications are polymethyl methacrylate (PMMA), polystyrene (PS), polycarbonate (PC), cyclo olefin plastic Zeonex E48R, alicyclic methacrylate copolymer Optorez 1330 and Bayer which is a polycarbonate material. Measured refractive indices at eight wavelengths are presented in table 1. Columns 1, 2, 3 and 5 refer to the indices obtained by means of the PR2 instrument and the rest of the columns – the results received on the goniometric set-up. Column 4 presents laser measurements at emission wavelength 632.8 nm. Additional refractometric data at the d-line is included in table 2 to make more convenient the comparison to optical glass types with similar refraction.

Table 1. Measured refractive indices of OPs.

OPs	Wavelength [nm]							
	435.8	486.1	546.1	632.8	656.3	703	879	1052
	1	2	3	4	5	6	7	8
PMMA	1.5025	1.4973	1.4934	1.4890	1.4890	1.4863	1.4835	1.4813
PS	1.6171	1.6056	1.5963	1.5872	1.5862	1.5820	1.5756	1.5718
PC	1.6117	1.5994	1.5896	1.5802	1.5793	1.5749	1.5683	1.5645
Optorez 1330	1.5219	1.5163	1.5119	1.5075	1.5065	1.5054	1.5021	1.4984
Zeonex E48R	1.5431	1.5376	1.5333	1.5284	1.5282	1.5256	1.5224	1.5204
Bayer	1.6121	1.5998	1.5905	1.5814	1.5803	1.5765	1.5698	1.5660

Presented data reveals a monotonous decrease of refractive indices with increasing wavelength of all studied polymers which confirms their normal dispersion behaviour in the measuring spectral region from 435 to 1052 nm. On the base of the experimentally obtained results we analyse the dispersive properties of OPs and comparison to optical glass is possible.

The range of refractive index values of plastics is much more restricted compared to glass. Glass catalogues propose materials with refraction usually between 1.49 up to 1.90 at the d-line but special products with n_d of about 1.43 and 2.01 are also available [9, 10]. Measured data of n_d of all of the examined OPs varies in a limited interval from 1.47 for the Cellulose material [5] up to 1.59 for the PS polymer (table 2). Recently, developing the technological processes some companies report for special plastics with n_d as high as 1.74 [11]. Nevertheless, designers have at their disposal not as many optical-grade polymers as compared to glass.

In table 2 we present optical characteristics of measured OPs (columns 1-5). Column 6 refers to the one of the most frequently used in ophthalmology CR39 plastic which is an allyl diglycol carbonate material produced by PPG Industries [12]. The included types of optical glass in columns 7 and 8 are also for ophthalmic use and column 9 presents cited data of the popular SCHOTT glass N-BK10 [9]. HC-Weiss 0290 and Unicrown materials are trademarks of spectacle glass types of Barberini [13] and Corning [14] companies, respectively.

Table 2. Optical characteristics of OPs and selected glass types.

Characteristics	PMMA	PS	PC	Optorez 1330	Zeonex E48R	CR39	HC-Weiss 0290	Uni-crown	N-BK10
	1	2	3	4	5	6	7	8	9
n_d	1.4914	1.5917	1.5849	1.5094	1.5309	1.4985	1.6008	1.5230	1.4978
$n_F - n_C$	0.0083	0.0194	0.0201	0.0098	0.0094	0.0084	0.0136	0.0089	0.0074
ν_d	59.2	30.5	29.1	52.0	56.5	59.3	44.2	58.8	66.95
$P_{g,F}$	0.626	0.593	0.612	0.571	0.585	–	0.573	0.590	0.530
$P_{d,C}$	0.289	0.283	0.279	0.296	0.287	0.345	0.301	0.303	0.309
$\Delta n_e/\Delta T \times 10^{-6}$/K	-130	-131	-100	-120	-126	–	–	–	3.4
UV cut-off [nm]	360 [15]	390	380[16]	–	340[17]	355[12]	330[13]	320[14]	270[9]

The significant influence of temperature on refraction of polymers is demonstrated by the presented in table 2 values of the calculated gradient $\Delta n_e/\Delta T$ for the range 20÷40 °C. Measured n_g, n_F, n_e, n_d, n_C values of studied polymers decrease with increasing temperature [18]. Thermal properties of plastics are rather different compared to glass and this fact should be carefully regarded in medical vision applications. The thermo-optic coefficients of OPs are negative and contrast to the positive coefficients of almost all optical glass types. Their absolute values are with about two orders of magnitude larger than those for glass [9-11]. It seems that integration of plastic and glass materials in a hybrid component may result in its thermal stability suitable for high performance medical imaging systems. A polymer layer can be laid on the glass body for stable thermal behaviour of the hybrid.

3.2. Dispersion

Important optical parameters in the design of vision systems are Abbe numbers ν_d or ν_e, principal dispersion $n_F - n_C$ and relative partial dispersions $P_{x,y}$ at selected wavelengths x and y. In table 2 these characteristics are presented for studied OPs while the respective values for the CR39 ophthalmic material and considered glass types are evaluated on base of catalogue data. Included relative partial dispersions are calculated by the following relations:

$$P_{g,F} = \frac{n_g - n_F}{n_F - n_C} \text{ and } P_{d,C} = \frac{n_d - n_C}{n_F - n_C}. \quad (1)$$

These characteristics are important for chromatic correction. In the design of multiple component optical systems partial dispersions should be closely matched while Abbe numbers differ substantially. As for example, in the VIS region the PMMA could be combined with the PS or PC material in a plastic doublet. The Zeonex material and PS plastic also form a suitable achromatic pair.

In figure 1(a) the optical scheme of all-plastic endoscopic micro-objective is proposed as an example of a medical vision application. The purpose of such objectives is to form and process qualitative colour images of investigated human tissues and organs. High performance endoscopes are expensive and their multiple usages require time-consuming and exhaustive cleaning procedures every time they are used. This problem might be solved by low cost but reliable disposable endoscopes. The presented wide-angle triplet is made of PMMA and PS plastics for better colour image correction. In figure 1(b) residual comma aberrations, calculated by means of OSLO optical program, are presented at different imaging field of view. Spot size increases from 0.009 mm in on-axis view to 0.015 mm for the full field of the triplet. Calculated diffractive point spread functions (DPSFs) and fractional energy versus spot radius are illustrated in figure 1(c) again at different angles of the incident rays. The charts show good aberration correction of the micro-objective in case of on-axis view in the focal plane.

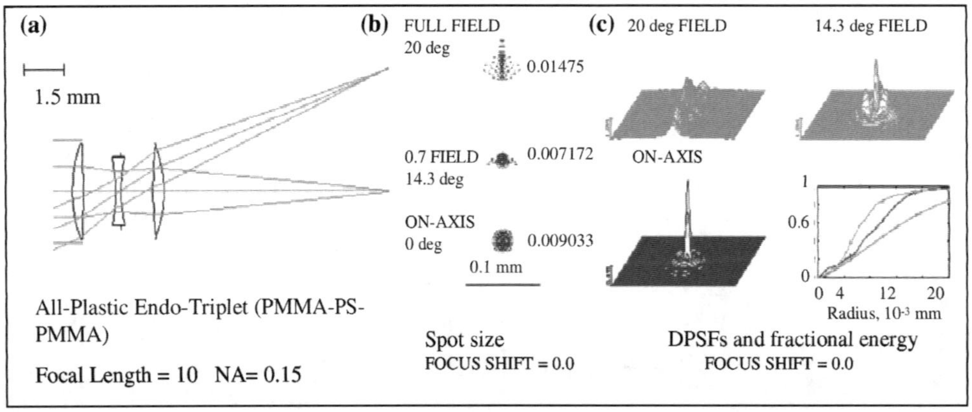

Figure 1. Optical scheme and aberrations of endoscopic triplet.

Our results show that OPs transmit well in the VIS and NIR regions and have transmittance better than 90% in the considered spectral range from 430 to 1100 nm [8]. Requirements of ISO standards for ophthalmic materials include information on the ultraviolet (UV) cut-off point. Therefore, in table 2 we present this data, since the UV radiation contributes to the development of various ocular disorders. As it can be seen, high refractive lens materials generally absorb more UV light. Ordinary glasses absorb UV rays with wavelengths under 300 nm and IR light over 3500 nm but some ophthalmic types as HC-Weiss 0290 and Unicrown have a little bit higher UV cut-off. Plastic lenses do not protect the eye from UV rays unless properly tinted. Different UV inhibitors should be added to polymers or tinting by organic dyes of lens surfaces is carried out to decrease the UV absorption.

Obtained transmission spectra as well as presented refractive data of studied OPs reveal their normal dispersion properties. We have applied the Cauchy–Schott's formula involving six dispersion coefficients which assures calculation accuracy better than ±0.0001 in the VIS and NIR regions from 435.8 to 1052 nm [5]:

$$n_\lambda^2 = A_1 + A_2\lambda^2 + \frac{A_3}{\lambda^2} + \frac{A_4}{\lambda^4} + \frac{A_5}{\lambda^6} + \frac{A_6}{\lambda^8}, \qquad (2)$$

where $A_1,\ldots, A_6$ are the Cauchy's coefficients. A program OptiColor was realized on the base of equation (2) that allows us calculation of dispersion coefficients, charts, Abbe numbers and random values of refractive indices. The input data consists of six measured indices at selected wavelengths and a linear system of equations is solved to compute the Cauchy's coefficients. Obtained dispersion charts of some of the studied polymers are illustrated in figure 2(a) in the whole considered spectral region. Higher refractive materials have steeper curves and therefore exhibit greater dispersion.

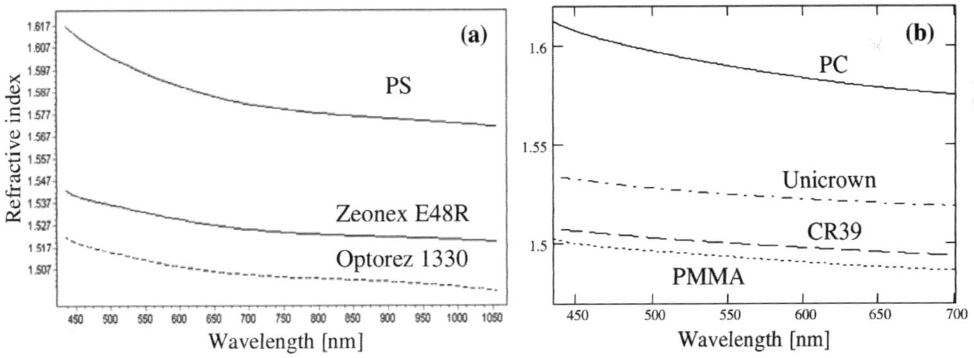

Figure 2. Dispersion charts of (a) principal OPs; (b) ophthalmic OPs and glass.

In figure 2(b) graphs of some ophthalmic OPs and the Corning Unicrown type glass are shown. Because of lack of sufficient refractometric data for the CR39 plastic, which is most frequently used for vision corrections, dispersion coefficients and charts have been calculated in the VIS light using only three terms of the series in equation 2. The curves of PMMA and CR39 as well as their Abbe numbers (table 2) confirm their low dispersion. However, the CR39 material is preferred in ophthalmic applications because it is harder and more resistant to scratches than other plastic lens materials. The Unicrown spectacle glass is also with similar dispersion. Our investigations on more than twenty optical polymers prove that OPs are higher dispersive materials in VIS light but may have similar or lower dispersion in NIR spectrum [19]. This fact makes them preferable materials in night vision photonic instruments.

Medical lasers appear in surgical and many non-surgical treatments in cosmetics, dentistry, therapy and vision diagnostic or biomedical analysers. In all cases light is transmitted by means of all-plastic or hybrid glass-plastic optics. In table 3 calculated refractive indices at emission wavelengths of some commercial medical lasers, including diode lasers, are presented. Data may be useful for designers of laser devices.

Table 3. Calculated refractive indices at laser emission wavelengths.

OPs	Laser emission wavelength [nm]								
	430	457.1	488	515	532	670	785	867	980
PMMA	1.504	1.499	1.497	1.486	1.495	1.487	1.484	1.483	1.482
PS	1.619	1.611	1.605	1.601	1.599	1.584	1.578	1.576	1.574
PC	1.614	1.605	1.599	1.595	1.592	1.577	1.571	1.569	1.566
Optorez 1330	1.523	1.519	1.516	1.514	1.513	1.506	1.504	1.502	1.500
Zeonex E48R	1.544	1.540	1.537	1.364	1.534	1.527	1.524	1.523	1.522
Bayer	1.614	1.605	1.599	1.954	1.592	1.577	1.571	1.569	1.566

4. Conclusions

Our investigations on optical properties of OPs reveal their appropriate characteristics and compatibility with glass for application in all-plastic or hybrid glass-plastic optics for medical vision devices and systems. Refractive, as well as dispersive parameters of polymers conform to their implementation in well corrected and high performance medical imaging components. Because of safety and twice lower weight OPs are preferable materials in consumer optics. Ophthalmic plastics allow for production of thinner, lightest lenses possible at reduced price. Some of their drawbacks as low scratch resistance and UV absorption can be surmounted by special coatings or tinting. Attention should be drawn to the influence of temperature on refraction and dispersion of polymers.

Plastic optics offers additional design freedoms that are not achievable or economical with glass optics.

References

[1] Beich W 2002 *Phot. Spectra* **36** 127
[2] Doushkina V 2011 *BioPhotonics* **18** 32
[3] 1991 *Handbook of Ophthalmic Optics* (Oberkochen: Carl Zeiss)
[4] Sultanova N, Nikolov I and Ivanov C 2003 *Opt. Quant. Electron.* **35** 21
[5] Kasarova S, Sultanova N, Ivanov C and Nikolov I 2007 *Opt. Mater.* **29** 1481
[6] 1976 *The Manual of PR2* (Jena: Carl Zeiss JENA)
[7] Nikolov I, Sultanova N and Kasarova S 2005 *Proc. SPIE* **5830** 511
[8] Sultanova N, Kasarova S and Nikolov I 2012 *J. Phys.:Conf. Ser.* **356** 012049
[9] SCHOTT Glass Technologies http://www.us.schott.com
[10] HOYA Corporation USA http://www.hoyaoptics.com
[11] MITSUI Chemicals Inc. http://www.mitsuichem.com
[12] PPG Industries http://www.ppg.com
[13] Barberini GmbH http://www.barberini.eu
[14] Corning Incorporated http://www.corning.com
[15] Steeneken S, Buma A and Gieskes W 1995 *Photochem. Photobiol.* **61** 276
[16] Meister D and Sheedy J 2008 *Introduction to Ophthalmic Optics* (San Diego: Carl Zeiss Vision) p 109
[17] ZEON Corporation, http://www.zeonex.com
[18] Kasarova S, Sultanova N and Nikolov I 2010 *J. Phys.: Conf. Ser.* **253** 012028
[19] Sultanova N, Kasarova S and Nikolov I 2009 *ACTA PHYS. POL. A* **116** 585

Nanoscale surface modification of plastic substrates for advanced tissue engineering applications

N Donkov[1], V Safonov[2], A Zykova[2], J Smolik[3], R Rogovska[3], A Goltsev[4], T Dubrava[4], I Rossokha[4], V Georgieva[5]

[1]Emil Djakov Institute of Electronics, Bulgarian Academy of Sciences, Sofia, Bulgaria
[2]NSC „Kharkov Institute of Physics and Technology, Kharkov, Ukraine
[3]Institute for Sustainable Technologies, NRC, Radom, Poland
[4]Institute for Problems of Cryobiology and Cryomedicine NASU, Kharkov, Ukraine
[5]Georgi Nadjakov Institute of Solid State Physics, Bulgarian Academy of Sciences, 72, Tzarigradsko Chausse Blvd., 1784, Sofia, Bulgaria

E-mail: nikolaidd@abv.bg

Abstract. Modified surface properties such as composition, nano roughness, wettability have effect on the most important processes at biomaterial interface. The research of stem cells (MSCs) adhesive potential, morphology, phenotypical characteristics on oxide coated and plastic substrate with different surface parameters was made. The oxide coatings deposition on plastic substrates shifts the surface properties at the more hydrophilic region and results in next positive cell/ biomaterial response in vitro tests. The MSCs marker number increases on the oxide nano structural surface of plastic substrates.

1. Introduction.

Recently, various stem cells, including mesenchymal stem cells (MSCs), have great potential in the tissue engineering area due to their biological capacity to differentiate into specific lineages and next advanced therapies applications [1-3]. Functional coatings deposition is an effective way of surface modification with direct control of stoichiometry, impurity elements, functional groups and surface charges. Coatings based on Ti, Al, Zr, Ta oxides exhibit unique properties: high inductivity, density, bio- and chemically inertness, which are very important for next implant and tissue engineering applications [4,5]. As the characteristic size scale of cellular receptors interaction is a few 100 nm, the surface relief modifications should be also controlled at a nanometer-submicronic level [6,7]. The separation of different factors influenced on cell/surface interaction such as nano topography, surface chemistry and surface free energy is of great interest for next tissue engineering applications.

2. Materials and Methods

At the present study the samples were plastic substrates (Petri dishes), uncoated and oxide coated (Al_2O_3, ZrO_2 (MS)) magnetron sputtered with different roughness parameters – 20, 200 and 400 nm. Al_2O_3 and ZrO_2 (MS) coatings deposition was performed in high vacuum pumping system with the base pressure about 10^{-3} Pa. The main process parameters were magnetron discharge power 1-8 kW, oxygen source power 1kW, deposition rate 8 μm/hour [8]. The deposition process in the case of fusible materials (plastic Petri dishes) has some difficulties - in the standard regimes the plastic Petri dishes melting. To optimize the regimes of deposition (the low temperature of samples and high rate of coating deposition process) the additional research of magnetron plasma and inductive discharge parameters that influence on the temperature of substrate was made. The temperature of substrates depends both on the parameters of magnetron and induction discharges, substrate location and method of cooling. By means of thermocouple the temperature dependences and the parameters of discharge, distance to the substrate, temperature distribution at the deposition area were measured. The radial temperature distribution in the plane of substrate holder was presented (figure1). Figure 2 demonstrates the temperature dependences for different regimes of magnetron sputtering oxide coatings deposition: magnetron power -5 kW, 2,5 kW and distance - 30 cm The discharge power decrease results in significant temperature decreasing. The distance to the target shift from 25 cm to 30 cm also leads to temperature decreasing.

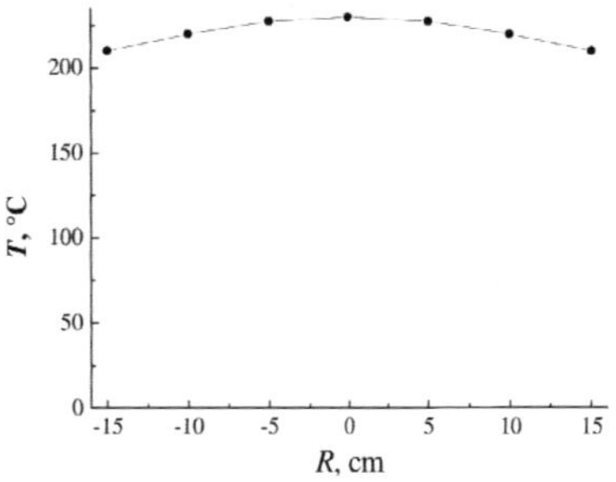

Figure 1. The temperature radial distribution: magnetron power -5 kW, distance - 25 cm

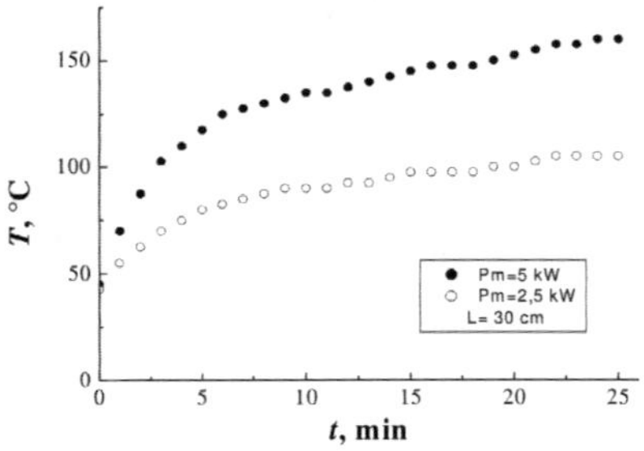

Figure 2. The temperature dependences of different regimes of magnetron sputtering coating deposition process on plastic

The results allow to optimize the process of coatings deposition, to avoid the excessive heating of substrates and to select the most effective conditions for deposition process on plastic substrates.

The structure of Al_2O_3, ZrO_2 (MS) magnetron sputtered coatings were investigated by means of XPS and XRD methods. X-ray diffraction profiles of Al_2O_3 (MS) were observed by means of diffraction device "DRON-3" with filtered Cu-Ku radiation. X-ray photoelectron spectroscopy was carried out using ESCALAB MkII (VG Scientific) electron spectrometer at a base pressure in the analysis chamber of $5x10^{-8}$ Pa (during the measurement $1x10^{-6}$ Pa), using Al Kalpha X-ray source (excitation energy hv=1486.6 eV). The instrumental resolution measured as the full width at a half maximum (FWHM) of the Ag3d5/2, photoelectron peak is 1 eV [9].

The surface roughness was estimated by profilometer Hommel T-2000 measurements. The surface topography was investigated by means of Interferometric Microscope Talysurf CCI (Taylor Hobson) and AFM (Quesant Instrument Corporation, USA) methods.

The contact angles were measured by means of tensiometric method in Kruss K12 Tensiometer at temperature 20° C. The standard liquids with well-known values of surface tension, component of dispersion and polar interaction such as water, formamide, diiodo methane, ethylene glycol, α-bromo naphthalene and glycerol were used [9].

The others parameters such as surface free energy (SFE), its polar and dispersion components and fractional polarity were determined by means of Wu, Neumann-Kwok methods for plastic substrates.

Integral index of structural and functional state of cells reflecting their functional activity is adhesion. Therefore, experimental studies in vitro were started from the investigation of adhesive potential of bone marrow cells (BM), cultured on plastic uncoated and oxide coated substrates with different roughness parameters. MSCs phenotypical characteristics were determined by cytofluorimeter FACS Calibur with using fluorochromal monoclonal antibodies to CD73, CD106, CD44 structures. The CD105 fractures were extracted from cultivated MSCs by means of antimouse CD105 MultiSort Kit (PE) for next phenotype evaluation by cytofluorimetry.

3. Results and Discussion.

X-ray diffraction profiles of Al_2O_3 (MS), ZrO_2 (MS) as-deposited coatings .demonstrate the amorphous nature, no peaks were observed. The tailored surface chemistry, stoichiometric composition and surface states of coatings are very important for next cell/biomaterial response. The structural analysis of Al_2O_3 (MS) and ZrO_2 (MS) oxide coatings by means of XPS method was presented. The photoelectron spectra of Al_2O_3 (MS) and ZrO_2 (MS) oxide demonstrate that oxide coatings have strong stoichiometric composition with binding energy peaks E(Al2p) – 74.4 eV, E(O1s) – 531.3 eV for Al_2O_3 (MS) films (figure 3 a, b). The binding energy peaks E(Zr3d) – 182.4 eV (figure 3 c), E(O1s) – 530.2 eV for ZrO_2 (MS) coatings were observed. At the Zr3d XPS spectra exist 2 peaks, demonstrated the energy shift 2,43 eV of 3d level due to spin orbit coupling [10].

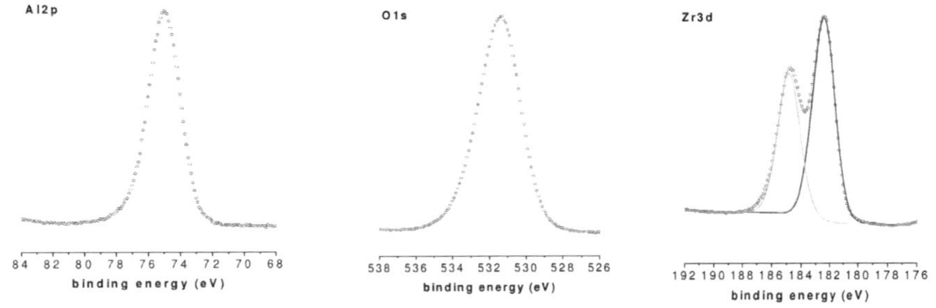

a) b) c)

Figure 3. Al2p XPS spectra of Al_2O_3 (MS) coatings (a), O1s XPS spectra of Al_2O_3 (MS) coatings (b) and Zr3d XPS spectra of ZrO_2 (MS) coatings (c).

The surface topography of investigated coatings deposited on the plastic substrates with different parameters of surface roughness was evaluated by means of Interferometric Microscope Talysurf CCI (Taylor Hobson) and AFM (Quesant Instrument Corporation, USA) methods. (figure 4).

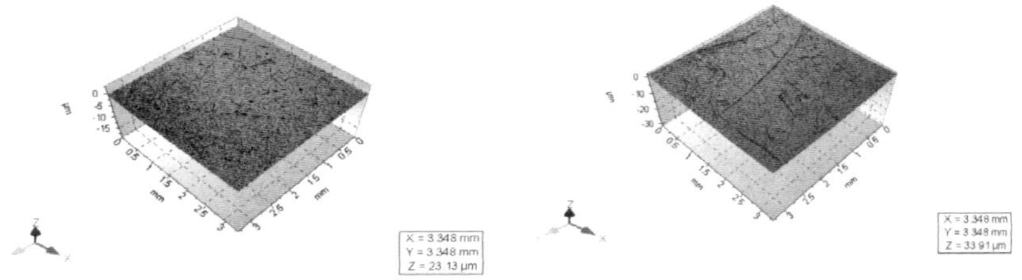

a) **b)**

Figure 4. The surface topography. a) plastic substrate (Petri dishes), b) plastic substrates /Al_2O_3 (MS)

The surface free energy SFE and its polar and dispersion parts, fractional polarity estimations were made by Wu method for two liquids system (ethylene glycol – water at temperature 20°C) and by Neumann-Kwok methods for plastic substrates. with various roughness parameters. The SFE was in the range 30-32 [mN/m] for plastic substrates and 43-45 [mN/m] for ZrO_2 and 54-55 for Al_2O_3 oxide coated on plastic substrates by Wu method.

The most hydrophobic nature, maximum values of advancing contact angles and minimum values of SFE for various roughness parameters in the range 20-400 nm possess plastic substrates. The hydrophobic properties of plastic substrates change in the more hydrophilic region and surface free energy increase for oxide coated plastic substrates (figure 5).

The influence of coatings roughness and composition parameters on BM adhesive potential was analyzed. The adhesive potential of BM cells was statistically and significantly different depending on substrate material properties. The best results were obtained on substrate surfaces with minimum roughness parameters in the range 20-400 nm.

The study of the expression of MSCs genes was made. An important component of immune modulating activity of MSCs is enzyme indoleamine 2,3-dioxygenase (IDO) which is produced by cells. The rise in the expression rate of IDO gene in BM after culturing on plastic/Al_2O_3 (MS) substrate in comparison with BM cultured on plastic substrate after the 1st and the 2nd passage was observed (figure 6).

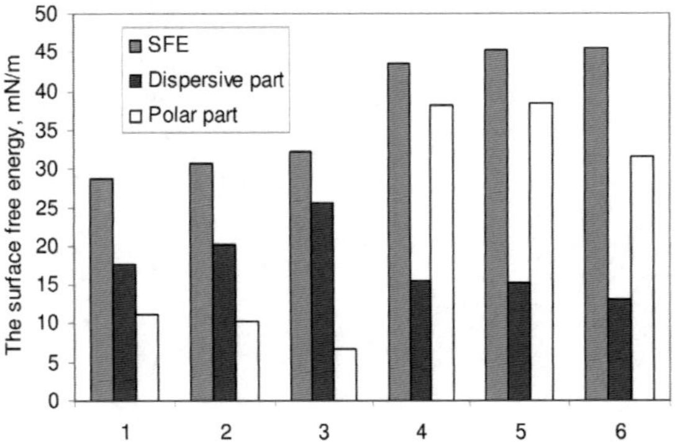

Figure 5. The surface free energy SFE and its polar and dispersion parts [mN/m] 1-3 – plastic with roughness 20, 200, 400 nm, 4-6 – plastic 20, 200, 400 nm with oxide coatings

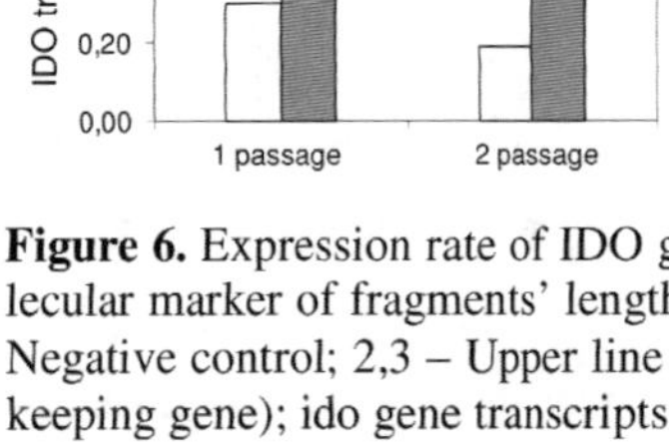

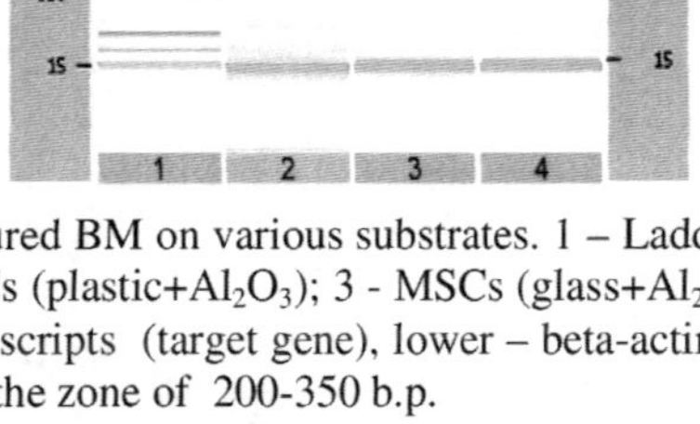

Figure 6. Expression rate of IDO gene in cultured BM on various substrates. 1 – Ladder (molecular marker of fragments' length); 2 – MSCs (plastic+Al_2O_3); 3 - MSCs (glass+Al_2O_3); 4 – Negative control; 2,3 – Upper line – IDO transcripts (target gene), lower – beta-actin (house keeping gene); ido gene transcripts located in the zone of 200-350 b.p.

Due to the increase of the cell content with MSCs markers during culturing on oxide nano structural surface, rise in the expression rate of IDO gene is logic since in the culture of BM cells, namely MSCs are IDO producers.

The rise in the mentioned transcripts of the cells with MSCs markers during BM culturing on oxide nano structural surface was demonstrated both for plastic and glass substrates.

4. Conclusions.

The best MSCs adhesive potential and phenotypical characteristics were obtained in the case of magnetron sputtered oxide Al_2O_3 (MS), ZrO_2 (MS) coatings with minimum parameters of roughness in the range 20-400 nm and intermediate values of surface free energy in the range 45-55 mN/m, the greater part of SFE polar components and fractional polarity calculated by means of Wu, Neumann-Kwok methods for plastic substrates with various roughness parameters. The changes at molecular-genetic apparatus MSCs (IDO gene expression degree) and MSCs marker number increasing on the oxide nano structural surface of plastic substrates were observed. The results show the significant effect of surface parameters modification on the regularities and mechanisms of nano materials interaction with mesenchymal stem cells (MSCs) in vitro tests.

Acknowledgment
The study was supported by the Program for Cooperation between Bulgarian Academy of Sciences and the National Ukrainian Academy of Sciences.

References

[1] Pittenger M F, Plence P and Muzic R F 1999 *Science* **284** 143
[2] Anselme K 2000 *Biomaterials* **21** 667-81
[3] Fibbe W E and Noort W A 2003 *Ann. NY Acad. Sci.* **996** 235
[4] Dhayal M, Cho S I, Moon J Y, Cho S J and Zykova A 2008 *J. Applied Surface Sci* **254** 3331
[5] Links J, Boyan B D, Blanchard C R, Lohmann C H, Liu Y and Cochran D 1998 *Biomaterials* **19** 2219
[6] Li D, Ferguson S J, Beutler T, Cochran D L, Sittig C, Hirt H P and Buser D. 2002 *J. Biomed. Mater. Res.* **60** 325
[7] Hammer D A and Lauffenburger D A 1987 *J. Biophys.* **52** 475
[8] Zykova A, Safonov V, Luk'yanchenko V, Walkowicz J and Rogovska R 2007 *Prob. At. Sci. Tech. Series:Plasma Physics* **13** 200
[9] Donkov N, Mateev E, Zykova A, Safonov V and Luk'yanchenko V 2010 *J. Phys. Conf. Series* **223** 012030
[10] Safonov V, Zykova A, Smolik J, Rogovska R, Donkov N, Goltsev A, Dubrava T, Rassokha I and Georgieva V 2012 *J. Phys. Conf. Series* **356** 012046

Low dose X –ray effects on catalase activity in animal tissue

R Focea, C Nadejde, D Creanga, T Luchian
"Alexandru Ioan Cuza" University, Faculty of Physics, Blvd. Carol I, No.11, 700506 Iasi, Romania

E-mail: ramona_focea@yahoo.com, mdor@uaic.ro

Abstract. This study was intended to investigate the effect of low-dose X ray-irradiation upon the activity of catalase (CAT) in freshly excised chicken tissues (liver, kidney, brain, muscle). The tissue samples were irradiated with 0.5Gy and 2Gy respectively, in a 6 MV photon beam produced by a clinical linear accelerator (VARIAN CLINAC 2100SC). The dose rate was of 260.88cGy/min. at 100 cm source to sample distance. The catalase level was assayed spectrophotometrically, based on reaction kinetics, using a catalase UV assay kit (SIGMA). Catalase increased activity in various tissue samples exposed to the studied X ray doses (for example with 24 % in the liver cells, $p<0.05$) suggested the stimulation of the antioxidant enzyme biosynthesis within several hours after exposure at doses of 0.5 Gy and 2 Gy; the putative enzyme inactivation could also occur (due to the injuries on the hydrogen bonds that ensure the specificity of CAT active site) but the resulted balance of the two concurrent processes indicates the cell ability of decomposing the hydrogen peroxide–with benefits for the cell physiology restoration for the chosen low dose radiation.

1. Introduction

Since the discovery of X-ray over 100 years ago, radiation has been used increasingly in medicine and industry to help for diagnosis, treatment, and technology purposes. However, radiation hazards present an enormous challenge for the biological and medical safety. The deleterious effects of ionizing radiation in biological systems are mainly mediated through the generation of reactive oxygen species (ROS) in cells as a result of water radiolysis [1]. Among them, particularly, the highly damaging hydroxyl radical ($^{\bullet}OH$) can cause injury by reacting with biomolecules [2, 3]. ROS and oxidative stress may contribute to radiation-induced cytotoxicity and to metabolic and morphologic changes in animals and humans during radiotherapy, experimentation, or even space flight [4]. Against oxidative stress, cells are equipped with several natural enzymatic and non-enzymatic antioxidant defenses [5]. The exposure of human body to ionizing radiation could lead, however, to depletion of these endogenous antioxidants [6] and ultimately to the development of systemic diseases. Irradiation up to 14.4 Gy caused marked decrease in serum melatonin and its pineal biosynthesis three and five days after radiation exposure [7]. In addition, total antioxidant capacity of plasma was reduced in patients exposed to whole body irradiation for the purpose of reducing tumor growth. Consequently, the cellular antioxidant capacity was decreased and the organs become more susceptible to the deleterious effects of ROS [8]. On the other hand, more than 35 countries worldwide allow the irradiation of food. In the USA, in addition to raw beef, pork and lamb, spices, wheat, flower, potatoes, pork, fruits and vegetables, have been irradiated. Irradiation was approved to control *Trichinea* in pork in 1990, *Salmonella* in chickens in 1994, and *E. coli* in row meat in 1999. In Canada, irradiation is used to prepare meals for hospital patients. French food processors also use irradiation. The World Health Organization designated food irradiation as a perfectly sound food preservation technology that is needed in the world where food borne diseases are on the increase, and where in between 1/4 to 1/3 of the global food supply is lost post – harvest. The approved doses for food preservation are: 4 kGy for chilled beef meat, 7 kGy for frozen beef meat, maximum 30 kGy for spices, and less than 1 kGy for fresh vegetables. Exposure of living organisms to low-dose ionizing radiation – still considered a new radiobiological concept - could trigger various metabolic reactions that help cells to defend against oxidative stress such as that related to hydrogen peroxide - enhanced following water radiolysis, often generating the indirect effect of radiations. Hydrogen peroxide is highly deleterious to the cell and its accumulation causes oxidation of cellular targets such as DNA, proteins, and lipids leading to mutagenesis and cell death [9-11]. In the context of food preservation, this study was intended to investigate the effect of low-dose X ray-irradiation upon the activity of catalase (CAT) in freshly excised chicken tissues (liver, kidney, brain, muscle).

2. Materials and methods

2.1. X –ray exposure

The biological material was represented by freshly excised chicken tissues (kidney, muscle, liver, brain).

The tissue samples were irradiated with 0.5 Gy and 2 Gy respectively, in a 6 MV photon beam produced by a clinical linear accelerator (VARIAN CLINAC 2100SC – Regional Institute of Oncology, Iasi, Romania). The dose rate was of 260.88c Gy/min. at 100 cm source to sample distance. The exposure time for the studied doses delivery was calculated using the following formula [12]:

$$t = D/\dot{D}\ (z_{max},\ 10,\ h\nu) \times RDF\ (A,\ h\nu) \times PDD\ (1{,}5cm,\ A,\ h\nu) \times 0.005029 \qquad (1)$$

were t is the irradiation time, D is the absorbed dose in the sample; $\dot{D}\ (z_{max})$ is the dose rate at the point z where it reaches its maximum value on the central axis of a 10 x10 cm^2 photon beam; $RDF\ (A)$ is the relative dose factor (depending on the radiation type); *PDD* (1.5 cm, A, $h\nu$) is the percentage depth dose at 1.5 cm depth, for a *A* field size. The parameters for the given irradiation geometry (6 MV photon beam) were obtained after beam calibration procedures (in accordance with dosimetric standard IAEA TRS- 398) using a 3D Blue Water Phantom and a PTW Unidose Electrometer.

2.2. Catalase activity

The catalase activity was determined using UV analysis method described by Hugo Aebi [13], using a catalase UV assay kit (SIGMA). The assay involves spectrophotometrically survey of the decrease in absorbance of hydrogen peroxide at 240 nm with kinetic recording software. The assay was performed within a short time period (60 seconds), because catalase reaction is very fast.

Unit definition: One unit of catalase will decompose 1.0 micromole of hydrogen peroxide to oxygen and water per minute at pH 7.0 at 25 °C at a substrate concentration of 10 mM hydrogen peroxide.

2.3. Statistical analysis

To ensure statistical significance, three replays of every sample and control were analyzed in identical conditions. Average values and standard deviations were considered for graphic representations. Student *t*-test was applied to assess the statistical significance of differences between controls and irradiated samples, with the threshold of 0.05 in all experiments.

3. Results and discussions

The study on the action of low-dose X-rays on the antioxidant defense system in the kidney tissue is represented by monitoring the absorbance decrease of hydrogen peroxide and calculating the enzymatic activity of catalase. The absorbance dynamics of hydrogen peroxide for 60 s for the kidney samples is shown in figure 1, where the downward trend of the absorbance of hydrogen peroxide as a result of its decomposition into O_2 and water by the action of catalase can be seen.

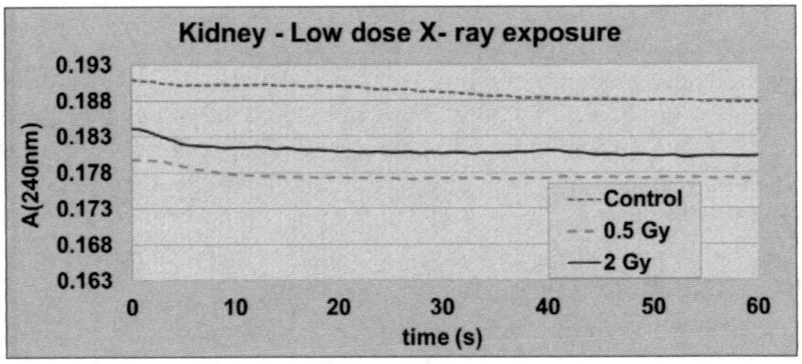

Figure 1. Hydrogen peroxide absorbance dynamics at 240 nm for control and irradiated chicken kidney tissue

The catalase activity (figure 2) for the kidney tissue revealed a significant increase in the enzymatic activity, with about 40% and 100% ($p<0.05$) for 0.5 Gy and respectively 2 Gy, compared to control.

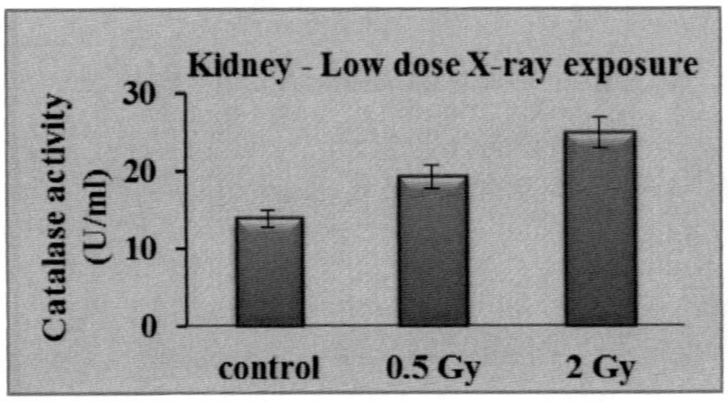

Figure 2. Catalase activity in chicken kidney tissue *vs*. X-ray dose

The exposure of brain tissue to low doses of X-ray led to an increase in catalase activity in this tissue, as shown in figure 3. Thus, in brain tissue samples irradiated with 0.5 Gy the levels of catalase activity were increased by 25%, while for samples irradiated with 2 Gy the catalase activity was increased almost 2.4 times, compared the non-irradiated samples (figure 4).

Next sets of samples were consistent with muscle tissue. Their exposure to the chosen radiation doses also led to a decrease in absorbance of H_2O_2 as a result of increased catalase activity

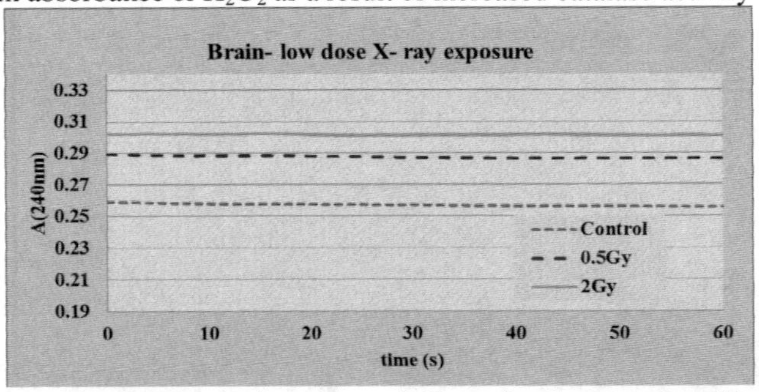

Figure 3. Hydrogen peroxide absorbance dynamics at 240 nm for the control and irradiated chicken brain

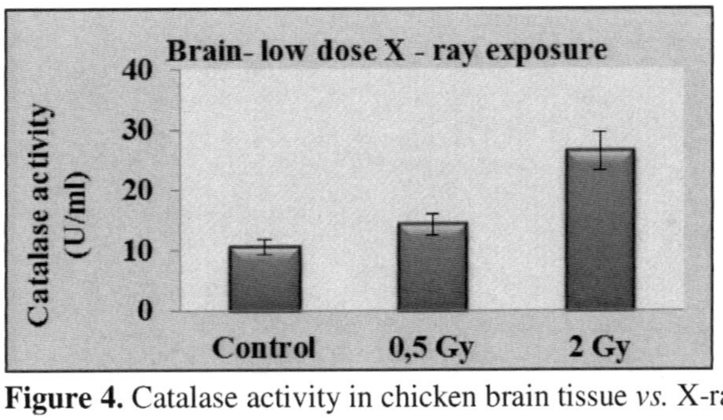

Figure 4. Catalase activity in chicken brain tissue *vs.* X-ray dose

Figure 5 illustrates the dynamics of H_2O_2 absorbance spectra in the chicken muscle tissue.

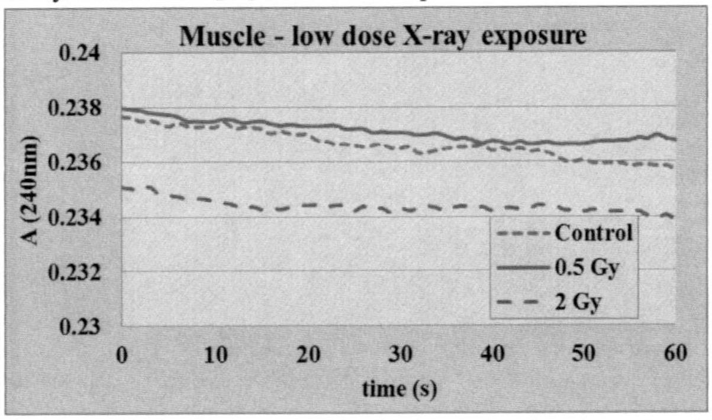

Figure 5. Hydrogen peroxide absorbance dynamics at 240 nm for the control and irradiated muscle tissue

The experimental data illustrated in figure 6 show a stimulation of antioxidant defense system activity in muscle samples after exposure to low doses of radiation - after 0.5 Gy X-ray exposure, the catalase activity being intensified by up to 20% compared with non-irradiated tissue ($p < 0.05$).

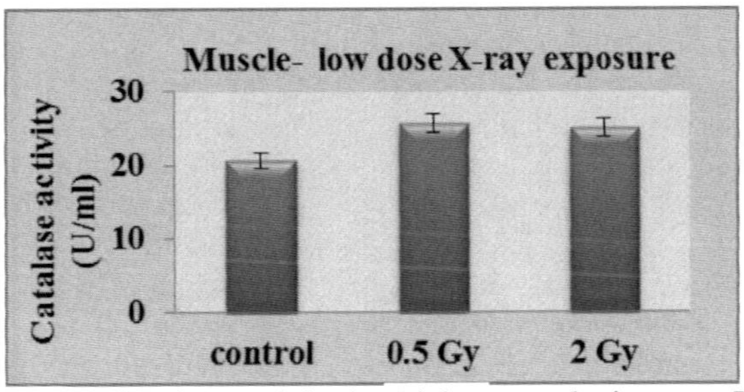

Figure 6. Catalase activity in chicken muscle tissue *vs.* X-ray dose

A major defense mechanism involves antioxidant enzymes such as superoxide dismutase (SOD) and catalase (CAT), converts active oxygen molecules into non-toxic compounds. The liver has the highest content of antioxidants and antioxidant enzymes, indicating the important role that it plays in antioxidant detoxification. Figure 7 presents the dynamics of hydrogen peroxide absorbance at 240 nm recorded for the extracts obtained from irradiated and non-irradiated liver tissue processing.

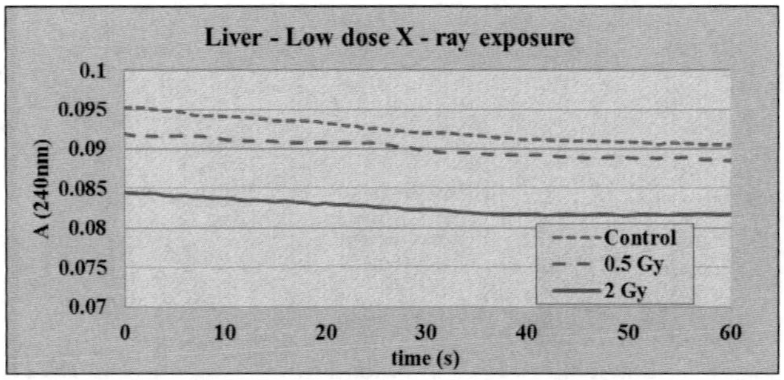

Figure 7. Hydrogen peroxide absorbance dynamics at 240 nm for the irradiated chicken liver tissue

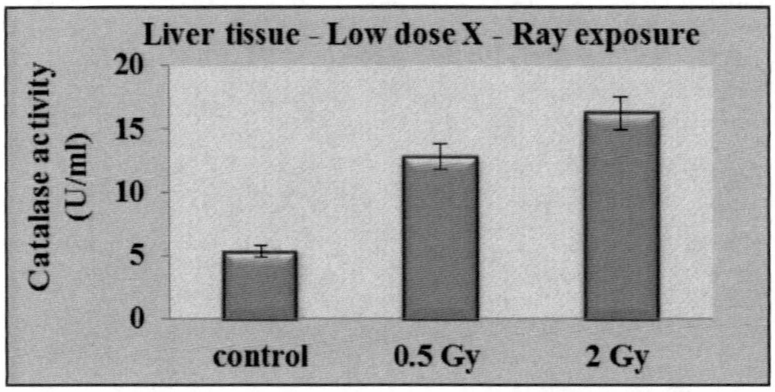

Figure 8. Catalase activity in chicken muscle tissue *vs.* X-ray dose

From figure 8 a significant increase in catalase activity in liver tissue irradiated compared to the non-irradiated can be seen. An exposure to a low dose of 0.5 Gy X radiation caused an increase in hepatic antioxidant defense process of about 2.3 times compared with non-irradiated samples ($p < 0.05$) while the exposure to 2 Gy resulted in tripling the intensity of catalase activity compared to non-irradiated samples.

4. Conclusions

In biological systems, in addition to normal metabolism, ionizing radiation is a source of active oxygen radicals. Cells can be damaged and even can lead to cell death under considerable exposure to ionizing radiation when reactive oxygen species content cannot be controlled by cellular antioxidants. In this study we observed the stimulation of antioxidant defense system status in different types of chicken tissues after low-dose (0.5Gy) and medium dose (2Gy) of X radiation. It can be concluded that under moderate oxidative stress (exposure to low and medium doses of ionizing radiation) cells are able to regulate a variety of physiological mechanisms in an attempt to cope with oxidative stress. On the other hand, exposure to higher doses caused a decrease in catalase activity, possibly due to inhibition or oxidative inactivation of the enzyme protein caused by reactive oxygen species generation. However, catalase is one of the three main families of enzymes in mammalian cells critical for removing peroxide. Reduction of catalase activity may be due to increased consumption of this enzyme to prevent lipid peroxidation

Acknowledgements: This work was partially supported by the European Social Fund in Romania, under the responsibility of the Managing Authority for the Sectorial Operational Program for Human Resources Development 2007-2013 [grant POSDRU/88/1.5/S/47646].

References

[1] Kamat J P, Boloor K K, Devasagayam T P A, Venkatachalam S R 2000 *J. Ethnopharmacol.* **71** 425–435
[2] Breen A P, Murphy J A, 1995 *Free Radical Biol. Med.* **18** 1033–1077
[3] Jagetia G C, Rajanikant G K, Shaival K, Rao M, Baliga S 2003 *Clin. Chim. Acta* **332** 111–121
[4] Fang Y, Yang S, Wu G 2002 *Nutrition* **18** 879–887
[5] Halliwell B 1992 *J. Neurochem.* **59** 1609
[6] Koc M, Taysi S, Buyukokuroglu M E, Bakan N 2003 *J. Radiat. Res.* **44** 211–215
[7] Ahlersova E, Pastorova B, Kassayova M, Ahlers I, Smajda B 1998 *Physiol. Res.* **47** (2) 133–136
[8] Karbownik M, Reiter R J 2000 *Proc. Soc. Exp. Biol. Med.* **225** 9–22
[9] Yousri R, Noaman E, El Shawi O 2011 *J. Cancer Therapy* **2** 616-628
[10] Amundson S A, Lee R A, Koch-Paiz C A, Bittner M L, Meltzer P, Trent J M, Fornace Jr A J 2003 *Mol. Can. Res.* **1** 445-452
[11] Lee W J, Majumder Z R, Jeoung D I, Lee H J, Kim S H, Bae S, Lee Y S, 2006 Radiat. Res. **165** 562-569
[12] Podgorsak E B 2005 *Radiation Oncology Physics: A Handbook for teachers and students* (Vienna: International Atomic Energy Agency) p 212
[13] Aebi H 1974 *Catalase. Methods of Enzymatic Analysis* 2nd ed. (Germany: Verlag Chemie Weinheim, Bergmeyer, H. U. eds.) p 673-684

Journal of Physics: Conference Series **398** (2012) 012033 doi:10.1088/1742-6596/398/1/012033

Rapid screening of pharmaceutical drugs using thermal desorption – SALDI mass spectrometry

A A Grechnikov[1], A E Kubasov[1], V B Georgieva[2], A S Borodkov[1], S M Nikiforov[3], Ya O Simanovsky[3], S S Alimpiev[3],

[1] Vernadsky Institute of Geochemistry and Analytical Chemistry of RAS, Kosygin str., 19, 119991 Moscow, Russia
[2] Georgi Nadjakov Institute of Solid State Physics, Bulgarian Academy of Sciences, 72 Tzarigradsko chaussee blvd., 1784 Sofia, Bulgaria
[3] General Physics Institute of RAS, Vavilov str., 38, 119991 Moscow, Russia

E-mail: lazarova@issp.bas.bg

Abstract. A novel approach to the rapid screening of pharmaceutical drugs by surface assisted laser desorption-ionization (SALDI) mass spectrometry with the rotating ball interface coupled with temperature programmed thermal desorption has been developed. Analytes were thermally desorbed and deposited onto the surface of amorphous silicon substrate attached to the rotating ball. The ball was rotated and the deposited analytes were analyzed using SALDI. The effectiveness of coupling SALDI mass spectrometry with thermal desorption was evaluated by the direct and rapid analysis of tablets containing lidocaine, diphenhydramine and propranolol without any sample pretreatment. The overall duration of the screening procedure was 30÷40 sec. Real urine samples were studied for drug analysis. It is shown that with simple preparation steps, urine samples can be quantitatively analyzed using the proposed technique with the detection limits in the range of 0.2÷0.5 ng/ml.

1. Introduction

The development of laser-assisted mass spectrometric techniques has resulted in new facilities, approaches and methods in the field of organic mass spectrometry. Surface-assisted laser desorption-ionization (SALDI) [1] is an excellent example, which holds great promise for analytical applications. In SALDI the gas-phase ions are formed from molecules deposited on a particular surface substrate that is irradiated with a pulsed laser. That process does not require the entrainment action of an added matrix compound for desorption [2]. Wide variety of solid state materials, such as graphite [1], activated carbon [3], titanium oxide (4) and zinc oxide [5], have been evaluated as a possible platform for SALDI with different degrees of success. However silicon materials, in particular, porous, nanocrystalline and amorphous silicon (a-Si) are so far the most commonly used substrates for SALDI.

The traditional approach to the analysis of organic and bioorganic samples by SALDI includes deposition of liquid solution of analytes as a droplet on the SALDI-active substrate. The droplet is dried and the sample support inserted into the mass spectrometer for analysis. The main problems with this approach are very poor repeatability from substrate to substrate, inhomogeneous distribution of analyte on the substrate surface and strong influence of matrix effects.

The implementation of gas-phase deposition SALDI as a gas chromatography/mass spectrometry (GC/MS) instrument was recently demonstrated [6]. In this method, individual, GC-separated compounds are adsorbed on the SALDI substrate and subsequently ionized and desorbed by the laser pulse. Gas-phase introduction of analytes was shown to be associated with high ionization efficiency, up to 1%. However, applying GC/SALDI MS for screening of pharmaceutical drugs is difficult because the analysis is time-consuming.

This work presents a novel approach for reproducible and quantitative analysis of solid and liquid organic and bioorganic samples by SALDI using gas-phase analyte introduction. Such approach is based on the combination of SALDI mass spectrometer with rotating ball interface [7] and thermal desorption chamber. Legal pharmaceutical products containing lidocaine, diphenhydramine and propranolol were used to demonstrate the effectiveness of proposed technique for the rapid screening of pharmaceutical drugs without a sample pretreatment or chromatographic separation step. Also, real urine samples were studied for drug analysis.

2. Experimental

The experimental setup is schematically shown in figure 1.

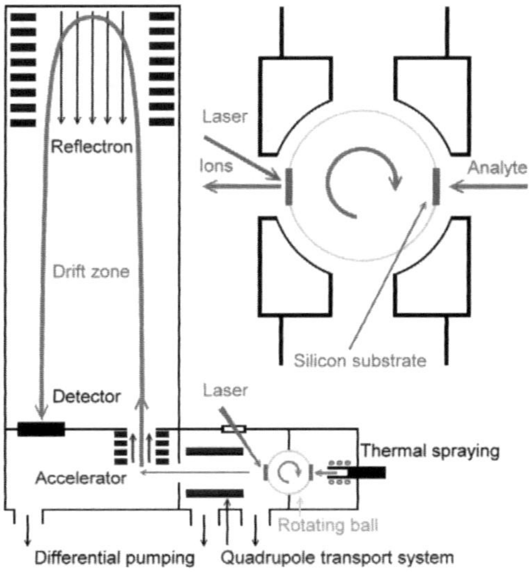

Figure 1. Schematic of SALDI - time of flight mass spectrometer coupled with temperature programmed thermal desorption. The inset shows a rotating ball interface.

The mass spectrometer was a lab-built, time of flight reflectron (0.9 m) with the orthogonal ion injection via gas-filled radiofrequency quadrupole (3 MHz, 1 kV). The ions were detected by a secondary electron multiplier ETP 14882 (ETP Pty Ltd). A 40-mm-diameter, rotating stainless-steel ball forms the interface between the sample deposition chamber, where analytes were deposited on the SALDI active substrate, and the ion source, where the deposited analytes were ionized and desorbed by laser irradiation for the subsequent mass spectrometric analysis. Rotation of the ball is actuated by a computer-controlled stepping motor.

A diode-pumped Nd:YAG laser with near-diffraction limited beam quality, 0.5 ns pulse duration, was operated at 300 Hz. A frequency-tripling crystal yielded a 355 nm laser pulse of about 100 μJ. The laser pulse energy was varied in the range of 10-30 μJ using an attenuator. In order to more efficiently utilize the analyte, the laser focus was scanned over a 1×2 mm^2 sample deposition area using a computer-controlled, dual-mirror scanner. The laser focus moved approximately 0.08 mm between successive laser shots, and about 300 individual mass spectra were obtained from different points on the probed area of the SALDI active substrate.

Amorphous Si (α-Si) substrates were deposited on monocrystalline silicon by standard RF sputtering of Si in a low pressure (10^{-3} Torr) Ar atmosphere, using a turbo-molecular pumped, lab-built apparatus [2]. The thickness of the deposited films measured using quartz crystal microbalances was approximately 0.3 μm. Two α-Si substrates were installed on opposite sides of the rotating ball. Analytes were deposited using thermal desorption on the substrate, the ball was rotated 180°, and the analytes were ionized and desorbed by laser irradiation for the subsequent mass spectrometric analysis. For temperature programmed thermal desorption, a resistively heated nickel-chromium wire wrapped around the stainless steel vaporizer tube was used to slowly warm the vaporizer. The temperature was monitored using a chromel-alumel type K thermocouple placed in contact with the vaporizer. Vaporizer was positioned at 2 mm distance from the surface of α-Si substrate.

Pharmaceutical tablets containing lidocaine, propranolol and diphenhydramine were obtained commercially. Small amount (less than 1 μg) of a tablet, removed using a scalpel, was used for the rapid screening of pharmaceutical drugs. To obtain liquid solution of analyte a tablet was firstly grinded to a powder. After dissolution in high-purity water with added sodium hydroxide the drug was extracted with ethyl acetate. Then the extract was filtered and dried, and analyte was purified by crystallization and dissolved in water methanol (1:1) mixture.

3. Results and discussion

3.1. Screening of pharmaceutical drugs

Lidocaine, propranolol and diphenhydramine were selected as test analytes because of their high proton affinities, differences in chemical structure, and their interest as widespread drugs, that can give a good indication of the utility of thermal desorption - surface assisted laser desorption ionization. In the rapid screening mode, the vaporizer with studied samples was rapidly heated to 170°C followed by the ball rotation and the SALDI/MS analysis of deposited probes. The preliminary study showed that 10 mJ/cm^2 of laser fluence gives the best performance in the rapid screening mode.

Propranolol was detected as the protonated molecule at m/z 260, with no fragment ions. Figure 2 shows typical SALDI mass spectrum of lidocaine. Mass spectrum contains the protonated molecule, MH$^+$, at m/z 235, and major fragment ion at m/z 86.

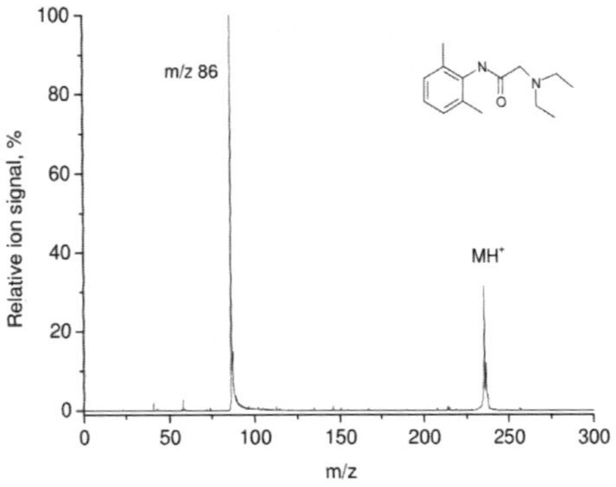

Figure 2. SALDI mass-spectrum of lidocaine.

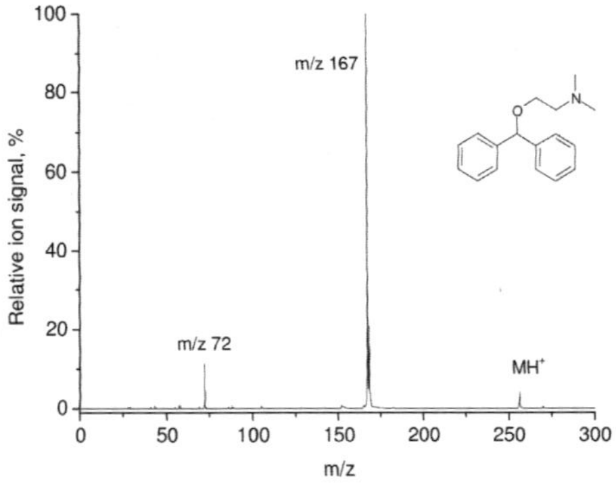

Figure 3. SALDI mass-spectrum of diphenhydramine.

The fragmentation of protonated lidocaine molecule is formed by the C–C bond cleavage, accompanied by hydrogen transfer from the amino group to the carbonyl group, yields a protonated amine with m/z 86 as it is seen in the Scheme 1.

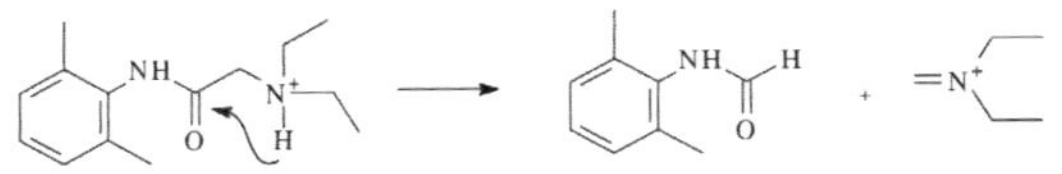

Scheme 1.

Figure 3 demonstrates typical SALDI mass spectrum of diphenhydramine. It is seen that diphenhydramine is more label compound, as compared with propranolol and lidocaine and gives three main peaks due to the protonated molecule, MH^+, at m/z 256, and two fragment ions. The proposed fragmentation pathways are illustrated in scheme 2 and scheme 3. Major fragment ion at m/z 167 is formed by the cleavage of the bond between oxygen and the carbon of the amino group, while minor fragment ion at m/z 72 is formed by the cleavage of the bond between oxygen and the carbon of the diphenylmethyl substituent.

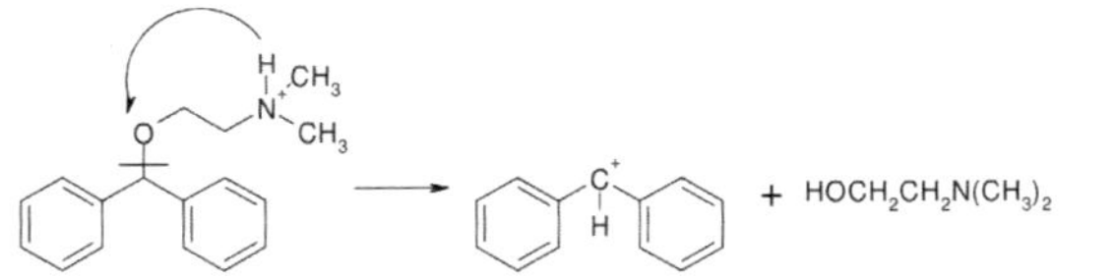

Scheme 2.

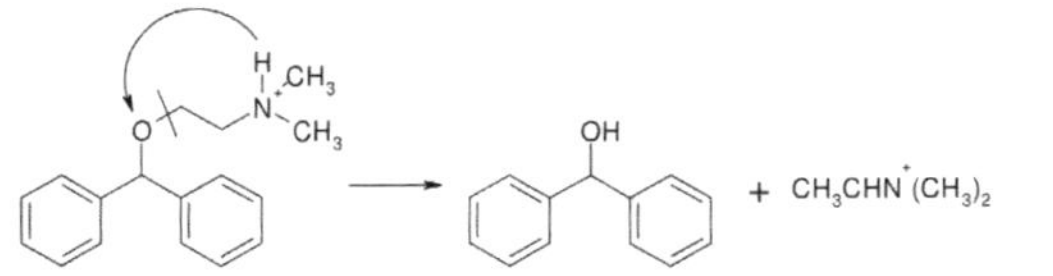

Scheme 3.

The resulting SALDI mass spectra are seen to contain structural information about analyte molecules. The simultaneous monitoring of fragment ion peaks and the protonated molecule peaks increases the ability to discriminate and identify drugs when complex probes are analyzed.

The overall duration of the screening procedure using thermal desorption – SALDI analysis with the rotating ball interface is 30÷40 sec in comparison with the 10–20 min GC/MS runs routinely used in forensic analysis.

3.2. Urine analysis

With simple preparation steps, urine samples can be directly analyzed using thermal desorption/ SALDI MS. Liquid solution of analyte was added to the urine. Then 5 mg of $NaHCO_3$ and 0.5 mL of chloroform isobutanol (6:1) mixture were added to 0.1 mL of urine. Samples were vortexed for 2 min and centrifuged at 10000 rpm for 3 min. The extraction efficiency was found to be 95÷98%. Some µL of the bottom organic layer was transferred into the vaporizer. The temperature programmed thermal desorption mode was used for analyte sampling. In this mode the temperature of the vaporizer was programmed for rapid heating to 120°C, then to 200°C with a rate of 10°C/min. Laser fluence was 25 mJ/cm^2.

Figure 4 shows the ion mass thermogram obtained for propranolol protonated molecules. The mass thermogram was a record of ion current response for one mass unit (m/z 260) over the time of analysis. It is seen that propranolol thermally desorbs over a relatively narrow temperature range after about 3 min from starting programmed thermal desorption. This is reflected in the ion signal trace, which has well-resolved maximum. Figures 4b and 4c present mass spectra in the region of mass-spectral peak of protonated molecule before the beginning of propranolol desorption and at the maximum of ion current. These were obtained by integrating the mass spectra over the one cycle of analyte deposition and laser desorption-ionization.

Evidently, peak desorption temperatures can be used as a guide for identification of the drugs, along with mass spectral information. It is especially important when analyte ions have near equal m/z values, because a vapor pressure difference gives separation of analytes for accurate mass spectral identification.

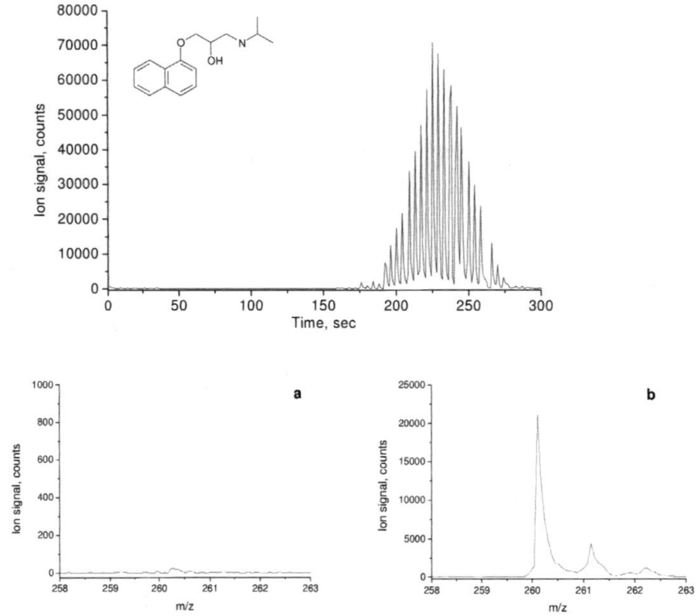

Figure 4. Mass thermogram of propranolol $(MH)^+$ ion. The lower insets show mass spectra obtained a) after 150 sec, b) after 220 sec. Concentration of propranolol in urine was 2 μg/mL.

The linearity of response was studied by varying the concentration of drugs in urine samples. Figure 5 demonstrates the dependence of analytical signal on the amount of propranolol in extract solution placed in the vaporizer. It is seen that the linear dynamic range extends over at least 3 orders of magnitude. The correlation coefficient found to be above 0.995. The relative standard deviation for five replicate analyses was less than 10%. Very similar results were obtained with lidocaine and diphenhydramine. The detection limits for studied drugs was estimated to be 0.2÷0.5 ng/ml.

The obtained data clearly indicates the capability of proposed technique for the quantitative determination of low molecular weight drugs in urine samples.

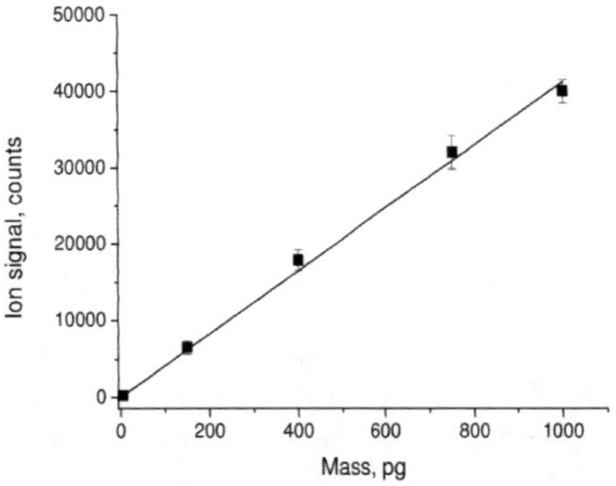

Figure 5. Analyte ion signal plotted against the amount of propranolol placed in the vaporizer.

Conclusions

Temperature programmed thermal desorption is coupled to a SALDI mass spectrometer with rotating ball interface for the first time. The application of such approach for drug detection is demonstrated. Lidocaine, propranolol and diphenhydramine were studied as test analytes. It is shown that the presented technique provides a simple and rapid screening method for tablets and other pharmaceutical products with total analysis time as short as 40 sec or less. Tested analytes are detected as protonated molecules with no or very structure-specific fragmentation. The schemes of fragmentation are proposed.

It is also demonstrated that the presented technique is applicable to the quantitative analysis of liquid biological materials such as urine. The detection limits of studied drugs were estimated to be 0.2÷0.5 ng/ml.

Acknowledgement

This work was supported by the Program for basic research of the Presidium of RAS №9 and by bilateral project (2012-2014) between GEOKHI, RAS and IFTT, BAS.

References

[1] Sunner J, Dratz E and Chen Y C 1995 *Anal.Chem.* **67** 4335

[2] Alimpiev S S, Grechnikov A A, Sunner J, Karavanskii V A, Zhabin S N, Simanovsky Ya O and Nikiforov S M 2008 *J.Chem.Phys.* **128** 014711

[3] Han M and Sunner J 2000 *J.Am.Soc.MassSpectrom.* **11** 644

[4] Yuan M, Shan Z, Tian B, Tu B, Yang P and Zhao D 2005 *Microporous Mesoporous Mater.* **78** 37

[5] Grechnikov A A, Georgieva V B, Alimpiev S S, Borodkov A S, Nikiforov S M, Simanovsky Ya O, Dimova-Malinovska D and Angelov O I 2010 *JOP:Conf.Ser.* **223** 012038

[6] Alimpiev S S, Grechnikov A A, Sunner J, Borodkov A S, Karavanskii V A, Simanovsky Ya O and Nikiforov S M 2009 *Anal.Chem.* **81** 1255

[7] Alimpiev S S, Grechnikov A A, Sunner J, Karavanskii V A, Simanovsky Ya O and Nikiforov S M 2011 *Rapid Commun. Mass Spectrom.* **25** 140

Journal of Physics: Conference Series **398** (2012) 012034 doi:10.1088/1742-6596/398/1/012034

A study on the interaction of nanoparticles with lipid membranes and their influence on membrane fluidity

P B Santhosh[1], S Penič[2], J Genova[3,4], A Iglič[4] , V Kralj-Iglič[5]and N P Ulrih[1,6]

[1]Department of Food Science and Technology, Biotechnical Faculty, University of Ljubljana, Jamnikarjeva 101, 1000 Ljubljana, Slovenia.
[2]Laboratory of Bioelectromagnetics, Faculty of Electrical Engineering, University of Ljubljana, 25, Trzaska, 1000 Ljubljana, Slovenia.
[3]Institute of Solid State Physics, Bulgarian Academy of Sciences, 1784 Sofia, Bulgaria
[4]Laboratory of Biophysics, Faculty of Electrical Engineering, University of Ljubljana, 1000 Ljubljana, Slovenia.
[5]Biomedical Research Group, Faculty of Health Sciences, University of Ljubljana.
[6]Centre of Excellence for Integrated Approaches in Chemistry and Biology of Proteins (CipKeBiP), Jamova 39, 1000 Ljubljana, Slovenia.

E-mail: poorni_balaji@yahoo.com

Abstract. In recent years, liposomes encapsulated with nanoparticles have found enormous scopes in various biomedical fields such as drug design, transport, imaging, targeted delivery and therapy. These applications require a clear understanding about the interaction of nanoparticles with cell membranes. The present work aims to investigate the effect of encapsulation of uncharged and positively charged nanoparticles in three different types of lipids such as1-stearoyl-2-oleoyl-*sn*-glycero-3-phosphocholine (SOPC),1-stearoyl-2-oleoyl-*sn*-glycero-3-phosphocholine and1-palmitoyl-2-oleoyl-*sn*-glycero-3-phospho-L-serine(SOPC-POPS) mixture and archaeal lipids. Through the temperature dependent fluorescence anisotropy measurements, we have found that the entrapment of nanoparticles in the bilayer has decreased the lipid transition temperature and increased the membrane fluidity of all three types of lipid vesicles. The results were more predominant in SOPC-POPS mixture because of high density encapsulation of nanoparticles in the vesicles due to electrostatic interaction between negatively charged membrane and positively charged iron oxide nanoparticles.

1. Introduction

Liposomes, the self-assembled lipid structures have received extensive attention due to their potential application in various fields. Because of versatile organization, well-defined physicochemical properties and ability to mimic membrane scaffolding, they are widely studied as model membranes. The encapsulation of nanoparticles in liposomes provides a biologically inspired route in designing therapeutic agents and as a means of reducing nanoparticle toxicity. The hybrid lipid/nanoparticle conjugates have diverse biomedical applications including imaging of cancer cells, drug/gene delivery, targeted therapy, immunoassay, cell/protein separation, biosensing etc. Currently, little is known about the influence of nanoparticles on physicochemical properties of lipid vesicles such as stability, elasticity, membrane fluidity and bilayer phase behavior [1].

Interest in the synthesis of metal nanoparticles (NPs) is steadily growing due to their unique properties and potentialities. Nanoparticles are highly effective to penetrate the plasma membrane and to alter the natural processes within the cell. They serve as excellent carriers of therapeutic cargos through the membrane, likely due to internalization mechanisms like physical rupturing, membrane mediated transport, pore formation, etc. Recent studies on the effects of different metal NPs on membrane stability and/or deformation have revealed that the incorporation of metal NPs within membrane have altered the phase behavior of the lipids by decreasing the phase transition temperature and increasing fluidity of the bilayer. Since the polymorphic phase behavior of lipids influence different membrane related processes, it has become very important to study the effect of nanoparticle interaction with different lipid membranes [2].

Enormous research has been carried out with homogeneous bilayers consisting of zwitterionic phospholipids, but very less work has been done to understand the electrostatic attraction between the negatively charged lipid bilayers and positively charged nanoparticles. Therefore we intend to study this property in detail and prepared negatively charged lipid vesicles by mixing SOPC and POPS lipids in the ratio of 4:1, respectively. Due to opposite charges, cationic iron oxide NPs are electrostatically attracted towards the negatively charged phosphate group of phospholipids and gets adsorbed. The adsorption process of NPs onto lipid molecules was thermodynamically favorable, and enhanced due to the ultrafine size of the nanoparticles [3].

We have also analyzed the influence of nanoparticles on archaeal lipids. The domain Archaea represents a third evolutionary form of life and their ability to survive in extreme environmental conditions is attributed to their unique lipid composition. The presence of ether linkages and highly branched isoprenoid side chains offers more stability archaeal membranes [4, 5]. We have grown the Aeropyrum pernix K1 archaeal cells in our lab [6] and extracted the lipid from them to study their membrane properties. We encapsulated uncharged Cobalt Ferrite ($CoFe_2O_4$) NPs and positively charged Iron Oxide (Fe_2O_3) NPs in archaeosome and studied their influence on membrane fluidity. The polar lipids of *A. pernix*K1 consist solely of $C_{25,\,25}$-archaeol (2, 3-di-sesterpanyl-*sn*-glycerol), with $C_{25,\,25}$-archetidyl (glucosyl) inositol (AGI) accounting for 91mol%, and the remaining 9 mol% by $C_{25,25}$-archetidylinositol (AI) [4, 5].

In lipid vesicles, nanoparticle encapsulation can be achieved by trapping the particles within the aqueous core or in the hydrophobic bilayer. To be embedded in the lipid bilayers, the nanoparticles must possess two important features. They should be smaller in size to fit within a lipid bilayer and should have a hydrophobic surface (by coating with appropriate agents such as sterylamine). When the nanoparticles are entrapped within bilayers, it can lead to changes in lipid packing and may disrupt lipid-lipid interactions amongst the head groups and/or acyl tails. Disruption of such interlipid interactions can result in changes in lipid bilayer phase behavior, which is related to the degree of lipid ordering and bilayer viscosity [8]. When some charged proteins or nanoparticles are adsorbed onto cell surface, the membrane undergoes deformation and lipids in the constituent bilayers will be reorganized due to electrostatic interaction between the lipids and nanoparticles/proteins. Since the membrane is negatively charged, positively charged nanoparticles are attracted more towards the surface of cell-membrane and show higher levels of internalization when compared to uncharged and negatively charged particles. Hence, depending on their size and surface chemistry, embedded nanoparticles may influence the stability and function of hybrid vesicles, domain formation, phase separation etc [9].

1.1 Modes of nanoparticle interaction in lipid bilayer

Depending upon the size, electrostatic charge and hydrophobicity, the nanoparticle may be partly or fully trapped in the bilayer. Three different possibilities of nanoparticle entrapment are shown in the figure 1. The first possibility indicates a structure in which a nanoparticle is partially embedded in the bilayer or resting on the vesicle surface. The second possibility shows that the nanoparticle spans the hydrophobic region making equivalent contact with the two quencher populations. The third mode depicts the entrapment of nanoparticle perfectly in middle of the bilayer due to hydrophobic interactions [10].

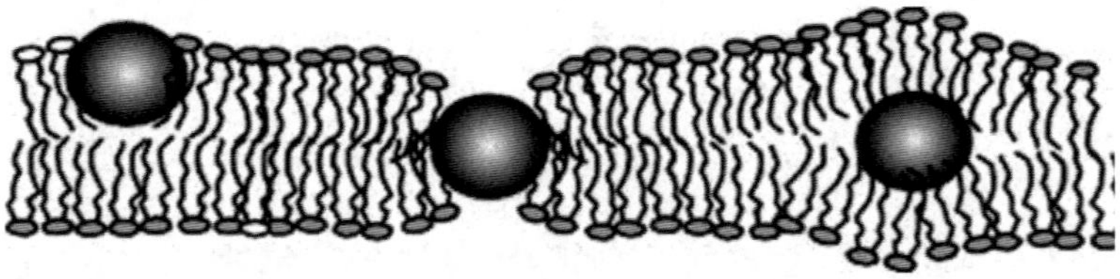

Figure 1. Sketches of possible nanoparticle/bilayer interactions

2. Materials and methods

2.1. Synthesis of nanoparticles

The superparamagnetic maghemite nanoparticles (γ-Fe_2O_3) are synthesized through a controlled chemical coprecipitation method. An aqueous mixture of ferric, ferrous salts and sodium hydroxide were prepared as alkali stock solutions. The corresponding metal hydroxides were precipitated during the reaction between the alkaline precipitating reagent and the mixture of metal salts and subsequently oxidised in air to form γ-Fe_2O_3. To achieve purification and uniform size distribution, the iron oxide nanoparticles were precipitated out from the solution using ethanol and separated from the supernatant by differential centrifugation method. The purified sample was dried under argon and redispersed in double distilled water. Their surfaces were stabilised electrostatically with positive charge by adsorbtion of citric acid on their surface. Citric acid provides strong negative surface charge and creates repulsive forces which prevents their aggregation and ensures the stability. The nanoparticles were characterised using X-ray diffractometry and Transmission Electron Microscopy (TEM). The size of the synthesized γ -Fe_2O_3 nanoparticles was found to be 10±2 nm by TEM analysis [7].

The cobalt ferrite nanoparticles were purchased from Sigma Aldrich. They were synthesized by co-precipitating the stoichiometric mixtures of $Fe(NO_3)_6.9H_2O$ and $Co(NO_3)_2.6H_2O$ in aqueous solutions. The pH was maintained between 9.5-11 using 10% NaOH solution and the temperature was set between70-95°C for 4-5 hours under vigorous magnetic agitation. The resulting mixture was then centrifuged for fifteen minutes at 3000 rpm. The supernatant was then decanted and centrifuged rapidly until a thick black precipitate was obtained. The precipitate was then washed thoroughly with water and acetone for purification and dried overnight at 100°C in hot air oven. The dried samples were then dispersed in double distilled water. In order to avoid the nanoparticle agglomeration in aqueous solutions, strong surface charges are applied by varying the pH of the solution. This results in high zeta potential value and increased nanoparticle stability. The size of $CoFe_2O_4$ nanoparticles were found to be in the range of 5-15 nm by TEM and the zeta potential value was estimated to be ± 34 using DLS.

2.2. Isolation and purification of Archaeal lipids

Aeropyrum pernix K1 was purchased from Japan Collection of Microorganisms (number 9820; Wako-shi, Japan) and the archaeal cells are cultivated in our lab. They were grown in 800 mL growth medium in 1000 mL heavy-walled flasks, with a magnetic stirring hot plate and forced aeration (0.5 $L{\cdot}min^{-1}$) at 92°C. After 40 h, the suspensions were cooled and centrifuged at 11,000×g for 10 min at 10°C. The cell pellets were washed twice with the corresponding buffer (20mM Hydroxyethyl-piperazineethanesulfonic acid (HEPES), pH 7.0, containing 3% NaCl). Later the archaeal cells were lyophilized to extract polar-lipid methanol fraction (PLMF) containing approximately 91% $C_{25,25}$-archetidyl (glucosyl) inositol (AGI) and 9% $C_{25,25}$- archetidylinositol (AI). The lipids were fractionated using adsorption chromatography and analysed by Thin Layer Chromatography (TLC) with chloroform/methanol/acetic acid/water (85/30/15/5) solvent. The methanol fraction containing the polar lipids (PLMF) was used for further analysis. This lipid solution was dried by slow evaporation under a constant flow of dry nitrogen, followed by vacuum evaporation of solvent residues [3, 4].

2.3. Preparation of liposome – nanoparticle conjugates
Adequate volumes of SOPC, SOPC-POPS (both purchased from Avanti Polar Lipids) and Archaeal lipids were dissolved in chloroform and transferred into round-bottomed glass flasks. The solvent from the lipid samples were evaporated using a Rotavapor under reduced pressure (17 mbar). The dried lipid films were then hydrated with the aqueous nanoparticle solutions so that the final concentration of the lipids was made to 1 $mg \cdot mL^{-1}$. Multilamellar vesicles (MLVs) were prepared by vortexing the lipid suspensions vigorously with glass beads for 10 minutes. The MLVs were further transformed into small unilamellar vesicles (SUVs) by sonication for 30 minutes with 10 s on-off cycles at 50% amplitude with a Vibracell Ultrasonic Disintegrator VCX 750 (Sonics and Materials, Newtown, USA). To separate the debris from SUVs after sonication, the sample was centrifuged for 10 min at 14,000 rpm (Eppendorf Centrifuge 5415C). The control lipid vesicles without nanoparticles were prepared in a similar way but diluted with 1 ml of 20 mM HEPES buffer instead of nanoparticle solution.

2.4. Bilayer melting and fluidity: Fluorescence anisotropy
Bilayer melting temperatures and fluidity were examined by fluorescence anisotropy measurements using 1,6-diphenyl-1,3,5-hexatriene (DPH) and trimethyl-ammonium-6-phenyl-1,3,5-hexatriene (TMA-DPH) in control liposomes and nanoparticle encapsulated liposomes in a 10 mm-path-length cuvette using a Cary Eclipse fluorescence spectrophotometer (Varian, Mulgrave, Australia), in the temperature range from 20°C to 90°C. Varian autopolarizers with slit widths of 5 nm for both excitation and emission were used. Here, 10 µL DPH or TMA-DPH (Sigma- Aldrich Chemie GmbH, Steinheim, Germany) in dimethyl sulphoxide (Merck KGaA, Darmstadt, Germany) was added to 2.5 mL of 100 µM SUV solution in the relevant buffer, to reach a final concentration of 0.5 µM DPH and 1.0 µM TMA-DPH. DPH and TMA-DPH fluorescence anisotropy was measured at the excitation wavelength of 358 nm, with the excitation polarizer oriented in the vertical position, while the vertical and horizontal components of the polarized emission light were recorded through a monochromator at 410 nm for both probes. The anisotropy <*r*> was calculated using built-in software of the instrument using below formula:

$$r = \frac{I_{\parallel} - I_{\perp}}{I_{\parallel} + 2I_{\perp}}$$

where $I_{\parallel}$ and $I_{\perp}$ are the parallel and perpendicular emission intensities, respectively.

2.5. Characterization of nanoparticle encapsulated liposomes
Liposomes were further analyzed by X-ray photoelectron spectroscopy (XPS) to determine their chemical composition [3].

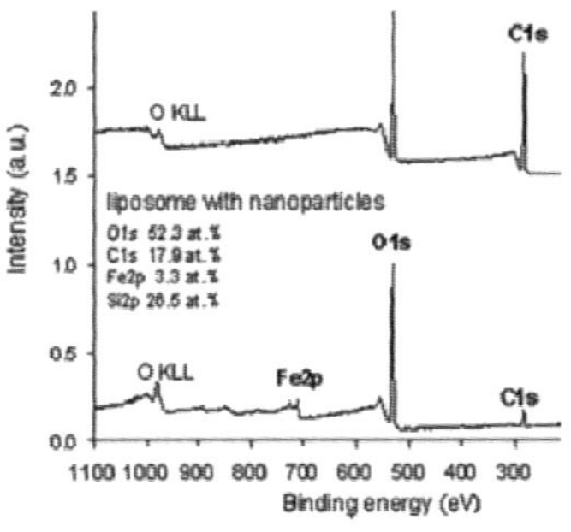

Figure 2. XPS survey depicting the difference in chemical composition (at.%) between 2 spectrums. Upper spectra: Control magnetosomes without Fe_2O_3 NPs, Lower spectra: magnetosomes encapsulated with Fe_2O_3 NPs.

3. Results and discussion

Figure 2 shows the chemical composition of set of liposomes containing magnetic nanoparticles and control liposomes without nanoparticles. In the first case, we can see oxygen and carbon originating from phospholipids as well as high peaks due to Si and Fe which are constituents of magnetic nanoparticles, while in the second case, we can observe only oxygen and carbon originating from phospholipids.

Given that the nanoparticles had diameters exceeding the thickness of a bilayer, this work suggests that lipid bilayers can distort to accommodate such particles and this distortion reduces lipid ordering. This result is consistent with the ability for a cell membrane to accommodate large transmembrane proteins. Since the colloidal particles in bilayer are in dynamic state, the particles are moving and vibrating continuously above absolute zero temperature. Also, the motion of colloidal particles increases with temperature due to bilayer melting. Thus, the movement of colloidal particles would disturb the crystalline structure of bilayer, resulting in decrease of phase transition temperature and increase of the membrane fluidity.

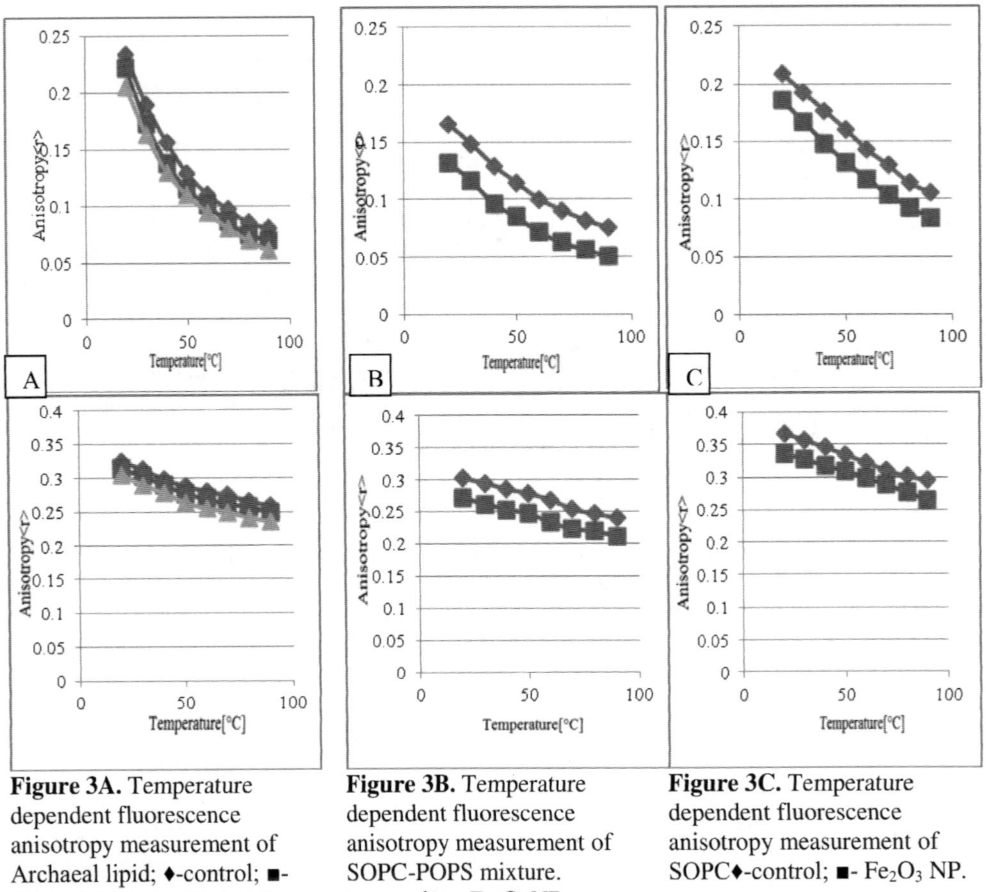

Figure 3A. Temperature dependent fluorescence anisotropy measurement of Archaeal lipid; ♦-control; ■- $CoFe_2O_4$ NP; ▲- Fe_2O_3 NP. DPH with head groups.

Figure 3B. Temperature dependent fluorescence anisotropy measurement of SOPC-POPS mixture. ♦control; ■- Fe_2O_3 NP.

Figure 3C. Temperature dependent fluorescence anisotropy measurement of SOPC♦-control; ■- Fe_2O_3 NP.

The upper graph depicts the interaction of DPH and lower graph TMA DPH

DPH and TMA-DPH are the widely used fluorescent probes to study the membrane properties. The results of anisotropy measurements of DPH and TMA DPH in three different types of lipids are shown in the figure 3. Anisotropy is a measure of lipid ordering and the bilayer microviscosity. It is inversely related to the membrane fluidity, that is, lower anisotropy values indicate an increase in the membrane fluidity. In order to study the influence of size of nanoparticle in bilayer uptake, we also encapsulated $CoFe_2O_4$ NP (20-30 nm) in archaeal lipids. These NPs are almost three times bigger in size compared to Fe_2O_3 NPs. The initial values of the order parameter of DPH at 20°C were: 0.23±0.01 for control archaeosome; 0.22±0.01 for $CoFe_2O_4$ encapsulated archaeosome; and 0.20±0.01 for Fe_2O_3 incorporated archaeal lipids 0.08±0.01, 0.07±0.01, and 0.06±0.01 respectively. In case of SOPC, the DPH anisotropy values at 20°C were: 0.20±0.01for control and 0.18±0.01 for Fe_2O_3 loaded liposome. For SOPC-POPS mixture the initial values were 0.24±0.01for control and 0.19±0.01 for Fe_2O_3 loaded liposome. The results have shown that the NPs show lower anisotropy values in all the three types of lipids when compared with the control liposomes without nanoparticles. The differences were significant in DPH rather than TMA-DPH. Due to electrostatic attraction, the cationic iron oxide NPs are attracted more towards the negatively charged SOPC-POPS lipid membrane. The observed encapsulation ratio is increased comparing the uncharged archaeal and SOPC lipids.

4. Conclusions

Fluorescence anisotropy of DPH and TMA-DPH gradually decrease with increasing temperature in all three types of liposome-nanoparticles conjugates. Though there was a gradual decrease in the anisotropy values of all these samples, the difference was predominant in the case of negatively changed lipid mixture (SOPC-POPS) containing positively charged iron oxide NPs. This result is in good agreement with the understanding that electrostatic interactions promote the encapsulation process. It is well known that the fluorescent probe DPH locates primarily in the core of the hydrophobic tails whereas TMA-DPH is anchored within the head group region close to the lipid-water interface. The fact that stronger quenching and significant decrease in the anisotropy values in case of DPH when compared to TMA-DPH confirms that the nanoparticles are trapped predominantly in the middle of the bilayer rather than partially embedding on the membrane surface or equally spanning the bilayer. The results from XPS spectra also confirm that the nanoparticles are successfully encapsulated in the liposomes.

Acknowledgements

This study was supported by Slovene Human Resources Development and Scholarship Fund.

References

[1] Al-Jamal W, Kostarelos K 2007 *Nanomed.* **2** 85
[2] Zhang Land Granick S 2006 *Nano Lett.***6**694
[3] Elersic K, Pavlic J, Iglic A, Vesel A and Mozetic M 2012 *Chem. Phys. Lipids***165** 120
[4] Gmajner D, Ota A, Šentjurc M and Ulrih N P 2011 *Chem. Phys. Lipids***164** 236
[5] Ulrih N P, Gmajner D and Raspor P 2009 *Appl.Microbiol.Biotechnol.***84**249
[6] Milek I, Cigić B, Skrt M, Kaletunç G and Ulrih N P 2005 *Canad. J. Microbiol.* **51** 805
[7] Park S H, Oh S G, Mun J Y and Han S S 2006 *Coll. Surf. B* **48** 112
[8] Bothun G D 2008 *J. Nanobiotechn.* **6** 1
[9] Bhandary S, Sultana P, Basu R, Das S and Nandy P 2011 *Adv. Sci. Eng. Med.***3** 1
[10] Jang H, Pell L.E, English D.S 2003 *J. Photochem.Photobiol. A:* **158** 111

Optical properties of dimeric liquid crystals doped with single-walled carbon nanotubes

M Petrov[a], B Katranchev[a], P M Rafailov[a], H Naradikian[a], U Dettlaff-Weglikowska[b] and E Keskinova[a]

[a] Institute of Solid State Physics, Bulgarian Academy of Sciences, 72 Tzarigradsko Chaussee Blvd., 1784 Sofia, Bulgaria.
[b] School of Electrical Engineering, Korea University, Anam-Dong, Sungbuk-Gu, 136-713 Seoul, Korea.

E-mail: mpetrov@issp.bas.bg

Abstract. Well oriented distinct local single crystals were grown in smectic C liquid crystal (LC) (heptyloxybenzoic acid - 7OBA) using predefined orientation of the single-walled carbon nanotubes (SWCNT). The SWCNT/SiOx/ITO/glass surface is found to cause a substantial enlargement of the smectic C distinct local single crystals and to reinforce both surface memory effect and surface anchoring. We synthesized and examined a set of nanocomposites, built of LC with hydrogen-bonded in dimers molecules (7OBA) and SWCNT. We found that the surface orientation strength and the interaction of the dimeric molecules with the carbon nanotubes play a decisive role for the type (symmetry), thermal stability and chirality of the LC states induced in the nanocomposites. As a result, a set of phase transitions and new phases (chiral smectic C, reentrant nematic, reentrant chiral nematic and a unique low-temperature chiral biaxial fluid smectic phase C_G) not typical for pristine 7OBA were found.

1. Introduction

In the last few years, a wide range of nanosize materials have been the subject of current research due to their interesting properties and technological applications. Nanomaterial as a guest, in liquid crystals (LC) material may open a new era for the LC based display devices because of their unique structural, electrical and mechanical properties [1]. The proper selection of nanomaterials, concentration, homogeneity and their compatibility with liquid crystals show a remarkable impact on structural, switching, optical and dielectric properties of LC. Much of the works were focused on the dispersion of carbon nanotubes in liquid crystals for lower threshold voltage, faster switching time and better optical responses [2,3]. Single walled carbon nanotubes (SWCNT) show high conductivity along the tube length and are also known to align parallel to each other in low concentration solutions. These spontaneous self-alignment properties of SWCNT make them an attractive alternative for surface and bulk alignment in LC. It was indicated that the trapping of ions, present as impurities in any LC material, due to attracting of those to carbon nanotubes.

Investigations of SWCNT as an alignment substrate for LC, as well as material for bulk orientation improvement were so far concentrated entirely on nematic and ferroelectric LC. Here we focus on the surface alignment of the smectic C (S_C) liquid crystals and on the bulk properties of mixture of SWCNT and nematics with short range smectic C order.

The goal of the present work is to further explore the anisotropy of SWCNT for formation of an oriented S_C. As a second stage we study the optical and phase state properties of mixture SWCNT and nematics with short range S_C order. In the two cases we used as a LC the seventh homologue of p-n-alkyloxybenzoic acids (7OBA), which characterizes with hydrogen bonded in dimers molecules.

SiO_x layer, deposited on an ITO treated glass surface induces distinct local single crystals (DLSCs). We investigate its orientation, light polarization characteristics as well as their sizes. We also analyze the memorization strengths of these orienting substrates and compared them with those at SiO_x- and ITO- treated surfaces. We prepare mixtures of 7OBA with SWCNT in a wide range of concentrations and explore conditions for phase thermal stability, new phase induction, compatibility of the SWCNT with the nematic matrix, chirality and phase transitions in these nanocomposites.

To test the imposed by the SWCNT deposition surface orientation responsible for DLSCs texture enlargement and reinforcement of the memorization of an oriented S_C texture in the temperature range of the N phase, known as surface memory effect (SME), we use spatially resolved polarized Raman spectroscopy [4,5]. Both the optical properties and phase states of the SWCNT treated LC cell surface and doped by SWCNT LC matrix we investigate by optical microtexture polarization analysis.

2. Experimental results and discussion

The experiments have been carried out with 4, n-heptyloxybenzoic acid (7OBA) which has following phase transitions:

$$Cr \xleftrightarrow{92^o C} S_C \xleftrightarrow{98^o C} N \xleftrightarrow{146^o C} I$$

Its chemical structure in dimeric form is demonstrated below:

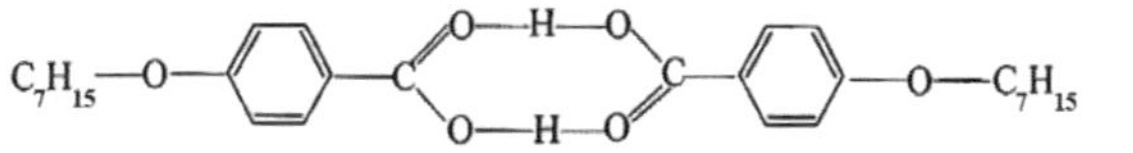

The LC material in its isotropic phase was filled in the cell using the capillary force action. The gap between the glass plates, defining a cell thickness of 20 μm, was maintained with Mylar spacers. The temperature of the sample was varied with a rate of 0.2 °C min^{-1} by a hot stage temperature controller Linkam TMS 90, and the chosen experimental temperature was stabilized with an accuracy of ± 0.1°C.

A microscope Zeiss NU2 was used for the texture analysis.

The latent heat of 7OBA is 10.8 $kJ.mol^{-1}$. Under bulk or surface impact the closed dimers transform in open dimers or monomers. The most typical external influence causing such transformation is the temperature variation. As a result a peculiar temperature T^*=117°C inside the N temperature range (98-146°C) exists for 7OBA, where a structural transition occurs [6]. Above T^* the nematic behaviour resembles that of the ordinary nematics, while below this temperature strong smectic C fluctuations appear, thus forming nematic with short-range smectic C order.

Purified single-walled carbon nanotubes with diameters in the range 1.2 – 1.4 nm and an aspect ratio of 20 – 2000, assembled in thin bundles of > 2 μm length, produced by arc discharge at Hanhwa NanoTech Co. Ltd. have been used as purchased. The sonication and centrifugation procedure was applied to remove the big SWCNT agglomerates and most of metallic impurities. The remaining well dispersed SWCNT suspension in a form of droplet was placed on a series of substrates and blown with an air stream in the "rubbing" direction of the substrates until the solvent evaporated. This procedure pushes the suspension on the surface and leads to the self assembly of the SWCNT bundles along the blowing direction while the solvent evaporates, so that some alignment of the SWCNT is expected.

The Raman spectra were measured on a Jobin Yvon Labram spectrometer equipped with a CCD detector and entrance slits set to 1.5 cm^{-1} spectral width. The 632.8 nm line of a He-Ne laser was used for excitation, the absolute accuracy being about 0.5 cm^{-1}. The laser beam with a power of about 1mW was focused on a spot of about 2 μm in diameter on the sample surface using microscope optics. In this way, a possible alignment of nanotubes lying on the sample surface should be detected due to the strong antenna-effect arising from the combination of a strongly resonant nature of the Raman effect in SCWNTs with the polarization effect due to the large nanotube aspect ratio, which renders the

Raman tensor approximately proportional to the polarizability tensor. This makes the resonance Raman signal from a carbon nanotube strongly orientation dependent. The scattering intensity of all Raman modes in a nanotube, regardless of their vibrational pattern, has then the same orientational dependence, approximately proportional to cos4α, where α is the angle between the incident laser polarization and the nanotube axis. Thus the strongest signal is obtained when the laser polarization is parallel to the nanotube axis (α=0; ±180°), while the Raman intensity should nearly vanish for α=90°. Spot-resolved measurement series performed as described above revealed an uniform orientational dependence of the intensity of all bands, as depicted in figure 1, with a period of 90°, closely resembling that of the function cos2α or cos4α. We thus conclude that the SWCNT deposited on the SiO_x/ITO/glass surface have indeed a preferred orientation which can be determined with polarized Raman spectroscopy.

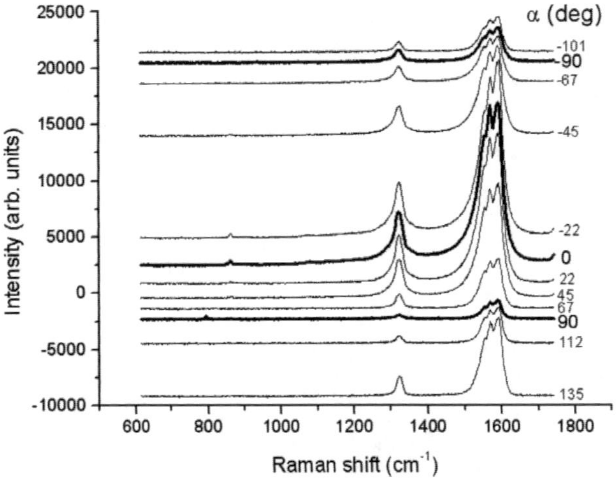

Figure 1. Spatially resolved polarized Raman spectra at various orientation angles α.

Starting from the isotropic phase with very slow cooling, we performed consecutive growth of a planar N phase, and upon further cooling in the temperature interval between 92 and 98°C, we obtained a S_C phase, containing DLSCs. Such DLSCs texture grown at 96.5°C on SWCNT/SiO_x/ITO/glass surface is presented in figure 2(a). As seen from this figure, the layers inside the black and white single crystals are directed at ±45° with respect to the horizontal direction. From observations by crossed polarizers we found these layers to be optically equivalent showing the same four extinction positions upon turning the sample by 2π, allowing us to depict the S_C layers topology (figure 2(b)): due to the imposed 'easy' direction in the xy plane, the smectic layers follow ±45° directions, the projections of the tilted molecules being parallel to the $\boldsymbol{n}_o \equiv x$ axis. The inset shows a S_C layer with a normal $\boldsymbol{N}$ and a cone (formed by the tilt angle ω≈45°) expressing the rotation around $\boldsymbol{N}$ of the director $\boldsymbol{n}$ (or equivalently the smectic C director $\boldsymbol{c}$), thus minimizing the free energy of the system. Well ordered and enlarged (with average spatial lateral extent of the single crystals ≈0.2mm^2) DLSCs are formed with surface anchoring, provided by the SWCNT/SiO_x/ITO/glass surface. For comparison, the sizes of DLSCs grown on SiO_x or ITO/glass surfaces are 0.1 and 0.08mm^2 respectively. We assume that this enlargement of the DLSCs results from a decrease of the deviation of $\boldsymbol{n}_s$ from the "easy" direction, forcing φ to tend to zero, thus improving the bulk orientation.

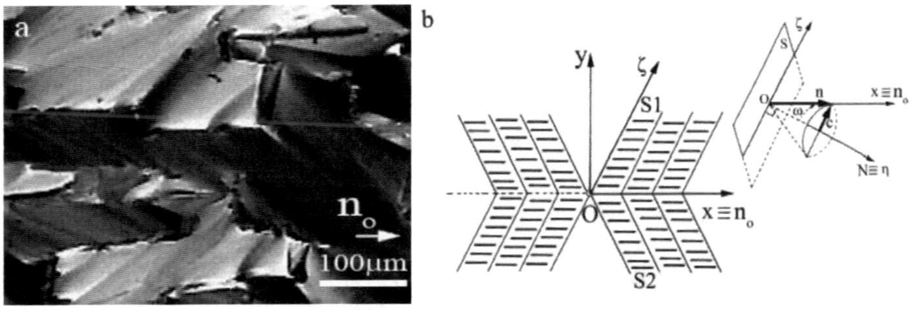

Figure 2. (a) The smectic C DLSCs texture in crossed polarizers. (b) A scheme of the S_C layers' topology. The inset shows a smectic layer, with the layer's normal ***N*** and the cone (formed by the tilt angle *ω)* expressing the rotation of the director ***n*** around ***N***. The layer's locations denoted by S1 and S2 are compatible.

We verify the ability of the $SWCNT/SiO_x/ITO$/glass surface to reinforce surface anchoring and SME strength accordingly. The following memorization procedure is applied: we keep the cell at T_{rec}=96.5°C for the recording time t_{rec} to achieve an imprint of the S_C texture like the one shown in figure 2(a) on the SWCNT-coated surface. Then we heat the cell up to a sufficiently high temperature T_{er} in the N state, to prevent emerging of any S_C precursors. Even at this high temperature, the recorded DLSC-texture imprint still imposes its shape on the N-phase volume thus temporarily maintaining a similar texture in the N phase. Then we apply static erasure through annealing by keeping the cell at T_{er}, for the so-called erasure time t_{er}, during which the memorized S_C texture is extinguished. In our experiment, the recording time was fixed at t_{rec}=3min and the achieved erasure time for T_{er}=135°C, was 150 min. To gain more insight in SME mechanism, we prefer static erasure through annealing as described above. The SME-imprint on the SWCNT-coated surface results from a reorientation of the LC molecules in the surface monolayer induced by the bulk LC alignment in the S_C phase. Within the rotational diffusion model [7], the reorientation of the molecules in this monolayer, at recording and erasure, can be viewed as jumping over potential barriers between the local wells in conditions of reversibility of the SME. This can be approximated with an exponential Arrhenius equation [8] $t_{er}=\tau\exp[-Q/(k_BT)]$. Here Q is an energy barrier, which has to be surpassed by the thermal activation k_BT for effective destruction of the picture through rotational diffusion during a time interval t_{er} and τ is an extrapolation parameter. The dependence ln $t_{er}(T^{1})$ is linear, and from the curve slope we calculate the erasure activation energy Q. We have also measured the erasure time t_{er} of the DLSC texture as a function of the temperature.

The obtained values of Q for the three orienting surfaces: $SWCNT/SiO_x/ITO$/glass, SiO_x/ITO/glass and ITO/glass respectively are 470 kJ mol^{-1}, 192 kJ mol^{-1} and 160 kJ mol^{-1}.

Further we study the bulk action of the SWCNT, but using the most popular surface orientation - ITO/glass suitably rubbed. The important point now is the preparation of the compatible SWCNT/LC mixtures. For preparation of a composite 0.6 mg SWCNT powder was mixed in a mortar with 620 mg of 7OBA during heating. After some technological procedures a homogeneous gray powder has been obtained.

The prepared SWCNT/7OBA mixture was examined to check the existence of LC states starting with SWCNT concentration of 0.1 weight %. Such concentration does not display LC phase. The most general inference at mixing is: varying the concentrations and the type of surface influence (non-oriented or oriented LC cells) we can obtain new phases, which have not been observed until now in the pristine 7OBA.

Richer variety of phases is observed at SWCNT concentration of 0.01 weight % in a LC cell with d=8μm: after forming a chiral N (N^*) at 142°C, the system enters at 93°C into a chiral smectic C (S_C^*)

phase (in figure 3(a) the texture of S_C^* is depicted). The S_C^* texture is characterized by the typical stripes indicating the helix direction, which in this case lies in the *xy* cell plane or is tilted with respect to it. Upon further cooling, at 90°C the reentrant chiral nematic (N_r^*) phase is reached. At 86 °C we observe a transition from N_r^* to a low-temperature smectic phase (figure 3(b)), which is fluid (well indicated by small colored glass particles), not smooth, and colored in red and green mozaic domains, thus closely resembling the G phase [9,10], which is, as we will indicate by polarization analysis, a fluid biaxial smectic G (C_G).

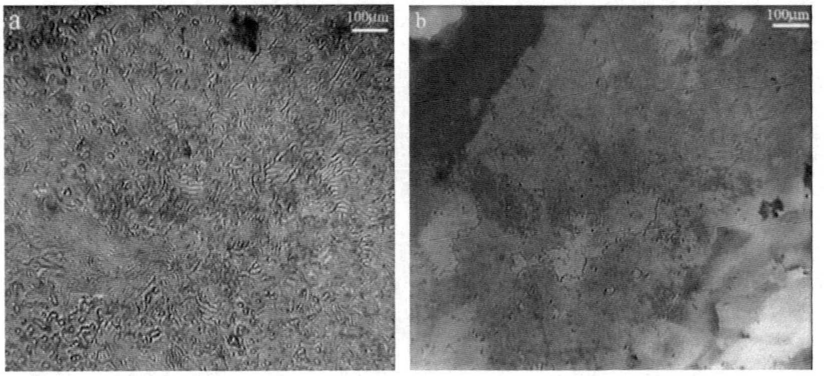

Figure 3. The textures of 0.01 wt % SWCNT/7OBA in a LC cell with d=8µm: (a) Chiral smectic C at T=91°C. (b) Low-temperature smectic phase C_G at T=86°C

Following the symmetry variation activated in the bulk by both thermal influence and by the interaction with the SWCNT, we find that the phase-transition and phase-stabilization sequence is as follows: the chiral N^* starts immediately below the I phase and upon further cooling it turns into a S_C^* (figure 3(a)), thus decreasing the symmetry to C_2 (twofold symmetry axis) and inducing an in layer's plane polarization typical for the ferroelectric LCs. Yet further cooling brings the system into the reentrant chiral nematic state N_r^*, below which the system transforms to C_G (figure 3(b)) and the symmetry decreases to the lowest possible one: C_1.

We focus on the smectic C_G phase (see figure 3(b)). Applying a microtexture polarization analysis, we observe two red and green mosaic domains. We find that in crossed polarizers, upon sample rotation in either direction, the red color turns into green and vice versa. Furthermore, conventional chiral smectic phases are optically uniaxial because of their helix structure, whereas the C_G phase observed here turned to be biaxial. To confirm this we perform special polarization analysis. The transition from N_r^* is not only connected with a change in fluidity but also with the sudden appearance of birefringence (red and green colors) just below 92°C and the texture loses its smoothness. From this we infer the biaxial nature of the C_G phase, in contrast to conventional cholesterics and chiral S_C^* which exhibit no change in color nor in intensity of the transmitted light upon rotating the film between crossed polarizers. Thus the phase observed here should be a C_G biaxial fluid smectic phase. Certainly, this C_G phase possesses a net permanent polarization and can be switched in a bistable way in an electric field, which we plan to perform in future.

The anisotropic in-plane fluidity of C_G (like that of achiral S_C and unlike the smectic A, where this in-plane fluidity is isotropic) requires that none of the principle axes of the second rank tensor [13], characterizing the orientational order, to make an angle of 0° or 90° with the smectic layers planes. Thus, its polarization vector *P*, is tilted at an angle to the smectic layers. The polar vector now is expressed by its components $P(P_x,P_y,P_z)$. Its direction can not be determined by simple symmetry considerations. Also in contrast to S_C^*, C_G has locally C_1 and globally C_∞ symmetry. This implies that, unlike to S_C^* the direction of macroscopic polarization is not fixed in a special symmetry direction (i.e.

there is no specific direction set by a mirror-plane or a twofold axis). Thus the C_G is helielectric in the plane of its layers and ferroelectric perpendicular to them or it should be ferroelectric in the bulk.

3. Conclusion

A smectic C liquid crystal texture consisting of well oriented distinct local single crystals was grown using predefined orientation of the single-walled carbon nanotubes adsorbed on a SiO_x/ITO/glass surface. The SWCNT/SiO_x/ITO/glass surface is found to cause a substantial enlargement of the smectic C distinct local single crystals as compared to SiO_x/ITO/glass and ITO/glass surfaces. It was found that the surface memory effect and surface anchoring induced by the SWCNT/SiO_x/ITO/glass surface was many times stronger than those promoted by conventional SiO_x/ITO/glass and ITO/glass surfaces, normally used for liquid crystals orientation. We synthesized and examined a set of nanocomposites, built of liquid crystal with hydrogen-bonded dimers in its molecules (7OBA) and single-walled carbon nanotubes. We found that the surface orientation strength and the interaction of the dimeric molecules with the carbon nanotubes play a decisive role for the type (symmetry), thermal stability and chirality of the induced liquid crystal states (not typical for pristine 7OBA): chiral smectic C, reentrant nematic, reentrant chiral nematic and the unique low-temperature chiral biaxial fluid smectic phase C_G.

Acknowledgments

This study was supported by Grant BK-2/11 from the Institute of Solid State Physics, Bulgarian Academy of Sciences. The Korean Ministry of Education, Science and Technology is acknowledged for support from the World Class University Project (WCU, R32-2008-000-10082-0).

References

[1] Dierking I, Scalia G and Morales P 2005 *J. Appl. Phys.* **97** 044309
[2] Podgornov F, Suvorova A, Lapanik A and Haase W 2009 *Chem. Phys. Lett.* **479** 206
[3] Russel J M, Oh S, LaRue I, Zhou O and Samulski E T 2006 *Thin Solid Films* **509** 53
[4] Gommans H, Alldredge J W, Tashiro H, Park J, Magnuson J and Rinzler A G 2000 *J. Appl. Phys.* **88** 2509
[5] Rafailov P M, Thomsen C, Gartsman K, Kaplan-Ashiri I and Tenne R, 2005 *Phys. Rev. B: Condensed Matter and Materials Phys.* **72** 205436
[6] Petrov M 2010 Optical *and Electrooptical Properties of Liquid Crystals Nematic and Smectic Phases* (New York: Nova Science Publishers)
[7] Ouchi Y, Feller M B, Moses T and Shen Y R 1992 *Phys. Rev. Lett.* **68** 3040
[8] Myrvold B O 1995 *Liq. Cryst.* **18** 287
[9] Brand H R, Cladis P E and Pleiner H, 1998 *Eur. Phys. J. B* **6**(3) 347
[10] Jakli A, Kruerke D, Sawade H and Heppke G 2001 *Phys. Rev. Lett.* **86** 5715

Journal of Physics: Conference Series **398** (2012) 012036 doi:10.1088/1742-6596/398/1/012036

Smectic C liquid crystal growth through surface orientation by $Zn_xCd_{1-x}Se$ thin films

B Katranchev, M Petrov, I Bineva, Z Levi and M Mineva

Institute of Solid State Physics, Bulgarian Academy of Sciences, 72 Tzarigradsko Chaussee Blvd., 1784 Sofia, Bulgaria.

E-mail: mpetrov@issp.bas.bg

Abstract. A smectic C liquid crystal (LC) texture, consisting of distinct local single crystals (DLSCs) was grown using predefined orientation of ternary nanocrystalline thin films of $Zn_xCd_{1-x}Se$. The surface morphology and orientation features of the $Zn_xCd_{1-x}Se$ films were investigated by AFM measurements and micro-texture polarization analysis. The $Zn_xCd_{1-x}Se$ surface causes a substantial enlargement of the smectic C DLSCs and induction of a surface bi-stable state. The specific character of the morphology of this coating leads to the decrease of the corresponding anchoring energy. Two new chiral states, not typical for this LC were indicated. The physical mechanism providing these new effects is presented.

1. Introduction

The liquid crystals (LCs) are highly anisotropic liquids. Their molecules tend to orient parallel to each other along one direction characterized by the director $\boldsymbol{n}$ ($\boldsymbol{n}^2=1$) [1]. This direction is arbitrary in space and imposed by the anisotropic interactions between liquid crystal molecules and a boundary wall. Uniform and homogeneous orientation of the liquid crystal, in particular nematic (N) LC, was obtained by using various surface treatments including mechanical striations or rubbing of thin polymeric layers [2]. The same orientation, using the same treatments, was reached also for the more complex layered smectic LC and specially for smectic C (S_C) LC, where the director $\boldsymbol{n}$ is tilted on an angle ω (tilt angle) with respect to the layer‘s normal $\boldsymbol{N}$. Deposition of SiO layer by evaporation at oblique incidence has been long ago known to provide nematic liquid crystal alignment

For an angle of evaporation incidence $45°<\alpha<80°$, from the normal (z direction) of the LC cell substrate (xy plane), the N LC orientation is planar. For grazing incidence $80°<\alpha<90°$, the N LC orientation is oblique in the incidence plane with a tilt angle toward the evaporation direction [3]. The orientation properties of LC molecules on the surface were explained by elastic models: the LC bulk is deformed to minimize its elastic energy in response to the anisotropic structure of the surface treatment, which could be SiO obliquely evaporated rough surface, providing well controlled films, or polyimide unidirectional rubbed surfaces [2]. Very often the rough orienting surfaces like that of SiO obliquely evaporated cause a continuous transition between the planar and the oblique orientations [3]. Such complicated state of the director, caused by the anisotropic interactions between LC molecules and a boundary wall, could be understand by the correlation between a complicated topography and the LC molecules orientation. This requirement is still more necessary when the LC molecules are those of S_C LC, where the director $\boldsymbol{n}$ is restricted to rotate (or reorient) simultaneously around two cones: one with apex angle 2ω within each layer and another one on the surface, coaxial with the substrate's surface normal *z*. Being put under these two constrains the director $\boldsymbol{n}_s$ near the surface thus predetermines the orientation of the grown S_C texture. Unlike the NLC, and due to these constrains the alignment of the smectic C characterizes with polycrystal textures, known as distinct local single crystals (DLSCs) [4]. Obtaining of large and well oriented DLSCs, that are convenient for experimentation, as well as for the needs of the contemporary display techniques, practically is a tough problem, since one must control independently the molecular and the layer orientation and to avoid fragmentation of the S_C texture. Due to its strong elasticity, however, S_C ensures comparatively fast electro-optical response (decay times of only few ms) [5], and is of practical interest for the fast LC displays. This motivates the study of the surface forces driving the DLSCs growth and efforts to grow comparatively large single crystals. Since the details of LC surface alignment are not well understood, the search for more easily quantifiable alignment materials continues, seeking to achieve adequate control of the surface orientation forces. Such new materials must be well acquainted, on a nano-scale dimension, in order to deduce the physico-chemical mechanism controlling the surface orientation of the S_C LC system. Till now it is not known whether that the strong or the weak surface anchoring could provoke growth of large and well oriented DLSCs. Furthermore, a bistable surface state, which could be induced at rough surfaces, is especially important for the display and other LC device techniques. Since the S_C is rather faster system than the nematic one the surface bistability in S_C must be a desired effect. This determines the goal of the present work: to obtain large, oriented and able to ensure fast bistable switching smectic C DLSCs. We expect that using as orienting surface film nanocrystalline layers of $Zn_{0.59}Cd_{0.41}Se$ with thickness of 400 nm, we will be able to find an electrically controllable bi-stabile texture. The understanding of the physical mechanism of this texture will be an important fundamental goal of the present work. We also observed two new states, not typical for the here investigated heptyloxybenzoic acid (7OBA) substance: chiral smectic C (S_C^*) and biaxial fluid low-temperature chiral smectic phase (C_G). These states due to the externally forced symmetry decrease. The orientation, light polarization characteristics as well as the size of the grown DLSC will be verified using predefined orientation of $Zn_{0.59}Cd_{0.41}Se$ treated glass surface. The anchoring strength will be estimated and compared with those at SiO_x- and ITO- treated surfaces.

2. Experimental results and discussion

The experiments have been carried out with 4, n-heptyloxybenzoic acid (7OBA) which has following phase transitions:

$$Cr \xleftrightarrow{365.2K} S_C \xleftrightarrow{371.2K} N \xleftrightarrow{419.2K} I$$

The LC material in its isotropic phase was filled in the cell using the capillary force action. The gap between the glass plates, defining a cell thickness, was maintained with Mylar spacers. The temperature of the sample was varied with a rate of 0.2 K min^{-1} by a hot stage temperature controller Linkam TMS 90, and the chosen experimental temperature was stabilized with an accuracy of $\pm$ 0.1K.

Thermal evaporation of ZnSe and CdSe powders (Merk, Suprapure) from two independent, simultaneously heated tantalum crucibles was applied for the preparation of $Zn_xCd_{1-x}Se$ layers with x=0.59 and thickness of 400 nm. The crucibles were disposed in the bottom of two cylindrical screens (not intentionally heated) whose top is close to the substrates. Thus, the deposition of each material was carried out in a quasi-closed volume.

The layers of $Zn_xCd_{1-x}Se$ were deposited on crystalline Corning 7059 glass substrates maintained at room temperature. During the film deposition the substrates were rotated at a rate V_r=20 turns/min and during each turn pass they spent over each crucible 1/12 of the turn time. The time Δt between the consecutive deposition of CdSe and ZnSe is 5/12 of the turn time. The substrate rotation was started after fixing the desired deposition rates for both CdSe and ZnSe. The rates were controlled by two independent preliminary calibrated quartz microbalance systems MIKI FFV. The layer composition was varied by changing the ratio of the deposition rates of both materials; a ratio 1.5/1.5 (V_d^{CdSe} (nm/s)/V_d^{ZnSe} (nm/s)) was used. Thus, $Zn_xCd_{1-x}Se$ single layers were prepared by alloying of the consecutively deposited small portions of CdSe and ZnSe with equivalent layer thicknesses d_e=0.37nm (at 15nm/s) [6,7].

The surface morphology of the thin films was investigated by AFM measurements that were carried out using a Multimode V (Veeco, Santa Barbara, CA). Imaging was performed in tapping mode and height, amplitude, and phase images were recorded. The scan rate was in the range 1-2 Hz, the image resolution was 512 lines per scan direction. At least three different points on the sample surface were explored. Silicon cantilevers for soft tapping mode (TAP150-Al-G, BudgetSensors Innovative Solutions Bulgaria Ltd., 1113 Sofia, Bulgaria), with a nominal resonance frequency of 150 kHz and a typical force constant of 5 N/m were used. The tip radius for these probes is less than 10 nm. Image processing was performed by means of the Nanoscope 7.30 programme. All the images were flattened and no further processing was carried out.

Monitoring of the optical textures, in the LC cell *xy* plane, prepared using $Zn_xCd_{1-x}Se$ coated substrates, was carried out using a video camera Hitachi on a microscope Zeiss NU2.

In figure 1 the bi-stable texture of grown DLSCs in LC cell prepared by two different plates with respect to the coating (hybrid LC cell): one coated with $Zn_xCd_{1-x}Se$ (upper plate) and the other one coated with rubbed ITO (lower plate). By rotation (on right) of the cell in crossed polarizers on an angle of ≈45°, one notes the change of the black into white DLSCs and conversely the change of white into black at left rotation. This is interesting for the fast display techniques due to the fast response of the smectic C to external electric field, as well as for the observed good contrast. The AFM morphology of upper, covered with fresh $Zn_xCd_{1-x}Se$, plate is indicated in figure 2 in its planar and 3D images. We do not found any difference in the orienting properties of the fresh and old (e.g. 4 years old) $Zn_xCd_{1-x}Se$ coatings, although nanosized grains of the last coating seem ordered in one direction.

Figure 1. Bi-stable smectic C texture for planarly oriented hybrid LC cell. T=368.3K. Crossed polarizers.

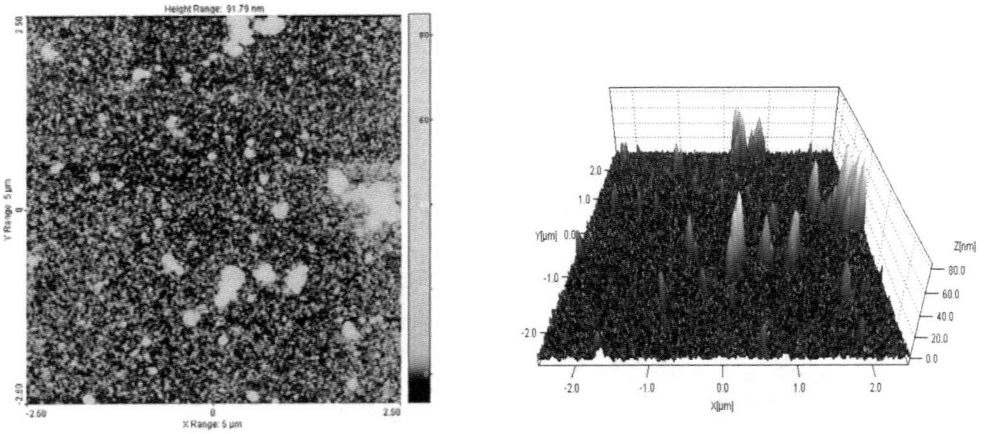

Figure 2. 2D and 3D AFM images of fresh $Zn_xCd_{1-x}Se$ film.

Using two equal $Zn_xCd_{1-x}Se$ coated plates as upper and lower for LC cell preparation we found transformation of achiral smectic C, typical for 7OBA, in chiral smectic C^*. One notes that such texture is only S_C^* fluctuation and exists with grate difficulty 2 degrees. At further cooling this chiral state transforms in a chiral biaxial, fluid in the layer planes, smectic C_G phase. These chiral states are seen in figure 3. We confirmed the biaxiality and fluidity, as well as the double tilting character of this state, by suitable microtexture polarization analysis. The S_C^* phase does not change the character of the achiral nematic phase above it.

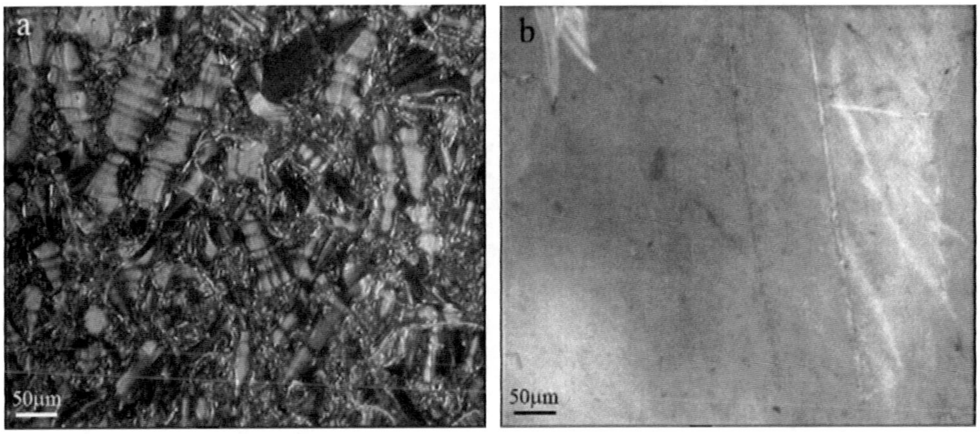

Figure 3. Texures of (a) Chiral smectic C^* state at T=369.4K in crossed polarizers and (b) Smectic C_G phase state at T=360.6K in crossed polarizers.

Considering the N phase (figure 4), in such prepared non hybrid LC cell, after cooling from the isotropic phase, we found well underlined Neel walls allowing the calculation of the surface anchoring energy. Hence we can use the formula [8] $W_s=\pi^2Kd/2l^2$, where d and l are the cell and wall thicknesses and K is the elastic constant (10^{-11}J m^{-1}) considered in the one-constant approximation [1]. We measured the wall width at temperature 408.2K in the N phase for two cell thicknesses d=8 and 12µm to be l=14 and 15μm respectively. The wall thickness l was derived from a study of interference fringes for normally incident monochromatic light. Using the measured l values and the Rapini-Papoular formula [9] $W_s=K(\mathbf{n}.\mathbf{n}_o)^2L^{-1}$, we determine the extrapolation length L which is a measure for the anchoring strength. We estimated the anchoring energy for the basic surface treatments and found the extrapolation lengths for the two thicknesses to be 5 and 3.85 μm respectively, corresponding to anchoring energies 0.2×10^{-5} J m^{-2} and 0.26×10^{-5} J m^{-2} respectively. These values indicate that the anchoring strength at $Zn_xCd_{1-x}Se$ coating is about one order weaker than those of the anchoring induced by SiO or ITO orientations (see for comparison [10])

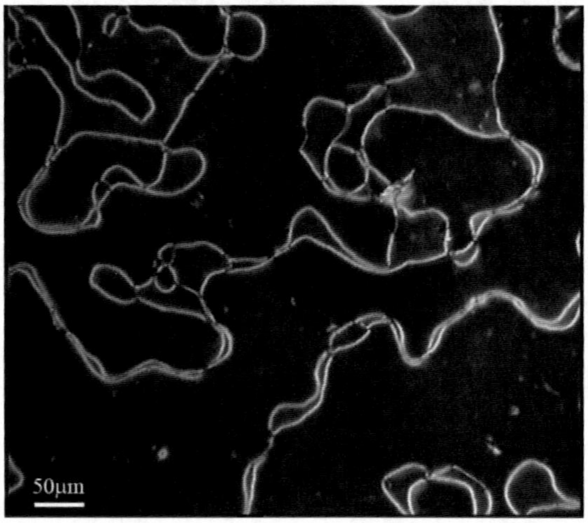

Figure 4. The Neel walls in the nematic phase in non hybrid LC cell at T=408.2K in crossed polarizers.

The weak anchoring at the interface $Zn_xCd_{1-x}Se$ coating – LC allows modeling of the physical mechanism responsible for the observed bistability and the chiral phases induction.

Varying the coating layer thickness, one can observe a column to needle nanostructure transition, with increasing surface roughness. The $Zn_xCd_{1-x}Se$ surface roughness is expected to induce a nematic surface order decrease (surface melting).

Therefore, we indicate the effect of the surface topography and roughness created by the $Zn_xCd_{1-x}Se$ surface on the N LC molecules orientation. The results could be quantitatively explained by an "order electric" model. In this model, the $Zn_xCd_{1-x}Se$ roughness destroys the LC order at the surface, inducing a gradient of this order towards the ordered bulk [11]. An order electric polarization proportional to this gradient rises to an electrostatic term in the LC orienting surface free energy. The surface roughness induces decrease of the surface order parameter to zero i.e. the LC phase near the surface would melt (goes to the isotropic state). This order parameter variation creates an order – electricity polarization $\mathbf{P}$ near the surface which tends to orient the molecules on the surface toward the magic angle θ_m. The competition between the van der Walls interaction with the surface and the order-electric effect may give the zenithal molecular orientation. We can present this effect using the scalar form of the order parameter $S=\langle(3cos^2\theta-1)\rangle/2$. When, due to the gradient of S along the normal to the substrate, the S value goes to zero, then the $\theta=\theta_m\approx54°$ is magic angle defined by $cos^2\theta=1/3$. The pre surface molecules are tilted on this angle and only the horizontal component of the polarisation keeps $P_x=3\nabla S sin\theta cos\theta/2$. ∇S is the ratio of the reduction in the order parameters S between the surface and the bulk. The surface horizontal order electric polarization P could provoke twist rotation of the surface director $\mathbf{n}_s$. On the $Zn_xCd_{1-x}Se$ grains P produces electric charges, which create a depolarizing electric field E_D parallel to the plate surfaces (xy plane). This electric field must disappear on the conductive surface as that of ITO but is active on that low-conductive $Zn_xCd_{1-x}Se$ surface and is able to create director rotation, dominating on the conventional van der Walls interaction with the surface.

This horizontal polarization action on the upper $Zn_xCd_{1-x}Se$ plates and corresponding rotation creates the observed bi-stability seen in figure 1. The *P* surface action on the two plates for LC cell constructed by two $Zn_xCd_{1-x}Se$ plates could create a deformation of the dimmer ring of 7OBA leading to decrease of the symmetry and induction of both chiral phases: S_C^* and C_G. The effect of decrease of symmetry and in turn induction of a low-symmetric phase (transformation of achiral smectic C into S_C^* and C_G phases) is unique and due to the possibility of the dimmer molecules to vary (at external action) between open dimmer state and monomers. The study of this new effect as well as the electrical control of the detected in this work bistability will be our important future task.

3. Conclusion

The surface morphology and orientation features of the $Zn_xCd_{1-x}Se$ films were investigated by AFM measurements and micro-texture polarization analysis. The $Zn_xCd_{1-x}Se$ surface causes a substantial enlargement of the smectic C liquid crystals monodomains in the 7OBA. Using of a hybrid LC cell, built by two different substrate plates, treated by $Zn_xCd_{1-x}Se$ and rubbed ITO, we detected grown of a surface bi-stable smectic C texture. Two new chiral states, namely ferroelectric smectic C^* and the unique chiral, biaxial and fluid in the layer's plane smectic C_G phases, (not typical for 7OBA) were indicated when the two liquid crystal cell's substrates were treated only with $Zn_xCd_{1-x}Se$. A model explaining the observed bi-stability on the base of the surface induced ordo-electric polarisation was presented. The specific character of the morphology of this coating leads to the decrease of the corresponding anchoring energy. The physical mechanism providing these new effects is presented.

Acknowledgments

This study was supported by the Bulgarian Ministry of Education, Youth and Science under grant DMU 03-91.

References

[1] de Gennes P G and Prost J 1993 *The Physics of Liquid Crystals* second edition (New York: Oxford University Press)
[2] Jerome B. 1991 *Rep. Frog. Phys.* **54** 391
[3] Monkade M, Boix M and Durand G 1988 *Europhys. Lett.* **5** 697
[4] Petrov M, Katranchev B, Rafailov P M, Naradikian H, Dettlaff-Weglikowska U and Keskinova E 2012 *J. Mol. Liq.* submitted
[5] Petroff B, Petrov M, Simova P and Angelov A 1978 Ann. Phys. (Paris) **3** 331
[6] Nesheva D, Aneva Z, Scepanovic M J, Bineva I, Levi Z, Popovic Z V and Pejova B 2010 *J. Phys.: Conf. Ser.* **253** 012035
[7] Nesheva D, Aneva Z, Scepanovic M J, Levi Z, Iordanova I and Popovic Z V 2011 *J. Phys. D: Appl. Phys.* **44** 415305
[8] Vitek V and Kleman M 1975 *J. de Phys.* **36** 59
[9] Rapini A and Papoular M 1969 *J. Phys. Colloq.* **30** 54
[10] Blinov L M and Chigrinov V G 1994 *Electropotic effects in liquid crystal materials* (New York: Springer)
[11] Barbero G, Dozov I, Palierne J F and Durand G 1986 *Phys. Rev. Lett.* **56** 2056

Journal of Physics: Conference Series **398** (2012) 012037 doi:10.1088/1742-6596/398/1/012037

Influence of Cholesterol on the Elastic Properties of Lipid Membranes

J Genova[1,2], V Kralj-Iglič[3], A Iglič[2], R Marinov[1] and I Bivas[1]

[1]Institute of Solid State Physics, Bulgarian Academy of Sciences, 72, Tzarigradsko Chaussee Blvd., 1784 Sofia, Bulgaria.
[2]Laboratory of Biophysics, Faculty of Electrical Engineering, University of Ljubljana, 25, Trzaska, 1000 Ljubljana, Slovenia.
[3]Biomedical Research Group, Faculty of Health Sciences, University of Ljubljana, 1000 Ljubljana, Slovenia.

E-mail: ulia@issp.bas.bg

Abstract. Thermally induced shape fluctuations of giant quasi-spherical lipid vesicles are used to study the influence of the cholesterol, incorporated in the lipid membranes, on the bending elasticity modulus k_c of the lipid membrane. The influence of cholesterol is investigated throughout a considerably wide interval of concentrations. The values of the bending elastic modulus for 10, 20, 30 and 50 mol% of cholesterol in the SOPC membrane are obtained as a mean weighted value of 6-11 vesicles for each system. The dependence of the bending elasticity modulus on the concentration of cholesterol in the lipid membrane is obtained. At low concentration of cholesterol in the SOPC membrane (10 mol %) a decrease of the bending elasticity modulus is observed, compared to pure SOPC membrane. At high cholesterol content (50mol% and above) a twofold increase of the bending modulus is obtained. The data for k_c for mixed SOPC - cholesterol membrane is compared to the results obtained by different methods on different lipid matrices.

1. Introduction

The simplest model of the biological cell is the so-called lipid vesicle (figure 1). It is a closed structure formed by a double lipid layer (bilayer). Such objects are prepared from natural or synthetic lipids in laboratory conditions using different formation techniques [1, 2]. By means of adding proteins, carbohydrates, cholesterol etc. to the lipid matrix the studied object is subsequently complicated in order to obtain model system resembling the real membranes.

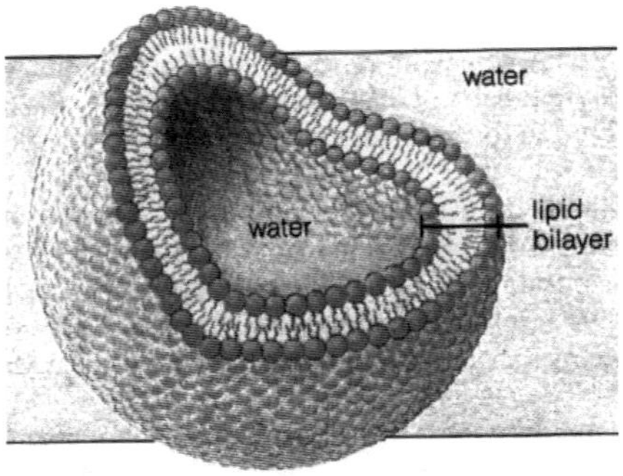

Figure 1. Scheme of a vesicle

Cholesterol (figure 2) is a vital component of the living cell membranes varying up to 50 mol % of the total lipid content [3]. It is necessary for the synthesis of hormones of the adrenal cortex - cortisone, corticosterone, aldosterone, male hormones - testosterone, androsterone, and female sex hormone - estrone, estriol, estradiol, progesterone, vitamin D and bile acids. Cholesterol is considered as a key molecule in the formation of small raft domains, providing a favorable nanoenvironment for the biochemical function of membrane proteins. At the high concentrations it is found in the cell membranes (close to 50 mol %) cholesterol separates the phospholipids so that the fatty acid chains can't come together and decreases the temperature of lamellar liquid crystal – gel phase transition.

These are just some of the reasons for the constant scientific interest towards this molecule and its influence on different physicochemical properties of the lipid membranes.

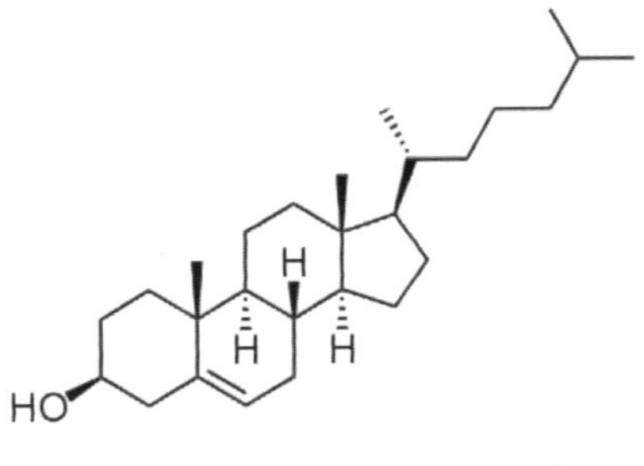

Figure 2(a). Structure of cholesterol.

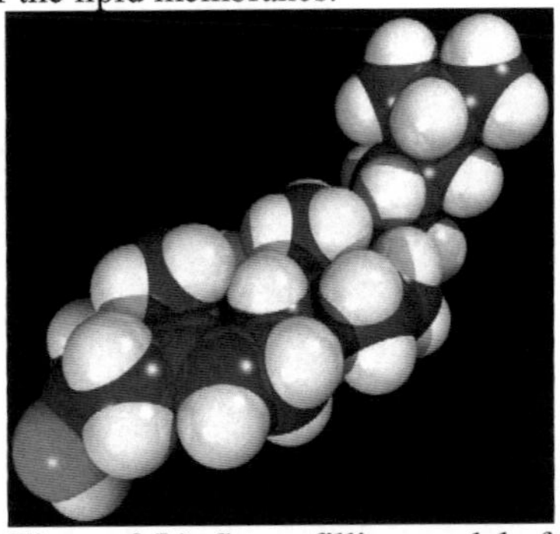

Figure 2(b). Space filling model of cholesterol.

There are many studies available in the literature, investigating the influence of cholesterol on the mechanical properties of lipid membranes using different experimental methods as micromanipulation [4], tether pulling [5], vesicle deformation induced by electric fields [6], nuclear magnetic resonance (NMR), X-ray diffraction [4, 8-10], etc. These studies demonstrate that cholesterol has significant ordering effect on the lipid hydrophobic chains and its presence in the cell membrane leads to an increase of the lipid bilayer thickness and decrease of the area per molecule.

It is also shown in the literature that the effect of cholesterol is lipid-specific. For example, the bending modulus of dioleoylphosphatidylcholine (DOPC) membranes does not change significantly after addition of cholesterol, but the sphingomyelin membranes become more flexible [6]. The presence of cholesterol strongly increases bending modulus when both chains are fully saturated, but not at all when there are two unsaturated chains [7, 8].

NMR measurements of acyl-chain order parameter of lipids with perdeuterated chains show that cholesterol disorders the lipid chain below the main transition and orders the chain above it [11] Ipsen).

In this work we focus our attention on the influence of cholesterol on the elastic properties of SOPC lipid membranes in pure water environment. Using the non-invasive method - thermally induced shape fluctuations of nearly spherical vesicles (liposomes) we measure the bending elastic modulus k_c of lipid vesicle's membrane in presence of different percentage of cholesterol in it.

2. Theory

The first theoretical models for the mechanical properties of lipid membranes proposed by Helfrich [12] and Evans [13] describe the elastic energy per unit area of lipid membrane, F_c by the expression:

$$F_c = \frac{1}{2}k_c\left(c_1 + c_2 - c_0\right)^2 + \bar{k}_c c_1 c_2 \tag{1}$$

where: c_1 and c_2 are the membrane principal curvatures, c_0 is the spontaneous curvature, and k_c and $\bar{k}_c$ are bending and saddle bending elasticity moduli of lipid bilayer, respectively. The spontaneous curvature of a symmetric membrane in a symmetric environment vanishes, $c_o = 0$.

After the first detailed theoretical model of thermally induced shape fluctuations has been proposed by Milner and Safran [14], the experimental procedures, based on the analysis of thermally induces shape fluctuations of quasi-spherical vesicles were developed for the precise measurements of the bending elastic modulus [15, 16]. The fundamental expression used by the authors is [14]:

$$\left\langle \left| U_n^m(t) \right|^2 \right\rangle = \frac{k_B T}{k_c} \frac{1}{(n-1)(n+2)[\bar{\sigma} + n(n+1)]} \tag{2}$$

where $\left\langle \left| U_n^m(t) \right|^2 \right\rangle$ is the mean squared amplitude of the membrane fluctuations' decomposition in spherical harmonics $Y_n^m(\theta,\varphi)$, k_B is the Boltzmann's constant, T is the absolute temperature, m and n are the numbers, characterizing the given mode and $\bar{\sigma} = \sigma R^2 / k_c$ is the dimensionless membrane tension (an adjustable parameter depending on the membrane tension and the difference of the lipid molecules between the inner and the outer layer of the lipid bilayer).

In fact what is measured in an experiment of fluctuating quasi-spherical giant vesicle is the equatorial cross section radius in 128 equidistant directions from the center of the vesicle for every recorded image. In spherical coordinates the radius of the vesicle in given direction can be written by the expression:

$$\rho(\varphi,t) = R[1 + u(\frac{\pi}{2},\varphi,t)] \tag{3}$$

where R represents the radius of a sphere with equal volume and $u(\theta = \pi/2,\varphi,t)$ is a function describing the membrane fluctuations, $\theta = \pi/2$ represents the equatorial cross section of the vesicle with the plane, passing through the vesicle centre, parallel to XY plane of the coordinate system.

It is assumed that the amplitudes of the fluctuations are small compared to the vesicle radius, $|u(\theta,\varphi,t)| << 1$.

The normalized angular autocorrelation function of the vesicle radius is given by the expression:

$$\xi(\gamma,t) = \frac{1}{R^2}[\frac{1}{2\pi}\int_0^{2\pi} \rho(\varphi+\gamma,t)\rho^*(\varphi,t)d\varphi - \rho^2(t)] \tag{4}$$

It is shown in [15] that the time averaged angular autocorrelation function can be decomposed into Legendre polynomials with amplitudes B_n, related to the mean squared amplitudes of spherical harmonics:

$$B_n = \frac{2n+1}{4\pi}\left\langle \left|U_n^m(t)\right|^2 \right\rangle \qquad (5)$$

Taking into account the relations (4), (5) and (2) we can calculate the bending elasticity modulus of the membrane from the decomposition of the angular autocorrelation function of the equatorial cross-section radius as following:

$$k_c = \frac{1}{B_n}\frac{k_B T}{4\pi}\frac{(2n+1)}{(n-1)(n+2)[\bar{\sigma}+n(n+1)]} \qquad (6)$$

In all the experimental data provided in this work stroboscopic illumination was used to remove the artifact due to the video camera integration time. Stroboscopically illuminated sample presents an instant picture of the object to the observer [17-19].

3. Materials and Methods

All the experiments are performed with vesicles composed of SOPC (1-stearoyl-2-oleoyl-*sn*-glycero-3-phosphocholine (C18:0/C18:1)), Avanti Polar Lipids Inc., USA). The cholesterol (5-Cholesten-3β-ol) used is product with high purity > 99% (GC): (Sigma - Aldrich). All the chemicals are used without any further purification.

The giant vesicles, studied in these experiments are prepared via modified electroformation method. The lipid is dissolved in chloroform 1 mg/ml. The cholesterol is dissolved in methanol 2 mg/ml. The final lipid –cholesterol solution for electroformation is prepared by mixing SOPC and cholesterol in the desired proportion for every cholesterol content. The electroformation cell, used for all experimental procedures is shown on figure 3. Glasses, coated with transparent conductor, indium tin oxide (ITO; thickness of 100 ± 20 nm, resistively of 100 Ω/□) are acting as electrodes.

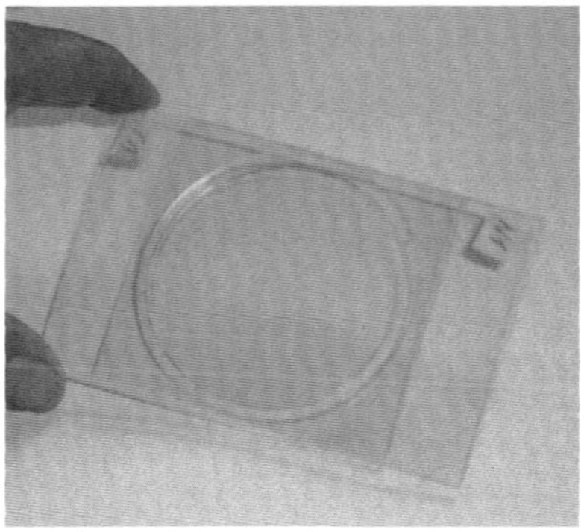

Figure 3 (a). An image of the electroformation cell.

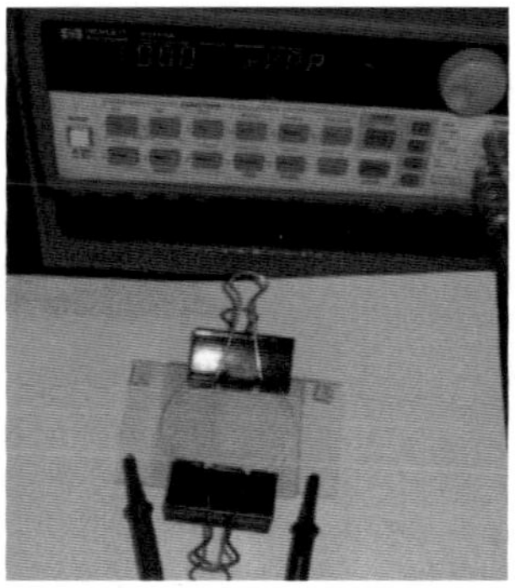

Figure 3 (b). An image of the electroformation set-up (electrofomation cell connected to the generator.

A number of small drops of the final lipid solution are laid on the surface of the glasses of the experimental cell in order to obtain as much as possible lipid depots for vesicle formation. The prepared in this way glasses are put under vacuum for at least 30 minutes. After the entire evaporation of the solvent the experimental cell is filled with double distilled water. We establish the most suitable for our experiment regime of electroformation. A low frequency (10 Hz) sinusoidal alternative voltage was applied (1.5 V PP) to the conductive glasses overnight. This procedure leads to the formation of vesicles, appropriate for our experiment. We choose giant (diameter of the order of 20-40 μm) fluctuating vesicles without any visible defects.

The samples of the fluctuating giant vesicles were observed under phase contrast microscope (Axiovert 100, Zeiss, Germany, oil immersion objective Ph3 100x magnification).The experimental equipment was improved using stroboscopic illumination, based on xenon flash lamp L6604 Hamamatsu, Japan) with damping vibration system, short light pulses (less than 3-4 μs long full width at half maximum) and high input energy (2 J) [19].

The obtained digital data was further recorded on the hard disk drive of the PC. Every second an image was acquired and recorded till the total number of images reaches a preliminary given value (about 400).

Further details on the contour determination, mean squared amplitudes calculation and fitting procedure to determine the bending elastic modulus, k_c, and the dimensionless membrane tension $\bar{\sigma}$, can be found in the article of Faucon et al. [15].

4. Results and Discussion

The analysis of thermally induced shape fluctuations of giant vesicles is used to study the influence of cholesterol on the elasticity of lipid membrane.

Table 1. Bending elasticity modulus of SOPC membrane, containing different concentration of cholesterol. The first column shows the molar percentage of cholesterol in the membrane, the second column shows the mean weighted value of the bending elasticity modulus and the third column shows the number of vesicles over which the mean is calculated.

Cholesterol concentration (mol %)	Weighted mean value of the bending elasticity modulus $\overline{k_c} \times 10^{-19} (J)$	Number of vesicles
0	1,7±0,1	10
10	1,56±0,05	11
20	1,98±0,11	11
30	2,02±0,09	10
50	2,67±0,13	6

The obtained experimental data for the bending elasticity modulus, k_c for different concentration of cholesterol in the lipid membrane (0 - 50 mol %) are presented in table 1. The values for the bending elasticity modulus for every cholesterol concentration in the lipid membrane are calculated as a weighted average value of about 6-11 vesicles. The value for the bending elasticity modulus of pure SOPC membrane in double distilled water, measured by the same method is given for comparison. All the experiments are performed in double distilled water environment.

As it can be seen from the obtained data for low cholesterol concentration (10 mol %) the bending elastic modulus decreases in comparison with that of pure lipid membrane. The theoretical studies for the influence of different admixtures on the elastic properties of lipid membranes [20] predict that for all the admixtures, influencing the elastic modulus of lipid membranes al low concentrations in the membranes reduction of the bending elastic modulus should be observed. For the substances that gradually reduce the bending elastic properties with the increase of its concentration in the membrane this result is obvious and experimentally proven. For the case of admixtures causing an increase of the elastic modulus of lipid membranes at high concentrations in the membrane (such as cholesterol) this result is specific and proves the theoretical predictions. Up to our knowledge this is the first experimental prove of the theory for membranes, containing substances increasing the bending elasticity at high concentration in the membrane.

At higher cholesterol content (20mol % and above) the value for the bending elastic modulus increases with the increase of cholesterol percentage in the lipid membrane. At the highest cholesterol content, studied by us (50 mol %) the value for the bending elasticity modulus is approximately two times higher than that of pure SOPC membrane. The obtained result is in very good agreement with the data, achieved by fitting the liquid crystalline theory with LAXS diffuse scattering data for SOPC membranes, containing cholesterol [21] as well as with the experimental data, obtained by different working groups for the influence of POPC (1-Palmitoyl-2-oleoyl-sn-glycero-3-phosphocholine) [10] and DMPC (1,2-Dimyristoyl-sn-glycero-3-phosphorylcholine) [22] on the elastic properties of lipid membranes.

5. Conclusion

The thermally induced shape fluctuation study of the elastic properties of cholesterol containing membranes shows that cholesterol gradually increases the bending elastic modulus of SOPC lipid membrane at concentrations above 20 mol % in the membrane. This result is in good agreement with the existing in the literature data provided by different experimental techniques and for other types of lipid membranes. At low cholesterol content in the membrane (10 mol %) the value of the bending elastic modulus is reduces. This result is in good agreement with the theoretical predictions.

Acknowledgements

This study was supported by Internal Grant VK-03-11 of The Institute of Solid State Physics, Bulgarian Academy of Sciences, Bulgaria and by the Grant 11013-9/2012-6 from Slovene Human Resources Development and Scholarship Fund.

References

[1] Angelova M I, Soleau S, Méléard P, Faucon J F and Bothorel P 1992 *Prog. Coll. Pol. Sci.* **89** 127
[2] Pavlič J, Genova J, Popkirov G, Iglič A, Kralj-Iglič V and Mitov M D 2011 *Chem. & Phys. Of Lipids* **164** 727
[3] Emmelot P 1977 *Mamalian Cell Membranes* (Butterworths, London)
[4] Needham D, Nunn R S 1990 *Biophys. J.* **58** 997
[5] Song J, Waugh R E 1993 *Biophys. J.* **64** 1967
[6] Gracia R S, Bezlyepkina N, Knorr R L, Lipowsky R and Dimova R 2010 *Soft Matter* **6** 1472
[7] Pan J, Mills T T, Tristram-Nagle S and Nagle J F 2008 *Phys. Rev. Lett.* **100** 198103
[8] Henriksen J, Rowat A C, Brief E, Hsueh Y W, Thewalt J L, Zuckermann M J and Ipsen J H 2006 *Biophys. J.* **90** 1639
[9] Chen Z, Rand R P 1997 *Biophys. J.* **73** 267
[10] Henriksen J, Rowat A C, Ipsen J H 2004 *Eur. Biophys. J.* **33** 732
[11] Ipsen J H, Moritsen O G, Bloom M 1990 *Biophys. J.* **57** 405
[12] Helfrich W 1973 *Z. Naturforsch.* **28c** 693
[13] Evans E A 1973 *Biophys. J.* **13** 941
[14] Milner S T, Safran S A 1987 *Phys. Rev. A* **36** 4371
[15] Faucon J F, Mitov M D, Méléard P, Bivas I and Bothorel P 1989 *J. Physique* **50** 2389
[16] Mitov M D, Faucon J F, Méléard P, Bivas I and Bothorel P 1992 *Adv. Supramol. Chem.*. G. W. Gokel (Eds.), JAI Press Inc.: Greenwich **93**
[17] Genova J, Vitkova V, Aladjem L, Méléard P, Mitov M D 2005 *J. of Optoel. & Adv. Mat.* **7** 257
[18] Méléard P, Faucon J F, Mitov M D and Bothorel P 1992 *Europhys. Lett.* **19** 267
[19] Genova J and Pavlic J 2012 *Bulg. J. Phys.* **39(1)** 65
[20] Bivas I and Meleard P 2003 *Phys. Rev. E* **67** 012901
[21] Pan J,Tristram-Nagle S, and Nagle J 2009 *Phys. Rev. E* **80(2)** 021931
[22] Meleard P, Gerbeaud C, Pott T, Fernandez-Puente L, Bivas I, Mitov M D, Dufourcq J and Bothorel P 1997 *Byophys. J.* **72** 2616

Trans-cis photoizomerization-induced tilted anchoring in photoactive guest-host liquid crystalline systems

Y G Marinov[1], G B Hadjichristov[1], A G Petrov[1], S Sridevi[2], U S Hiremath[2], C V Yelamaggad[2] and S K Prasad[2]

[1] Institute of Solid State Physics, Bulgarian Academy of Sciences, 72 Tzarigradsko chaussee, Sofia 1784, Bulgaria
[2] Centre for Liquid Crystal Research, Jalahalli, Bangalore 560 013, India

E-mail: ymarinov@issp.bas.bg

Abstract. We report on UV light-induced tilted anchoring in a mixture of the nematic liquid crystal N-(4-methoxybenzylidene)-4-butylaniline (MBBA) with a photoactive liquid crystalline azo dye (4-hexyloxybenzoloxy-4'-cyanoazobenzene) at 1 wt.% concentration. The surface anchoring found in this guest-host liquid-crystal system confined in homeotropic thin films is due to *trans-cis* photoizomerization of the guest azo molecules upon UV light.

1. Introduction

Of great interest for various applications of liquid crystals (LCs) are highly aligned and well ordered LC layers [1-3]. For the nematic LCs (as most exploited LCs) three basic types of LC alignment (known as anchoring) can be achieved depending on the boundary conditions: homeotropic, planar and tilted [4]. In order to control the LC surface anchoring, temperature-driven or light-driven layers acting as command surfaces, are commonly applied [5-10]. In all cases, both anchoring and orientational order are affected. The substantial depression of the order near the LC surface attracts a considerable attention [11-14] since this unwanted effect destroys the surface-induced order, and thereby the homogeneity of the anisotropic properties of the LCs [15].

Here we report on a kind of photo-induced anchoring at the surface of LCs, namely *trans-cis* photoizomerization-induced tilted anchoring, in a photoactive guest-host LC system. In particular, a homeotropic layer of a LC mixture consisting of a nematic LC as the host, and a photo-responsive LC azo dye as the guest, was considered.

2. Experimental

In the investigated mixture, the nematogen *N*-(4-methoxybenzylidene)-4-butylaniline (MBBA) was the host compound. The nematic-isotropic transition of this LC (from Reahim) is at 43° C. The azo dye 4-hexyloxybenzoloxy-4'-cyanoazobenzene (UI-147, the chemical structure is given in figure 1) was the photoactive guest. In the guest-host system, this liquid-crystalline compound synthesized at the Centre for Soft Matter Research in Bangalore, India, was present at a concentration of 1 wt.% (not eutectic mixture).

For the investigation, a sandwich type cell was prepared in a manner similar to that described in [16]. The cell of dimensions 1.5 cm × 2 cm, was assembled from two parallel 1 mm thick glass plates, and had a gap of 2 mm, defined using cooper spacers. They also served as electrodes. Initial homeotropic orientation of the LC layer was achieved using self-assembled layers of lipid lecithin [17]. The employed procedure of dipping the glass plates in a lecithin/chloroform solution of 20μM concentration resulted in loosely packed orienting layers [18]. The cell was filled with the LC mixture in its isotropic phase, and subsequently cooled slowly to the nematic phase to achieve good homeotropic orientation of the molecules. An identical cell, filled with MBBA only, was used as a reference. The samples were placed between two crossed polarizers (figure 2).

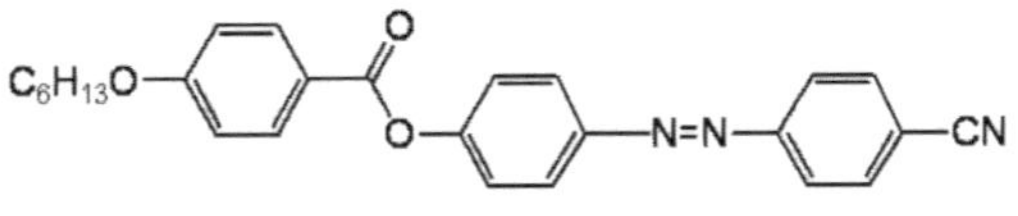

4-hexyloxybenzoloxy-4'-cyanoazobenzene (UI-147)

Cr 365,2 K **SmC** 398,5 K **N** 547,4 K

Figure 1. Molecular structure of the LC azobenzene dye UI-147. The phase transition temperatures are also given.

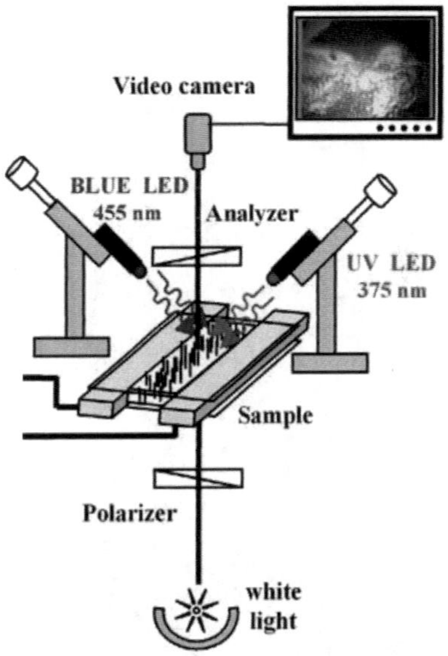

Figure 2. Schematic representa-tion of the experimental set-up used in this work.

3. Results and Discussion

Figure 3 presents the changes in the texture of the LC mixture upon light illumination. In the absence of the illumination by the LEDs, the field of view under the polarizing microscope in the crossed polarizer configuration is that of a good homeotropic alignment, proved also by conoscopy. In the following, the time elapsed after switching the UV source on, is denoted as t. It must be emphasized that once turned on, the UV source remained in the on condition for the rest of the experiment. At the beginning (t = 0, before the UV is on), a dark image, corresponding to the nematic phase with homeotropic orientation, is seen. Upon illumination by the two LEDs, the field of view changed as presented in figure 3. After t = 45 s, the texture starts changing from the cell periphery with the appearance of defects (figure 3(a)) spreading to the center. The bright part on the bottom right corner in the micrograph shown in figure 3(a) represents the change in LC anchoring to a tilted orientation.

Since the change is from a homeotropic to a tilted nematic orientation, a new state of anchoring starts to develop at the surface. The observed tilt-angle texture implies that the degree of homeotropic orientation within the sub-surface region of the LC layer decreases. The UV light-induced effect is continuously accumulated thus encompassing the whole cell area (figure 3(b)). The umbilicus-like texture visible in figure 3(b) indicates the presence of a conical degeneration of the LC director [4].

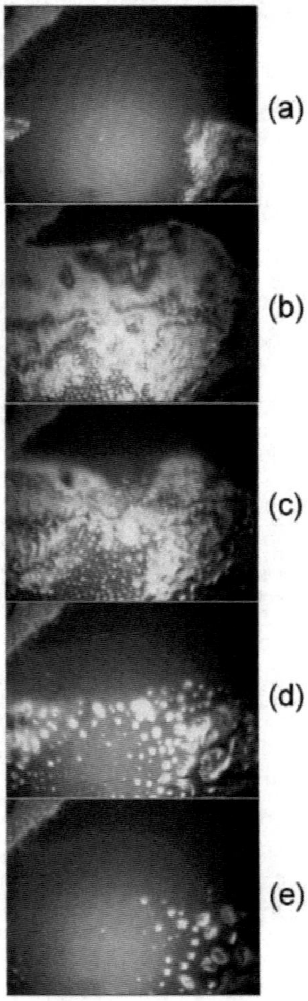

Figure 3. Polarizing microscope images of the textures observed for the LC mixture upon UV illumination after a lapse of:
(a) 45 s;
(b) 1 min 30 s;
(c) 2 min 15 s;
(d) 3 min;
(e) 4 min.

The blue light illumination starts after the instance (b).

The light intensities are:
UV LED - 11 mW/cm^2;
blue LED - 15 mW/cm^2.

At t = 1 min 30 s, the blue LED was turned on, while keeping the UV LED also on. The effect from the blue light is the gradual shrinking of the tilted area, from the periphery to the center (figure 3(c)). The reduction of the tilted anchoring at the surface upon blue LED illumination results in a complete diminishing of the UV-induced effect, as seen in figure 3(d) and (e). Such reversibility suggests that the observed texture changes are not a consequence of the sample heating.

The spontaneous return to the homeotropic state is achieved by switching the UV light off. This indicates that the observed UV light-induced anchoring modification is monostable unlike the bistable reported in the literature [7]. This process is relatively slow, occurring on the timescale of an hour. At higher intensity of the LEDs, the observed texture changes become faster. All these features, as well as the texture changes as given in figure 3, are exactly reproducible upon illumination over repeated cycles. We have to point out that the above changes are not observed in pure MBBA sample under the same experimental conditions, at higher intensities of the UV LED, even up to 23 mW/cm^2. This rules out pure thermal effects (and heat transport) as the cause for the observed features.

Our results on UV light-driven anchoring can be compared to temperature-driven anchoring transitions reported for nematics [5,6], even in the presence of photoactive agent, in particular, an azo compound [8]. The theory of temperature-driven anchoring transitions explains them as due to: (i) decrease of the surface order parameter and (ii) the competition of aligning effects of both surface and subsurface region (the surface melting region). The first effect occurs at low temperatures while the second is expected close to the nematic-isotropic (N-I) transition. Specific for our system is the ability of blue light to overcome the UV-driven effect and to quench the anchoring change. The non-thermal nature of the observed tilted anchoring is supported also by direct proof of the magnitude of the eventual temperature change (the heating) of the cell due to the UV light illumination in our experiment. For the purpose, the LC cell was replaced by a cholesteric LC thermographic sheet. The temperature variation estimated by thermal mapping with this sensitive thermoindicator was less than 1 K. Such a temperature change is negligible to bring about the nematic-isotropic phase transition, thereby proving that the UV-driven effect of tilted anchoring at the LC surface is not thermally induced (in any form, either non-local or local, e.g. by UV light absorption-induced thermal effects).

The tilted anchoring observed in our sample can be explained as due to a conformational change of the UI-147 azo compound. The loosely packed lecithin layer contains a lot of cavities. Those vacancies are naturally accommodated by LC molecules resulting in pronounced homeotropic anchoring. In the absence of the UV radiation, the UI-147 molecules are in the *trans*-conformation. Since the UI-147 molecule is polar, the concentration of *trans*-conformers in the near(sub)surface region is very high due to preferred adsorbance of polar azo molecules. Moreover, being highly polar in contrast to MBBA molecules, the azo-molecules presumably replace majority of them from the surface cavities. The molecular length of *trans*- and *cis*- conformers are respectively: 2.62 nm and 1.81 nm [19]. Now taking into account the average length of lecithin molecules to be ca. 3 nm [20], which scales with those of azo-conformers, we can assume formation of rugged interface upon UV photoizomerization. The latter modifies the LC surface leading to change of the surface anchoring similar to the case of surface-induced disorder [12]. In this sense, the effect observed here resembles the surface-melting transition (an effect due to depression of the order close to the surface) [11-14].

One of the related phenomena is the surface polarization. In particular it plays an important role for the flexoelectricity in LCs [21]. Thus, the near(sub)surface region enriched with highly polar azo-molecules serves as a 'command surface' for some pronounced photo-induced effects, e.g. it is responsible for the photoflexoelectricity found in homeotropic layer of a mixture of nematic LC with the same photoactive LC azo dye upon UV light illumination (in particular, at the wavelength of 365 nm) [22]. Upon UV illumination, the *trans*-conformers absorbing the light are transformed to *cis* form, and in the process losing their rod-like shape. As a consequence of anisotropy change, the nematic order in the sub-surface region is disturbed, leading to the umbilicus-like texture observed in our experiment (figure 3(b)). The observed photo-induced effect is a *trans-cis* photoizomerization-induced tilted anchoring at the LC surface. Conversely, the back recovery process takes place due to conformer *cis*-to-*trans* photoizomerization when the blue light is introduced.

The *trans-cis* photoisomerization of azobenzene dye UI-147 in the MBBA/UI-147 mixture under the conditions of the present experiment can be evidenced by monitoring the photo-induced change of the absorption of light passing through the LC layer. Figure 4 shows the absorption spectra in the wavelength range 420 – 500 nm obtained without and with illumination of the sample by the LEDs inserted within the spectrophotometer instrument, thus reproducing the experimental conditions close to those by the inspection of the tilted anchoring effect. The observed photo-induced spectral changes (figure 4) are indicative of the efficient interaction of the light of both UV and blue LEDs with the MBBA/UI-147 guest-host system, in spite of the relatively low light intensities. On the other hand, no change in the MBBA absorption (the reference cell) was present upon illumination with the LEDs under the same experimental conditions.

The change of the near(sub)surface region of the layer of the examined MBBA/UI-147 guest-host system upon UV light can be detected by reflectance measurements. Figure 5 plots reflectance spectra (recorded over the wavelength range from 200 nm to 500 nm) for MBBA/UI-147 without and with UV light illumination, when no cover glass plate was present for the LC sample. Even small, the difference in the optical response of MBBA/UI-147 system in both cases indicates the photo-induced change directly at the surface of the LC sample.

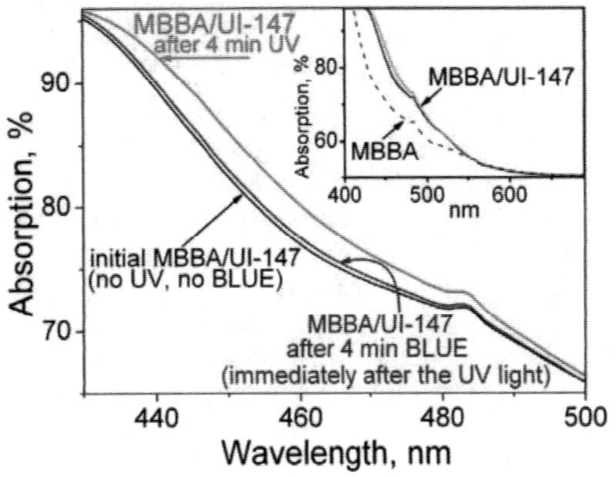

Figure 4. Absorption spectra recorded for the MBBA/UI-147 guest-host system upon illumination with the LEDs used in the present study. MBBA absorption is given in the inset (dashed line).

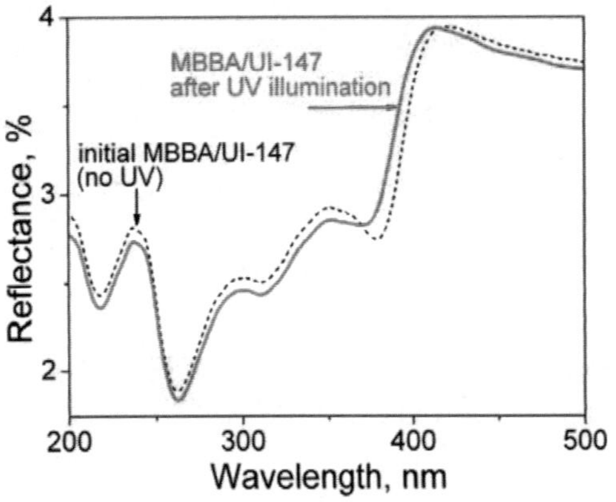

Figure 5. Reflectance spectra of MBBA/UI-147 guest-host system: data obtained without UV light illumination (dashed line), and with UV light illumination of the LC mixture (solid line).

4. Conclusion
We have reported an UV light-induced (not thermally-induced) effect of tilted anchoring in a photoactive guest-host LC system (1 wt.% guest) upon a relatively low intensity UV-light illumination (of the order of 10 mW/cm^2). This surface-induced effect is based on *trans-cis* photoizomerization of the guest azo-dye molecules. The near (sub)surface region of the LC layer rich with azo molecules is responsible for the observed *trans-cis* photoizomerization-induced surface anchoring. The tilted anchoring close to the surface is closely related to the LCs applications and has to be taken into account, especially at high light intensity, when this effect can limit and even frustrate the LCs laser-driven applications based on *trans-cis* photoizomerization.

Acknowledgments
This study was carried out as a part of Indo-Bulgarian joint research project supported by the Department of Science and Technology, New Delhi and the Ministry of Education and Science, National Science Fund of Bulgaria.

References
[1] Geary J M, Goodby J W, Kmetz A R and Patel J S 1987 *J.Appl. Phys.* **62** 4100
[2] Bahadur B 1995 *Liquid Crystals Applications and Uses* (Singapore: World Scientific)
[3] Paek S H, Durning C J, Lee K W and Lien A 1998 *J. Appl. Phys.* **83** 1270

[4] de Gennes J P and Prost J 1993 *The Physics of Liquid Crystals* (Oxford: Oxford University Press).
[5] Ryshenkov G and Kleman M 1976 *J. Chem. Phys.* **64** 404
[6] Hiltrop K and Stegemeyer H 1981 *Ber. Bunsenges. Phys. Chem.* **85** 582
[7] Aoki K, Seki T, Suzuki Y, Tamaki T, Hosoki A and Ichimura K 1992 *Langmuir* **8** 1007
[8] Shonova N, Petkova J I and Petrov A G 1993 *Compt. Rend. De l'Acad. Bulg. Sci.* **46** 45
[9] Ichimura K, Tomita H and Kudo K 1996 *Liq. Cryst.* **20** 161
[10] Fedorenko D, Ouskova E, Reshetnyak V and Reznikov Y 2006 *Phys. Rev. E* **73** 031701
[11] Barbero G and Durand G 1991 *Mol. Cryst. Liq. Cryst.* **203** 33
[12] Barbero G and Evangelista L R 2006 *Adsorption Phenomena and Anchoring Energy in Nematic Liquid Crystals* (Boca Raton: CRC Press)
[13] Biscari P and Turzi S 2007 *SIAM J. Appl. Math.* **67** 447
[14] Blinov L M 2011 *Structure and Properties of Liquid Crystals* (Dordrecht: Springer)
[15] Rasing T and Musevic I 2004 *Surfaces and Interfaces of Liquid Crystals* (Berlin: Springer)
[16] Kumar P, Marinov Y G, Hinov H P, Hiremath U S, Yelamaggad C V, Krishnamurthy K S and Petrov A G 2009 *J. Phys. Chem. B* **113** 9168
[17] Marinov Y, Shonova N, Naydenova S and Petrov A G 2000 *Mol. Cryst. Liq. Cryst.* **351** 411
[18] Naydenova S, Petrov A G and Yarwood J 1995 *Langmuir* **11** 3435
[19] Sridevi S, Hiremath U S, Yelamaggad C V, Prasad S K, Marinov Y G, Hadjichristov G B and Petrov AG 2011 *Mat. Chem. Phys.* **130** 1329
[20] Aileen C and Moy V T 2000 *Biophys. J.* **78** 2814
[21] Petrov A G 2001 Measurements and interpretation of flexoelectricity *Physical Properties of Liquid Crystals: Nematics* ed D Dunmur et al (London: IEE INSPEC) pp. 251-264
[22] Petrov A G, Marinov Y G, Hadjichristov G B, Sridevi S, Hiremath U S, Yelamaggad C V and Prasad S K 2011 *Mol. Cryst. Liq. Cryst.* **544** 3

17ISCMP IOP Publishing
Journal of Physics: Conference Series **398** (2012) 012039 doi:10.1088/1742-6596/398/1/012039

Performance Analysis of Multi-Wavelength Transmission Scanner for Polarized NIR Light

K Atroshchenko[1], C L Fontana[1], M De Rosa[1], M Bello[2], G Moschini[2], N M Uzunov[2, 4] and P Rossi[1, 3]

[1] Dept. of Physics "Galileo Galilei", University of Padua, 8 Francesco Marzolo str., 35131 Padua, Italy
[2] National Laboratories of Legnaro, INFN, Viale dell'Universita 2, 35020 Legnaro (PD), Italy
[3] INFN section of Padua, 8 Francesco Marzolo str., 35131 Padua, Italy
[4] Faculty of Natural Sciences, Shumen University "K. Preslavsky", 115 Universitetska str., 9712 Shumen, Bulgaria

E-mail: nikolay.uzunov@lnl.infn.it

Abstract. A system for small-object imaging, comprising a multiple-wavelength scanner for Near Infra-Red (NIR) light is under development in the Laboratory of Radiopharmaceuticals and Molecular Imaging (LRMI) at the National Laboratories of Legnaro, INFN, Italy. The System performs scanning of biological objects using NIR light in the interval of 900nm – 1700nm. The scanned region is a rectangular with dimensions of 50mm x 80mm and is performed by consecutive positioning of InGaAs linear image sensor sliding close to the scanned object. The scanning is carried out in two different modes. The first mode is performed in transmitted linearly polarized NIR light using a set of five light emitting diodes with fixed wavelengths. The process of scanning is realized by a consecutive positioning of the NIR sensor and signal acquisition at the corresponding position. In the second scanning mode the fluorescence emission of nanoparticles such as single-walled carbon nanotubes (SWCNTs), administered in the imaged object, is excited by NIR lasers with different wavelengths. Spatial resolution of the system for transmitted linearly polarized NIR at five fixed wavelengths has been determined. Polarimetric measurements of some optically active sugars such as fructose and lactose were conducted at some fixed wavelengths in the range of 900 - 1200nm. The system sensitivity with respect to the concentrations of these agents has been estimated.

1. Introduction

Near infra-red (NIR) imaging is a noninvasive method which recently gains popularity in biology and medicine because of its very useful properties. The wavelength region of NIR is situated in the interval spanning from 750 to 2500nm of electro-magnetic field spectrum. The range of 750 – 1400 nm, also known as a diagnostic window [1], belongs to the region of the spectrum where the absorption has lowest values that determines the deepest tissue penetration. The maximum energy value carried by NIR is around 150 kJ/mol, while the bonding energy between atoms in molecules is between 200 – 700 kJ/mol [2]. It is clear that the NIR light has insufficient energy to break up intra-molecular bondings of the organic molecules and hence does not introduce measurable effects in the tissue.

A number of techniques such as NIR bio-fluorescence imaging [3, 4], NIR tissue polarimetry analysis [5, 6], NIR transmission analysis [7, 8], etc., enabling *in vivo* imaging of physiological, metabolic, and molecular function, have emerged exploiting the advantages of NIR. The development of these techniques is stimulated by the number of recently made available targeted biological agents for diagnosis and basic medical research, enabling *in vivo* analyses and imaging in NIR. The specific physical properties of the new agents allow imaging by absorption, by scattering, and by excitation (such as fluorescence). On the other hand the search for better imaging of physiological, metabolic, and molecular function has lead to new imaging techniques such as multimodal imaging which exploit combinations of different physical properties of the imaged objects and requires nanoparticles or supramolecular assemblies of quantum dots, obeying advantages in faster circulation time and brighter images with respect to single molecule imaging agents [9].

A system for small-object imaging, based on a multiple-wavelength scanner for Near Infra-Red (NIR) light is under development in the Laboratory of Radiopharmaceuticals and Molecular Imaging (LRMI) at the National Laboratories of Legnaro, INFN, Italy. The System performs object scanning in the range of 900nm – 1700nm and is designed to operate in two different modes. In the first mode the scanning is performed in transmitted linearly polarized NIR light, traversed through the object. In the second scanning mode the fluorescence emission of nanoparticles administered in the imaged object, is excited using lasers with different wavelengths.

In this article we present the experiments conducted to determine some of the optical characteristics of the system such as the spatial resolution of the system for linearly polarized transmitted NIR light as a function of the distance to the scanned object, as well as polarimetric measurements, conducted on some optically active sugars such as fructose and lactose at wavelength range 900nm - 1200nm.

2. Experimental setup

A block-diagram of the scanner for small-animal NIR imaging is shown in figure 1. The scanner for NIR is realized on the basis of 256 pixels InGaAs array sensor Hamamatsu G9203-256D, positioned in a socket of a NIR detector board, which is fixed inside the scanner carcass. There is a compartment with NIR light emitting diodes (LED) attached to a support on the scanner carcass (figure 2). A system of NIR lasers (Laser1 and Laser 2) for imaging of reflected NIR light and the excited fluorescence is also fixed to the system. The scanned object is placed in a platform with a thin CaF_2 window and the scanning is performed while the platform with the object is moving in X and Y directions of a plane.

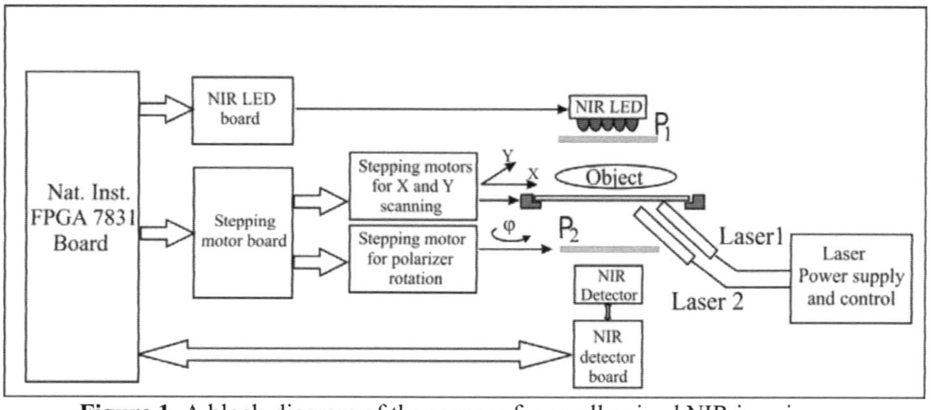

Figure 1. A block-diagram of the scanner for small-animal NIR imaging

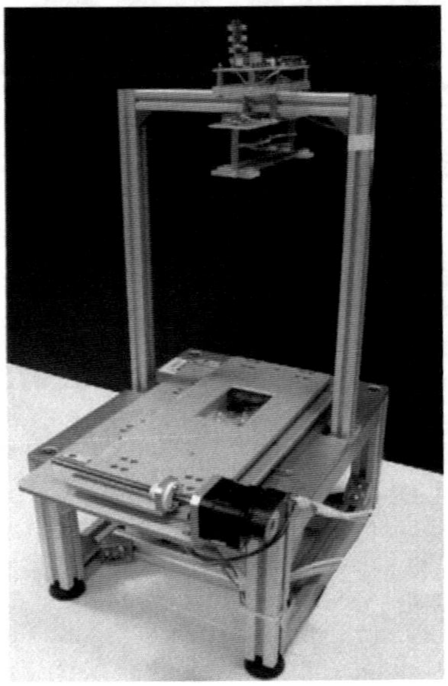

Figure 2. A photograph of the multifunctional NIR scanner with removed cover of the NIR LED compartment on the top. The NIR LED compartment and the NIR sensor are fixed. The platform with the object on the CaF_2 rectangular window is moving in XY plane.

The NIR LED compartment contains a set of five high-power diodes, emitting NIR light with different wavelengths. A linear polarizer P_1 is situated below the diodes in the NIR LED compartment. A second linear polarizer P_2, situated in a ring rotated by stepping motors, is placed in front of the NIR sensor and below the CaF_2 window.

The scanned region has a rectangular shape with dimensions of 50mm x 80mm. The scanning in both directions can be performed in steps as small as 10^{-3}mm. At each step the NIR light, transmitted through the object, is collected separately for each of the five wavelengths by switching consecutively the NIR light-emitting diodes and the acquisition. The scanning procedure thus produces five NIR images of transmitted light. The smallest rotation step of the analyzer P_2 is 1^0. The stepping motors guidance, the NIR diodes emission as well as the image acquisitions are controlled by a PC-based multifunction intelligent board National Instruments FPGA 7831R. A graphical user interface based on the developing environment LabView 8.5 was created to control the scanner functions as well as to visualize and process the acquired images.

3. Spatial resolution measurements

The spatial resolution of the scanner as well as the spatial resolution of the scanner, optically coupled with a fibre optical plate (FOP) model FOP(D35T20.5), have been determined along two perpendicular directions using USAF 1951 resolution standard, placed at different distances from the sensor and from the sensor with the FOP. For convenience the scan direction perpendicular to the sensor array was assigned X and the direction parallel to the sensor array was assigned Y. The spatial resolution in X and Y was measured for different distances to the scanned object for each of five NIR wavelengths: 940nm; 1070nm; 1200nm; 1300nm and 1550nm. Overall value of the spatial resolution was calculated averaging the spatial resolutions at all five wavelengths at each sensor-to-object distance [10]. The overall spatial resolution dependence of the system along X direction is shown in figure 3 (sensor without FOP). The corresponding overall spatial resolution dependence of the system along direction Y is shown in figure 4 (sensor without FOP). The corresponding overall spatial resolutions (assigned as sensor with FOP) along X and Y directions are shown in figure 3 and figure 4.

4. Polarimetric measurements of fructose and lactose

The rotation of the plane of polarization of linearly polarized NIR light passing through optically active solutions such as fructose and lactose in equilibrium with their chiral components was studied.

Besides the necessity to measure experimentally the values of the specific rotation of water solutions of these two compounds, our idea was to conduct system performance studies to assess the lowest detectable quantities of the system with respect to these sugars.

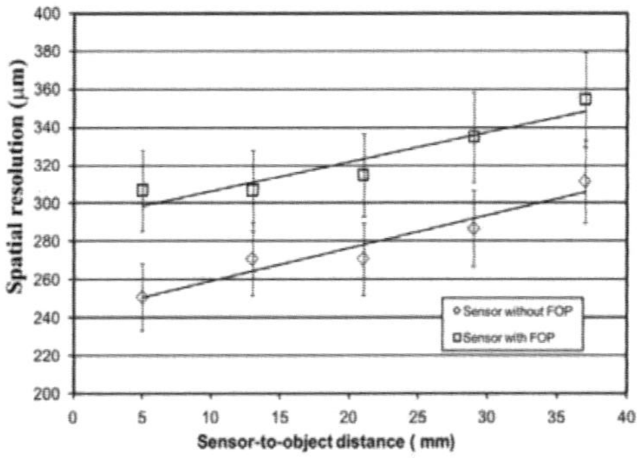

Figure 3. Overall sensor-to-object spatial resolution along direction X, calculated from the spatial resolution at five wavelengths of NIR light.

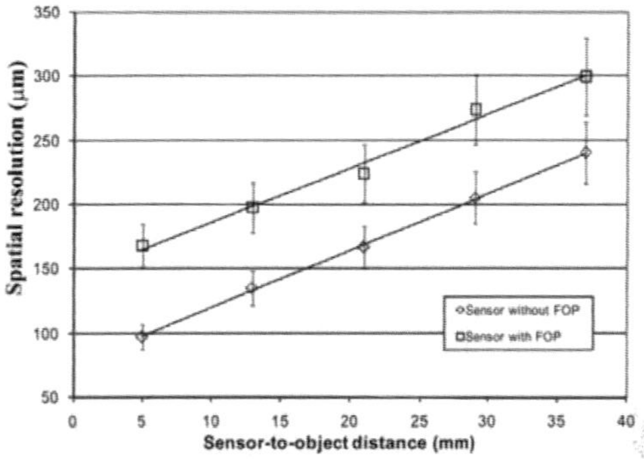

Figure 4. Overall sensor-to-object spatial resolution along direction Y, calculated from the spatial resolution at five wavelengths of NIR light.

Fructose ($C_6H_{12}O_6$) is a monosaccharide which is an isomer of glucose but has a different structure. The crystalline fructose has cyclic six-membered structure known as D-fructopyranose. When dissolved in water, because of mutarotation, it reaches equilibrium for about 30 minutes at ambient temperature where the β-D-pyranose is the most abundant of the anomers (68 – 72%). The next most abundant anomer is β-D-fructofuranose (28 – 32%), and finally there are less abundant quantities of α-D-glucopyranose (about 4%) [11,12]. Water solution of fructose in equilibrium possess a specific levorotation $[\alpha]_D^{20} = -93,78^0$.

Lactose ($C_{12}H_{22}O_{11}$) is a disaccharide, compounded by galactose and glucose. The water solution of lactose at ambient temperature causes degeneration of all crystalline modifications of the lactose to alpha- and beta- lactose [13]. Due to the process of mutarotation the anomeric ratio evolves towards equilibrium. This process normally takes about 30-40 minutes. Both anomers are dextrorotatory and the water solution of lactose in equilibrium exhibits a specific dextrorotation $[\alpha]_D^{20} = 55,48^0$ [13].

Measurements of the rotation of the plane of polarization in fructose and lactose were conducted using a cylindrical-shaped large-diameter quartz cuvette model 35I (Madatec Srl.), with diameter 50mm and 50mm path length (thickness of the analyzed layer). Solutions with different concentrations of fructose and lactose were prepared and scanned using linearly polarized NIR light in three wavelengths: 940nm; 1070nm and 1200nm. A diaphragm with a small opening (diameter of 6mm) was placed on the top of the cuvette to prevent reflections from its cylindrical walls. The analyzer P_2 was rotated from -90^0 to $+90^0$ with a step of 10^0. Images at each analyzer's angle were acquired and stored. NIR images of distilled water were acquired at the same analyzer angles for the same wavelengths in order to determine the zero position of the polarimeter. All images were saved as matrixes which elements were corresponding to the pixels of the image and with values corresponding to the intensity of the registered NIR light. The intensity of NIR light of a chosen region was calculated using averaging the value of the corresponding matrix elements.

Plots of the measured light intensity at all positions of the analyzer P_2 for each wavelength and for each sugar concentration were obtained. Following the idea that the observed NIR intensity would obey Malus' law, a 4-parametric fit with a periodic function has been conducted:

$$I = A\cos^2\left(B\varphi + C\right) + D, \qquad (1)$$

where: I is the observed intensity; parameter A is a parameter, corresponding to the amplitude of the sine function; B corresponds to the observed frequency; C is the phase shift and D is a correction parameter for the vertical shift of the sine function (the dark current parameter). The angle of rotation of the polarized light for given concentration and wavelength is deduced from the comparison with the fitted data for distilled water at the same wavelength. The measured dependence of this angle on the concentration of fructose and lactose is shown in figure 5 and figure 6. The values of the calculated specific angle of rotation $[\alpha]_\lambda^T$ are shown in Table 1.

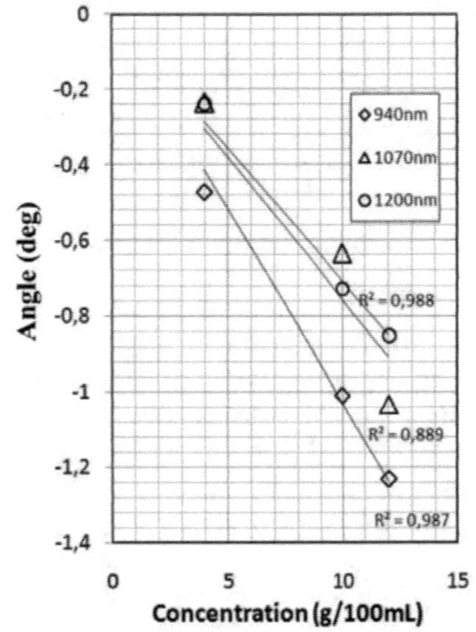

Figure 5. Rotation of angle of polarization for fructose as a function of the concentration

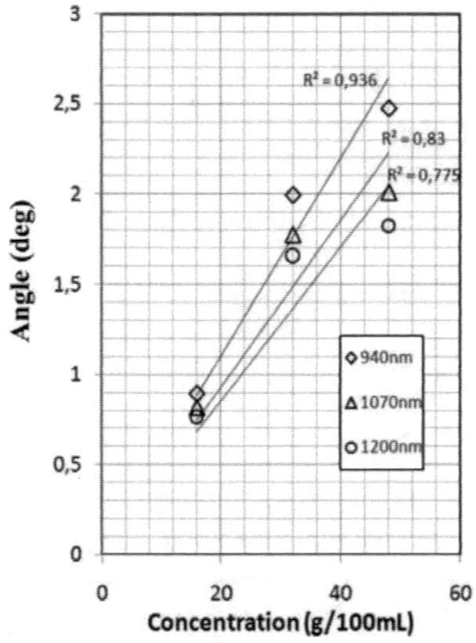

Figure 6. Rotation of angle of polarization for lactose as a function of the concentration

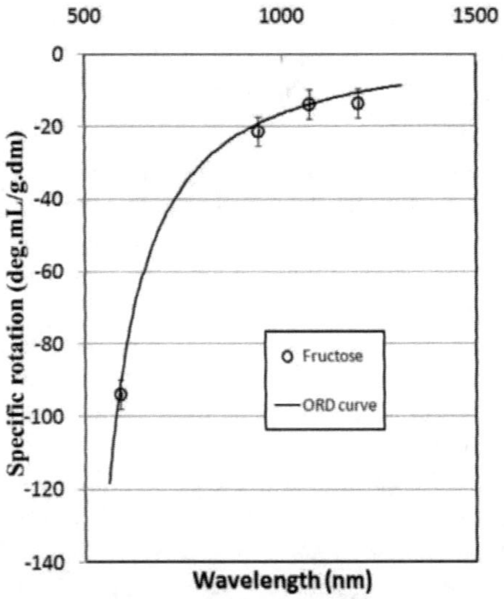

Figure 7. Measured specific rotation of fructose compared with the optical rotatory dispersion (ORD) curve

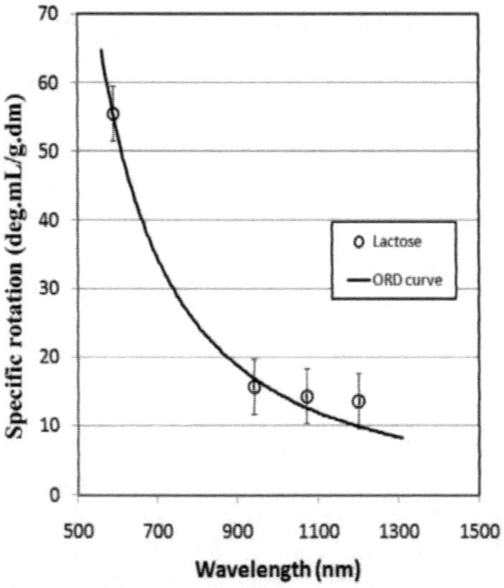

Figure 8. Measured specific rotation of lactose compared with the optical rotatory dispersion curve

Exploiting Drude's formula concerning the optical rotatory dispersion (ORD) [14], a two-parametric approximation has been applied to the obtained experimental data for fructose and lactose:

$$[\alpha]_\lambda^T = A/\left(\lambda^2 - B^2\right), \tag{2}$$

where parameter A is the rotation constant and B is the dispersion constant. The resulted ORD curves and the experimental values of $[\alpha]_\lambda^T$ for fructose and lactose are shown in figure7 and figure 8.

Table 1. Measured specific angle of rotation for fructose and lactose

	Specific angle of rotation (deg.mL/dm.g)		
	940 nm	1070 nm	1200 nm
Fructose	-21 ± 2	-14 ± 1,3	-13 ± 1,3
Lactose	16 ± 1,6	14 ± 1,4	13 ± 1,3

5. Conclusion

The system possess a good spatial resolution for analyzing layers of tissues in transmitted NIR light as well as for imaging using bioluminescence of small biological objects. The differences in the spatial resolution of the scanner along two perpendicular directions of the scanned surface are consequence of the different pixel dimensions of Hamamatsu G9203-256D: the pixel size is 50μm x 500μm oriented with its longer size along the slab. The use of FOP slightly deteriorates the spatial resolution along both directions, resulting in a shift of about 50μm in both graphs.

The measured values of the specific rotation for fructose and lactose are in agreement with the optical rotatory dispersion curves passing through the initial values for the specific rotation at wavelength λ=589nm. The system's lowest measurable concentrations, estimated experimentally lowering the sugar concentrations, were 500mg/dL for fructose and about 1000mg/dL for lactose and glucose.

References

[1] Awathy R G, Yoshida Y, Maekawa T and Kumar D S 2010 *Anal. Bioanal. Chem.* **397(4)** 1417

[2] Moor J *Application of NIR imaging and multivariate data analysis for pharmaceutical products* (Diploma thesis Univ. of Appl. Sci. München 2010)

[3] Sevick-Muraca E M, Houston J P and Gurfinkel M 2002 *Curr Opin Chem Biol.* 6(5) 642

[4] Frangioni J V 2003 *Curr Opin Chem Biol.* 7(5) 626

[5] Ghosh N and Vitkin I A 2011 *Journal of Biomedical Optics* 16(11) 110801

[6] Giakos G C, Valluru K, Adya V, Ambadipudi K, Pituri S, Bathini P, M Becker M, Farajipour P, Marotta S, Paxitzis J, Mandali B, Zervakis M and Livanos G 2009 *Meas. Sci. Technol.* **20** 104003

[7] Zhang X, Bloch Sh, Akers W and Achilefu S 2012 *Current Protocols in Cytometry* Publ. Online

[8] Tao H, Yang K, Ma Zh, Wan J, Zhang Y, Kang Zh and Liu Zh 2012 *Small* Vol. 8 Issue 2 281

[9] Pansare V J, Hejazi S, Faenza W J and Prudhomme R K 2012 *Chem. Mater.* **24 (5)** 812

[10] Uzunov N M, Bello M, Moschini G, Rossi P, Rosato A, Rondina M B, Montagner I M, Boldrin D and Muzzio P C 2010 *Conference record 2010 NSS and MIC* 3820

[11] Shallenberger R S 1978 *Pure & Appl. Chem.* **Vol. 50** 1409

[12] Domingues J R A *Glucose isomerase action on acid whey lactose hydrolisate and other sugars* (PhD Thesis University of Arizona 1984)

[13] Hargreaves J *Characterization of lactose in the liquid and solid state usin nuclear magnetic reaction NMR and other methods* (PhD Thesis Massey University 1995)

[14] Shechter E and Blout E R 1964 *Proc Natl Acad Sci USA* **V 51(4)** 695

Monte Carlo evaluation of water equivalency of some plastic materials for realistic electron IORT beams

M Oprea[1,2], D Mihailescu[1], C. Borcia[1]
[1]Faculty of Physics, "Al. I. Cuza" University of Iasi, Romania
[2]Radiotherapy Department, The Oncology Institute Iasi, Romania

E-mail: dmihail@uaic.

Abstract. The water equivalency of some commercially available water substitute materials for high energy electron beams dosimetry (PMMA, polystyrene and solid water WT1) has been investigated in this work for electron beams generated by the IORT linear accelerator NOVAC 7. The beams were simulated by the BEAMnrc/EGSnrc Monte Carlo code, while the dose distributions in water and plastic phantoms were calculated using DOSXYZnrc. The stopping power ratios were evaluated using SPRRZnrc user code. The results obtained for the depth- and fluence-scaling factors have been compared with the values recommended by the TRS-398 IAEA code of practice for absorbed dose determination in external beam radiotherapy. Due to the significant differences observed (sometimes more than 1%) and to the dependence of the scaling factors on the beam quality we can conclude that every time when plastic phantoms are used in electron IORT dosimetry, a theoretical or experimental investigation of the water equivalency of the water substitute materials must be done.

1. Introduction

IAEA TRS-398 [1] and other international dosimetry protocols recommend the use of water as standard phantom material for the dosimetry of high-energy electrons. For electron energies $E_0 < 10$ MeV, and only in special circumstances (when reference dosimetry measurements in a water phantom is difficult to perform), water-equivalent plastic materials may be used. In these circumstances, the dose-distribution in the solid phantom must be converted into appropriate dose-distribution in water by means of a depth-scaling factor, c_{pl}. Moreover, the dosimeter reading $M_{Q,pl}$ at any depth in plastic should be scaled to the equivalent reading in water, M_Q, using a fluence-scaling factor, h_{pl}.

In the case of conventional radiotherapy with electron beams, the use of c_{pl} and h_{pl} values given by international dosimetry protocols guarantees an acceptable level of accuracy. However, these protocols should be used with care in the case of electron beams generated by linear accelerators for Intra-Operative Radiation Therapy (IORT) because these accelerators have different dosimetric characteristics. The mobile accelerators are equipped with long cylindrical applicators. Due to their length, these applicators have a major contribution to the energy degradation, as well as to the spatial and angular distributions of the electrons at the phantom/patient surface [2]. Important parameters (as stopping power ratios and perturbation factors) used for absorbed dose determinations, depend on the electron beam characteristics. The correction factors and physical parameters recommended in the international dosimetry protocols [1, 3] are experimentally determined or calculated for conventional electron beams. Applying those data to beams such as those used for IORT could be, in principle, not appropriate if the characteristics of IORT and conventional beams are too much dissimilar from each other. Taking into account that, in the calibration procedures of electron beams, the accuracy of the dosimetric measurements must be as good as possible, such errors must be avoid.

In the present work is investigated the water equivalency of some plastic materials for realistic IORT beams generated by NOVAC7® linear accelerator calculating the values of scaling factors by a Monte Carlo method. The modelling of the NOVAC7® IORT accelerator is described in our previous work [2]. In the present calculations, only 3 MeV and 9 MeV electron beams and applicators having the extreme sizes (diameters of 4 and 10 cm) are used. The electron beams were obtained using BEAMnrc Monte Carlo code and dose distributions were calculated with DOSXYZnrc. The scaling factors have been evaluated through a technique described in our earlier works in which mono-energetic beams in interaction with polystyrene, PMMA and solid water WT1 were employed [4, 5]. Agreements until ~1% with TRS-398 international code of practice were obtained. However, a theoretical investigation of the water equivalency of water substitutes materials has never been reported for realistic beams generated by an IORT accelerator. It is well-known that the IORT beams are far to be mono-energetic, in fact being more degraded than the electron beams obtained with conventional accelerators. It is interesting to know if the values recommended by the IAEA TRS-398 dosimetry protocol for the scaling factors can be applied to electron beams generated by an IORT accelerator such as NOVAC7®. The main goal of our work is to give an answer to this question.

2. Theoretical considerations

According to the IAEA TRS-398 code of practice [2], solid phantoms in slab form such as polystyrene, PMMA, and certain epoxy resin "solid water" (water substitute) phantom materials such as solid water, plastic water, virtual water, etc. (see table 1 from the reference [5]) may be used for low energy electron beam dosimetry (under 10 MeV) and are generally required for low energy X rays dosimetry. However, the dose determination must always be referred to the absorbed dose to water at the reference depth in a homogeneous water phantom, the scaling procedure described in the TRS-398 code of practice being mandatory.

Depth dose distributions in plastic phantoms can be converted to depth dose distribution in water by means of depth-scaling. For a measurement made at a depth z_{pl} (g ·cm^{-2}) in a plastic phantom, the appropriate depth in water z_w (g ·cm^{-2}) is given by:

$$z_w = z_{pl} c_{pl} \tag{1}$$

where c_{pl} is the *depth-scaling factor* which can be calculated [6] by:

$$c_{pl} = \frac{z_{av}^{w}\rho_w}{z_{av}^{pl}\rho_{pl}} = \frac{R_{50}^{w}\rho_w}{R_{50}^{pl}\rho_{pl}} \tag{2}$$

In the above equation, z_{av}^{w} and z_{av}^{pl} are the average penetration depths (i.e. the half-value depths, R_{50}) in water and plastic phantom, respectively, and ρ_w, ρ_{pl} are the material densities.

In the case of identical irradiation conditions, when the absorbed dose to water is D_w and the absorbed dose in the solid (plastic) phantom is D_{pl}, the *fluence-scaling factor* can be calculated using the following formula:

$$h_{pl} = \frac{M_Q}{M_{Q,pl}} = \frac{D_w}{D_{pl}} s_{pl,w} \tag{3}$$

where, according to IAEA TRS-398 [2], M_Q and $M_{Q,pl}$ are the dosimeter readings at the reference depth

$$z_{ref} = 0.6R_{50} - 0.1 \text{ g/cm}^2 \quad (R_{50} \text{ in g/cm}^2) \tag{4}$$

in water and in the plastic phantom, respectively; $s_{pl,w}$ is the plastic material-to-water mass stopping power ratio.

The IAEA recommended values for the depth-scaling factors c_{pl}, fluence-scaling factors h_{pl} and the nominal densities ρ_{pl} for certain commercially available water substitute plastic materials are given in the table 21 (chapter 7) of the IAEA TRS-398 code of practice [2].

When using a plastic phantom to determine the depth dose distribution, each measurement depth in plastic must be scaled using equation (1) to give the appropriate depth in water. The dosimeter reading at each depth must also be scaled using equation (3).

3. Modelling

The NOVAC7® system is a robotic mobile intra-operative electron beam unit. This accelerator produces pulsed electron beams with four different nominal energies: 3, 5, 7 and 9 MeV. The accelerator is equipped with a 3D movable arm that can be pointed on the operating field. The basic system includes four types of Perspex (PMMA) cylindrical applicators with inner diameters 4, 6, 8 and 10 cm, wall thickness 0.5 cm and lengths 69, 67, 67 and 87 cm, respectively. The source-to-surface distance (SSD) is 80 cm, except for the applicator with the diameter of 10 cm for which the SSD is 100 cm. There are no scattering foils or flattening filters inside the treatment head, as the spatial uniformity of the treatment field is obtained by the scattering processes of electrons on the applicator wall [7].

The geometry of the NOVAC7® IORT accelerator (i.e. all of the essential components in the treatment head) was built using EGSnrc/BEAMnrc [8 - 10], The accelerator was modelled as a series of simple BEAMnrc component modules with cylindrical symmetry centred on the *z*-axis (see the figure 1 from our previous work [2]). The shape, dimension and material of these components were simulated according to the information provided by the manufacturer.

A number of four IORT beams have been simulated in this work corresponding to the combination of the lowest and highest nominal energies (3 MeV and 9 MeV) with the applicators having the extreme sizes (4 cm and 10 cm).

For each simulated beam, the complete information (energy, position, direction, charge, etc.) about any particle that crosses a given plane perpendicular to the beam axis (scoring plane) was stored in a data file (the *phase-space* file) further used as input file for the calculation of dose distributions in water and plastic phantoms. The correctitude of the simulations was verified by comparing the dose distributions calculated in the water phantom with the experimental dose distribution in water, as described in the reference [2]. An agreement of about 2% was obtained.

Depth dose distributions on the central axis of the four investigated beams where calculated in 20 x 20 x 5 cm^3 water, PMMA, polystyrene and WT1 phantoms of Cartesian voxel geometry using EGHSnrc user-code DOSXYZnrc [11]. Dose scoring grid was set to 1 cm x 1 cm x 0.1 cm. A number of $4 \cdot 10^7$ source electrons were using, ensuring a statistical uncertainty less than 1%.

The phase-space files at the phantom surface obtained for the IORT clinical beams were also used as source inputs for the EGSnrc/SPRRZnrc code [12] to calculate Spencer–Attix stopping power ratios of plastic-to-water, $s_{pl,w}$, as a function of depth in phantom. The $s_{pl,w}$ values were calculated along the central axis of the beam in cylindrical regions with a thickness of 0.1 cm and a radius of 1 cm. The statistical uncertainties were less than 0.1%.

In all simulations, have been used default values for EGSnrc particle's transport parameters, PRESTA-I for boundary crossing algorithm and PRESTA-II as electron transport algorithm. The cross section data were created using PEGS4 [10] including Sternheimer density effect corrections from ICRU 37 [13]. The energy cut-offs for particle transport were set to ECUT = AE = 0,521 MeV (kinetic energy plus rest mass) and PCUT = AP = 0,010 MeV.

4. Results and discussion

Percentage depth dose distributions in the water phantom for the IORT beams investigated in this work are shown in figure 1a. Each distribution is normalized to the maximum dose value. As we can see, in the case of the beams having 9 MeV nominal energy, the applicator size has practically no influence on dose distributions.

The necessity of depth scaling is illustrated by figure 1b. The significant deviation of the dose distribution curves in PMMA is due to the higher density of this material (1.19 g/cm^3) in comparison to water (1 g/cm^3), polystyrene (1.06 g/cm^3) and WT1 (1.02 g/cm^3). Nevertheless, the depth-scaling factors calculated using equation (2) does not differ so much because the depths z_w and z_{pl} are expressed in g/cm^2.

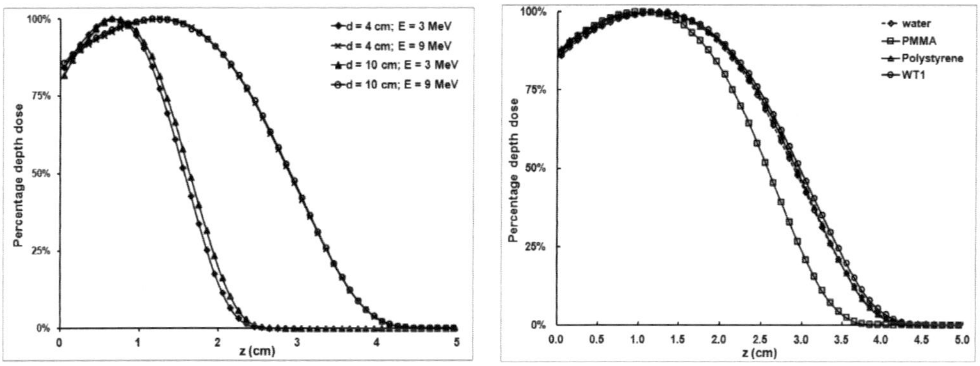

(a) (b)

Figure 1. Percentage dose distributions: (a) in the water phantom for the IORT beams investigated in this work (d = the applicator diameter, E = the nominal energy); (b) in different materials (d = 10 cm, E = 9 MeV).

The stopping power ratios of plastic-to-water, $s_{pl,w}$, as a function of depth in different phantom materials for the same beams as in figure 1b are shown in the next figure in which the solid markers indicate, approximately, the reference depths (z_{ref}) at which have been evaluated the dose and stopping ratios values. The $s_{pl,w}$values for 9 MeV electron beams are given in table 1. For 3 MeV beams were obtained slightly different values (up to ±0.1%).

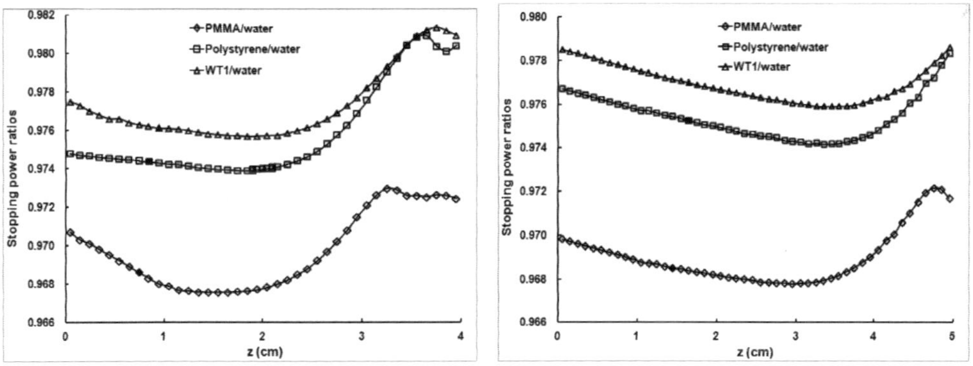

(a) (b)

Figure 2. Stopping power ratios as a function of depth:
(a) d = 10 cm, E = 3 MeV; (b) d = 10 cm, E = 9 MeV.

The values of scaling factors calculated by the Monte Carlo method are given in table 2 and compared with the values recommended by the IAEA TRS-398 code of practice. The statistical uncertainties are less than ±0.3% (1σ) for the small applicator and less than ±0.4% (1σ) for the large field size applicator. However, a final estimation of the uncertainties should also refer to the experimental dose distributions. Taking into account the ±2% agreement [2] between calculated and experimental depth dose distributions in the water phantom, previously mentioned, the range of uncertainty should be unacceptable. Fortunately, in the regions of half-value (R_{50}) and reference (z_{ref}) depths, the 1σ uncertainties do not exceed ±0.9%. Assuming the same uncertainty for the other water substitute materials, the overall 1σ uncertainties for the scaling factors are ±0.95% (d = 4 cm applicator) and ±1% (d = 10 cm applicator). Comparing with TRS-398 values, generally higher values have been obtained (up to ~1%, with some exceptions – the bold values in table 2).

Table 1. Spencer–Attix stopping power ratios plastic-to-water, $s_{pl,w}$ (for beams with 9 MeV nominal energy).

	PMMA	Polystyrene	WT1
$s_{pl,w}$	0.968	0.975	0.977

An interesting case is that of the most degraded beam, i.e. the 3 MeV beam generated using the smallest applicator (d = 4 cm). In this case, in which the scattered component is larger than the direct component, the depth scaling factors are much smaller, negative deviations from the TRS-398 values (up to -5.7%) being obtained. For this beam, the same problem also arises in the case of the fluence-scaling factors, but deviations are positives this time and not so high (up to 2.4%). It is obvious and understandable that for such low energy beams, used only in the Intra-Operative Radiation Therapy, the TRS-398 values cannot be applied in practice without introducing unacceptable errors. For the rest of the beams, with some exceptions, we generally found an agreement in the limit of about ±1%, fact that suggest the applicability of the TRS-398 values.

Table 2. Scaling factors for water substitute materials investigated in this work. In the first (left) column the values in parentheses are those recommended by the IAEA TRS-398 code of practice. Deviations from these values are given below the Monte Carlo calculated values. The bold values are deviations significantly exceeding the overall uncertainties of 0.95% and 1%.

	d = 4 cm E = 3 MeV	d = 4 cm E = 9 MeV	d = 10 cm E = 3 MeV	d = 10 cm E = 9 MeV	Mean values[(*)]
			Depth-scaling factors, c_{pl}		
PMMA	0.925	0.938	0.944	0.948	0.943
(0.941)	**(-1.7%)**	(-0.3%)	(+0.3%)	(+0.7%)	(+0.2%)
Polystyrene	0.871	0.931	0.932	0.937	0.933
(0.922)	**(-5.5%)**	(+1.0%)	(+1.1%)	**(+1.6%)**	**(+1.2%)**
WT1	0.895	0.954	0.951	0.957	0.954
(0.949)	**(-5.7%)**	(+0.5%)	(+0.2%)	(+0.8%)	(+0.5%)
			Fluence-scaling factors, h_{pl}		
PMMA	1.024	1.006	1.024	1.015	1.015
(1.009)	**(+1.5%)**	(-0.3%)	**(+1.5%)**	(+0.6%)	(+0.6%)
Polystyrene	1.051	1.036	1.051	1.036	1.041
(1.026)	**(+2.4%)**	(+1.0%)	**(+2.4%)**	(+1.0%)	**(+1.5%)**
WT1	1.029	1.022	1.029	1.020	1.027
(1.011)	**(+1.8%)**	(+1.1%)	**(+1.8%)**	(+0.9%)	**(+1.6%)**

[(*)]Excepting the most degraded beam (d = 4 cm, E = 3MeV).

Taking into account that only the extreme cases have been chosen for the Monte Carlo investigation (the smallest and biggest applicator, the lowest and highest nominal energies) we can conclude that for other combinations applicator-nominal energy the scaling factors should have values situated somewhat between those obtained in this work (see the last column from table 2). However, without experimental validation (which is the purpose of a future work) we cannot suggest the application of these results in the real measurements.

5. Conclusions

Water equivalence of PMMA, polystyrene and solid water (WT1) for realistic beams generated by the NOVAC 7 IORT accelerator was investigated by the Monte Carlo method. Only four IORT beams have been used: a combination of the lower and higher nominal energies (3 and 9 MeV) with the thinnest and the thickest applicator (4 and 10 cm). The depth- and fluence-scaling factors were calculated and compared with the values recommended by the IAEA TRS-398 dosimetry protocol. The results obtained for the four beams investigated in this work show a clear dependence on the beam quality, especially in the case of depth-scaling factors. However, in most cases, deviations ~ 1% from the TRS-398 values have been obtained, excepting the beam with a nominal energy of 3 MeV obtained with an applicator of 4 cm in diameter for which the deviations go until -5.7%. Taking into account these results, we conclude that for an accurate dosimetry a preliminary investigation of the water equivalence of the water substitute materials should be performed. Otherwise, the use of these materials - not recommended by the international dosimetry protocols but permitted in some special circumstances - could introduce larger and unknown errors, at risk being especially the low energy beams.

Acknowledgments

This work was supported by the Europen Development Fund through the Project "Transnational network of integrated postdoctoral research in science communication. Institutional building (school postdoctoral) and scholarship program (CommScie)" POSDRU/89/1.5/S/63663.

References

[1] International Atomic Energy Agency – IAEA TRS-398, IAEA Vienna 2000 85.
[2] Pimpinella M, Mihailescu D, Guerra A S, and Laitano R F 2007 *Phys. Med. Biol.* **52** 6197–6214.
[3] International Atomic Energy Agency - IAEA TRS 381, IAEA Vienna 2000.
[4] Mihailescu D and Borcia C 2006 *Rom. Rep. Phys.* **58** No. 4, 415–425.
[5] Borcia C and Mihailescu D 2008 *Rom. J. Phys.* **53** Nos. 7–8, 851–863.
[6] Saitoh H, Tomaru T, Fujisaki T, Abe S, Myojoyama A and Fukuda K, 2002 Proceedings of the Tenth EGS4 Users Meeting in Japan, *KEK Proceedings*, **18**, 55-64.
[7] Picardi L, Ronsivalle C, Tata A, Vignati A, The Italian "IORT PROJECT", *Proceedings of EPAC*, Vienna, Austria, 2000 (http://accelconf.web.cern.ch/Accelconf/e00/PAPERS/WEP5B04.pdf).
[8] Rogers D W O, Faddegon B A, Ding G X, Ma C-M, We J and Mackie T R, 1995 *Med. Phys.* **22** 503–524.
[9] Rogers D W O, Walters B and Kawrakow I, 2005 "BEAMnrc Users Manual" *National Research Council of Canada Report PIRS-0509 (A) revK* (Ottawa: NRC).
[10] Kawrakow I and Rogers D W O, 2003 "The EGSnrc Code System: Monte Carlo Simulation of Electron and Photon Transport" *National Research Council of Canada Report PIRS-701* (Ottawa: NRC).
[11] Walters B, Kawrakow I and Rogers D W O, 2004 "DOSXYZnrc Users Manual", *National Research Council of Canada Report PIRS-794 revB* (Ottawa: NRC).
[12] Rogers D W O, Kawrakow I, Seuntjens J P, Walters B R B and Mainegra-Hing E 2003 "NRC User Codes for EGSnrc" *National Research Council of Canada Report PIRS-702 (revB)* (Ottawa: NRC).
[13] International Commission On Radiation Units And Measurements 1984 *Stopping Powers for Electrons and Positrons*, Rep. 37, ICRU, Bethesda, MD.

Journal of Physics: Conference Series **398** (2012) 012041 doi:10.1088/1742-6596/398/1/012041

Dynamics of narrow solitons in two Ablowitz-Ladik chains with different interchain couplings

M T Primatarowa and R S Kamburova

Georgi Nadjakov Institute of Solid State Physics, Bulgarian Academy of Sciences, 1784 Sofia, Bulgaria

E-mail: prima@issp.bas.bg

Abstract. We study the evolution of discrete (narrow) solitons in a system of coupled Ablowitz-Ladik (AL) chains. Two types of interchain coupling are investigated: one which admits reduction of the system to the standard (integrable) AL model and one which couples opposite sites of the chains and does not admit reduction to the AL model. The condition for a perfect soliton switching between the two chains is obtained and the characteristics of the different couplings are analyzed.

1. Introduction

The study of nonlinear waves is receiving much attention due to the potential application within different branches of physics from nonlinear optics to Bose-Einstein condensate. The interplay between discrete diffraction and nonlinearity leads to the formation of discrete solitons. They promise an efficient way to control switching of optical signals in a system of coupled waveguides. So waveguide-based devices have received considerable attention in literature and this field has been extensively explored theoretically and experimentally. A discrete coupler involving two waveguides which exchange power as a result of weak overlap of their evanescent fields is the basic realization of a waveguide switch. The rate of power swapped back and forth between waveguides depends on the strength of the coupling, the degree of similarity of the waveguides and the initial pulse energy [1-5]. Waveguide arrays are particularly interesting because of their possible applications in signal processing [6-10]. All considerations regard discrete soliton switching mainly in homogenous waveguides with constant or smoothly varying coupling between them. However in practical applications the properties of inhomogeneous waveguides are more interesting and rather inevitable for switching [11-15].

Widely investigated are the standard discrete nonlinear Schrödinger (NLS) equation, as well as the completely integrable discrete Ablowitz-Ladik (AL) equation [16-18]. Although the two equations have the same linear properties and yield the same NLS equation in the continuum limit, their nonlinear properties are different. This leads to differences in the dynamics of narrow solitons (bright or dark) for the two models. Soliton solutions in two coupled discrete nonlinear chains were found and their stability was investigated in [19-21]. In the present paper we study the soliton dynamics in two coupled Ablowitz-Ladik chains with a complicated coupling that includes linear, nonlinear and dispersive interactions between the chains.

2. The model

We shall consider two parallel chains of particles described by the following system of coupled Ablowitz-Ladik equations:

$$
\begin{aligned}
i\frac{\partial \alpha_n}{\partial t} &= M(\alpha_{n+1}+\alpha_{n-1})(1+\gamma|\alpha_n|^2)+[d_1(\beta_{n+1}+\beta_{n-1})+2d_2\beta_n](1+\gamma|\alpha_n|^2) \\
i\frac{\partial \beta_n}{\partial t} &= M(\beta_{n+1}+\beta_{n-1})(1+\gamma|\beta_n|^2)+[d_1(\alpha_{n+1}+\alpha_{n-1})+2d_2\alpha_n](1+\gamma|\beta_n|^2) \qquad (1)
\end{aligned}
$$

$\alpha_n(t)$ $[\beta_n(t)]$ is the amplitude of an excitation at site n of the first (second) chain. M is the coupling interaction between neighboring particles in one and the same chain. The two chains are coupled to each other through the real parameters d_1 and d_2. The parameter d_2 governs the interchain coupling between opposite sites (nondispersive), while d_1 is a coupling of the AL type and allows reduction of the system to the AL model which is completely integrable. Both interactions include linear and nonlinear terms. The parameter γ determines the type of the soliton solution (bright for $\gamma > 0$ and dark for $\gamma < 0$) of the AL equation. In what follows we consider only bright solitons and set $\gamma = 1$ due to the scaling property of the AL system.

We shall study the influence of the two different couplings on the soliton properties. (1) is a generalization of the models considered in [19] and [20] and can be derived from the Hamiltonian

$$
\begin{aligned}
H &= \sum_n [M(\alpha_n\alpha^*_{n-1}+\alpha^*_n\alpha_{n-1}+\beta_n\beta^*_{n-1}+\beta^*_n\beta_{n-1}) \\
&+ d_1(\alpha_n\beta^*_{n-1}+\alpha^*_n\beta_{n-1}+\alpha_{n-1}\beta^*_n+\alpha^*_{n-1}\beta_n)+2d_2(\alpha_n\beta^*_n+\alpha^*_n\beta_n)] \qquad (2)
\end{aligned}
$$

using the deformed Poisson brackets [17,18]

$$
\begin{aligned}
\{\alpha_n,\alpha^*_m\} &= i(1+|\alpha^*_n|^2)\delta_{n,m} \\
\{\alpha_n,\alpha_m\} &= \{\alpha^*_n,\alpha^*_m\} = 0 \qquad (3)
\end{aligned}
$$

and the equations of motion

$$
\frac{\partial \alpha_n}{\partial t} = \{H,\alpha_n\}\,,\quad \frac{\partial \beta_n}{\partial t} = \{H,\beta_n\}\,. \qquad (4)
$$

The system (1) is nonintegrable but has two integrals of motion, the Hamiltonian H and the total number of particles N

$$
N = \sum_n [\ln(1+|\alpha_n|^2)+\ln(1+|\beta_n|^2)]\,. \qquad (5)
$$

For $d_2 = 0$ and the symmetric reduction $\alpha_n(t) \equiv \beta_n(t)$ the system (1) turns in an AL equation with the well known soliton solution:

$$
\begin{aligned}
\alpha_n(t) = \beta_n(t) &= \sinh\frac{1}{L}\,\mathrm{sech}\frac{n-vt}{L}\, e^{i(kn-\omega t)} \\
v = -2(M+d_1)L\sinh\frac{1}{L}\sin k\,, &\quad \omega = 2(M+d_1)\cosh\frac{1}{L}\cos k \qquad (6)
\end{aligned}
$$

The parameters k (wavenumber) and L (width) determine the velocity v and frequency ω of the soliton. In this case the conserved quantities have the form

$$
H = 8(M+d_1)\sinh\frac{1}{L}\cos k\,, \quad N = 4/L \qquad (7)
$$

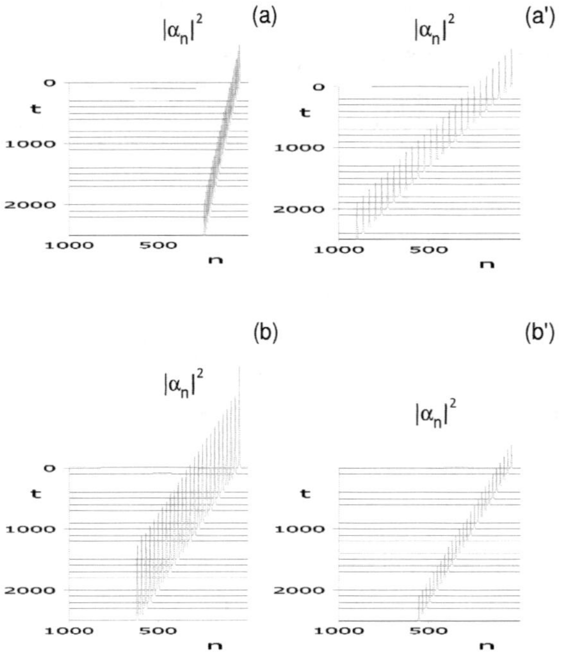

Figure 1. Propagation of equal narrow ($L = 2$) solitons in the two chains with $M = -1$, $k = 0.1$ and different coupling constants. (a): $d_1 = 0.628$, $d_2 = 0$; (a′): $d_1 = -0.628$, $d_2 = 0$; (b): $d_1 = 0$, $d_2 = 0.628$; (b′): $d_1 = 0$, $d_2 = -0.628$. The time is in units of $1/|M|$.

and it holds $\omega = \partial H/\partial N$.

Figure 1 shows the propagation of the soliton (6) when the amplitudes in the two chains are equal for different coupling constants. As can be expected the dispersive coupling d_1 influences the velocity of the soliton and according to the expression in (6) $v = 0.077$ for $d_1 = 0.628$ [figure 1(a)] and $v = 0.339$ for $d_1 = -0.628$ [figure 1(a′)]. The coupling between opposite sites in the two chains does not change significantly the velocity and $v = 0.208$ for positive as well as negative values of d_2 [figures 1(b) and (b′)]. The value of d_2 has mainly influence on the soliton's amplitude through the factor $\sqrt{M/(M+d_2)}$.

In the continuum limit $\alpha_n(t) \to \alpha(x,t)$, $\beta_n(t) \to \beta(x,t)$ which holds for wide solitons ($L \gg 1$) and for $\alpha(x,t) \equiv \beta(x,t)$ the system (1) reduces to the standard NLS equation of the form

$$i\frac{\partial \alpha}{\partial t} = G\alpha + (M+d_1)\frac{\partial^2 \alpha}{\partial x^2} + G|\alpha|^2\alpha\,, \quad G = 2(M+d_1+d_2)\,. \tag{8}$$

3. Numerical results

We shall investigate the propagation of an AL soliton which at the initial time is launched in one of the chains

$$\alpha_n(0) = \sinh\frac{1}{L}\,\mathrm{sech}\frac{n-n_0}{L}\,e^{ikn}\,, \quad \beta_n(0) = 0 \tag{9}$$

solving numerically the system (1). The simulations are carried out for 1000 sites of each chain, $n_0 = 500$ and periodic boundary conditions.

In the pure linear case the excitation will transfer from one chain to the other and back with a period $t_0 = \pi/|d|$, where d is the total linear coupling.

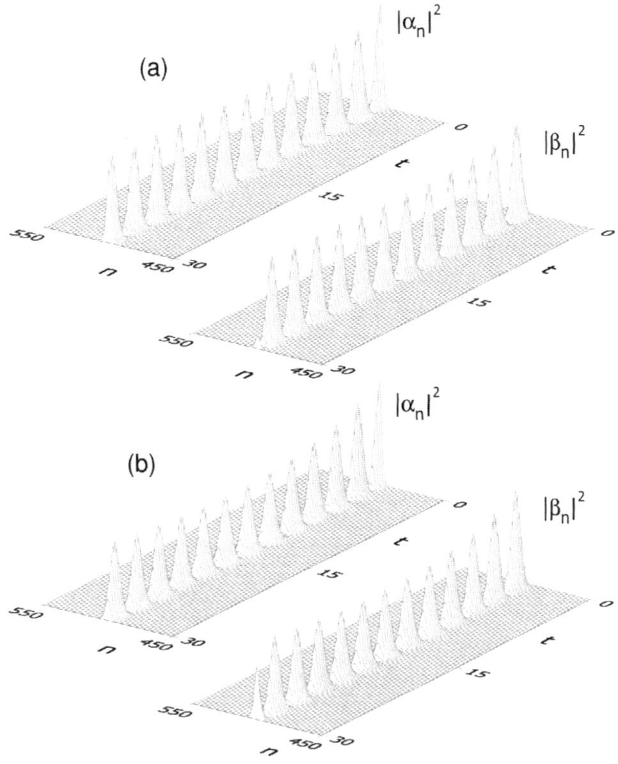

Figure 2. Switching of a narrow soliton with $k = 0$ between two chains for $|d_1| = 0.628$, $d_2 = 0$ (a) and $d_1 = 0$, $|d_2| = 0.628$ (b). The other parameters are the same as in figure 1.

For our complicated model (which is not linear) we observe that an energy exchange between the two chains take place with nearly the same period

$$t_0 = \pi/|d_1 + d_2| \tag{10}$$

and the energy exchange rate depends on the strength of the coupling. For small values of the coupling constants the soliton is only partially transferred. When the coupling increases the transferred rate grows. This behavior is due to the nonlinear coupling terms. We obtained that a soliton can transfer (perfect soliton switching) when the simple condition

$$4|d_1 + d_2|L^2 \gg 1 \tag{11}$$

is fulfilled. Figure 2 shows the soliton transfer when the coupling between the two chains is of type d_1 [figure 2(a)] or d_2 [figure 2(b)]. The process does not depend on the sign of the coupling. We observe that in the case of propagating solitons the soliton can be switched and its velocity remains the same as the initial one.

Figure 3 demonstrates the interplay between the two types of coupling. Their action is additive. When the constants d_1 and d_2 are of the same sign [figure 3(a)] the soliton switching is similar to that for one coupling's type but with the greater strength (figure 2). When the two couplings are of different signs the resulting coupling decreases and can be set to nearly zero as shown on figure 3(b). We observe that the soliton α_n remains in the first chain where it was initially excited and in the second chain $\beta_n \simeq 0$.

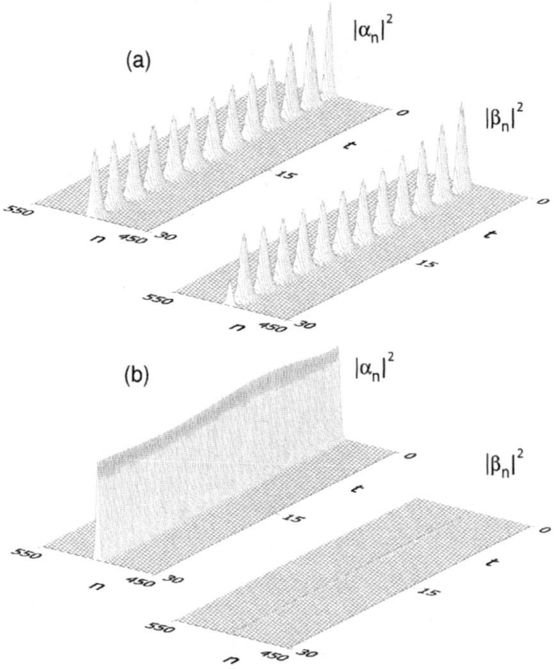

Figure 3. Evolution of a soliton with $k = 0$ excited in one of the chains for $d_1 = d_2 = 0.314$ (a) and $d_1 = -d_2 = 0.314$ (b). The other parameters are the same as in figure 1.

Figure 4 illustrate the differences between the two types of interchain coupling. For large values of the coupling constants the soliton switching is preserve for the coupling between opposite sites of the chains d_2 [figure 4(b)]. For the dispersive coupling d_1 an energy exchange with the same period t_0 takes place but the soliton formation is destroyed [figure 4(a)]. Obviously to achieve soliton switching $|d_1| < |M|$ has to be fulfilled. We have observed that for wide solitons ($L \gg 1$) the dispersive character of the coupling d_1 vanish and the solitons in the two chains are stable.

4. Conclusion

We have studied the evolution of discrete solitons in a system of coupled Ablowitz-Ladik chains. Two types of interchain coupling are introduced and compared. The coupling of the AL type admits reduction of the system to an AL model. Perfect soliton switching can be obtained for large enough values of the coupling constant in both cases. In the case of dispersive coupling the soliton is destroyed when the coupling constant exceed a critical value which depends on the soliton width and the interaction between the neighboring particles in the chains.

Acknowledgments

This work is supported by the National Science Fund of Bulgaria under Grant No. DO 02-264.

References

[1] Jensen S M, 1982 *IEEE J. Quantum Electron.* **18** 1580-83
[2] Trillo S, Wabnitz S and Stegeman G 1989 *IEEE J. Quantum Electron.* **25** 1907-16
[3] Paré C and Florjańczyk M 1990 *Phys. Rev.* A **41** 6287-95

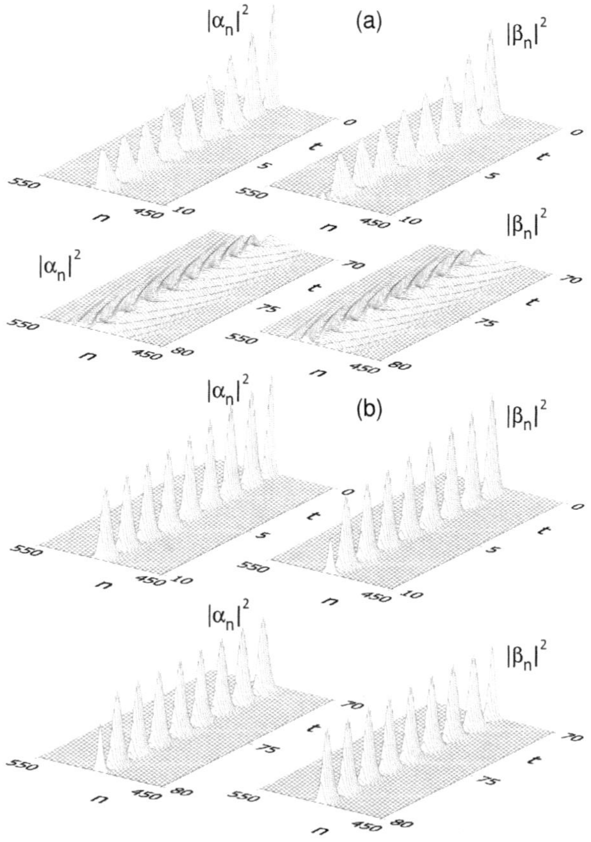

Figure 4. Switching of a narrow soliton with $k = 0$ between two chains for $|d_1| = 1.256$, $d_2 = 0$ (a) and $d_1 = 0$, $|d_2| = 1.256$ (b). The other parameters are the same as in figure 1.

[4] Kivshar Yu S 1993 *Optics Lett.* **18** 7-9
[5] Chu P L, Peng G D and Malomed B A 1993 *Optics Lett.* **18** 328-30
[6] Aceves A B, De Angelis C, Peschel T, Muschall R, Lederer F, Trillo S and Wabnitz S 1996 *Phys. Rev.* E **53** 1172-89
[7] Christodoulides D N and Eugenieva E D 2001 *Phys. Rev. Lett.* **87** 233901(4)
[8] Peschel U, Morandotti R, Arnold J M, Aitchison J S, Eisenberg H S, Silberberg Y, Pertsch T and Lederer F 2002 *J. Opt. Soc. Am.* A **19** 2637-44
[9] Vicencio R A, Molina M I and Kivshar Yu S 2003 *Optics Lett.* **28** 1942-44
[10] Molina M I, Vicencio R A and Kivshar Yu S 2006 *Optics Lett.* **31** 1693-95
[11] Królikowski W and Kivshar Yu S 1996 *J. Opt. Soc. Am.* B **13** 876-87
[12] Longhi S 2006 *Phys. Rev.* E **74** 026602(9)
[13] Morales-Molina L and Vicencio R A 2006 *Optics Lett.* **31** 966-68
[14] Goodman R H and Weinstein M I 2008 *Physica* D **237** 2731-60
[15] Belmonte-Beitia J, Pérez-García V M and Torres P J 2009 *Nonlin. Science* **19** 437-51
[16] Ablowitz M J and Ladik J F 1976 *J. Math. Phys.* **17** 1011-18
[17] Scharf R and Bishop A R 1991 *Phys. Rev.* A **43** 6535-44
[18] Cai D, Bishop A R and Grønbech-Jensen N 1996 *Phys. Rev.* E **53** 4131-36
[19] Bülow A, Hennig D and Gabriel H 1999 Phys. Rev. E **59** 2380-92
[20] Malomed B A and Yang J 2002 *Phys. Lett.* A **302** 163-70
[21] Kamburova R S and Primatarowa M T 2010 *AIP Conf. Proc.* **1203** 261-266

Crystallization of nickel nanoclusters by molecular dynamics

H Chamati and K Gaminchev

Institute of Solid State Physics, Bulgarian Academy of Sciences,
72 Tzarigradsko Chaussée, 1784 Sofia, Bulgaria

E-mail: chamati@issp.bas.bg

Abstract. We investigated the melting properties of bulk nickel and the crystallization of nickel nanocrystals via molecular dynamics using a potential in the framework of the second moment approximation of tight-binding theory. The melting behavior was simulated with the hysteresis approach by subsequently heating and cooling gradually the system over a wide range of temperatures. The crystallization of nickel nanoclusters consisting of 55, 147 and 309 atoms was achieved after repeatedly annealing and quenching the corresponding quasicrystals several times to avoid being trapped in a local energy minimum. The time over which the global minimum was reached was found to increase with the cluster size.

1. Introduction

Metallic clusters, with sizes ranging from 1 to 100 nm, has received considerable attention during the last three decades [1, 2]. These materials are arranged in structures that are completely different from their bulk parents. It has been found both experimentally and theoretically that they may exist in many different structures: icosahedral, decahedral, Marks polyhedra, etc. The change in the behavior of nanosized particles can be traced back to their large surface to volume ratio. Thus the major contributions to the physical and chemical properties emanates from their surfaces [3, 4].

Nickel, a transition metal with many technological and industrial applications especially when combined with other chemical elements to form alloys, has always attracted the attention of materials researchers, experimentally and theoretically. On the theoretical side various potentials have been proposed to model the interaction among the atoms that are able to reproduce the experimental measurements known so far and to explore unknown physicochemical properties that might lead to important technological applications.

Applying quantum mechanical methods, e.g. *ab initio*, to get insights into the physical properties of materials is computationally expensive, restricting the simulations to very small samples on short time scales. For systems with large number of particles over large time scales it is more appropriate to utilise semiempirical potentials, that incorporate the main features of the quantum mechanical properties of the system under study. There exist several efficient semiempirical potential formalisms depending on the way their corresponding expressions are derived from quantum mechanics. These potentials take into account the local density profiles of materials. They include the embedded atom method [5], Finnis-Sinclair potentials [6], second-moment approximation of tight-binding theory [7, 8], to name a few. In particular in the second moment approximation, the expression of the total energy is based on a small set of adjustable parameters, which are usually determined by matching the relevant experimental data [8] or first-principles results [9].

In a recent work [10] one of the present authors has proposed an improved version of the interatomic potential of reference [11] for the paramagnetic nickel, within the framework of the second moment approximation of the tight-binding theory. While the original potential [11] was obtained by fitting the relevant adjustable parameters to augmented plane wave results of the total energies of nickel and the stable alloys it forms with Aluminum, the potential of reference [10] was adapted solely for nickel. It was modified in such a way to improve on the results obtained with its original counterpart. The named potential was fitted only to the experimental values of the lattice parameter and the cohesive energy of Ni. This potential was used to estimate several properties, such as the bulk modulus, elastic constants, vacancy and surface formation energies. Its predictions for the elastic constants showed a satisfactorily good agreement with experimental data, except for C_{12}. The calculated surface energies were in good agreement with the experimental results and showed the correct anisotropies. Furthermore molecular dynamics simulations were carried to compute the temperature dependence of the thermal lattice expansion and the mean–square displacements, the phonon-dispersion curves along the high symmetry axes and the radial distribution function above the theoretical melting temperature. In general it was found that this potential was superior to its original counterpart in predicting the physical properties of nickel.

In this paper we continue to test the ability of the potential of reference [10] to reproduce known experimental results and to check its transferability to different atomic environments. In addition we use it to find the equilibrium shapes of quasi-crystals with free surfaces containing a well defined number of atoms by minimizing the total energy. Here the number of atoms is chosen to correspond to magic numbers identifying two well defined atomic structures: icosahedra and decahedra [12]. It is worth mentioning that nickel should crystallize in an icosahedral structure [3].

The plan of this paper is as follows. In Section 2 we give details of the computational procedure. Our results and their comparison to experimental, along with relevant discussions are presented in Section 3. Our conclusions are drawn in Section 4.

2. Computational methodology

In the second moment approximation of the tight-binding theory the total energy of the system reads

$$E = \sum_i \left[A \sum_{j \neq i} e^{-p\left(\frac{r_{ij}}{r_0}-1\right)} - \xi \sqrt{\sum_{j \neq i} e^{-2q\left(\frac{r_{ij}}{r_0}-1\right)}} \right] \tag{2.1}$$

The first term is a of the Born–Mayer type pairwise potential describing repulsive contributions, while the second term represents the band–structure term. To make contact with the embedded atom formalism we identify the sum under the root square as the density function and the root square itself as the embedding function. In expression (2.1) r_{ij} is the distance between a pair of atoms i and j. Here interactions up to fifth neighbors were taken into account. The adjustable parameters A, ξ, p, q, and r_0 were part of a large set of parameters that have been determined in reference [11] through fitting to the total energy of the system as computed by first–principles augmented plane wave calculations and taking into account the experimental values of the cohesive energy, the lattice constant and the structural stability of the ground state of Al, paramagnetic Ni (both fcc and bcc structures) and their alloys $NiAl_3$, Ni_3Al ($L1_3$ structure), as well as NiAl (B2 structure). It is worth noticing that the constructed potential for Alumimum was found to be accurate in predicting a number of thermodynamic and structural properties [13]. The corresponding values of the first four parameters in equation (2.1) are taken from reference [11] and are given by $A = 0.07415$ eV, $\xi = 1.4175$ eV, $p = 13.8297$ and $q = 2.2448$, while the fifth parameter, instead of $r_0 = 2.4307$ Å of [11], we have proposed in reference [10] the value $r_0 = 2.502$ Å. This value was chosen to give better fit to the equilibrium lattice constant a_0 and the cohesive energy E_0 of fcc nickel. For the sake of comparison we report here some results obtained with both potentials. These are $E_0 = 4.435$ eV/atom and $a_0 = 3.421$ Å [11], against $E_0 = 4.43$ eV/atom and $a_0 = 3.52$ Å [10]. The later values are in accord with the experimental result [14].

Notice that despite the relatively simple form (2.1) and the small number of parameters used to fit the total energy, this potential has proven its ability to reproduce quite well some experimental data in comparison to the embedded atom method potential of reference [15], whose total energy was fitted to 17 experimental values and *ab initio* results.

The above interatomic potential is used in conjunction with molecular dynamics simulations in the canonical isothermal (NVT) and isobaric (NPT) ensembles. The used simulation box was made up of 4000 atoms arranged on a fcc lattice with 40 layers with 100 atoms each. Simulations were carried out using periodic boundary conditions on the three space directions in order to neglect the emergence of possible surface effects. The equations of motion were integrated by means of the Verlet algorithm and a time step $\delta t = 5 \times 10^{-15}$ s. This value ensures energy conservation of the trajectories during the simulation process. In the NVT ensemble, the system was equilibrated at a target temperature during 1000 time steps, which were sufficient to obtain stationary values for the kinetic and the potential energies. The thermodynamic averages were computed satisfactorily over trajectories lasting 50 ps. In the NPT ensemble used to explore the hysteresis by gradually heating and cooling the sample at intervals of 10 K, the simulation box was equilibrated at a given temperature for over 6000 time steps and averages were computed over the next 1000 steps. Here we have monitored the change in the simulation box volume and the energy per atom.

During the simulation, the value of the lattice constant at each temperature was chosen so as to result in zero pressure in the system. The mean–square displacements at each temperature were computed in order to get insights into the melting behavior of nickel within the framework of the present model.

The structural properties are identified via the computation of the radial distribution function expressed via

$$g(r) = \frac{1}{4\pi N}\left\langle \frac{1}{r^2\rho\Delta r}\sum_{i=1}^{N} n_i(r)\right\rangle, \tag{2.2}$$

where $n_i(r)$ stands for the number of particles in a spherical shell of thickness Δr and at a distance r from the i–th atom, ρ is the number density of the studied material and N – the number of molecular dynamics steps. The brackets $\langle \cdots \rangle$ in equation (2.2) indicate the thermodynamic average. The system was first equilibrated at a higher temperature than the targeted one for 2000 times steps and then cooled and equilibrated at the named temperature for another 2000 steps. Statistical averages for the calculation of the structural properties were obtained from the data of the next 4000 steps.

Energy minimization leading to the most stable shape of nanoclusters was achieved, with the aid of the Wulff construction, by starting from fcc structures having free surfaces, with a well defined number of atoms corresponding to the one of the magic numbers according to the formula [12]

$$\varsigma(k) = \frac{10}{3}k^3 - 5k^2 + \frac{11}{3}k - 1, \tag{2.3}$$

identifying nanoparticles with five–fold symmetry structured in k shells. So for k a natural number the series of magic numbers is 13, 55, 147, ... The free surfaces were produced by fixing the dimensions of the computational box at a length large enough to avoid interactions between the periodic slabs in all directions. The initial fcc quasicrystals were annealed at high temperatures about the melting one and then were the subject of successive molecular dynamics quenching procedures (similar to the quasidynamic minimization procedure) that led to the final configurations.

3. Results and Discussion

In reference [10] results computed with the potential (2.1) for various structural and thermal properties were presented. In particular the melting temperature extracted from the behavior of the mean-square displacement extracted from the atomic density profile and based on the Lindemann criterion

[16] was established at 1880 K. For the purpose of this paper we have calculated anew the mean-square displacement at various temperatures using an approach taking the atomic positions. A linear extrapolation of our atomic mean-square displacement results yields a melting temperature of about 1930 K, showing that the present potential overestimates the melting temperature of nickel, compared to the experimental result of 1728 K [14], but comparable to the theoretical results of reference [8]. To check the validity of this value having in mind that the potential is transferable to different atomic environments, such as the liquid phase, we computed the pair distribution functions (see Section 2 for details) at 1900 K and 2000 K. The results are shown in figure 1. This clearly shows that at 1900 K the potential still describes a crystalline phase, while nickel should be liquid at this temperature. At 2000 K, however we have a behavior characteristic of a liquid phase. These findings are a clear confirmation that the melting temperature is most likely to be at 1930 K. On the other hand it is possible that the relaxation time at 2000 K is much longer than the simulation length.

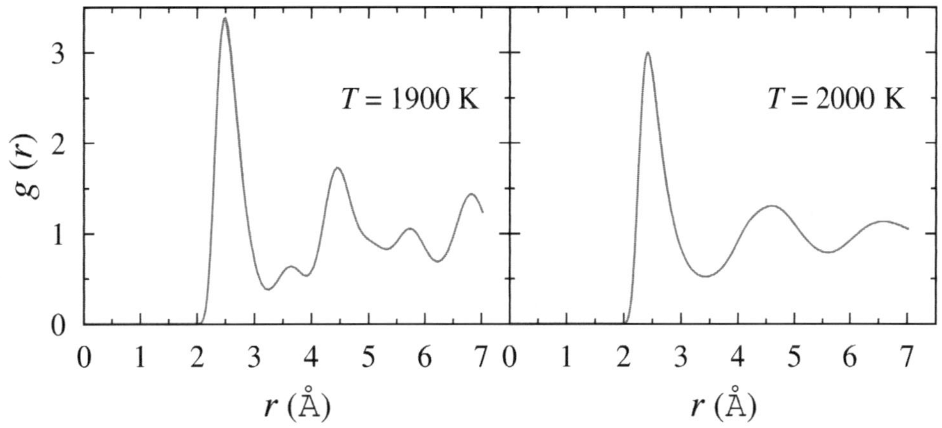

Figure 1. Radial distribution functions of nickel at T = 1900 K and T = 2000 K, obtained using molecular dynamics simulation.

Another way to find the equilibrium melting temperature from molecular dynamics is the use of the hysteresis approach. In this approach the simulation box is heated gradually beyond melting and then cooled in a similar way until crystallization in the NPT ensemble at zero pressure. The temperature dependence the lattice constant are depicted in figure 2. During the process of heating we observe an abrupt increase in both the energy (not shown here) and the volume upon melting. The inverse process (cooling) leads to a sudden drop at a lower temperature i.e. where nickel solidifies. Following reference [17] we deduce the equilibrium melting point T_m from the expression $T_m = T_+ + T_- - \sqrt{T_+T_-}$, where T_+ and T_- are the temperatures for the maximum degree of superheating and supercooling, respectively. According to this expression and using $T_- = 1210$ K and $T_+ = 2180$ K, we obtain $T_m = 1766$ K, which slightly above the temperature reported in reference [17] obtained with a quntum corrected Sutton-Chen potential.

A final test of our model consists in the crystallization of nanoparticles with numbers of atoms corresponding to magic numbers. We have started our simulations with 55, 147 and 309 atoms arranged in fcc structures with free surfaces as described above. The lowest energy configuration nanoclusters was obtained by employing a simulated annealing and quenched molecular dynamics technique. This procedure is commonly used to find the global and local minima in complicated situations. The results were tested by varying cooling rates and melted configurations. In general this approach does not guarantee that the energy minimum found is mandatory the global one. Usually the procedure stops at the first local minimum reached. Consequently, various different verifications are necessary to make

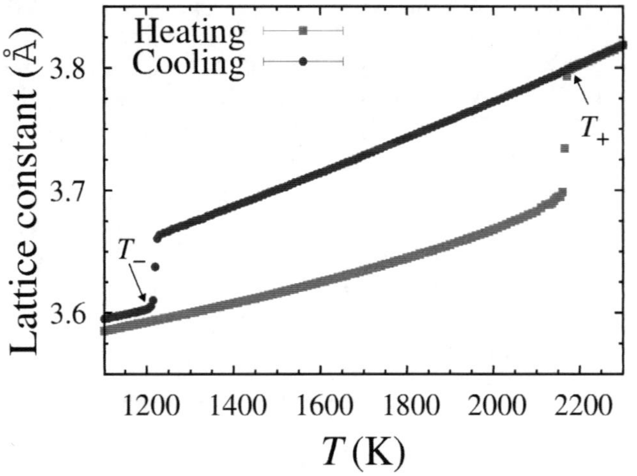

Figure 2. Dependence of the lattice constant of bulk fcc nickel on the temperature in the isobaric canonical ensemble via molecular dynamics simulation. The statistical errors are smaller than the symbols size.

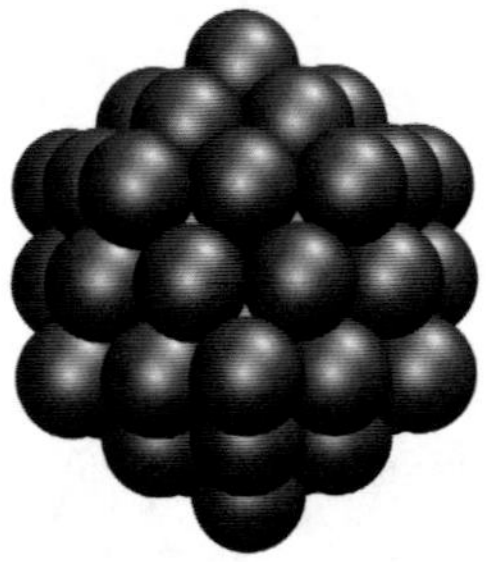

Figure 3. Icosahedral structure composed of 55 atoms of nickel obtained using molecular dynamics simulation.

sure that after a series of annealed and quenched molecular dynamics runs the minimum found is likely to correspond to the global one. In addition, icosahedral clusters have five-fold symmetry and therefore they are metastable structures. Nevertheless, it is found in several cases that icosahedral metallic clusters have the lowest energy compared to other structures. We started with fcc clusters consisting of a number of atoms corresponding to a magic number [given by equation (2.3)] of the icosahedral structure. We heated the initially constructed clusters up to temperatures that are higher than the melting point. The resulting liquid is subsequently quenched with different cooling rates in order to minimize the energy. The procedure is repeated several times and the coordinates of the quenched cluster were saved. A comparison of the energies of the saved configurations showed that indeed those with the lowest energy have an icosahedral-like structure. In figure 3 we show the shape of the lowest energy nanocluster with 55 atoms. To show its degree of symmetry we present three different projections on two dimensional planes. We have also noticed that changing the cooling rates during simulations leads to polycrystalline structures. It was found the time needed to reach the global minimum increases with the particle size. This is in agreement with the conclusions of references [18, 19], where the influence of the cooling rates in the crystallization of nanoclusters was investigated.

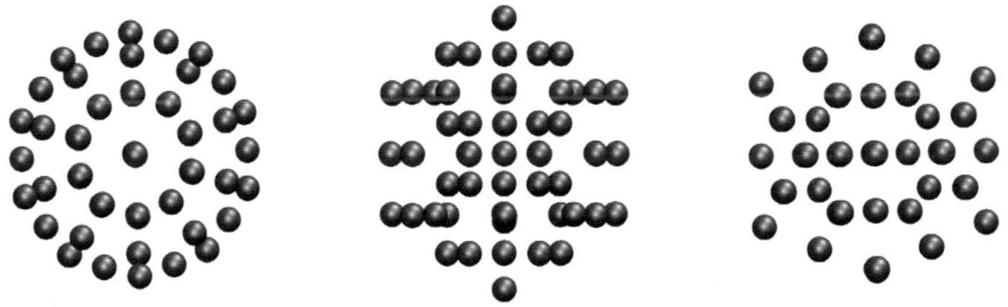

Figure 4. two-dimensional projections of the icosahedral structure shown in figure 3.

4. Conclusions

We presented an extensive molecular dynamics study of the melting properties of bulk nickel and the crystallization of quasicrystals in the nanometer range. The simulations were performed with the help of an interatomic potential in analogy to the tight-binding scheme in the second moment approximations. This potential was proposed in reference [10] and has proven its ability to reproduce with a fair accuracy the experimental data of a number of thermodynamic, as well as structural properties of nickel.

To gain insight into the melting properties of nickel we computed the mean-square displacement over a wide range of temperature. According to the Lindemann criterion [16] we estimated the melting temperature at 1930 K. This result is higher than the experimental value of 1728 K [14], but in agreement with most of the potentials designed to describe to the properties of nickel. Our result was checked by exploring the behavior of the pair distribution function at 1900 K and 2000 K. We found a fcc structure at 1900 K and a liquid phase at 2000 K. Further we constructed the hysteresis of nickel by heating and cooling gradually a simulation box over a wide range of temperatures. We found a large hysteresis with two abrupt jumps of the volume of the box at two different temperatures: upon heating at $T_+ = 2180$ K and upon cooling at $T_- = 1210$ K.

The transferability of the model was checked further by investigating the crystallization of nickel quasicrystals using a series of alternating annealing and quenching molecular dynamics simulations till the lowest energy shape is reached. Our procedure was applied to quasicrytals composed of a number of atoms, corresponding to magic numbers characteristic of icosahedra and decahedra. Starting from perfect fcc structures, containing 55, 147 and 309, with free surfaces above melting and repeating the procedure several times we end up with almost perfect icosahedral structures. The global minimum of the stable configuration was attained after a certain number of cycles that depended upon the size (number of atoms) of the nanocrystal.

We conclude that despite its relative simplicity the potential of reference [10] is capable of producing the main characteristics of nickel, yet it has to take into account the magnetic nature of the real element.

Acknowledgments
We thank Professor N. Papanicolaou for a careful reading of the manuscript. This paper was partially supported by grant No BK-01/2012.

References

[1] Contescu C I and Putyera K (eds) 2009 *Dekker encyclopedia of nanoscience and nanotechnology* 2nd ed (Boca Raton: CRC Press)

[2] Cao G and Wang Y 2011 *Nanostructures & nanomaterials : synthesis, properties, and applications* (*World Scientific Series in Nanoscience and Nanotechnology* vol 2) (New Jersey: World Scientific)
[3] Mitin V V, Sementsov D I and Vagidov N Z 2010 *Quantum Mechanics for Nanostructures* (Cambridge University Press)
[4] Roduner E 2007 *Nanoscopic Materials: Size-Dependent Phenomena* (Cambridge: Royal Society of Chemistry)
[5] Daw M and Baskes M 1984 *Phys. Rev. B* **29** 6443–6453
[6] Finnis M W and Sinclair J E 1984 *Phil. Mag. A* **50** 45–55
[7] Gupta R 1981 *Phys. Rev. B* **23** 6265–6270
[8] Cleri F and Rosato V 1993 *Phys. Rev. B* **48** 22
[9] Chamati H and Papanicolaou N I 2004 *J. Phys.: Condens. Matter* **16** 8399–8407
[10] Chamati H 2011 *J. Mater. Sci. Tech.* **19** 42–51
[11] Papanicolaou N I, Chamati H, Evangelakis G A and Papaconstantopoulos D A 2003 *Comput. Mater. Sci.* **27** 191–198
[12] Martin T P 1996 *Phys. Rep.* **273** 199–241
[13] Chamati H, Stoycheva M S and Evangelakis G A 2004 *J. Phys. Condens. Matter* **16** 5031–5042
[14] Kittel C 2005 *Introduction to solid state physics* 8th ed (Hoboken NJ: Wiley)
[15] Mishin Y, Farkas D, Mehl M J and Papaconstantopoulos D A 1999 *Phys. Rev. B* **59** 3393
[16] Lindemann F 1910 *Phys. Z.* **11** 609–612
[17] Luo F, Chen X R, Cai L C and Ji G F 2010 *J. Chem. Eng. Data* **55** 5149–5155
[18] Gafner Y, Gafner S and Entel P 2004 *Phys. Solid State* **46** 1327–1330
[19] Gafner S, Redel' L and Gafner Y 2007 *The Physics of Metals and Metallography* **104** 180–186

17ISCMP IOP Publishing
Journal of Physics: Conference Series **398** (2012) 012043 doi:10.1088/1742-6596/398/1/012043

Thermoluminescence properties of Eu and Li co-doped Gd_2O_3, induced by UV light

H Hristov[1], Y Baneva[2], D Nedeva[3], N Arhangelova[1], I Penev[4], V Velev[1] G Moschini[5], P Rossi[6] and N M Uzunov[1, 5]

[1] Faculty of Natural Sciences, Shumen University "K. Preslavsky",
115 Universitetska str., 9712 Shumen, Bulgaria
[2] Dept. of Physics and Biophysics, Medical University,
150 Tzar Liberator str., 9000 Varna, Bulgaria
[3] Faculty of Physics, Chemistry and Ecology, Technical University, Gabrovo,
4 H. Dimitar str., 5300 Gabrovo, Bulgaria
[4] Institute for Nuclear Research and Nuclear Energy, BAS,
72 Tzarigradsko chaussee Blvd., 1784 Sofia, Bulgaria
[5] National Laboratories of Legnaro, INFN,
Viale dell'Universita 2, 35020 Legnaro (PD), Italy
[6] Dept. Of Physics, University of Padua and INFN section of Padua,
8 Francesco Marzolo str., 35131 Padua, Italy

E-mail: nikolay.uzunov@lnl.infn.it

Abstract. For some specific biomedical applications, connected with in-situ measurements of the absorbed dose of ultraviolet (UV) light, we have developed materials, sensitive to the light emission with a wavelength up to 320nm. Thermoluminescence (TL) yield of Gd_2O_3, doped with Eu and Li has been analysed with respect to the quantity of Li co-dopant. Lithium has been added as Li_2CO_3 to a mixture of Gd_2O_3 with 10 wt% Eu_2O_3. Pellets with the mixture have been sintered at a temperature of 1000°C. The kinetic parameters of the phosphors thus obtained have been studied from the TL glow curves after irradiation with UV light. It has been demonstrated that the addition of 16 wt% of Li_2CO_3 to the Eu-doped Gd_2O_3 yields a maximum intensity of the peaks at 87°C and at 145°C. Studies on the kinetic parameters as well as the TL properties of Eu-doped Gd_2O_3 with the addition of 16% of Li_2CO_3 have been conducted. It has been measured that two of the TL peaks of this phosphor have relatively long fading. Analysis of the TL properties of the phosphors obtained from Gd_2O_3, doped with Eu and Li, shows that they possess good sensitivity to the UV emission and could be used as appropriate phosphors for detection and quantitative measurements of UV light.

1. Introduction

During the process of thermally stimulated luminescence (thermo-luminescence or TL) a release of luminescence light from some compounds (phosphors) is observed when they are heated. The emitted light is proportional to the accumulated energy from primary irradiation (x-rays, gamma rays, etc.), absorbed by the phosphor. The plots of the emitted light intensity with respect to the linearly increasing temperature are known as "glow curves" (GC). These curves usually have a complex shape, compounded by superimposed TL peaks. The intensity of these peaks is as a rule linearly correlated to the quantity of radiation absorbed by the material. TL is largely exploited at present for applications in dosimetry, in nuclear medicine, in archaeology in environmental studies, etc.

A very promising TL property of some materials is their ability to accumulate energy from photon sources with "softer" than the X-rays' and gamma-rays' energy such as ultraviolet (UV) and even visible light. It has been shown that some sintered materials from ZrO_2 doped with Er and Li could be used as appropriate detectors for ultra-violet (UV) light [1-3].

Studies about the influence of some co-dopants on the TL emission of some materials have shown that the presence of some ions in the matrix of the phosphors enhances the intensity of the TL. This effect is probably due to the presence of traps in the band gap created by the ions. The effect of co-doping with lithium ions is shown for a number of materials [4,5]. Following the idea to enhance the TL emission of the materials with addition of Li ions, in our previous publications we have demonstrated that quantities of lithium co-dopant, added as Li_2CO_3 to a mixture of ZrO_2 with 1wt% Er, improved the TL intensity of the sintered material [6]. In this article we present the enhancement of the TL emission of Gd_2O_3 doped with Eu and co-doped with Li (some preliminary results have already been reported in [7]). We discuss the opportunity such material to be exploited as a potential quantitative indicator, sensible to the ultraviolet light in the range of 290nm—340nm.

2. Materials and methods

To obtain the TL materials sensitive to the UV emission we have applied a simple sinterization procedure, similar to the one described in [1]. UV sensible phosphors have been made from sintered pellets of Gd_2O_3 doped with 10% Eu_2O_3. Quantities of Li have been added as Li_2CO_3, obtaining thus a set of samples with the following quantities of Li: 10%; 12%, 14%; 16% and 18%. The mixtures have been pressed in pellets with a diameter of 8mm and a thickness of about 1 mm using 5 ton press. Finally the pellets have been sintered at 1000°C for 8 hours.

A setup for precise TL measurements (TL reader) developed in the Laboratory for nuclear physics and radioecology (LNPR) at the University of Shumen has been used to obtain the glow curves of the phosphors [8]. The system consists of a setup for sample heating, a glow curve measuring device on the basis of photomultiplier tube, and a setup for sample irradiation with higher-intensity ultraviolet (UV) light. The operation and control of the system has been realized on the basis of PIC16F876 microcontroller as well as on original forward/backward feedback method, based on the so-called model predictive controllers, in order to achieve higher temperature stability and repeatability of the analyses conducted.

Trap parameters of the phosphors have been studied using initial irradiation with ultraviolet light from a xenon short-arc lamp type XBO 75 W/2 from OSRAM. The crystals have been placed at a distance of 10 cm and the time of irradiation was 5 min.

The intensity of the TL emission as a function of the linear increase of the temperature, known as TL glow curve (GC), is used to analyze some important parameters of the materials. Kinetic parameters such as the activation energy and the kinetic order are obtained applying glow curve decomposition analysis on each GC. Following [9] a single glow peak of a general order of kinetics is fitted using the formula:

$$I(T) = I_m b^{\frac{b}{b-1}} \exp\left[\frac{E}{kT}\cdot\frac{T-T_m}{T_m}\right]\left[(b-1)(1-\Delta)\frac{T^2}{T_m^2}\exp\left[\frac{E}{kT}\cdot\frac{T-T_m}{T_m}\right]+Z_m\right]^{-\frac{b}{b-1}}, \quad (1)$$

where I(T) is the glow-peak intensity, I_m is the maximum glow-curve intensity, E (eV) is the activation energy, k is the Boltzmann constant, T (K) is the temperature, T_m (K) is the value of the temperature at the peak maximum, with $\Delta = 2kT/E$, $\Delta_m = 2kT_m/E$; $Z_m = 1+(b-1)/\Delta_m$ and the parameter b ($1 < b \leq 2$) is the so-called kinetic order.

When the glow peak is a composite peak, compounded by m overlapping glow peaks, the method consists of a minimization of the χ^2 function:

$$\chi^2 = \sum_{i=1}^{n} \left(I_i - \sum_{j=1}^{m} I_j(T_j) \right)^2 , \qquad (2)$$

where I_i are the measured GC intensity values, $I_j(T_i)$ are the values of each partial GC at the temperature T_i, n is the number of the experimental points and m is the number of the partial glow peaks compounding the observed peak.

The fading of the peaks in the TL glow curve (GC) has been obtained calculating the intensity of the corresponding partial TL emission (the net peak area) from the phosphors, kept for a certain time after the irradiation in a dark storage at a temperature of 20°C. Glow curves, obtained with the TL reader, have been analyzed using a developed in the LNPR computer program implementing data smoothing using a Savitzky-Golay filtering and subsequently applying a minimization procedure based on conjugated gradients method for the TL peak decomposition technique.

3. Experimental results

Glow curves of Gd_2O_3, doped with 10% of Eu_2O_3 and co-doped with different quantities of Li from Li_2CO_3 have been obtained measuring the TL emission immediately after 5 minutes of irradiation with UV without annealing. The GC, presented with small circles in figure 1, have been obtained at a heating rate of 0,16°Cs^{-1}.

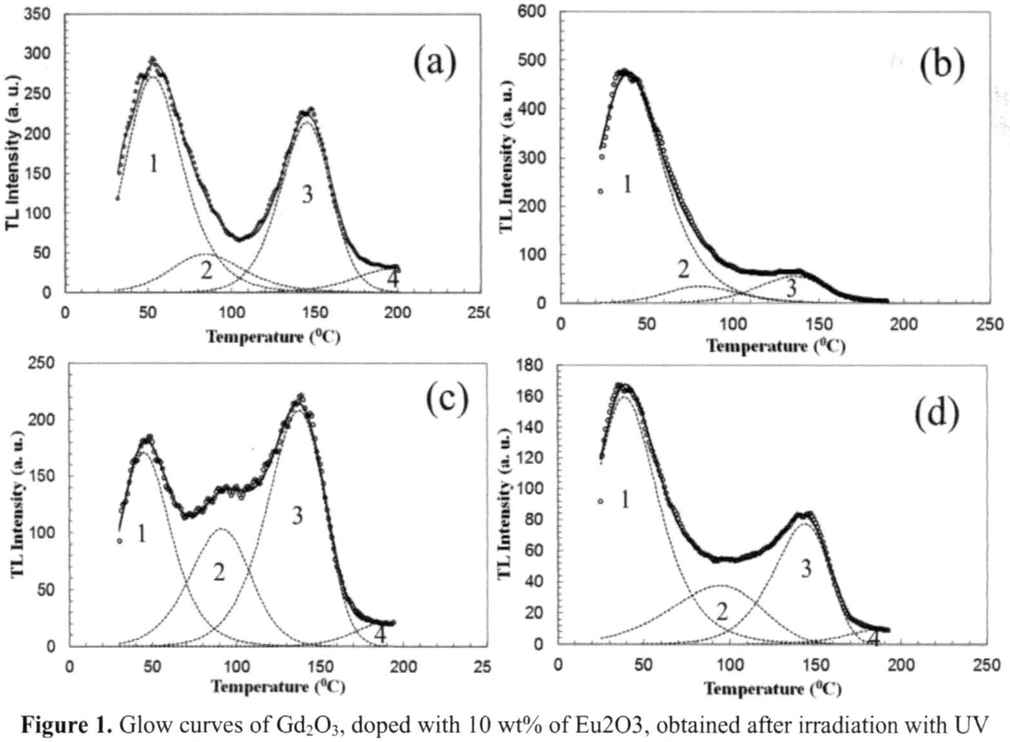

Figure 1. Glow curves of Gd_2O_3, doped with 10 wt% of Eu2O3, obtained after irradiation with UV light. The material is co-doped with: a) 12% Li from Li_2CO_3; b) 14% Li from Li_2CO_3; c) 16% Li from Li_2CO_3; d) 18% Li from Li_2CO_3.

Journal of Physics: Conference Series **398** (2012) 012043 doi:10.1088/1742-6596/398/1/012043

Three peaks, shown with a broken line in figure 1, have been revealed by the decomposition analysis of the glow curves (using formula (1) and (2)): peak 1 at 47°C, peak 2 at 87°C and peak 3 at 145°C, indicated as 1, 2 and 3 respectively. Another peak with higher values of the maximum temperature position, indicated as 4 in figure 1, has also been detected. The resulting fit, which is a sum of all peaks, is presented with thick line in the same figures.

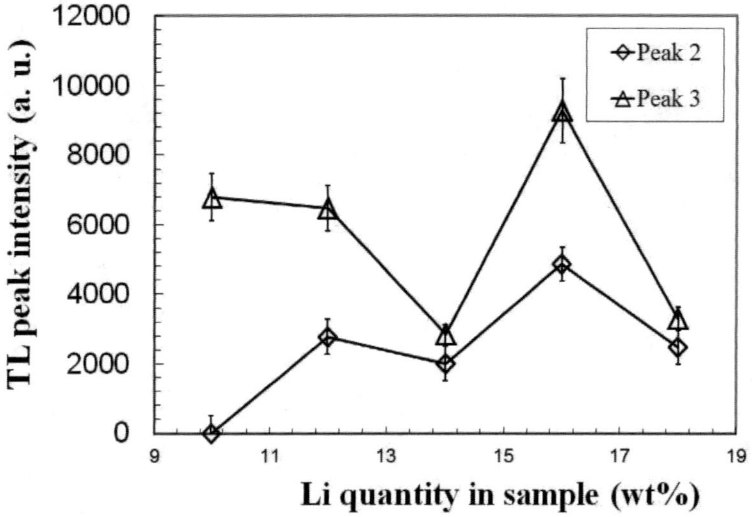

Figure 2. A plot of Peak 2 and Peak 3 intensities as a function of the quantity of Li co-dopant.

A plot of the calculated peak intensities (the integrated counts in the corresponding peak areas from figure 1(a) to figure 1(d)) as a function of the quantity of Li co-dopant is shown in figure 2. From the figure it is clearly seen that the highest TL emission of peak 2 and peak 3 is observed for the pellet co-doped with 16% Li from Li_2CO_3. The calculated kinetic parameters for the peaks appearing in the TL glow curves are shown in Table 1.

Table 1. Calculated kinetic parameters of the peaks in TL glow curves of Gd_2O_3 co-doped with Eu and Li

Peak reference	Maximum temperature position (°C)	Activation energy (eV)	Kinetic order
Peak 1	47 ± 2	0,6 ± 0,2	2,0 – 0,1
Peak 2	87 ± 1,8	0,7 ± 0,1	1,8 ± 0,1
Peak 3	145 ± 1,3	1,1 ± 0,08	1,2 ± 0,1

The dependence of the TL emission on the wavelength of the irradiating light (the spectral sensitivity) has been studied measuring the intensity of the TL glow curves for Gd_2O_3 doped with 10 wt% Eu from Eu_2O_3 and co-doped with 16wt% Li from Li_2CO_3. Interference optical filters for different wavelength were placed consecutively between the XBO lamp and the crystals during each irradiation. To perform these measurements the phosphors have been exposed to irradiation from the XBO lamp for a long enough time, ensuring thus saturation of the states. The calculated intensities for peak 1, peak 2 and peak 3 from the measured TL glow curves versus the wavelength of the irradiating light are shown in figure 3. The obtained thus spectral sensitivity of the material reveals a wavelength limit for peak 2 and peak 3 around 320 nm. The corresponding irradiation wavelength limit for peak 1 is about 400nm. The later fact shows that the appearance of peak 2 and peak 3 in the TL glow curve spectrum is caused by the UV emission from the XBO lamp only.

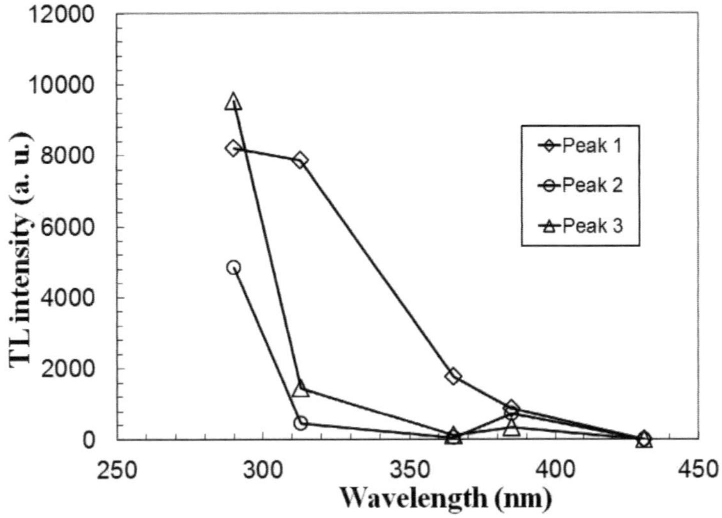

Figure 3. Spectral sensitivity of Gd_2O_3 doped with 10 wt% Eu from Eu_2O_3 and co-doped with 16wt% Li from Li_2CO_3.

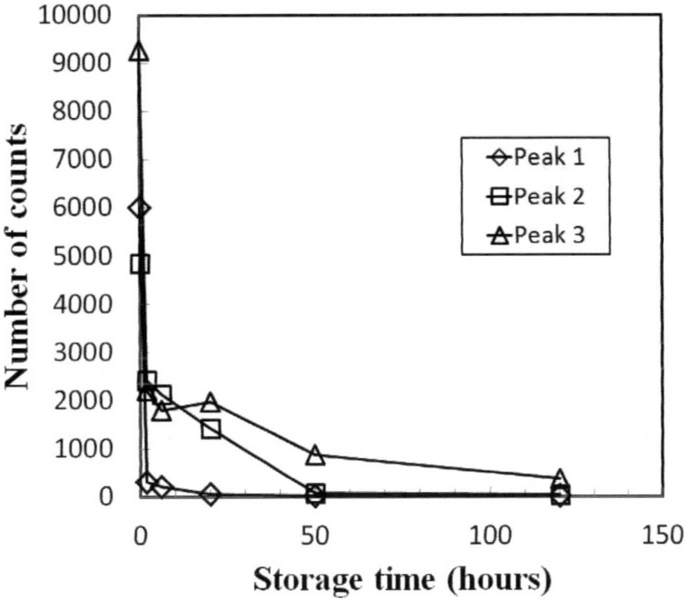

Figure 4. . Fading of Peak 1, Peak 2 and Peak 3 intensities as a function of the storage time.

Measurements to determine the fading of the observed peaks in fig 1 have been conducted for several sintered pellets, containing Gd_2O_3 doped with 10 wt% Eu from Eu_2O_3 and co-doped with 16wt% Li from Li_2CO_3. To perform this kind of measurements the phosphors have been irradiated each time with an equal amount of light (5 min) from the XBO lamp. After the irradiation the crystals have been kept for certain time periods in dark place at a temperature of 20ºC. A plot of the areas of the three peaks, calculated from the fits of the TL glow curves versus the storage time of the phosphors from is shown in figure 4. The maximum storage time interval was 120 hours.

4. Conclusions

The analysis of the fading of the three peaks of Gd_2O_3, co-doped with 10% of Eu and 16% of Li reveals that the low-temperature peak 1 disappears rapidly (in two hours). Peak 2 fades completely for about 50 hours, while peak 3 fades for about 120 hours. Both last peaks possess relatively good count-rate statistics. They have sufficiently long fading, allowing quantitative TL measurements within a couple of days.

We also compared the TL yield of Gd_2O_3 co-doped with 10% of Eu and 16% of Li with respect to the TL yield of ZrO_2 co-doped with 1% Er and 8% of Li from our previous studies [7]. Equal size pellets from both kinds of materials were tested using excitation with XBO lamp at equal other excitation conditions such as time of irradiation and source-to-detector distance. It was shown that the material from Gd_2O_3, co-doped with 10% of Eu and 16% of Li possesses higher TL yield. This fact and the relatively long enough fading of peak 2 and peak 3 open a reasonable opportunity to exploit such a material as a potential phosphor for short-term UV measurements.

Acknowledgements

One of the authors (N Uzunov) would like to acknowledge the support received from the ICTP, Trieste, Italy and in particular the Program for Training and research in Italian Laboratories.

References

[1] Hsieh W C anb Su C S 1994 *Journal of Physics D: Applied Physics* **27** 1763
[2] Grossi F H and Campos L L 2002 *Radiat Prot Dosimetry* **100(1-4)** 421
[3] Bitencourt J F S, Goncalves K A, Tatumi S H and Marcos P J B 2010 *Journal of Physics: Conference Series* **249** 012052
[4] Hatayama Y, Fukumoto S and Ibuki S 1992 *Jpn. J. Appl. Phys.* **31** 3383
[5] Prokic M, 2000 *Appl. Rad. and Isotopes* **52** 97
[6] H. Hristov H, Arhangelova N, Penev I, Bello M, Moschini G and Uzunov N 2010 *Journ. of Optoelectronics and Advanced Materials* **12(4)** 892
[7] Baneva Y I, Hristov H Y, Arhangelova N N, Penev I P, Rossi P, Moschini G and Uzunov N M Feasibility Studies of Eu + Li Co-doped GD_2O_3 as a Thermoluminescence Detector for UV Emission 2012 *Proc. Of The First Int. Conf. on Rad. and Dosimetry in Various Fields of Res. RAD2012* 83
[8] Hristov H Y, Arhangelova N, Penev I and Uzunov N M 2008 *Annual of Konstantin Preslavski University* **Vol. XVIII B1** 30
[9] Kittis G, Gomez-Ros J and Tuyn J 1998 *J. Appl. Phys.* **31** 2636

Influence Of The 4d And 5d Transition Metals On The Urbach's Parameters Of Doped $Bi_{12}SiO_{20}$

P Petkova
Shumen University "Konstantin Preslavsky", 115 Universitetska street, 9712 Shumen, Bulgaria

E-mail: Petya232@abv.bg

Abstract. In this work, the Urbah's rule region (1.77-2.88 eV) in the absorption spectra of doped $Bi_{12}SiO_{20}$ is established. The constant of the electron-phonon interaction and the value of the dynamic disorder in the crystal lattice are calculated in the cases of doping with Os, Re, Rh, and Ru. Urbach's energy and the cross-section of the impurity absorption are presented as function of the energy of photons hv.

1. Introduction

The basic application of the photorefractive $Bi_{12}SiO_{20}$ (BSO) single crystals is as holographic information storage [1] and these materials have found application in the field of nonlinear optics [2].

The interest many researchers took in the use of sillenites gave rise to the development of large and optically homogeneous crystal growth techniques. The main objective of optical experiments is to study the influence of different dopant elements on the absorption coefficient and to optimize the physical properties on the basis of practical application. The analysis of the impurity absorption in the

Urbach's rule region is very important because it can answer why some impurities improve the properties of the material they are inculcated in. This work establishes the influence of the impurities Os^{4+}, Re^{4+}, Rh^{4+}, and Ru^{4+} in the region of the Urbach's rule when the crystals are untreated and illuminated with UV light. Their influence is compared with the characteristic Urbach's rule parameters of untreated undoped $Bi_{12}SiO_{20}$.

2. Experimental Details

The crystals were prepared from a stoichiometric solution using the Czochralski method. The purity of the starting oxides used was 99.9999% for Bi_2O_3 and SiO_2. The dopants were placed in the melt solution in the form of oxides: RuO_2, RhO_2, PdO, Re_2O_7 and OsO_4.The concentration of the impurity ions in the crystal lattice was as follows: Os – $1.76x10^{20}$ cm^{-3}, Re – $1.8x10^{20}$ cm^{-3}, Rh – $1.63x10^{20}$ cm^{-3} and Ru – $3.23x10^{20}$ cm^{-3}. The diameter of the synthesized crystals was 30-45 mm and their lenght was 70-100 mm. The experimental set up for the measurement of the absorption coefficient in the visible region consisted of the following: a halogen lamp with a stabilized 3H-7 rectifier, a SPM-2 monochromator, a system of quartz lenses, a polarizer, a crystal sample holder, and a Hamamatsu S2281-01 detector.

3. Results and Discussion

The behaviour of the absorption coefficient α has been investigated at the absorption edge using the Urbach's formula $\ln\alpha$ = A+B($\hbar\omega$/T), where A and B are the constants, T is the temperature. The constant B is expressed by the dependence B = $\sigma(T)$/k, where $\sigma(T)$ is the parameter characterizing the slope of the absorption edge, k is the Boltzmann constant. The Urbach's rule or the experimental dependence $\ln\alpha(E)$ for the untreated and illuminated crystals is presented in the spectral region 1.77 – 2.88 eV (figure 1). The Urbach's energy for the undoped and doped sillenites is shown in Figure 2.

The value of $\sigma(T)$ is 0.05 for the untreated crystal, whereas $\sigma(T)$ is higher for the doped with Os, Re and Rh untreated samples (table 1). The parameter $\sigma(T)$ is the lowest for BSO:Ru – 0.04. When the doped samples are illuminated, $\sigma(T)$ for BSO:Re is equal to $\sigma(T)$ for BSO. This parameter decreases for BSO:Re, BSO:Ru and it increases for BSO:Os (table 2).

On the other hand, there is the equation $W_d = kT/\sigma$ [3]. The magnitude W_d describes the broadening of the absorption edge due to the dynamic disorder. When the temperature is higher, the absorption edge is wider due to the dynamic disorder in the crystal lattice. This disorder decreases in the untreated BSO:Os, BSO:Re and BSO:Rh (table 1). It is shown that W_d decreases for the illuminated crystals doped with Os, Re and it increases respectively for BSO:Rh and BSO:Ru (table 2). The parameter $\sigma(T)$ and W_d are calculated at room temperature (T = 300 K).

The approximation of the experimental data shows that $\sigma(T) = \sigma_0(2kT/h\nu_0)\mathrm{th}(h\nu_0/2kT)$, where $h\nu_0$ is the energy of the effective phonons, strongly interacting with photons, and σ_0 is the high temperature constant [4]. The magnitude $h\nu_0$ for the crystals under investigation corresponds to the energy $h\nu_0$ = 31.7 meV of the longitudinal optical phonons (ω = 257 cm^{-1}) which are observed in the IR absorption spectra of BSO [5].

In the case under investigation the observed values of σ_0 are as follows: σ_0 = 0.08 ($Bi_{12}SiO_{20}$), σ_0 = 0.07 – 0.12 (the untreated doped $Bi_{12}SiO_{20}$) (table 1) and σ_0 = 0.05 – 0.12 (the illuminated doped crystals). The comparison of the values of σ_0 for the samples under investigation with the values of the same constant in [6] shows that the undoped sillenite under investigation has surplus of Bi^{3+} and Bi^{5+} ions in the tetrahedral complexes in the crystal lattice. When this crystal is doped with 4d elements the surplus of Bi ions decreases there.

The dependence $g = 2/3\sigma_0$ determines the strength of the electron-phonon interaction [4]. When g > 1, the electron-phonon interaction is strong. The value of this strength for the undoped crystal is g = 8.32 (Table1). The electron-phonon interaction increases when the untreated crystal is doped with Ru and it decreases for BSO:Os, BSO:Re and BSO:Rh (table 1). When the doped samples are illuminated with UV light, the electron-phonon interaction decreases for BSO:Os and BSO:Re (table 2). The constant g increases for BSO:Rh and BSO:Ru (table 2).

The Urbach's rule region is shifted to the longest wavelengths for all doped samples in the untreated and illuminated state in comparison with the same region of $Bi_{12}SiO_{20}$ (figure 1).
Urbach's energy is connected with the carrier impurity interaction, the carrier-phonon interaction and the structural disorder [7]. That is why this energy is calculated by the formula $E_u = \alpha(E)/(d\alpha/dE)$. The investigation of the selected crystals leads to the conclusion that Urbach's energy is lower for all untreated and illuminated doped crystals than for the undoped sillenite in the spectral region 1.77-2.88 eV (Figure 2).

The next step in the calculations is the determination of the cross-section of the impurity absorption [8]. It is very important to establish how the radiation is absorbed by the impurity ions in the crystals. The total cross-section σ_a of the impurity absorption is defined by the integration within the absorption band of the impurity ions $\sigma_a = \frac{1}{N}\int_{E_1}^{E_2}\alpha(E)dE$, where N is the number of the impurity ions in the unit volume, α is the impurity absorption coefficient typical of an energetic interval from E_1 to E_2. For the investigated crystals here E_1 = 1.77 eV and E_2 = 2.88 eV. The cross-section σ_a can vary significantly from one absorption band to another. In the case under investigation the value of the cross section σ_a is the highest for the untreated BSO:Os and the lowest for the untreated BSO:Rh (table 1). The illumination with UV light proves that σ_a for the doped samples changes as follows: it decreases for BSO:Os, BSO:Ru and it increases for BSO:Re, BSO:Rh (table 2).

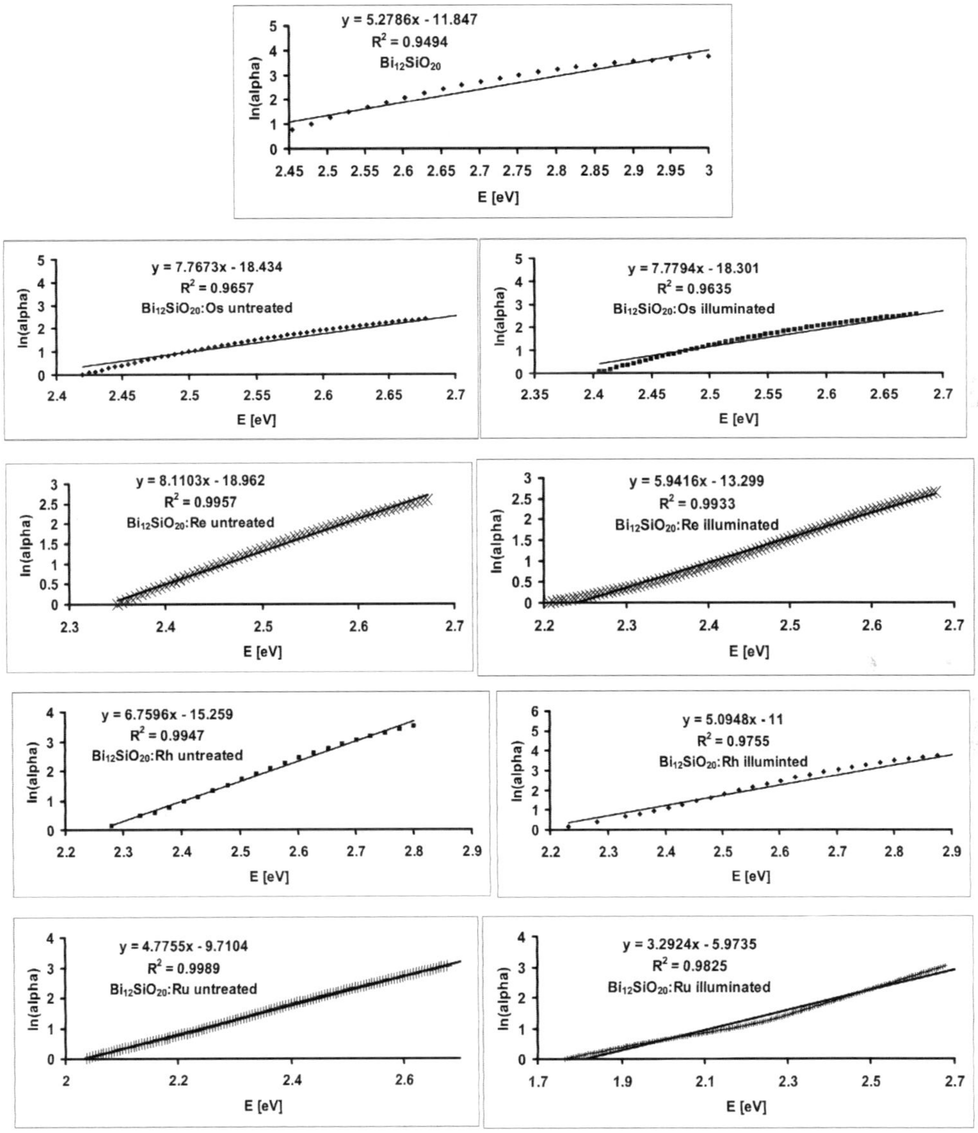

Figure 1. The Urbach's rule region of the untreated and illuminated doped sillenites compared with the same region of untreated undoped BSO.

Table 1. Urbach's parameters of the untreated undoped and doped sillenites.

Untreated Crystals	$\sigma(T)$	σ_0	g	W_d [meV]	σ_a [cm^2]
$Bi_{12}SiO_{20}$	0.0455	0.0801	8.3229	56 818	-
$Bi_{12}SiO_{20}$:Os	0.067	0.118	5.6497	38 585	1.2015x10^{-20}
$Bi_{12}SiO_{20}$:Re	0.0699	0.1231	5.4157	36 984	8.4597x10^{-20}
$Bi_{12}SiO_{20}$:Rh	0.0583	0.1027	6.4914	44 343	13.4601x10^{-20}
$Bi_{12}SiO_{20}$:Ru	0.0412	0.0726	9.1827	62 748	1.3479x10^{-20}

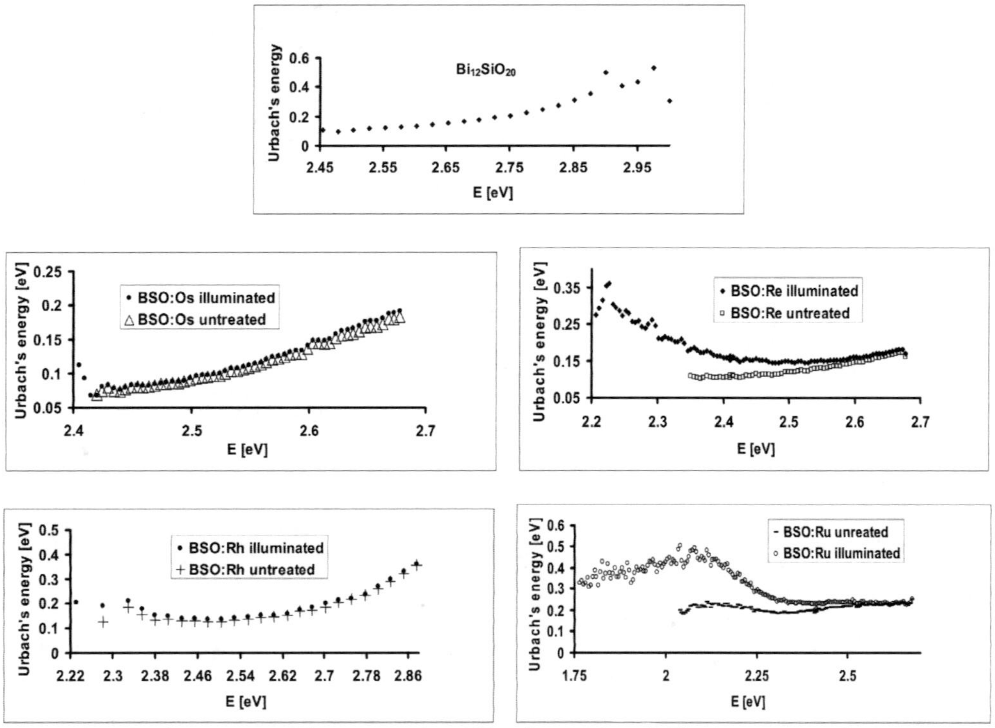

Figure 2. Urbach's energy of the untreated and illuminated doped sillenites compared with the same energy of untreated undoped BSO.

Table 2. Urbach's parameters of the illuminated doped sillenites.

illuminated doped crystals	$\sigma(T)$	σ_0	g	W_d [meV]	σ_a [cm^2]
$Bi_{12}SiO_{20}$:Os	0.067	0.118	5.6497	38 585	1.4136x10^{-20}
$Bi_{12}SiO_{20}$:Re	0.0512	0.0902	7.391	50 492	8.0206x10^{-20}
$Bi_{12}SiO_{20}$:Rh	0.0439	0.0773	8.6244	58 888	13.3483x10^{-20}
$Bi_{12}SiO_{20}$:Ru	0.0284	0.05	13.3333	91 028	1.5657x10^{-20}

Conclusions

The value of the parameter *σ(T)* is the highest for BSO:Re in an untreated state and for BSO:Os in the illuminated state. Thus the conclusion is that the number of the longitudinal phonons is the biggest for the same doped crystals. The value of the electron-phonon interaction constant is the highest for the Ru doped sample in the untreated and illuminated state. The dynamic disorder in the crystal lattice is again the biggest in BSO:Ru in the untreated and illuminated state. Urbach's energy of the untreated crystal does not change its curvature for BSO:Os and BSO:Re in the untreated and illuminated state.

The same energy changes the slope of its curvature for BSO:Re and BSO:Ru in the illuminated state. The cross-section of the impurity absorption σ_a is the highest for BSO:Os and its value is the lowest for BSO:Rh in the untreated and illuminated state.

Acknowledgement

Grants 5240, RD – 07-718/04.04.2012 from the University of Shumen, Bulgaria supported the present study.

References

[1] Georges M P Scauffaire V S and Lemaire P C 2001 *Appl. Phys.* B**72** 761
[2] Ramaz F Rakitina L Gospodinov M and Briat B 2005 *Optical Materials* **27** 1547-1559
[3] Kunets V Kulish N Kunets Vas and Lisitsa M 2002 *Semiconductor Physics, Quantum Electronics & Optoelectronics* **5** 9
[4] Kurik M 1971 *Phys. Status Solidi (a)* **8** 9
[5] Woidovsky W Lukasiewicz T Nazariwicz W and Zmija J 1979 *Phys. Status Solidi (b)* **94** 649
[6] Panchenko T Kopylova S and Osetskié Yu 1995 *Solid State Physics* **37** 1415
[7] Faruque Ahmed Sk Moon Myoung-Woon and Lee Kwang-Ryeol 2009 *Thin Solid Films* **517** 4035
[8] Optical Properties of Condensed Matter and Applications (Edited by Singh J Wiley J & Sons) 2006

Journal of Physics: Conference Series **398** (2012) 012045 doi:10.1088/1742-6596/398/1/012045

Jahn-Teller Effect And Spin-Orbit Interaction In $Bi_{12}SiO_{20}$:Ru

P Petkova
Shumen University "Konstantin Preslavsky", 115 Universitetska street, 9712 Shumen, Bulgaria

E-mail: Petya232@abv.bg

Abstract. The investigation of the optical properties of doped cubic crystals with the apparatus of molecular quantum mechanics is very important for their application in practice. This is the reason for the analysis of the experimental results of the Ru^{4+} and Ru^{5+} octahedral complex in the crystal lattice of doped $Bi_{12}SiO_{20}$. This paper makes a comparison between the manifestation of the Jahn-Teller effect and the spin-orbit interaction in the visible spectral region where the absorption structure of ruthenium is observed. In this work, the calculations are connected with the determination of Racah's parameters of the crystal field. The spin-orbit interaction and the exchange interaction are also analyzed.

1. Introduction

The sillenite-type $Bi_{12}SiO_{20}$ (BSO) crystals display optical activity and a strong photochromic effect. They are applied in dynamic holography, optical information processing, optical phase conjugation and real-time interferometry [1, 2]. This is the reason for the investigation of their optical properties.

The determination of the valence and the local symmetry of the involved dopants in the crystal structure is very important for the optimization of the synthesis conditions of these materials. That is why the aim of the present work is to present the energy level diagram of Ru^{4+} and Ru^{5+} in BSO:Ru.

2. Experimental Details

The crystals were prepared from a stoichiometric solution using the Czochralski method. The purity of the starting oxides used was 99.9999% for Bi_2O_3 and SiO_2. The structure of the sillenite is the final result from Si^{4+} tetrahedrons and Bi^{3+} octahedrons. The dopant was placed in the melt solution in the form of the oxide RuO_2.The concentration of the ruthenium ions in the crystal lattice was Ru – 3.23×10^{20} cm^{-3}. The maximum ruthenium concentration in the melt was established by several experiments which determined the optically homogeneous doped crystals without other phase inclusions. The diameter of the synthesized crystals was 30-45 mm and their lenght was 70-100 mm.

The experimental set up for the measurement of the absorption coefficient in the visible region consisted of the following: a halogen lamp with a stabilized 3H-7 rectifier, a SPM-2 monochromator, a The experimental set up for the measurement of the absorption coefficient in the visible region consisted of the following: a halogen lamp with a stabilized 3H-7 rectifier, a SPM-2 monochromator, a system of quartz lenses, a polarizer, a crystal sample holder, and a Hamamatsu S2281-01 detector. The investigation was carried out with an untreated sample that has been illuminated with UV light for 30 min.

3. Results and Discussion

The absorption coefficient is calculated using the formula: $\alpha = (1/d)\ln(I_0/I)$, where I_0 is the intensity of the incident light, I is the intensity of the passing light and d is the sample thickness. The absorption spectrum of the untreated and illuminated crystals is measured in the spectral region 1.6-1.8 eV (figures 1, 2). A well pronounced curvature can be observed around 1.70 eV when BSO:Ru is untreated (figure 1). This curvature is missing in the illuminated state of the crystal (figure 2). In order to explain this difference, the first derivative of the absorption coefficient must be calculated. Dα/dE provides information only for the number of 4d electron transitions in the Ru ion. So the second derivative of the absorption coefficient must be calculated. Thus information for the exact energetic position of 4d electron transitions is derived. It is seen that the number of the electron transitions in Ru^{4+} is two when the crystal is untreated (figure 1). The energy of these two electron transitions is as follows: 1.66 eV and 1.74 eV. The number of the electron transitions is three when the sample is illuminated and the valence of the ruthenium ion is 5+ (figure 2). This fact is connected with the colour changing of the crystal. The maximum around 1.73 eV in the first derivative of α(E) consists of two overlapping transitions (1.71 eV and 1.75 eV) in the second derivative of the absorption coefficient (figure 2). The energetic diagram of the 4d electron transitions in Ru^{5+} is presented in figure 3. The investigation of $Bi_{12}SiO_{20}$:Ru by Magnetic Circular Dichroism (MCD) proves that Ru^{4+} is inculcated in the Bi^{3+} octahedrons in the crystal lattice [3]. The symmetry of the octahedral complex changes to the tetragonal symmetry D_{4h}, because the ruthenium ions have smaller ionic radius than Bi^{3+}. The final result is expressed in the octahedral extension and it concludes in the appearance of the static Jahn-Teller effect in the basic energy level. The value of the crystal field parameter Dq is 14760 cm^{-1} for the Ru^{5+} ion. When there is more than one electron in an atom, there is certain electrostatic repulsion between those electrons. The strength of the repulsion varies from one atom to another depending upon the number and spin of the electrons and the orbitals they occupy. The total repulsion can be expressed in terms of the two parameters B and C which are known as the Racah parameters. The Racah parameters for the chromium are as follows: B = 1476 cm^{-1} and C = 6642 cm^{-1}. The literature [4] informs of the Nephelauxetic parameter β = B (complex)/B (free ion). If the value of β is in the interval 0.54-0.81, it manifests the metal ligand σ bond. The calculation of this parameter for BSO:Ru (β = 1.43) shows that the metal ligand π bond is present. The tetragonal distortion of the octahedral complex has full potential $V(D_{4h}) = V_{Oh} + V_{D4h}$. If the matrix elements are known, the value of the parameter Dq and the radial parameters Ds and Dt will be known. The paramerer Dt determines the difference between the equatorial and the axial center: Dt = 4/7[Dq(xy) – Dq(z)]. If the approximation of the theory of the ligand field is applied, then Dq ~ $(1/r)^5$, where r = 0,1466 cm^{-1} is the length of the bond M-O. Hence, r(xy) / r(z) = $[Dq(z) / Dq(xy)]^{1/6}$ = 0,895. For the complex under investigation, Dq(z) = 0.0049 cm^{-1} and Dq(xy) = 0,0095 cm^{-1}. The value of Dt is 0,0026 cm^{-1}. The parameter Ds = 0,014 cm^{-1} is defined by the equation Δ_2 = Ds + 10Dt (Figure 3). The symmetry of the complex is such that e_g and t_{2g} orbitals of the metal correspond only to σ – anti-connecting and π – anti-connecting orbitals [5]. The parameters $\delta\sigma$ and $\delta\pi$ are determined as follows: $\delta\sigma = \sigma_z - \sigma_{xy}$ = -(12/8)Ds – (15/8)Dt, $\delta\pi = \pi_z - \pi_{xy}$ = -(3/2)Ds + (5/2)Dt. The parameters σ and π are indirect indicators for the σ – anti-connecting and π – anti-connecting properties of the ligands. Finally, $\delta\sigma$ = -0,0161 cm^{-1} and $\delta\pi$ = 0,0275 cm^{-1}. The electron transitions in Ru^{5+} realize from the basic state $^4A_{2g}$ (^4F) to the first excited state $^4T_{2g}$ (^4F) (figure 3). These three electrons are situated on the d_{zx}, d_{xy} and d_z^2 respectively. This paper discusses the electronic structure of 3d valence electrons of Co which is embedded in a metallic glass host. It is convenient to introduce the exchange interaction J(dd) here, describing the intrashell exchange attraction between parallel spins [6]: J(dd) = $1/14(F^2 + F^4)$ and the additional parameter describing the angular part of the multiplet splitting is C(dd) = $1/14(9/7F^2 - 5/7F^4)$. J and C are especially useful when describing the lowest state of an l^n multiplet. For d electrons l = 2. According to Hund's rule, this is always the state with the highest possible spin quantum number. In the case under investigation, the values of J(dd) and C(dd) are as follows: 14 465 cm^{-1} and 6642 cm^{-1}. The next equations are fulfilled for the glass systems under investigation $F^2 = 49F_2$ and $F^4 = 441F_4$, where F_2 and F_4 are the Condon–Shortley parameters. The integral of the exchange interaction $K(z^2,xy) = 4B + C = 12\ 546\ cm^{-1}$. This exchange interaction is realized between the orbitals d_z^2 and d_{xy}. The orbital d_{xy} has very important role, because when the crystal is illuminated one electron is situated on this orbital. In the opposite case, when the sample is untreated, the d_{xy} orbital is empty (figure 3).

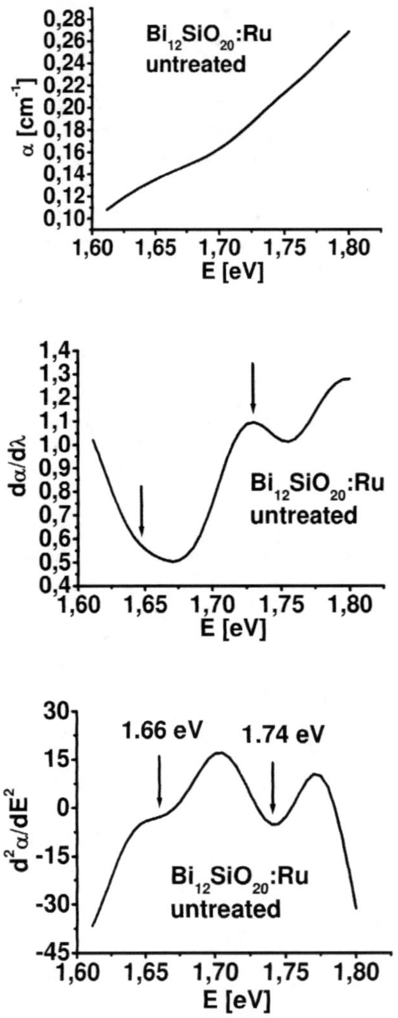

Figure 1. The absorption coefficient α(E), first derivative dα/dE and second derivative $d^2\alpha/dE^2$ for the untreated $Bi_{12}SiO_{20}$:Ru.

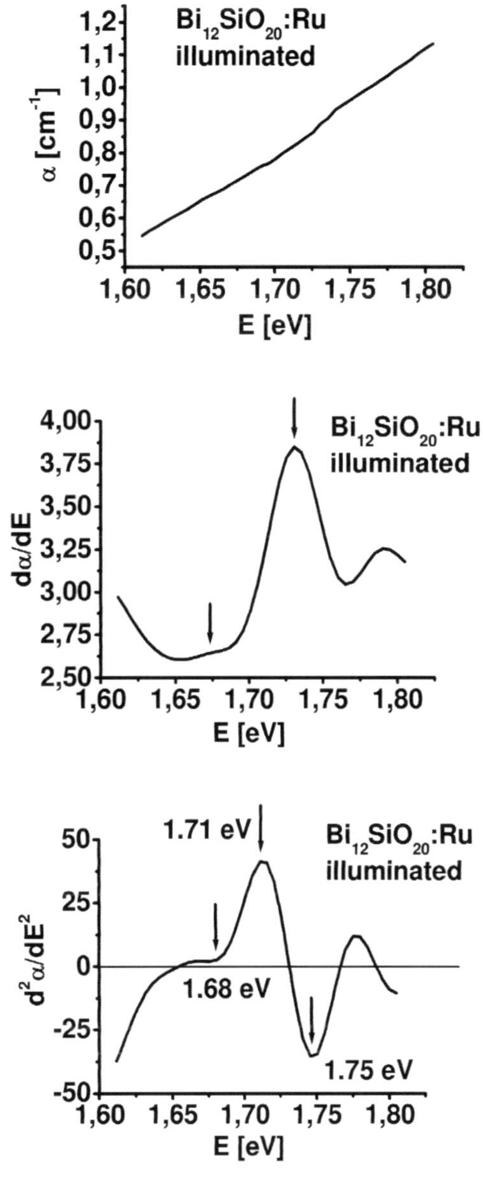

Figure 2. The absorption coefficient α(E), first derivative dα/dE and second derivative $d^2\alpha/dE^2$ for the illuminated $Bi_{12}SiO_{20}$:Ru.

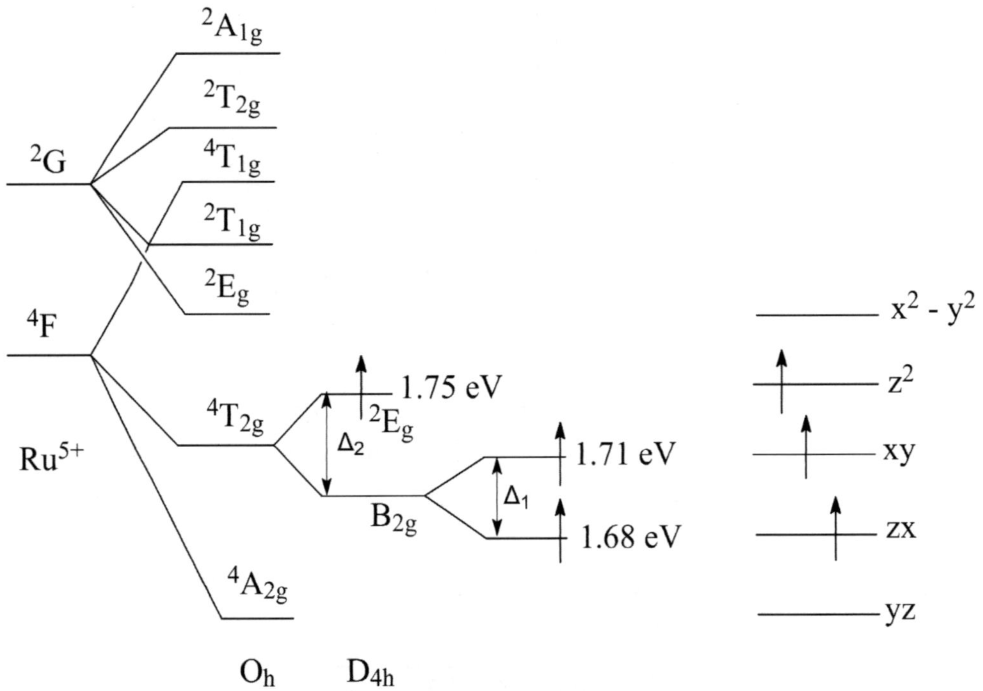

Figure 3. Energetic diagram of the 4d electron transitions in Ru^{5+}.

Conclusions

The illumination with UV light of the crystal $Bi_{12}SiO_{20}$:Ru clearly shows that $Ru^{4+} + e^- \rightarrow Ru^{5+}$ in the visible spectral region. This is very important for its application in practice as holographic information storage. The metal ligand bond in BSO:Ru is π, which means that Ru ions defy very big dynamic disorder in the crystal lattice. The intrashell exchange attraction between parallel spins is stronger than the angular part of the multiplet splitting in the case of Ru^{5+} ions.

References

[1] Raciukaitis G Gavryushin V Kubertavicius V Puzonas G 1993 *Jpn. J. Appl. Phys.* **32** 645
[2] Marinova V Veleva M Petrova D Kourmoulis I M Papazoglou D G Apostolidis A G Vanidhis E D Deliolanis N C 2001 *Journal of Applied Physics* **89** 2686-2689
[3] Ramaz F Rakitina L Gospodinov M and Briat B 2005 *Optical Materials* **27** 1547-1559
[4] Chandra S Gupta L 2005 *Spectrochimica Acta Part A* **62** 1125
[5] Mc Clure D S *Advances in the Chemistry of Coordination compounds* 1961 New York Macmillan 498
[6] van der Marel D Sawatzky G A *1988 Physical Review B* **37** 10674–10684

Improving Resonance Characteristics of Gas Sensors by Chemical Etching of Quartz Plates

Z Raicheva[1], V Georgieva[1], A Grechnikov[2], V Gadjanova[1], Ts Angelov[1], L Vergov[1], Y Lazarov[1]

[1] Georgi Nadjakov Institute of Solid State Physics, Bulgarian Academy of Sciences, 72, Tzarigradsko Chausse Blvd., 1784, Sofia, Bulgaria
[2] Vernadsky Institute of Geochemistry and Analytical Chemistry of RAS, Kosygin str., 19, 119991 Moscow, Russia

E-mail: zdravka_raicheva@yahoo.co.uk

Abstract. The paper presents the results of the influence of the etching process of AT-cut quartz plates on the resonance parameters and the QCM sensors. Quartz wafers (100 μm thick, with a diameter of 8 mm), divided into five groups, have been etched in $[NH_4]_2 F_2$: H_2O = 1:1 solution at temperatures in the range from 70°C to 90°C. The influence of etching temperature on the surface morphology of quartz wafers has been estimated by Atomic Force Microscopy (AFM). A correlation between the etching temperature and the dynamic characteristics is obtained. The optimal etching conditions for removing the surface damages caused by the mechanical treatment of the quartz wafers and for obtaining a clean surface were determined. The typical parameters of fabricated resonators on the quartz plates etched in the temperature range from 70°C to 90°C are as follows:

Frequency, Fs	16 MHz ± 100 kHz
Motional resistance, Rs	less *10* Ω
Motional inductance, Lq	higher than 3 mH
Motional capacitance, Cq	less 30 fF
Static capacitance, Co	around *5* pF
Quality factor, Q	from 46 000 to 70 000

Sorption properties of QCM - MoO_3 are evaluated at NH_3 concentrations in the interval from 100 ppm to 500 ppm.

1. Introduction

The surface state of a quartz plate is responsible for the resonators' quality. Its change leads to alteration of the resonators' parameters. Roughness effect on the frequency of a quartz-crystal resonator in contact with liquid is studied by K. Rechendorff [1]. The influence of nanometre scale surface roughness on the frequency of a quartz crystal microbalance operating in a liquid environment has been investigated by exposing quartz crystals, with a rough tantalum coating, to water solutions with varying concentrations of sucrose. A clear correlation between the measured surface roughness and the frequency shift is observed.

The different roughness of the quartz surfaces can be achieved by treatment with abrasive with different grain size or by chemical etching. The influence of the quartz roughness on the WO_3-QCM parameters and sorption properties was investigated [2]. Experiments were carried out using AT-cut

Published under licence by IOP Publishing Ltd

quartz crystal with surface roughness obtained by treatment with SiC abrasive with grain size from 3 µm to 20 µm. The correlation of the equivalent dynamic parameters of the structure and the surface roughness was determined. The surface morphology showed higher increase of the WO_3 surface roughness compared to the initial quartz surface. The sorption properties of WO_3-QCM system to NH_3 showed an increase of the sorption ability with the increase of the quartz roughness.

The authors in [3] investigated the lapping and polishing processes of quartz in relation to crystalline orientation. The measurements of the surface roughness were carried out in X-, Y-, Z- and AT-cut samples. It shows that the relationship found between material removal rate and stress depends on specimen orientation. Surface profiles and SEM micrographs show that roughness of lapped surfaces decreases with increasing the normal stress but an opposite behaviour is observed in polished surfaces. It is well known, that when the AT-cut lapped quartz is etched in saturated solution of ammonium bifluoride $[NH_4]_2\ F_2$ the surface roughness decreases. This way with increasing the depth of the etching, the plates became chemically polished. The same effect takes place for single rotated BT- cut and ST-cut, but not for double rotated IT and SC-cuts. With increasing the depth by the etching in $[NH_4]_2\ F_2$ solution, the surfaces of quartz plate become rougher as a result of the etching process. Fundamental studies of chemical etching of quartz were carried out by Watanable [4, 5]

The etch rate and surface roughness of polished and lapped AT-cut quartz subjected to hot (90, 110, and 130^{0}C), concentrated (50, 65, 80 wt %) $[NH_4]_2\ F_2$ have been investigated. Etch rates were found to correlate well, and possibly logarithmically, with temperature except for the hottest etching applied to lapped material. Roughness as a function of temperature and concentration behaved well for the lapped material, but lacked systematic variation in the case of the polished material. At the lowest temperature, concentration had no effect on the etch rate or roughness. Their future efforts are targeted at temperatures and concentrations closer to the solubility limit [6].

The aim of the work is to study the influence of etching process of AT-cut quartz in $[NH_4]_2\ F_2$ solution at low temperature (70^{0}C-90^{0}C) on quartz roughness, on resonator parameters and sorption sensibility of the QCM-MoO_3 structure.

2. Experimental

Quartz wafers (100 µm thick, with a diameter of 8 mm), divided into five groups, have been etched in $[NH_4]_2\ F_2$:H_2O = 1:1 solution at temperatures in the range from 70°C to 90°C.

The etching system was made of polypropylene. The etching bath was put in a thermostat, the temperature being stabilized in a water bath. The temperatures of the water and the etchant were controlled by a digital thermometer (Delta Instruments with accuracy of ±0.2°C) placed in a polypropylene tube. The typical temperature difference between the etchant and the water bath was 2°C. The plates were put vertically in a polypropylene holder, which was shaken occasionally in order to homogenize the etchant. Optimal time for etching is determined by the results, obtained in treatment of three series at temperatures 75°C, 80°C and 85°C that show the change of frequency as the result of etching duration.

Figures 2-4 show the crystal surfaces of the quartz plates measured with an Atomic Force Microscope (AFM). AFM measurements were carried out using a Multimode V (Veeco, Santa Barbara, CA) equipped with closed loop scanner. Imaging was performed in tapping mode and height, amplitude, and phase images were recorded. Scan rate was in range 1.0-1.5 Hz, the images resolution was 512 samples/ line. At least three different points on the sample surface were explored. Silicon cantilevers with a cantilever length of 125 µm and Al Reflective coating on the backside (RTESPA, Veecoprobes) were used in the experiments. RTESPA probes have a nominal resonance frequency of 300 kHz and a typical force constant of 40 N/m. The tip nominal radius for these probes is less than 10 nm. Image processing was performed by means of Nanoscope 7.30.

After etching of quartz plates, the samples were subjected to chemical treatment that consists of samples boiling for 30 min in $H_2SO_4 : H_2O_2 = 1 : 3$ and than followed rinsing in deionised water. Then Au electrodes (55 nm thick) with diameter of 4 mm were thermally evaporated on both sides of the plates. As a result of the above technological procedures, quartz resonators with basic frequency of 16 MHz ± 100 kHz were produced. After measuring the static capacitance Co, the serial resonance frequency Fs and dynamic resistance Rs, the dynamic capacity Cq, the dynamic inductance Lq and Q–factor were calculated. The average values of the equivalent dynamic parameters are marked with-(av) index.

Thin MoO_3 layers 200 nm were deposited by magnetron sputtering on both sides of the resonators. The sorption properties of the QCM-MoO_3 structures were measured at ammonia concentration of 100 ppm, 250 ppm and 500 ppm on a laboratory set up. The changes in the resonance characteristics of the quartz resonators, as a function of the surface roughness, were measured.

3. Results and discussion

The quartz plate frequency change versus etching time at different temperatures is shown on figure 1.

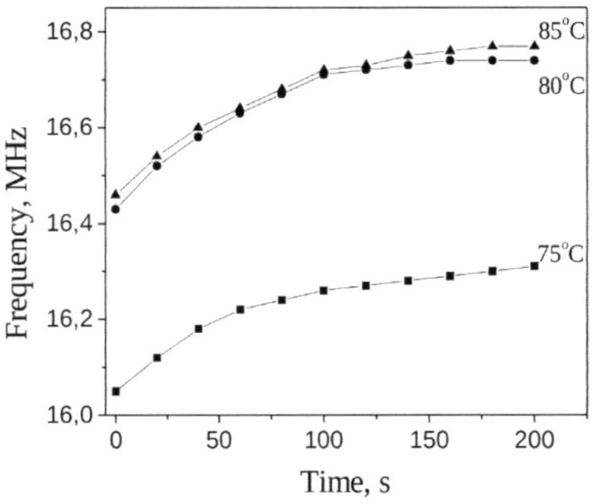

Figure 1. Frequency dependence versus etching time.

Figure 1 demonstrates the greatest change of the frequency which reaches up to 60 s. The high etching rate of 60s is caused by the great number of free chemical bonds on the quartz surface, resulting from the mechanical treatment. Negligible dependence changes are observed at etching time from 120s to 180s. The optimal time for etching - 120s was obtained. It shows that the frequency change occurs after the 120th second. The influence of the etching time and temperature of quartz surface relief and morphology was studied by Atomic Force Microscopy (AFM).

3.1. Surface roughness

The results from AFM measurements of the samples, etched in solution of ammonium bifluoride at different temperatures, are presented in figure 2-4.

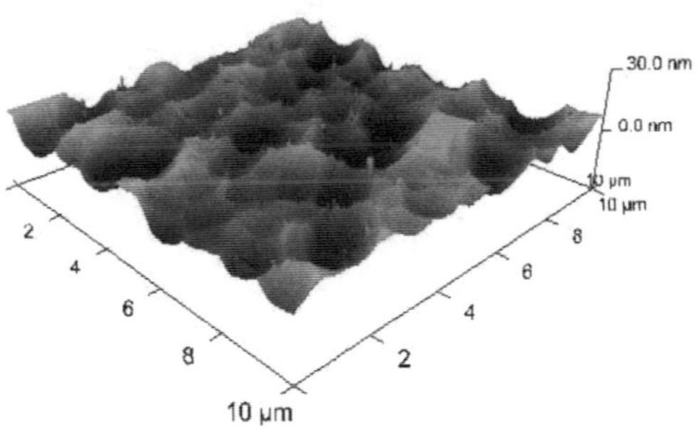

Figure 2. Quartz surface etched at 70°C. The root mean square roughness is 4.51 nm.

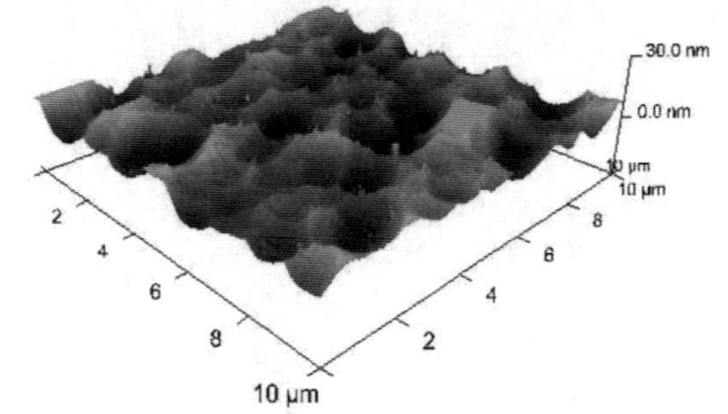

Figure 3. Quartz surface etched at 80°C. The root mean square roughness is 3.79 nm.

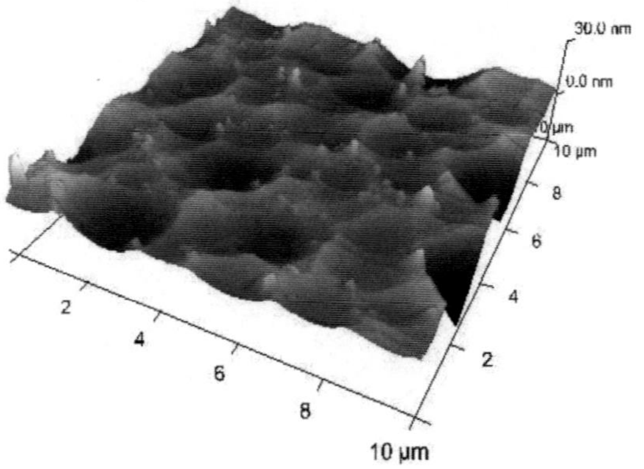

Figure 4. Quartz surface etched at 85°C. The root mean square roughness is 3.71 nm.

3.2. Motional parameters

Table 1. Dynamic parameters of experimental samples

Etching temperature, °C	Fs_{av}, Hz	Rs_{av}, Ω	Co_{av}, pF	Cq_{av}, F	Lq_{av}, mH	Q_{av}
Quartz resonators initial parameters						
25	15777136	8.67	5.634	2.57E-14	3.97	45 314
70	16105181	5.80	5.700	2.73E-14	3.57	62 341
75	16084578	5.20	5.715	2.72E-14	3.16	70 079
80	16118224	6.03	5.610	2.69E-14	3.63	60 862
85	16196423	7.41	5.720	2.84E-14	3.41	46 800
90	16509816	7.68	5.690	2.70E-14	3.45	46 537
Quartz resonators parameters after MoO_3 deposition						
25	15629965	15.23	5.657	2.54E-14	4.08	26 317
70	16010547	6.70	5.700	2.69E-14	3.68	55 163
75	15994868	5.66	5.724	2.59E-14	3.82	67 827
80	16029757	9.15	5.605	2.43E-14	4.06	44 662
85	16106289	9.00	5.727	2.42E-14	4.04	45 372
90	16416009	8.90	5.694	2.42 E-14	3.89	45 088
Quartz resonators parameters after NH_3 sorption						
25	15630164	14.10	5.646	2.52E-14	4.11	28 643
70	15995672	6.65	5.738	2.69 E-14	3.69	55 740
75	15995359	5.60	5.719	2.57 E-14	3.35	69 058
80	16030250	8.95	5.597	2.43 E-14	4.06	45 712
85	16107048	6.28	5.565	2.44 E-14	4.01	64 546
90	16416522	8.51	5.694	2.43 E-14	3.36	46 822

Table 1 shows that the quartz resonators created at room temperature (25^{0}C) have less favorable resonance characteristics (Rs_{av} and Q_{av} are respectively 8.67 Ω and 45 314) compare to the once with temperature treatment, which leads to the necessity of thermal treatment of the quartz plates.
After mass loading of resonators with MoO_3, their dynamic parameters get worse. For example: when etching quartz plates at temperature of 70^{0}C, Rs_{av} of the resonators gets from 5.80 Ω to 6.70 Ω and Q_{av} gets from 62 341 to 55 163.When etching temperature is 90^{0}C, Rs_{av} of the quartz resonators gets from 68 Ω to 8.90 and Q_{av} from 46 537 to 45 088.

Resonators' dynamic parameters deteriorate after mass loading with MoO_3. This is presented in table 1. It could be explained by the resonators' increased acoustic losses.

After sorption the values of the dynamic resistance and Q get better. For example: when etching quartz plates at temperature of 70^{0}C, Rs_{av} of the resonators gets from 6.70 Ω to 6.65 Ω, and Q_{av} changes from 55 163 to 55 740. When etching quartz plates at temperature of 90^{0}C, Rs_{av} of the resonators gets from 8.90 Ω to 8.51 Ω, and Q_{av} gets from 45 088 to 46 822. The reason for this is the excellent cleaning of quartz surface after it has been treated with dry air with high frequency (99.95%), which also contains a certain NH_3 concentration.

3.3. Sorption characteristic.

The change in QCM-MoO_3 resonance frequency as a function of quartz surface roughness is studied. The sorption sensitivity to NH_3 with concentration of 100 ppm, 250 ppm and 500 ppm is investigated. The results are presented in table 2 and in figure 5.

Table 2. Frequency change (Δ f) as a function of etching temperature and average roughness of the quartz surface at different NH_3 concentrations.

Etching temperature,°C	Roughness, nm	f, Hz at 100 ppm NH_3	f, Hz at 250 ppm NH_3	f, Hz at 500 ppm NH_3
70	4.51	47	205	294
75	3.88	8	23	40
80	3.79	6	20	29
85	3.71	6	18	28
90	3.55	6	16	26

The difference in the average roughness of plates etched at 70°C and 75°C is 0.63 nm, while the change in sensor sensitivity is 6, 9 and 7.4 times, respectively for NH_3 concentrations of 100, 250 and 500 ppm. At etching temperatures of 75°C to 90°C, the roughness decreases with 0.33 nm and the sensor sensitivity changes from 1 to 1.5 times for all concentrations respectively. As a result of the mechanical treatment a stress increases in the crystal could lead to micro cracks. Also, as a result of this treatment, a lot of free chemical bonds are retained, which can absorb moisture and gases from the environment. The aim of the etching is removing this disturbed surface layer. On the quartz surface reside silicon atoms that form single, double and triple bonds to the crystal lattice. In water soluble systems, the oxygen atoms on the quartz surface combine with protons and form hydroxyl groups. Water solutions of $(NH_4)_2F_2$ contain F and HF_2 - ions. The F- ions can replace the OH-groups. The process of substitution for double bonded to the lattice surface atoms involves a further sorption of HF_2-ions. This step in the reaction mechanism leads to releasing SiF_4, which has good solubility in the etchant by the following reaction:

$$SiO_2 + (NH_4)_2 F_2 \rightarrow SiF_4 + NH_4OH$$

The quartz surface relief became smoother as results of this mechanism.

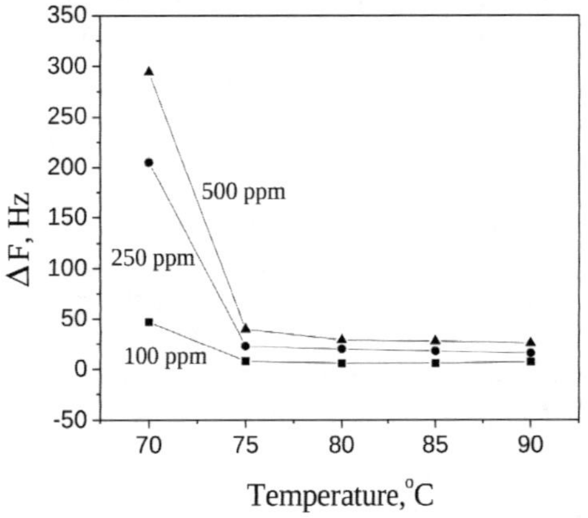

Figure 5. Frequency shift of the system QCM-MoO_3 vs. etching temperature at different NH_3 concentration.

The effective surface area is the highest at etching temperature of 70°C. This leads to a greater sorption of NH_3.

4. Conclusion.

The etching process of AT-cut quartz in $(NH_4)_2F_2$ aqueous solution at low temperature from 70°C to 90°C was studied. The created resonators on so treated quartz plates retain their equivalent dynamic parameters. The AFM analysis on quartz roughness shows that the greatest quartz roughness is obtained at 70°C. At etching temperatures from 75°C to 90°C the effective surface is reduced due to the reaction mechanism. It was shown that the difference in the average roughness of quartz plates etched at 70°C and 75°C of 0.63 nm, leads to sensitivity increase of QCM-MoO_3 in 6, 9 and 7.5 times, for NH_3 concentrations of 100, 250 and 500 ppm respectively.

The increase of effective sorption surface, obtained by chemical etching, can be successfully used to realize high sensitive gas sensors.

Acknowledgement
The work is supported partly by the bilateral project (2009-2011) IFTT, BAS and GEOKHI, RAS.

References

[1] Rechendorff K, Hovgaard M, Foss M and Besenbacher F 2007 *J. Appl. Phys.* **101** 114502
[2] Georgieva V, Raicheva Z, Grechnikov A, Gadjanova V, Atanassov M, Lazarov J and Manolov E 2010 *J.of Physics: Conference Series* **253** 012046
[3] Guzzo P, Daniel J and De Mello J 1999 *Proc. of the Joint Meeting of the European Frequency and Time Forum and the IEEE Int. Freq. Contr. Symp.* **1** 465
[4] Watanabe T 2001 *Proc. IEEE Int. Freq. Contr. Symp. and PDA Exhibition (Piscataway)*
[5] Watanabe T *European patent application* EP 1 127 965 A1, 2001
[6] Vallin O, Danielsson R, Lindberg U and Thornell G 2007 *IEEE Transactions on ultrasonics, ferroelectrics and frequency control* **54** 7 pp 1454 – 1462

Proton-exchanged optical waveguides in $LiTaO_3$: phase composition and stress

M Kuneva, K Christova and S Tonchev

Institute of Solid State Physics, Bulgarian Academy of Sciences
72 Tzarigradsko Chaussee Blvd., Sofia 1784, Bulgaria

E-mail: kuneva@issp.bas.bg

Abstract. Planar optical waveguide layers were obtained in Z-cut $LiTaO_3$ crystal substrates via proton exchange. Two different media were used as proton sources: benzoic acid melt and lithium hydrogensulphate vapors, controlling the thickness of the waveguides by duration and temperature of the proton exchange process and also by post-exchange annealing. The intrinsic stress caused by the penetration of the hydrogen ions into the crystal lattice was estimated by the optical integral method. The phase composition of proton-exchanged layers was analyzed based on the mode and IR-absorption spectra. An attempt to relate the level of stress to the level of proton doping has been made.

1. Introduction

Lithium tantalate (LT) is one of the attractive ferroelectric materials for integrated optics due to its strong nonlinearity, high electro-optical coefficients and optical resistivity, as well as to the transparency to shorter wavelengths compared to the most popular lithium niobate (LN).

The proton exchange (PE) method [1, 2] has been widely studied because of the easy and fast obtaining of waveguides with strong waveguide effect in electrooptical crystals like $LiNbO_3$ and $LiTaO_3$. The surface layer modified by hydrogen diffusion has a large extraordinary index change (Δn_e) and thus, a strong waveguide and polarizing effect. PE is a low-temperature process, compared to other technological methods of producing waveguides and is especially appropriate for LT, whose Curie temperature is lower. The $Li_{1-x}H_xTaO_3$ layers formed by Li-H ion exchange show complex phase behavior depending on the hydrogen concentration (value of x) [3, 4]. Studies have shown the presence of up to six different phases in PE LT [4, 5] originating as individual sublayers of several hundred nm thickness or less. In each of them one of them Δn_e is a linear function of proton concentration. Within a phase transition, the value of Δn_e and/or of the deformations perpendicular to the surface change by a leap.

Since the development of the PE technology, the main efforts have been directed to the study of phases coexisting in such waveguides and how technological conditions effect their formation. The proton presence causes changes in crystal lattice parameters which reflect on the vibrational spectra of exchanged layers. Using the results of optical profiles and IR-absorption spectra we could perform an analysis of the phase content of PE layers. The use of a combination of different methods to confirm the presence of a different phase or set of phases has been reported recently [6].

It is well known that PE is accompanied by accumulation and relaxation of internal strain [7]. An optical integral method [8] contributes to the characterization of the waveguide film from the point of view of mechanical stress (MS). Since the MS is a function of the film structure, by its change due to different film composition, an additional picture for the building units of the film could be gained.

2. Experimental

Proton-exchanged waveguide layers were obtained in Z-cut substrates of LT at the technological conditions described in table 1. Two technological methods were applied: PE in vapors (one-step process) and PE in a melt followed by annealing (APE two-step process). The LT substrates used in each method were of identical stoichiometry. The cooling after diffusion was slow (10 deg/min), so avoiding metastable phase formation and short-time instability of the waveguides as well.

Table 1. Technological and waveguide parameters of the proton-exchanged waveguides: (T – temperature, t – duration of the PE process, T_a and t_a – temperature and duration of the annealing, M – number of waveguide modes at λ = 633 nm, Δn_e – extraordinary refractive index change, d – waveguide depth, σ – intrinsic stress)

Sample	Proton source	T [^{o}C]	t [h]	T_a [^{o}C]	t_a [h]	M	d [μm)]	Δn_e	Phase composition	σ x10^9 [N/m^2]
TZ-1	$LiHSO_4$ (vapors)	250	10	-	-	1	0.89	0.0186	α, κ, δ	+22.80
TZ-2	$LiHSO_4$ (vapors)	250	20	-	-	1	1.23	0.0162	α, κ, δ	+3.41
TZ-3	$LiHSO_4$ (vapors)	220	48	-	-	1	1.44	0.0101	α, κ, δ	+2.50
TZ-4	$LiHSO_4$ (vapors)	200	72	-	-	1	2.09	0.0058	α, κ, δ	+23.00
T-1	Benzoic acid	240	8	265 295	1 1.5	2	2.18	0.0213	ß(γ)	+2.10
T-2	Benzoic acid	240	31	400	2	5	9.45	0.0155	α	-2.10
Z-1	Benzoic acid	240	8	295 350	1 1	4	4.25	0.0189	α, κ	+3.86

The refractive index profiles (figure 1) were reconstructed by an inverse WKB method after mode spectra measurements at a wavelength of λ=0.633 μm. For single-mode waveguides, the optical measurements were performed in two media, air and water, and two values of the effective mode index n_{eff} were obtained, allowing two mode-propagation equations to be written by assumption of a step-like optical profile. Then the depth and surface index change were calculated from both equations for the individually determined n_{eff}.

The IR absorption spectra (figure 2a) of protonated in vapors and as-grown LT samples were recorded in the wavenumber range of 2700-3700 cm^{-1} of OH-stretching modes with a Brucker LFS-113 V FTIR spectrometer with a resolution of 1 cm^{-1}. After measuring the IR-absorption spectra, a Gaussian-Lorentzian decomposition procedure was performed. The substrate spectrum (TZ-ref) is extracted from the spectra of protonated samples, so only specra of PE-layers are present on the hystograms. The results are shown in figure 2b.

The quantitative calculations of layer stress have been performed by an ex situ optical interference technique [8, 9]. The role of the substrate is assumed to be of LT with no waveguide on it. Since the deformation of the waveguide follows that of the substrate, the elastic parameters, Young modulus and Poison's ratio are taken for the substrates. The change in the interference fringe patterns between the substrate and an optical flat is used to measure the deformation of the layer-substrate system. The radius of curvature is actually the substrate radius. The error in the radii determined by curvature measurements was 10%. Since the temperature during the process of proton exchange is low, the thermal component of stress is negligible. Thus, the calculated stress is the intrinsic stress for the waveguide layers. It has to be noted that this method for determining stress gives an integral value over the waveguide thickness. Besides, the nature of the stress is determined by its sign. If the sign is negative, the stress is compressive whereas a positive sign corresponds to tensile stress.

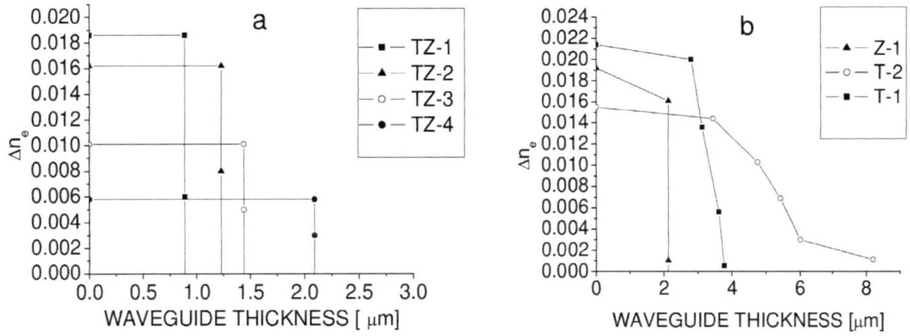

Figure 1. Optical profiles of the waveguides examined: (a) produced by direct PE in vapors; (b) produced by APE process.

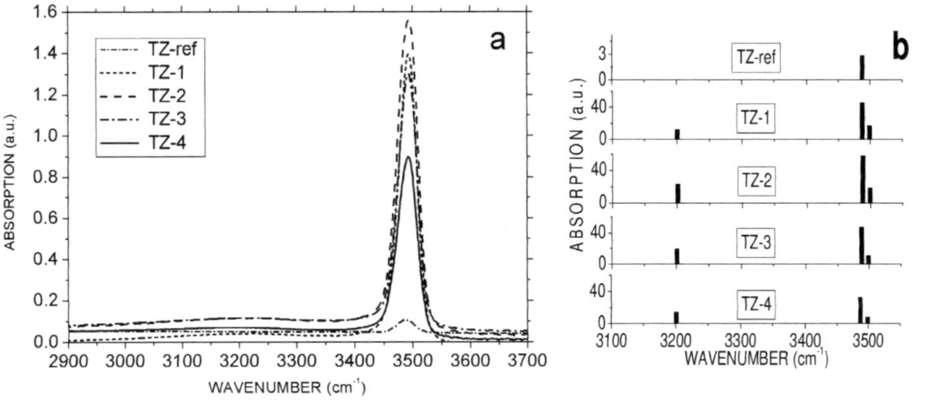

Figure 2. Infrared absorption spectra of the multiphase waveguides (a) and a histogram presentation of the integrated intensities of all bands after deconvolution (b).

3. Discussion

It was established earlier that the profile of the extraordinary index change $\Delta n_e(z)$ in $H:LiTaO_3$ layers has a step-like form preserved even after annealing at temperatures of up to 320-350 °C [10], so the shape of the optical profile could not be used for phase composition evaluation. An exception is the profile of single-α-phase waveguide which tends to a graded shape. The α phase which is the most important for different applications requiring preserving or restoration of the electro-optical properties could be obtained only by annealing of the waveguides obtained in ß, γ and δ phase.

Some results were reported for α-phase waveguides obtained in LT by direct proton exchange in benzoic acid vapors at 350 °C [11]. The essential detail of that experiment is the vapor temperature since the transition from high-protonated phases in LT to the lower-protonated α-phase starts at about

320 °C [10]. This way the diffusion temperature does not allow other phases other than the α-phase to be formed and the diffusion due to the annealing at such a temperature seems to be a reason for the decrease in Δn_e when the duration of PE is increased [11]. When PE in vapors (PEV) is used at an appropriate temperature instead of in liquid acid, it allows an annealing to be performed simultaneously at a vapor temperature higher than 350 °C, when phase transition to α phase takes place, and high-quality waveguides could be obtained by a one-step technological process. Since in the experiment described here we used vapor temperatures lower than 350 °C for obtaining waveguides by PEV, we could expect that the phase composition of the PE-layers is complex and contains phases with strongly differing lattice parameters. However, we believe that by using higher temperatures of $LiHSO_4$ vapors, α-phase waveguides could also be obtained.

A preliminary evaluation of the possible phases existing in the waveguide layers could be made on the basis of optical profiles of the investigated samples. The considerations performed are based on the phase model for Z-cut $Li_{1-x}H_xTaO_3$ suggesting that 5 different crystallographic phases could be formed in the PE-layer [4]. According to the phase diagram presented in [4], the determined values of Δn_e for samples obtained by PEV correspond to δ – phase (Δn_e= 0.01 – 0.02) except of TZ-4 having a value of Δn_e lower than 0.01, which corresponds to the α-phase. Since α-phase waveguides cannot be obtained by direct proton exchange [4], when the diffusion temperature is below 350 °C (350 °C is the temperature at which phase transition to α-phase takes place [10]) we could expect the waveguides TZ1-3 to have δ- or κ–phase as well. The fact that Δn_e decreases with increasing proton exchange duration (proton concentration respectively) also confirms the presence of the δ-phase. The other two possible phases (ß and γ) require larger values of Δn_e. Therefore, the determination of the crystallographic phase of the waveguide could not be performed on the basis of the extraordinary index change only, since it could correspond to both κ- and δ-phases [4].

The mobility of Li ions and their ability to be exchanged by protons in the lattice of LT is quite a bit lower than in the lattice of the widely used LN [12]. The amount of protons and the depth of their penetration is less than in the case of LN. That is why the PE-waveguides in LT obtained by direct PE are shallow and usually do not support more than one mode and need to be annealed.

Direct proton exchange in pure acid melts (benzoic, pyrophosphoric etc.) provides waveguides in the δ- phase, characterized by strongly reduced nonlinear coefficients in the waveguide region, and post-exchange annealing needs to be performed to restore the electro-optical and nonlinear properties when the α-phase is reached [13]. Samples obtained by APE treated over 350 °C undergo phase transformation to α phase (sample T-2), but sample T1 annealed at lower temperatures demonstrates a higher value of Δn_e characteristic for the ß- and γ- phases.

The IR-absorption spectra (figure 2) show three well-defined bands: at 3488 cm^{-1}, polarized perpendicularly to the Z-axis (α-phase), at 3500 cm^{-1} with the same polarization (ß- or γ- phase), and a large unpolarized band at 3200 cm^{-1}, which can be assigned to the much higher degree of substitution of Li^+ by H^+ [14] characteristic for the δ-phase. The shoulder at 3200 cm^{-1} suggests both high value of x (H^+ at Li^+ positions) and interstitial hydrogen.

The spectra deconvolution (figure 2b) allows some conclusions to be made about the rate of creation of the different monophase sublayers, as well as about the relative quota of each phase to the whole multiphase PE-layer. The fact that the band area of the α-phase is the largest proves the dominance of the α-phase in the waveguides obtained by PEV in $LiHSO_4$. The analysis of the phase composition confirms this way the presence of more than one phase in the waveguides obtained by PE in $LiHSO_4$ vapors at temperatures lower than 350 °C.

It is well known that the change in the refractive index of ferroelectric materials with doping is due to the changes in molecular refraction, in spontaneous polarization and to elastic deformations in the crystal [15]. When the proton concentration increases, the lattice extends in the direction normal to the surface in order to satisfy the boundary conditions, so in the horizontal plane the PE-layer and the substrate should have identical lattice parameters [16]. This way the deformations introduced by the dopant contribute to the change of the refractive index through the elasto-optic effect. In our case (Z-cut substrates) deformations should be parallel to the optical axis, and at high values of x the lattice should be under tensile stress, which is confirmed by stress measurements.

Phase identification by measurement of crystal lattice deformations of Z-cut samples in the optical axis direction was performed in [4, 17]. It has been shown that for each phase there is a linear dependence between the increase of the refractive index change and the deformations (the proton concentration respectively), but the coefficient of proportionality is different for each phase. The coefficient changes when a phase transition takes place.

It would be logical to assume that the higher level of the dopant causes stronger deformations of the lattice and therefore stronger mechanical stresses in the PE-layer. The authors of [18] have established that the most strained PE layers are those with a higher level of substitution when the surface damage takes place. The smallest internal stress calculated (table 1) is characteristic for homogeneous and less disturbed layers such as the α-phase (T-2) or single-phase ones (T-1).

From a technological point of view, it is interesting to note that the most strained layers have been obtained in two of the cases when PE was performed in vapors (TZ-1 and TZ-4). It is quite possible (especially at higher temperatures and short exchange times) that the kinetic energy of the substituting element transforms into energy of the lattice deformation. Longer diffusion (the temperature and proton source being the same, TZ-2) allows a redistribution of the diffused protons and lattice relaxation. Such low-temperature annealing does not lead to phase transitions and transformations, but probably causes homogenization of the existing phases, this way decreasing the MS in the layer [19]. In the case of TZ-4 the highest stress value is due to longer diffusion at lower temperature when a relatively thick multiphase layer is formed.

The only sample with compressive stress is T-2. It is the α-phase layer with the lowest value of MS measured. As it was discussed above, the strain introduced in the lattice after PE should be tensile and directed perpendicularly to the layer's surface. In the case of Z-cut samples, the Z direction coincides with the optical axis and it is quite possible that the sign „+" or „-" depends on the orientation of the optical axis towards the layer, probably due to the different orientation of the ferroelectric domains in the crystal.

It can be seen in table 2 that there should be a thickness d_s of the sublayer, responsible for the 3200 cm^{-1} band, at which the mechanical stress is minimal. As in the case of LN [19], that thickness is about 0.35 μm.

Table 2. The relative quota of the IR spectral bands of the multiphase waveguide layers and the thicknesses of the phase sublayers (corresponding phase quota respectively) compared to the measured stress.

SAMPLE	IR BAND INTENSITY [%] / PHASE-SUBLAYER THICKNESS [μm]			σ x10^9 [N/m^2]
	3200 cm^{-1}	3480 cm^{-1}	3500 cm^{-1}	
TZ-1	16.67/0.148	62.50/0.556	20.83/0.185	+22.80
TZ-2	22.00/0.271	58.00/0.713	20.00/0.246	+3.41
TZ-3	27.40/0.394	58.90/0.848	13.70/0.197	+2.50
TZ-4	28.30/0.591	62.28/1.302	9.43/0.197	+23.00

The phase model we referred to [4] is based on the fact that each phase has its own specific dependence of the refractive index on proton concentration, which is proportional to the lattice strain ε_{33} of the proton-exchanged layer [20]. The authors of [20], who have studied the electro-optical properties of all possible $H_xLi_{1-x}TaO_3$ phases have observed that the electro-optical effect (the value of the r_{33} coefficient respectively) decreases monotonically with the rise of mechanical stress (increase of x), but the authors of [21] have established that the electro-optical coefficient r_{33} has a minimal value at an intermediate value of x, corresponding to the κ- or ß-phase. This contradiction could be explained to some extent through the existence of d_s: by varying the highly protonated sublayer's thickness we could tune the internal stress and r_{33} respectively, not changing the value of x (for given value of x).

4. Conclusions

In summary, it could be concluded that the independent study of the phase composition of multiphase proton-exchanged waveguides by mode spectroscopy, IR absorption spectroscopy and mechanical stress measurements show that a correlation exists between mechanical stress and the phase composition (hydrogen presence) in the proton-exchanged layer. The stress depends substantially on the technological conditions for obtaining waveguides.

The use of various kinds of melts and technological regimes makes it possible to control to some extent the phase status of the layer obtained, which determines its quality in respect of optical losses and electro-optical properties. Thus, with an appropriate arrangement of the technology it is possible to avoid undesired effects such as phase mixture and formation of highly distorted layers.

The thickness of the phase sublayers having their own bands in the IR spectra could be assessed by spectral deconvolution and optical profile reconstruction. There is a certain thickness of the strongly protonated sublayer corresponding to a minimal stress for the entire PE layer.

References

[1] Jackel J L, Rice C E and Veselka J J 1982 *Appl. Phys. Lett.* **41** 607
[2] Spillman W B, Sanford N A, Sanford R A, Soref R A 1983 *Opt. Lett.* **8** 497
[3] Korkishko Yu N and Fedorov V A 1996 *IEEE J. Select. Top. Quant. El.* **2**1
[4] Haidi K E, Baldi P et al. 1995 *Opt. Lett.* **20** 1698
[5] Maring D B et al. 2002 *JOSA B* **19** 1575
[6] Kuneva M, Tonchev S, Nesheva D, Atanasov P 2010 *Solid State Sciences* **12** 1870
[7] Wu X-L, Zhang M S, Yan F, Feng D 1995 *J. Appl. Phys.***78** 1953
[8] Christova K 1993 PhD Thesis ISSP BAS, Sofia, Bulgaria
[9] Skordeva E, Christova K, Tzolov M and Dimitrova Z 1998 *Appl. Phys. A* **66** 103
[10] Ziling C C, Atuchin V V, Savatinova I and Kuneva M 1992 *Int. J. Optoelectron.* **7** 519
[11] Korkishko Yu N, Fedorov V A, Feoktistova O Y 2000 *J. Lightwave Technol.* **18** 562
[12] Salavcova L, et al. 2007 *Optical Materials* **29** 913
[13] Ahlfeldt A, Laurel F and G. Arvidsson G 1993 *Electron. Lett.* **29** 819
[14] Baldi P, De Micheli M P et al. 1998 *Opt. Eng.* **37** 1193
[15] Sugh K, Fucuma M and Iwasaki H, 1978, *J. Mat. Sci.*, **13** 523
[16] Chen S, Baldi P et al.1993 *Opt. Lett.* **18** 1314
[17] Fedorov V A and Korkishko Yu N 1994 *Ferroelectrics* **160** 185
[18] Wu X-L, Yan F, Zhang M-S, Jiang S-S and Feng D 1996 *J. Condens. Matter* **8** 2073
[19] Kuneva M, Christova K, and Tonchev S 2011 *EPL* **95** 67005
[20] Kostritskii S M, Yu.N.Korkishko Yu N and Fedorov V A 2004 *Proc. ICORS* **19** 477
[21] Ziling C C, et al. 1997 *J. Phys. D: Appl. Phys.* **30** 2698

Thermal effects on the Ga^+ ion beam induced structural modification of a-SiC:H

T Tsvetkova[1,2], C D Wright[2], M F Craciun[2], L Bischoff[3], O Angelov[4] and D Dimova-Malinovska[4]

[1] Institute of Solid State Physics, Bulgarian Academy of Sciences, 72 Tzarigradsko Chaussee, 1784 Sofia, Bulgaria
[2] University of Exeter, College of Engineering, Mathematics and Physical Sciences, Harrison Building, North Park Rd, Exeter EX4 4QF, UK
[3] Institute of Ion Beam Physics and Materials Research, Helmholtz-Zentrum Dresden-Rossendorf, P.O.B. 51 01 19, 01314 Dresden, Germany
[4] Central Laboratory of Solar Energy and New Energy Sources, Bulgarian Academy of Sciences, 72 Tzarigradsko Chaussee Blvd., 1784 Sofia, Bulgaria

E-mail: tania_tsvetkova@yahoo.co.uk

Abstract. The effects of implantation temperature and post-implantation thermal annealing on the Ga^+ ion beam induced optical contrast formation in hydrogenated silicon–carbon alloy (a-SiC:H) films and underlying structural modifications have been studied. The optical contrast formed (between implanted and unimplanted regions of the film material) has been made use of in the form of optical pattern formation by computer-operated Ga^+-focused ion beam. Possible applications of this effect in the area of submicron lithography and high-density optical data storage have been suggested with regard to the most widely spread focused micro-beam systems based on Ga^+ liquid metal ion sources. The implanted samples were structurally analysed using vibrational spectroscopies, like Raman and infra-red (IR) spectroscopy, to define optimum implantation conditions. The precise role of implantation temperature effects, i.e. the target temperature during Ga^+ ion irradiation, on the structural modification obtainable has been therefore a key part of this study. Appropriate post-implantation annealing treatments were also studied, since these are expected to offer further benefits in reducing the required ion dose and enhancing the optical contrast, thus increasing the cost-effectiveness of the method.

1. Introduction

The present work is concerned with a new approach to providing ultra-stable (>50 years), ultra-high density (> 1Tbit/sq.in.) data storage for archival applications. We used ion-implantation to write nanoscale data into hydrogenated amorphous silicon carbide (a-SiC:H) films [1-7]. The role of implantation conditions on the Ga^+ ion beam induced optical contrast formation in hydrogenated silicon–carbon alloy (a-SiC:H) films and underlying structural modifications have been studied, aiming to determine optimised conditions. The use of gallium as the ion implanted species is attractive since it is available in standard focused ion beam (FIB) machines, and in addition has been shown to be capable of generating large optical contrasts [8,9]. The fact that Ga has a very low melting point (Tm= 29.8°C) and an unusual feature of volume contraction on melting are factors which favour Ga incorporation upon ion-implantation as dispersed clusters, or small nanoparticles. It was previously noted that Ga precipitation into nanoparticles can vary dramatically (in terms of particle size) with Ga concentration and small changes in surface implant temperature [10, 11]. The implanted samples were structurally analysed using vibrational spectroscopies, like Raman and infra-red (IR) spectroscopy, to define optimum implantation conditions. The precise role of implantation temperature effects, i.e. the target temperature during Ga^+ ion irradiation, on the structural modification obtainable has been therefore a key part of this study. Appropriate post-implantation annealing treatments were also studied, since these are expected to offer further benefits in reducing the required ion dose and enhancing the optical contrast, thus increasing the cost-effectiveness of the bit-writing method.

Ga^+ broad-beam ion-implantation in a-SiC:H samples was carried out at different substrate temperatures and some of the implanted samples were annealed post-implantation at higher temperatures in vacuum. The expected benefit for the optical data storage method, which relies for readout on reduced optical transmission detected by scanning near-field optical microscopy (SNOM) [12,13], was that a lower implantation temperature would result in an increased amount of defects leading to an absorption increase, and hence further decrease in transmission (greater readout contrast). The expected effect of higher implantation temperatures, or high temperature post-implantation annealing, was that there would be an increase in the optical reflectivity and decrease in transmission due to Ga clusters coalescing into bigger Ga colloids.

2. Experimental

Wide bandgap a-$Si_{1-x}C_x$:H (x_1=0.18;x_2=0.35) thin film samples were deposited onto Corning glass substrates by RF (13.56 MHz) reactive magnetron sputtering. A composite target, composed of mono-crystalline (100) silicon wafer with chips of pure graphite placed on it, was sputtered in an Ar-20%H_2 gas mixture.The film thickness, defined by Talystep profilometer and optical measurements, was ~200 nm. Rutherford backscattering spectrometry was used to define the carbon content (x) of the films.

Ga^+ ion implantation was performed at Helmholtz-Zentrum Dresden-Rossendorf. A range of samples have been processed at different substrate temperature (typically from liquid nitrogen (LN_2) temperatures to around room temperature (RT), and also at a higher temperature T=+50°C). Some of the RT implanted samples were further annealed at several different temperatures in vacuum. The Ga^+ ion-beam intensity was I ~ 2 μA/cm^2, the ion energy was E = 30 keV, and ion doses in the range D = 10^{15} – 10^{17} ions.cm^{-2} were used (D1=1x10^{15} cm^{-2}, D2=5x10^{15} cm^{-2}, D3=2.5x10^{16} cm^{-2} and D4=1.25x10^{17} cm^{-2} (last dose only at RT). Furthermore, some additional samples were implanted with Ga^+ at 3 different temperatures (T1=RT, T2=LN_2 and T3=+50°C) with a single dose (D = 2xD3 = 5x10^{16} cm^{-2}). The samples implanted at RT were then thermally annealed in Ar atmosphere at 3 different temperatures: +50°C, +120°C and +250°C. All implanted samples were structurally analysed using vibrational spectroscopies, like Raman and infra-red (IR) spectroscopy, to define the optimum implantation conditions. IR transmittance measurements were performed with a Fourier Transform spectrophotometer Bruker (model IFS 113 V) in the region of 550–2200 cm^{-1}. Raman spectra are collected in ambient air and at different temperatures (T1=RT and T2=LN_2(liquid nitrogen)) with a Renishaw spectrometer. An excitation laser with a wavelength of 532 nm, focused to a spot size of 1.5 micrometers diameter and an x100 objective lens are used. To avoid sample damage or laser induced heating, the incident power is kept at 5 mW.

3. Results and discussion

3.1. Infra-red (IR) spectroscopy results on a-SiC:H/c-Si samples implanted with Ga^+ broad-beam implantation at different temperatures

The ion-implanted samples with Ga^+ broad-beam implantation at different temperatures (T1=RT, T2=LN_2(liquid nitrogen), and T3= +50°C), were studied by infra-red (IR) spectroscopy in the range 300 – 4500 cm^{-1}. Results are shown in figure 1.

Journal of Physics: Conference Series **398** (2012) 012048 doi:10.1088/1742-6596/398/1/012048

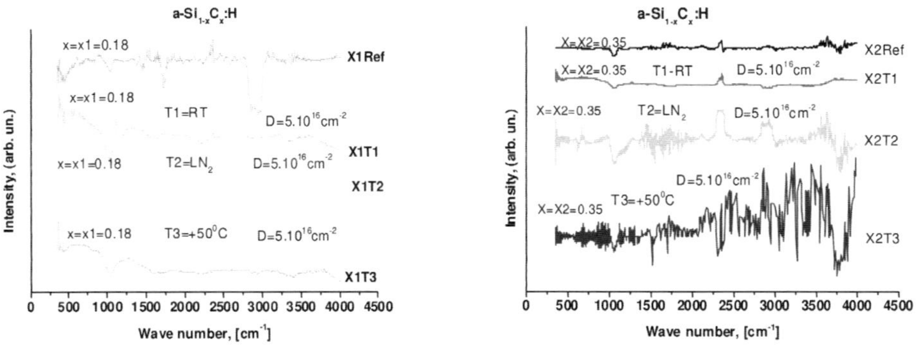

a) b)

Figure 1. IR spectra of Ga^+ implanted a-$Si_{1-x}C_x$:H (X1=0.18 **(a)**, X2=0.35 **(b)**) films with a dose D=$5x10^{16}$ cm^{-2}, at different temperatures: T1=RT, T2=LN_2(liquid nitrogen), and T3= +50°C.

The observed bands for the case of the lower carbon content samples (figure 1 a) are as follows:

- C-C, 460 cm^{-1}, at all samples
- Si-C, 780 cm^{-1}, stretching mode, pronounced at X1Ref, the band smears under implantation at all temperatures
- Si-O, stretching mode, at 1043 cm^{-1} at all samples
- C-C, at about 1358-1569 cm^{-1}, at sp^3 C-configuration at 1374 cm^{-1} (X1Ref) and 1358 cm^{-1} at T2 and T1, at T3 disappears; sp^2 C-configuration at 1550 cm^{-1} (T1), 1569 cm^{-1} (T2 and T3), but at X1Ref it is absent
- C-H and C-H2, at about 1739 cm^{-1} at sample X1Ref, under implantation at all samples this band smears
- C-Hn, stretching mode, in sp^2 (2860 cm^{-1}) and in sp^3 (2920 cm^{-1}) in sample XiRef, under implantation this band smears.

This analysis clearly shows that at all temperatures of ion implantation considerable additional disorder is introduced, as Si-C bond (780 cm^{-1}) is largely broken, while all C-H bonds (C-H, C-H2 and C-Hn) are also broken accompanied by the loss of the hydrogen from the films as a result of the ion beam bombardment. The results for the films with the higher carbon content (X2) also show the same trend (figure 1 b).

The obtained IR spectra were also studied for evidence of possible Si-Ga bond vibrations. Due to the considerable noise in the registered spectra in the range 350 - 400 cm^{-1}, where the main Si-Ga vibration modes appear, there is available reliable data only for the RT implanted samples with two different Ga^+ ion doses: D1=$5x10^{16}$ cm^{-2} and D2=$2.5x10^{16}$ cm^{-2} (figure 2).

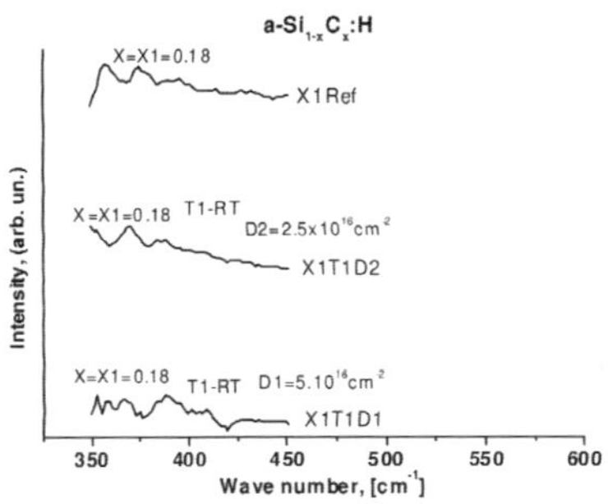

Figure 2. IR spectra of Ga^+ implanted a-$Si_{1-x}C_x$:H (X1=0.18) films with a dose D1=$5x10^{16}$ cm^{-2} and D2=$2.5x10^{16}$ cm^{-2} at RT.

The following bands for present Si-Ga bonds were observed:

at 358 cm^{-1} – present for both samples implanted with D1 and D2 (X1T1D1 and X1T1D2)

at 375 cm^{-1} – present for both samples implanted with D1 and D2 (X1T1D1 and X1T1D2)

at 398 cm^{-1} – present only in the case of D1 (X1T1D1); for D2 (X1T1D2) this band smears.

Only in these spectra the bands corresponding to vibrations of Si-Ga bonds can be initialized; in the other spectra due to the high level of noise signals such bands smear.

3.2. Infra-red(IR) spectroscopy results on a-SiC:H/c-Si samples implanted with RT Ga^+ broad-beam implantation and after treatments with post-implantation thermal annealing

The observed bands for the case of the lower carbon content samples (figure 3 a) are as follows:

- C-C, 456 cm^{-1} at all samples without shift after annealing;
- Si-C, stretching mode, 786 cm^{-1}, broad band (X1T1D2AsImpl), after annealing at 802 cm^{-1} (X1T1D2Ann250°C); 802 cm^{-1} (X1T1D1AsImpl), after annealing at 820 cm^{-1} (X1T1D1Ann250°C);
- Si-O, 1053 cm^{-1}, stretching mode, without shift after annealing;
- C-C, 1373 cm^{-1} sp^3 and 1538 cm^{-1} sp^2 C-configuration (X1T1D2AsImpl); after annealing the bands smear (X1T1D2Ann250°C); 1363 cm^{-1} at sp^3 and 1520 cm^{-1} sp^2 C-configuration (X1T1D1AsImpl); after annealing the bands smear (X1T1D1Ann250°C);
- C-H and C-H2 stretching mode, broad band at ~ 1750 cm^{-1} (X1T1D2AsImpl), it shifts with low intensity at ~ 1658 cm^{-1} after annealing (X1T1D2Ann250°C); less pronounced at 1628 cm^{-1} (X1T1D1AsImpl) and after annealing at ~ 1658 cm^{-1} (X1T1D1Ann250°C);
- C-Hn, 2875 cm^{-1} sp^2 and 2924 cm^{-1} sp^3 C-configuration (X1T1D2AsImpl), after annealing with low intensity of the band at 2860 cm^{-1} sp^2 and 2920 cm^{-1} sp^3 C-configuration (X1T1D2Ann250°C); 2845 cm^{-1} sp^2 and 2920 cm^{-1} sp^3 C-configuration (X1T1D1AsImpl), after annealing at 2870 cm^{-1} at sp^2 and 2920 cm^{-1} at sp^3 C-configuration (X1T1D1Ann250°C).

This analysis shows that the C-C bond at 456 cm^{-1} and Si-O bonds at 1053 cm^{-1} do not show changes after annealing at +250°C; however, the C-C bonds for both sp^3 and sp^2 C-configuration (i.e. the diamond-like and graphitic carbon bonds) are modified due to the influence of the presence of Ga nano-clusters, i.e. Ga colloid formation and growth.

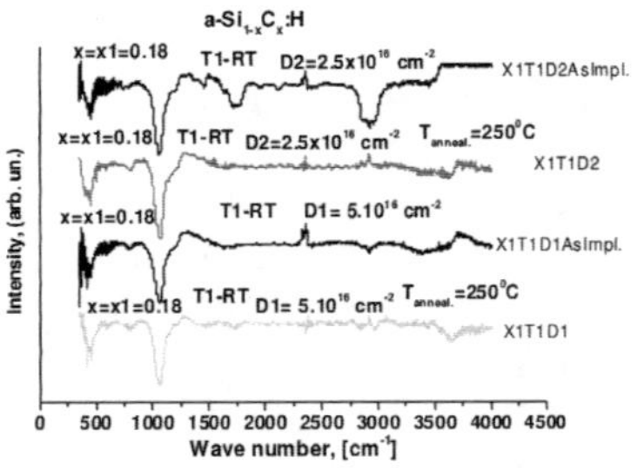

a)

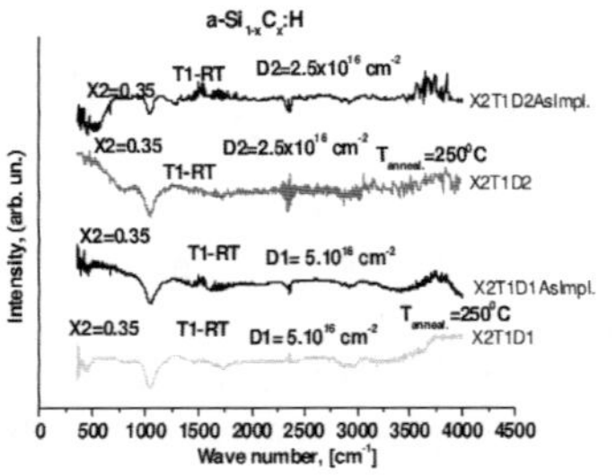

b)

Figure 3. IR spectra of Ga^+ implanted a-$Si_{1-x}C_x$:H (X1=0.18 (**a**), X2=0.35 (**b**)) films, with doses D1=5x10^{16} cm^{-2} and D2=2.5x10^{16} cm^{-2}, before and after thermal annealing at T= +250°C.

Similarly modified are also the main bonds present in the a-SiC:H films – the Si-C and C-H bonds - due to the change in the back-bone configuration of the a-SiC:H host lattice.

The results for higher carbon content films (figure 3 b) also show the same trends as above.

3.3. Raman spectroscopy results on a-SiC:H/c-Si samples implanted with Ga+ broad-beam implantation at different implantation temperatures

Wide bandgap a-$Si_{1-x}C_x$:H (x_1=0.18; x_2=0.35) thin film (d=200nm) samples deposited on c-Si substrates, as prepared by rf magnetron sputtering and ion-implanted with Ga^+ broad-beam implantation at different temperatures (T1 and T2), were studied by Raman spectroscopy in the range 100 – 3500 cm^{-1}. The Ga+ ion doses were in the range D = 10^{15} – 10^{17} ions.cm^{-2} (D1=1x10^{15} cm^{-2}, D2=5x10^{15} cm^{-2}, D3=2.5x10^{16} cm^{-2} and D4=1.25x10^{17} cm^{-2} (last dose only at RT). The Raman spectra were measured at different points on the sample (m_1, m_2, etc.).

The as-deposited samples show greater sp^2 - type carbon (graphitic type carbon-configuration) in the a-$Si_{1-x}C_x$:H films with the higher carbon content (x_2=0.35) as demonstrated in figure 4, where a clear band at ~1500 cm^{-1} appears for the case of X2 (x_2=0.35) (figure 4 b), which could be assigned to graphitic carbon, and which is not present in the case of X1 (x_1=0.18) (figure 4 a).

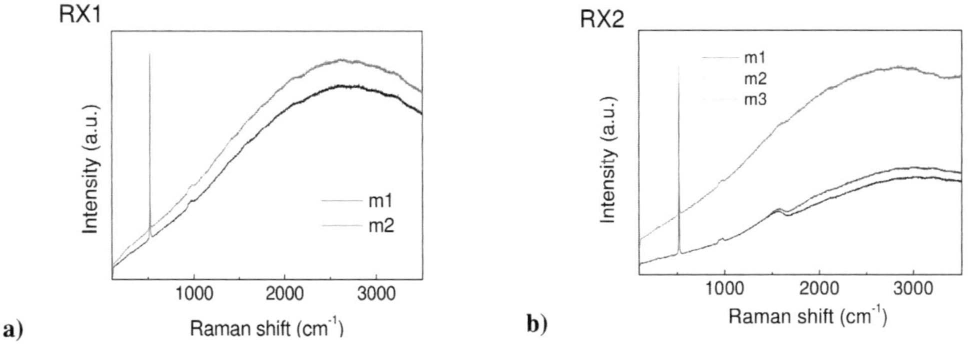

Figure 4. Raman spectra of as-deposited a-$Si_{1-x}C_x$:H (X1=0.18 **(a)**, X2=0.35 **(b)**) films.

In the case of the lower carbon content (x_1=0.18) (figure 5 a), the RT Ga^+ ion implantation results in graphitization of the sp^3 type carbon and the appearance of a graphitic (sp^2) type carbon related band at 1500 cm^{-1} even at relatively lower ion doses (D3) (figure 5 b), but is especially expressed for the highest dose (figure 5 c). However, at lower implantation temperatures (T2), this graphitic (sp^2) type carbon related band at 1500 cm^{-1} is smeared as a result of the ion bombardment (figure 5 d).

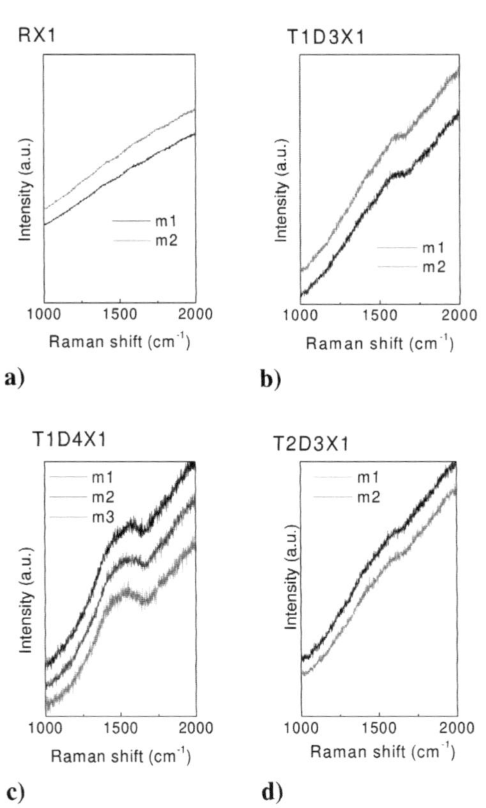

Figure 5. Raman spectra of a-$Si_{1-x}C_x$:H (x_1=0.18), as deposited and Ga^+ implanted at different D and T.

In the case of the higher carbon content (x_2=0.35) (figure 6 a), the RT Ga^+ ion implantation results in increased graphitization and better expressed graphitic (sp^2) type carbon related band at ~1500 cm^{-1} even at relatively lower ion doses (D3) (figure 6 b), but it is especially expressed for the highest dose D4 (figure 6 c). However, at lower implantation temperatures (T2), this graphitic (sp^2) type carbon related band at 1500 cm^{-1} is smeared as a result of the ion bombardment, due to the increased defects introduction and further breaking of the graphitic carbon bonds (figure 6 d).

4. Conclusion

Ga^+ broad-beam ion implantation in a-SiC:H samples was carried out at different substrate temperatures (T1=RT, T2=LN2(liquid nitrogen), and T3= +50°C) and some of the RT implanted samples were further annealed at higher temperatures in vacuum. The expected benefit for the optical data storage method, which relies on reduced optical transmission detected by SNOM, was that lower implantation temperature would result in an increased amount of defects leading to further absorption increase, and hence further decrease in transmission. This effect was indeed observed but was overridden by an effect of increased ion beam induced sputtering yield, which greatly reduced the film thickness and hence increased the transmission. Likewise, the expected increase in the optical reflectivity (due to Ga clusters coalescing into bigger Ga colloids which

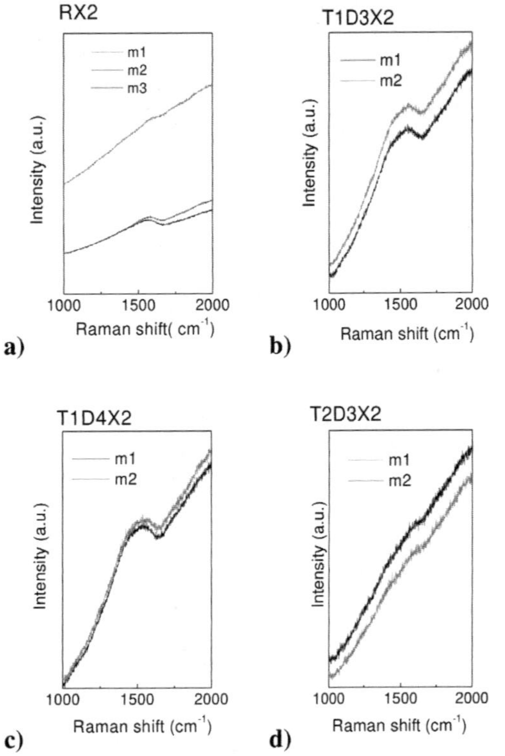

Figure 6. Raman spectra of a-$Si_{1-x}C_x$:H (x_2=0.35), as deposited and Ga^+ implanted at different D & T.

have higher reflectivity and hence should decrease the transmission) as a result of higher implantation temperature or high temperature post-implantation annealing was indeed observed but was overridden by an ion beam induced decrease of defect concentration (due to self annealing) that leads to an absorption decrease and hence further transmission increase. Therefore, it could be concluded that the best conditions for optical data storage for archival storage applications would be using Ga^+ ion implantation in a-SiC:H films with an optimal dose of D=5x10^{16} cm^{-2} at RT. The fact that the optimum implantation temperature for storage applications has been found to be that of room temperature is fortuitous since it means that the required ion-implantation process is simplified and reduced in cost.

Acknowledgements

This work has been supported by the European Community as an Integrating Activity 'Support of Public and Industrial Research Using Ion Beam Technology (SPIRIT)' under EC contract No. 227012. The Marie Curie Fellowship for T.Tsvetkova was also supported by the European Community under the contract PIEF-GA-2009-251845. The help of Dr. K.Gesheva with the measurement of the IR spectra is also gratefully acknowledged.

References

[1] Powell J A and Matus L 1989 *Amorphous and Crystalline Silicon Carbide*, ed G L Harris and C Y W Yang (Berlin: Springer)
[2] Ziegler J F 1988 *Ion Implantation* (New York: Academic Press)
[3] Ruttensperger B, Krötz G, Müller G, Derst G and Kalbitzer S 1991 *J Non-Cryst Solids* **137-138** 635
[4] Tsvetkova T 1996 *Beam Processing of Advanced Materials*, ed J Singh, S Copley and J Mazumder (Metals Park: ASM International) p 207
[5] Tsvetkova T, Takahashi S, Zayats A, Dawson P, Turner R, Bischoff L, Angelov O and Dimova-Malinovska D 2005 *Vacuum* **79** 94
[6] Tsvetkova T, Takahashi S, Zayats A, Dawson P, Turner R, Bischoff L, Angelov O and Dimova-Malinovska D 2005 *Vacuum* **79** 100
[7] Takahashi S, Dawson P, Zayats A V, Bischoff L, Angelov O, Dimova-Malinovska D, Tsvetkova T and Townsend P D 2007 *J Phys D: Appl Phys* **40** 7492
[8] Bischoff L, Teichert J, Kitova S and Tsvetkova T 2003 *Vacuum* **69** 73
[9] Tsvetkova T, Angelov O, Sendova-Vassileva M, Dimova-Malinovska D, Bischoff L, Adriaenssens G J, Grudzinski W and Zuk J 2003 *Vacuum* **70** 467
[10] Townsend P D, Chandler P J and Zhang L 1994 rev. 2006 *Optical effects of ion implantation* (Cambridge: Cambridge University Press)
[11] Hole D E, Townsend P D, Barton J D, Nistor L C and Van Landuyt J 1995 *J. Non-Cryst. Sol.* **180** 266
[12] Richards D and Zayats A V 2004 Nano-Optics and Near-Field Microscopy *Phil. Trans. R. Soc. London* **362** 699
[13] Takahashi S, Dickson W, Pollard R and Zayats A V 2004 *Ultramicroscopy* **100** 443
[14] Tsvetkova T, Wright C D, Hosseini P, Bischoff L and Zuk J 2012 *Acta Physica Polonica* in press

Influence of the Type of Metal Contact and Post-deposition Treatment on the Performance of P3HT:PCBM Organic Solar Cells

M Sendova-Vassileva[1], G Popkirov[1], P Vitanov[1], Ch Dikov[1], V Gancheva[2], D Tsocheva[2], P Mokreva[2]

[1]Central Laboratory of Solar Energy and New Energy Sources, Bulgarian Academy of Sciences, 72 Tzarigradsko Chausse, 1784 Sofia, Bulgaria
[2]Institute of Polymers, Bulgarian Academy of Sciences, 1113 Sofia, Bulgaria

E-mail: marushka@phys.bas.bg

Abstract. Bulk-heterojunction solar cells based on P3HT:PCBM with Al or Ag electrodes deposited either by magnetron sputtering or by thermal evaporation were produced and their performance compared. The best results were obtained with thermally evaporated Al electrodes. Post-production annealing was applied to the already encapsulated samples and significantly enhanced the efficiency of all four types of cells. For both Ag and Al electrodes deposited by thermal evaporation post-annealing improved the I-V characteristics eliminating the S-shape usually typical for as deposited samples.

1. Introduction

Organic solar cells based on the concept of the donor-acceptor heterojunction have attracted wide interest [1,2]. The efficiency of such solar cells depends strongly on the effectiveness of the photogeneration of excitons and their subsequent dissociation into free electrons and holes. To facilitate rapid dissociation of the excitons and fast transport of the released charges to the electrodes, interpenetrating networks of organic electron donor and acceptor materials are used as the absorbing layer in organic solar cells. Mixtures of semiconducting conjugated polymers serving as electron donor and fullerene derivatives serving as electron acceptor are the constituents of bulk heterojunction (BHJ) structures. The blend of poly(3-hexylthiophen) (P3HT) and the fullerene derivative [6,6]-phenyl-C61-butyric acid methyl ester (PCBM) is a popular donor-acceptor couple which has been successfully used according to the literature. In most reported cases, indium tin oxide (ITO), covered with a thin layer of poly(3,4-ethylenedioxythiophene):poly(styrenesulfonate) (PEDOT:PSS) as an intermediate hole-conducting layer, have been used as the transparent positive electrode either on glass or on flexible polyester substrates. Aluminium is often used as the negative electrode. Silver is more stable against oxidation and is also a prospective candidate for the role of the negative electrode. The almost standard method for cathode deposition is the thermal evaporation of Al in high vacuum [3]. Magnetron sputtering is better suited for large-scale production, but is usually avoided because of the potential sputter damages to the organic absorber layer. However, as shown in [4], magnetron sputter deposition can be an alternative deposition technique, if adequate post-annealing is applied.

The purpose of the present contribution is to test, characterize and compare polymer/fullerene bulk-heterojunction solar cells of identical structure and production technology avoiding production steps in glove box, and using Al or Ag electrodes deposited either by magnetron sputtering or by thermal evaporation.

2. Experimental

Test solar cells were produced using the substrate packs purchased from Osilla. A thin layer (≈ 90 nm) of PEDOT:PSS Clevious P, purchased from K.C. Stark was deposited from water dispersion onto pre-patterned ITO/glass substrates by spin coating at 1750 rpm for 60 s. The films were subsequently annealed for 30 min at 130°C in argon atmosphere. P3HT and PCBM, purchased from Rielke Metals and Solenne, respectively, were dissolved in chlorbenzene (1:1 weight ratio with concentration 12 mg/ml) by stirring for 48 hours at 50°C in argon. The photoactive layer was deposited onto the substrate by spin-coating at 400 rpm for 80 s and the thickness obtained was 80-90 nm as measured by a Talystep stylus profilometer. The samples were kept for 10 min in a closed vessel with solvent-vapor atmosphere to assure slow drying of the polymer for better thin-film morphology [5]. Finally, Ag or Al metal electrodes were deposited by thermal evaporation or sputtering through a steel shadow mask, thus yielding 6 small solar cells on each substrate with surface area of 0.045 cm^2 each. In the sputtering chamber the samples were subjected to thermal drying in Ar for 10 min at 100°C before sputtering and thermal annealing at 140°C for 15 min after the deposition. No thermal treatments were applied to the samples in the vacuum evaporation chamber. The active area of the samples was encapsulated using a thin glass plate and UV curable epoxy resin. Finally, electrical connection legs were mounted to provide reliable electrical contact for the measurements. Post-production annealing of some of the samples was performed by heating the already encapsulated samples on a hot plate at 150°C for 10 min.

Current-voltage (I-V) measurements with and without illumination were carried out by a computer controlled home-made potentiostat. The voltage scan rate was 20mV/s in the range from -0.5V to +0.6V. A 50W tungsten-halogen lamp powered by a regulated d.c. power supply served as illumination source. Impedance spectroscopy measurements were performed at different bias voltages using a fast Fourier transform (FFT) impedance spectrometer [6, 7] in the frequency range 1.5 Hz – 50 kHz with a 15 mVp-p perturbation voltage.

3. Results and discussion

3.1. Current-voltage measurements

The I-V characteristics obtained in a technological batch showed some deviations from cell to cell and from sample to sample, most probably due to inhomogeneities of the absorbing layers. Here data and results for typical of the respective batch representative single cells are presented and discussed.

I-V curves measured upon illumination of the four types of samples before post-annealing are depicted in figure 1, curves (1). For both metals evaporation leads to significantly higher photocurrents compared to sputtering with the used processing. Structure damages in the thin absorbing layer during the sputtering deposition cannot be excluded. The parameters obtained from the characterization of the cells are presented in table 1.

A look at the curves for the samples before post-annealing in figure 1 reveals not only low photocurrent values at short circuit, but also very low fill-factors and the existence of a characteristic S-shape close to the open-circuit voltage regime. This S-shape kink has been repeatedly mentioned in the literature [1,4,8-10] and several models have been proposed, but as of now this effect is not completely explained for both planar and bulk heterojunction organic solar cells. Often the existence of barriers at the metal contacts is considered. An attempt to model the I-V characteristics of BHJ organic solar cells is presented in [9], where the authors mention that non-ohmic contacts can strongly limit the current flow and lower the fill-factor. According to [8] the appearance of an S-shape I-V characteristic indicates slow charge transfer at one of the electrical contacts of the absorber layer. Further, these authors suggest an effect of possible corrosion of the Al electrode.

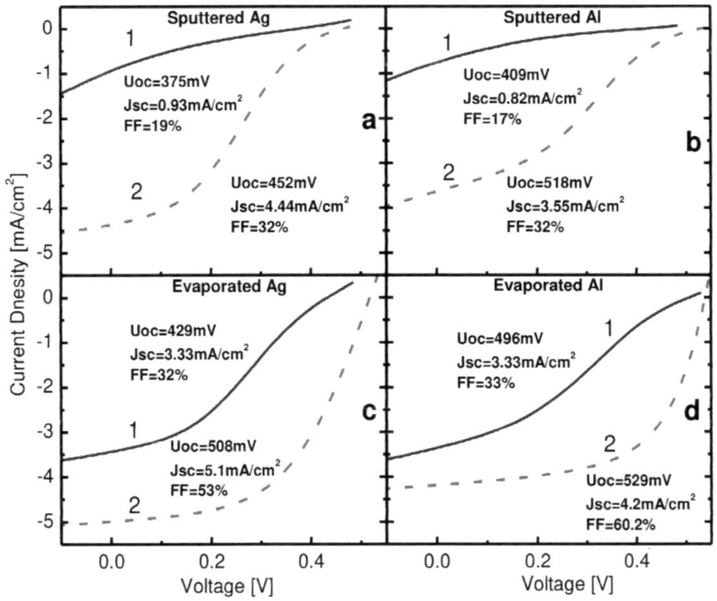

Figure 1. Current-voltage characteristics under illumination of the studied solar cells with sputtered Ag contacts (a), sputtered Al contacts (b), evaporated Ag contacts (c) and evaporated Al contacts (d). In each plot curve (1) is measured before post-annealing and curve (2) - after post-annealing.

The samples with evaporated Al and Ag electrodes worked better than the ones with sputtered electrodes but their photovoltaic performance was also disappointing, especially because of the low fill-factors and the S-shaped I-V characteristics. So the samples were annealed on a hot plate in air at 150°C for 10 min after encapsulation. Representative results are shown in figure 1, curves (2) and table 1. Very significant improvement of the photovoltaic parameters resulted for all of the samples. Short-circuit currents and open-circuit voltages increased markedly. The S-shape feature in the I-V characteristics of sputtered samples, unfortunately, remained, thus their fill-factors were still low. The improvement of the samples with evaporated metal electrodes was less pronounced, but their S-shape kink fully disappeared and fill-factors as high as 60% were calculated.

Using the data for the post-annealed cells with evaporated Al and Ag electrodes, which do not show S-shape near the open-circuit voltage, the series and the shunt resistances were calculated from the respective slopes of the I-V curves, as can be seen in table 1. If the I-V data sets which show an S-kink are evaluated in the same way series resistances in the range of kOhms would result, but the reasons behind the S-shape feature are more complex and not well understood, and thus, it is not correct to apply the slope-method for the estimation of the series resistance. Alternatively, we have tried to obtain some information about the series and shunt resistance values of the cells using impedance spectroscopy.

3.2. Impedance spectroscopy

The series resistance is an important parameter for solar cells of any kind. A number of methods for the estimation of the series resistance of semiconductor p-n junction solar cells are known from the literature [11], however, still further investigations are necessary to prove their validity for organic donor-acceptor solar cells.

Table 1. Parameters obtained from the I-V characterization of the cells and R_{ser} and R_{sh} from the impedance spectra.

Batch	U_{oc}, mV	J_{sc}, mA/cm^2	FF, %	R_{ser}, Ω R_{sh}, Ω (I-V)	R_{ser}, Ω R_{sh}, Ω (impedance)	S-shape
Ag, sputtered	375	0.93	19	-	-	yes
Ag, sputtered, post-annealed	452	4.44	32	-	-	yes
Ag, evaporated	429	3.33	32	-	-	yes
Ag, evaporated, post-annealed	508	5.1	53	760 31k	50 43k	no
Al, sputtered	409	0.82	17	-	-	yes
Al, sputtered, post-annealed	518	3.55	32	-	-	yes
Al, evaporated	496	3.33	33	-	-	yes
Al, evaporated, post-annealed	529	4.2	60	552 29k	100 31.7k	no

The series resistance of BHJ solar cells should include at least contributions from the ITO and PEDOT:PSS layers, the contact resistance in-between, plus the resistance of the metal electrode, plus the two contact resistances at both sides of the absorbing layer. At least partly, the bulk resistance of the active layer could also contribute to the estimated value of the series resistance.

Several authors have applied impedance measurements to characterize solar cells. Information on carrier lifetime and electron density-of-states [12], mobility [13], density and energetic position of defect states in the absorber [14], etc. can be obtained applying proper calculation models to measured impedance spectra in a wide frequency range, at different bias voltage, with or without illumination.

Often the impedance spectra of a solar cell exhibit a single semicircle in the complex plot of the imaginary part vs. the real part of the complex impedance. A simple equivalent circuit consisting of a series resistor, R_{ser}, connected to a capacitor, C_p, and a resistor, R_{sh}, in parallel to C_p, can be used to fit the measurement results obtained for different measurement conditions, i.e. illumination intensity and voltage bias. In most cases, however, the semicircle appears to be somehow depressed. A so called constant phase element (CPE) is then used to replace the capacitance and take account of the distributed nature of the non-ideal capacitance. More generally this effect is attributed to the distribution of time-constants, due to distributions of physical properties such as: structure, dielectric constants, and/or resistivity over the surface of an electrode and/or normally to the electrodes [15]. The impedance of the CPE is defined as $Z_{CPE} = 1/\left[Q_p.(j\omega)^{\alpha}\right]$, where $j = \sqrt{-1}$, ω is the circular frequency. For α=1, Q_p has capacitance dimension and the impedance spectra plot is a semicircle in the complex plane. For higher frequencies (up to 1 MHz), a second partial arc can appear and a more complex equivalent circuit has to be used [12, 16].

The impedance spectra of the samples were measured upon illumination in the range from 1.5 Hz to 50 kHz. The impedance spectra obtained depended strongly on the bias voltage, illumination level and markedly – on post-production annealing. Here only data for the best performing sample, the one obtained with thermal evaporation of Al are presented. Figure 2 shows data measured at reverse bias of -0.3V from the as-prepared sample (curve 1) and the one after post-production annealing (curve 2).

Figure 3 presents data acquired close to the open-circuit voltage. The spectra are shown in the complex plane (figure 2a and figure 3a) and as Bode-plots in figure 2b,c and figure 3b,c to show explicitly the frequency dependence of the real and imaginary parts of the impedance. The equivalent circuit as described above with a CPE instead of capacitance gave a good fit. Even without looking at the fit parameters it can be seen that for both bias voltages used the effect of post-production annealing is very strong.

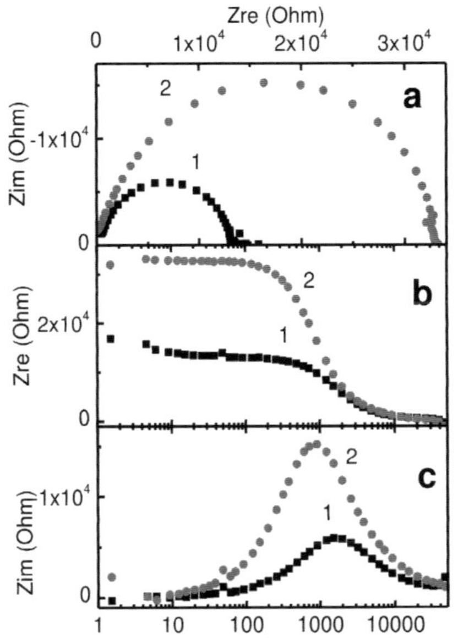

Figure 2. Impedance spectra obtained at reverse bias of -0.3V of the as-prepared sample with evaporated Al contacts (1) and after post-production annealing (2).

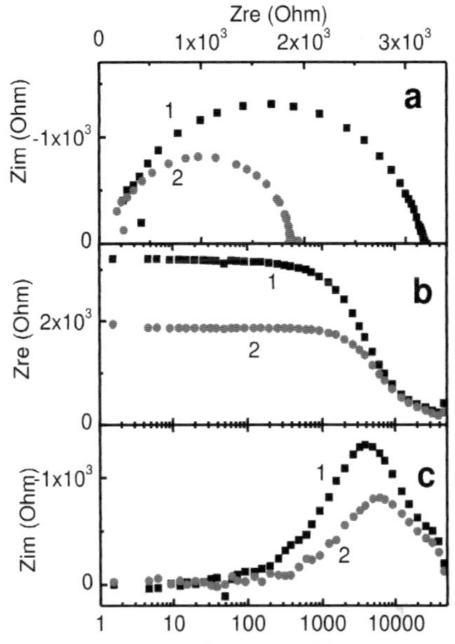

Figure 3. Impedance spectra obtained close to the respective open-circuit voltage of the as-prepared sample with evaporated Al contacts (1) and after post-production annealing (2).

The shunt resistance obtained from the fit of measured impedance data is representative of the bulk resistivity of the P3HT:PCBM layer due to recombination of charge carriers in the absorbing layer and possibly existing conductive paths between the two metal electrodes across the absorbing layer. The voltage and illumination dependence of R_{sh} is attributed to the photogeneration and/or injection of charge carriers. Measurements at reverse bias of -0.3V (cf. figure 2a) showed almost no significant change in the value of α (ca. 0.95) after post-annealing and the shunt resistance increased from 12.7 kOhm to 31.7 kOhm after annealing. Close to open circuit voltage the shunt resistance decreased from 3.07 kOhm to 1.76 kOhm after annealing. The depression of the semicircles (cf. figure 3a) decreased from α = 0.89 before to 0.96 after post-production annealing for measurement close to open-circuit conditions. These data are in good correspondence with the observed improvement of the short-circuit current, the open-circuit voltage and the fill-factor. The accepted opinion, that post-annealing leads to improvement of the morphology of the absorption layer (crystallinity, phase separation etc.) can also be confirmed by the increased value of α. The series resistance, as revealed from the fit, is in the range of 100 Ohm, a value much less than the one obtained from the I-V curve slope (552 Ohm, cf. table 1). Obviously, the slope of the I-V curve at open-circuit voltage depends also on other structural and/or interface factors. The value of R_{ser} obtained from the impedance spectra possibly does not account for any resistive effects due to the absorbing layer and its interface to the electrodes. Discussions concerning the nature of the CPE and its parameter Q_p would be very speculative, due to lack of further data for the absorbing layer morphology, and are beyond the scope of this note.

4. Conclusion

In this contribution bulk heterojunction organic solar cells with sputtered metal contacts are studied and compared to samples with evaporated metal contacts. The photoactive thin film consisting of mixed P3HT and the fullerene derivative PCBM was deposited by spin coating on patterned ITO covered glass plates coated with PEDOT:PSS. Metal electrodes of aluminium or silver were sputtered or evaporated through a shadow mask. The active area was encapsulated using a thin glass plate and UV curable epoxy resin. After encapsulation the samples were subjected to thermal annealing in air. The solar cells were characterized by current-voltage measurements and attempts were made to determine their effective series and shunt resistances by means of impedance spectra measurements. The results of the investigation indicate that cells with both sputtered and evaporated contacts improve their performance significantly after annealing but for sputtered contacts the S-shape of the I-V characteristic and resulting lower fill-factor values are not fully overcome.

Acknowledgement. The authors are grateful to the Bulgarian National Science Fund for financial support – Project 02-254/2008.

References

[1] Deibel C, Dyakonov V and Brabec C J 2010 *IEEE J. Sel. Top. Quant.* **16** 1517
[2] Wang Y, Wei W, Liu X and Gu Y 2012 *Sol. Energ. Mat. Sol. C.* **98** 129
[3] Singh V, Thakur A K, Pandey S S, Takashima W and Kaneto K 2008 *Org. Electron.* **9** 790
[4] Ahlswede E, Hanisch J. and Powalla M 2007 *Appl. Phys. Lett.* **90** 063513
[5] Chen F-C, Ko C-J, Wu J-L and. Chen W-C 2010 *Sol. Energ. Mat. Sol. C.* **94** 2426
[6] Popkirov G and Schindler R-N 1992 *Rev. Sci. Instrum.* **63** 5366
[7] Popkirov G and Schindler R-N 1993 *Electrochim. Acta* **38** 861
[8] Glatthaar M, Riede M, Keegan N, Sylvester-Hvid K, Zimmermann B, Niggemann M, Hinsch A and Gombert A 2007 *Sol. Energ. Mat. Sol. C.* **91** 390
[9] Kumar P, Jain S C, Kumar V, Chand S and Tandon R P 2009 *J. Phys.* D*: Appl. Phys.* 42, 055102
[10] Tress W, Petrich A, Hummert M, Hein M, Leo K and Riede *M* 2001 *Appl. Phys. Lett.* **98** 063301
[11] Bashahu M and Habyarimana A 1995 *Renew. Energ.* **6** 128
[12] Garcia-Belmonte G, Boix P P, Bisquert J, Sessolo M, Bolink H 2010 J *Sol. Energ. Mat. Sol. C.* **94** 366
[13] Garcia-Belmonte G, Munar A, Barea E M, Bisquert J, Ugarte I, Pacios R 2008 *Org. Electron.* **9** 847
[14] Khelifi S, Decock K, Lauwaert J, Vrielinck H, Spoltore D, Piersimoni F, Manca J, Belghachi A and Burgelman M 2011 *J. Appl. Phys.* **110** 094509
[15] Hirschorn B, Orazem M E, Tribollet B, Vivier V, Frateur I and Musianid M 2010 *J. Electrochem. Soc.* **157** C452
[16] Leever B J, Bailey C A, Marks T J, Hersam M C and Durstock M F 2012 *Adv. Energy Mater.* **2** 120

An experimental approach for synthesis of Fe-Al-O multiferroic fibrous material

N Starbov[1], K Starbova[1], Ph Vanderbemden[2], P Simeonova[1] and V Lovchinov[1]
[1] Institute of Solid State Physics, Bulgarian Academy of Sciences,
72 Tzarigradsko Chaussee Blvd., 1784 Sofia, Bulgaria
[2] University of Liège, S.U.P.R.A.T.E.C.S., Department of Electrical Engineering and Computer Science B28, Sart-Tilman, Liège

E-mail: lovcinov@issp.bas.bg

Abstract: Basic principles of the electro-hydrodynamics are applied for synthesis of solid state Fe-Al-O multi-ferroic fibrous material. For that purpose stable blend spinning solutions comprised of a high molecular assisting organic polymer and salts of iron and aluminum are developed. These solutions are tested under electrospinning conditions and synthesis of homogeneous as spun non-woven mats characterized by fibre mean diameters in the micro- and nano-size range is successfully demonstrated. Multi-step thermal procedure is applied for the consecutive solvent evaporation, polymer pyrolisis and final fibre calcination. Electron-optical imaging technique and XRD are applied for revealing the sample morphology and the phase composition correspondingly. The results obtained outline the wide possibilities for fabrication of multi-ferroics fibrous nano-materials on the basis of Fe-Al-O.

1. Introduction

In the last decade a class of materials named multi-ferroics has attracted great technological and academic interest [1]. The multi-ferroics are materials in which two or all three of the property "ferroelectricity", "ferromagnetism" and "ferroelasticity" occur simultaneously in the same phase. This multiplicity, combined with relatively simple structure of these materials, makes them suitable object for investigating the general principles that govern these properties. Moreover, the possibility of integration between the ferroelectromagnetic physical properties through the magnetoelectric effects can promote interesting practical advances in many electronic technologies. Therefore, the fabrication of single-phase multi-ferroics materials with appropriate properties for technological application is a provocative challenge for physicists and material scientists.

Among the variety of iron aluminium oxide systems $FeAlO_3$ is an attractive multi-ferroic material since it exhibits piezo-electricity, ferromagnetism and magnetoelectrical effects at low temperature [2, 3]. Moreover, $FeAlO_3$ shows an extremely high magnetic anisotropy at temperatures down to 4,2K up to 37T in pulsed magnetic fields [4]. Regardless of this R&D importance only a few activities on the synthesis of single-phase $FeAlO_3$ have been reported. This is due to the very narrow range of stability of the multi-ferroic phase in the binary equilibrium diagram of Fe_2O_3 - Al_2O_3 [5] and of the extremely high stabilities of both hematite - a-Fe_2O_3 and alumina - a-Al_2O_3, which hinders the reaction. Different experimental approaches are well known for synthesis of single-phase multi-ferroic $FeAlO_3$ [5], namely: solid-state, sol-gel, co-precipitation and the high-energy ball-milling methods.

In the present paper for the first time a disparate experimental approach for synthesis of $FeAlO_3$ multi-ferroic fibrous material is applied - electrospinning. Basic principles of the electro-hydrodynamics for synthesis of solid state $FeAlO_3$ multi-ferroic fibrous material are applied and discussed.

2. Experimental

A water soluble polymer polyethileneoxide – PEO, aluminium acetate - $AlOH(CH_3COO)_2$ and ferric nitrate - $Fe(NO_3)_3.9H_2O$ are chosen for the preparation of the spinning solution. All chemicals (Fluka - P.A.) are used without further purification. The solutions are prepared with distilled deionized water (σ < 0,5 μs/cm) from considerations of both purity and electrical conductivity. PEO with an average M_v = 800 000 (800 KD) as determined in distilled water at 30°C, using an Ubbelohde viscometer, by the equation $[\eta] = 1.25x10^{-4}\ Mv^{0.78}$ is used as high molecular polymer [6]. The starting polymer solution is prepared as 8,0 wt% aqueous PEO. The metal precursors $AlOH(CH_3COO)_2$ and $Fe(NO_3)_3.9H_2O$ are dissolved separately in distilled water, both salts being in equal concentrations. Afterwards, two $AlOH(CH_3COO)_2/Fe(NO_3)_3$ mixed solutions are prepared depending on their concentration dissolved in the water. A homogeneous translucent liquid is obtained if the starting solutions are diluted. Besides, an opaque solution – suspension, is produced from the supersaturated mixed solutions of both salts. Thus, two different blend solutions are prepared via drop wise addition of both laboratory prepared water solutions $AlOH(CH_3COO)_2/Fe(NO_3)_3$ to the corresponding, gently stirred PEO solution. The weight ratio of the blend spinning solutions used displaying optimal visco-elastic properties under electrospinning conditions is 1:1.

The electrospinning process is carried out under carefully controlled conditions. The blend solution is passed through a syringe with a stainless steel needle at the tip. The needle has been electrified using a high-voltage DC supply with an applied voltage of 15-20 kV. A grounded collector of thick aluminum foil served as a counter electrode. The solution has been pumped continuously using a compressed nitrogen and precise valve at an approximately flow rates of 0,1 mL/h. The applied field strength between the capillary tip and counter electrode was of the order of 1,0 kV/cm. Pieces of fused quartz glass arranged over the collector, served as substrates. Brownish colored mat from as-spun hybrid fibers of polymer/metal salts well adhered to these substrates are collected.

In order to obtain PEO free ceramic fibers the hybrid non-woven mats obtained from both blends solutions are further placed in a quartz tubular furnace and thermally modified via specially developed two-step calcination procedure. The latter is performed non-isothermally at 120°C - 360°C for water and PEO removal, in the presence of low pressure oil free compressed air, providing conditions for the efficient removal of the emanating gases from the heated space and controlled PEO pyrolysis [7, 8]. The samples are subjected further to calcination in the presence of low pressure oil free compressed air at 800°C maintaining a constant heating rate of the order of 1,0°C/min.

Scanning electron microscope Jeol SEM T-200 is used for studying the morphology of the thermally processed samples. The samples have been previously coated under high vacuum conditions with carbon layers. Two electron-optical modes of imaging are applied – SEI or BEI-compo for visualization of the samples in morphology and composition contrast respectively. The phase composition of the thermally processed mats is verified by conventional wide angle X-ray diffraction – WAXRD, by means of APD – 15 Philips X-ray diffractometer.

3. Results and discussion

The results obtained demonstrate that applying the electrospinning method and subsequently the two step thermal procedure for calcination of the as spun hybrid fibrous material a non-woven mats could be obtained. Figures 1 and 2 present the electron-optical micrographs of the inorganic fibres fabricated form both precursors. The morphology contrast of the secondary electron image – SEI, reveals the microstructure of the fibrous materials obtained (figures 1a and 2a). It is clearly seen from these micrographs that in both cases the spun non-woven mat is composed from long fibres with nearly homogeneous shape. The morphology of both fibre mats are similar but the fibres cross-section size differs slightly, that spun from suspension being 20% larger as compared to the fibres spun from homogeneous blend solution. Looking at the backscattered electron image - BEI-compo micrographs (figures 1b and 2b), one can see easily that in both cases the chemical composition of the fibres is uniform at least at the SEM magnifications used in this study.

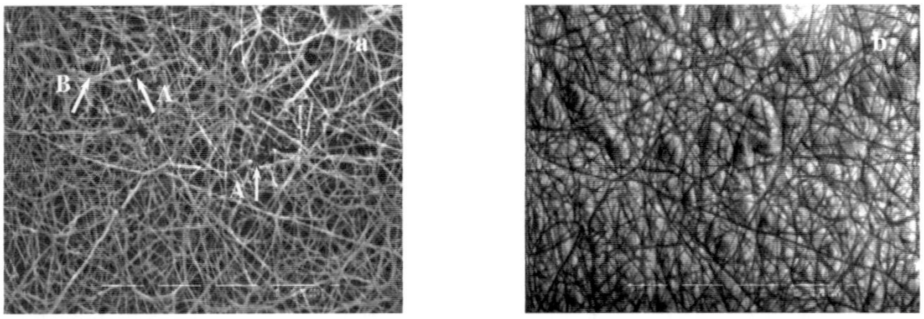

Figure1. SEI – a and b – BEI-compo micrographs of fibres spun from homogeneous blend solution; bar length 10 μm

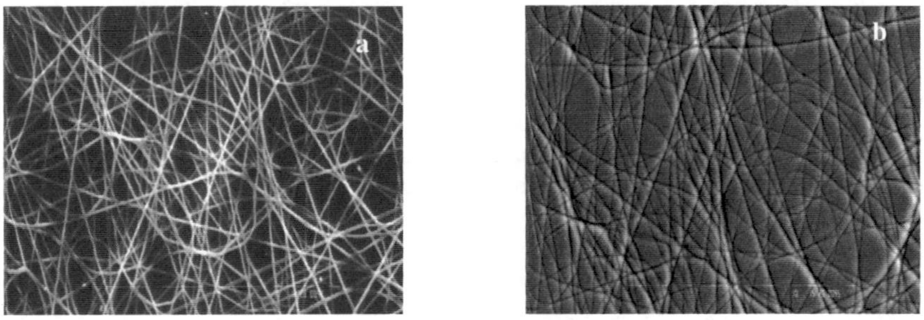

Figure 2. SEI – a and b – BEI-compo micrographs of fibres spun from suspension blend solution; bar length 10 μm

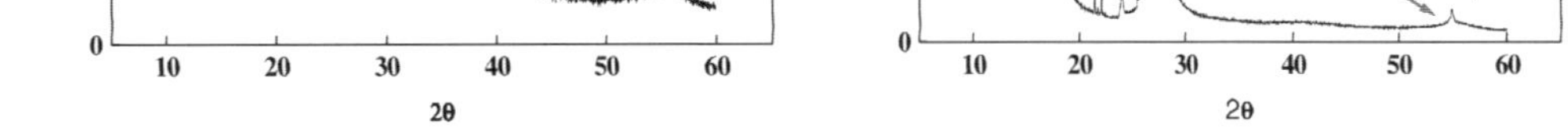

Figure 3. XRD spectra of the non-woven mats spun from homogeneous – a, or b – suspension blend solution

Figure 3 shows the XRD spectra obtained from both kinds of samples. As seen the most intensive peak at 2θ=26,64° as well as this at 2θ=54,87° belong to the quartz material used as substrate. The second intensive peak in both spectra and at 2θ=24,02° is very sharp and belongs to the $FeAl_2O_4$ phase and that at 2θ=18,83° - to $FeAlO_3$ phase. Only these phases are identified among the huge number of Fe-Al-O compounds known from the powder diffraction [8]. Besides, no traces from the very stable hematite α-Fe_2O_3 or alumina α-Al_2O_3 are dissolved in these XRD spectra which usually accompany the synthesis of $FeAlO_3$ with other methods as for instance the solid state reactions [5]. The results from XRD show that in both cases the fibrous material spun at optimal electrospinning conditions and thereafter calcinated during the two step thermal procedure is the multi-ferroic with practically same composition, e.g. – mixture of $FeAl_2O_4$ with small amount of $FeAlO_3$ phase.

It should be noted here that rarely some defects along the inorganic fibres spun from homogeneous blend solution are observed: broken – A, or fused B fibres (figure 1a). In some experiments a formation of different fibres defects spindles, knots or drops accompany the spinning process. This could be avoided selecting carefully controlled electrospinning conditions.

4. Conclusion

The present study demonstrates for the first time that applying the elctrospinning method it is possible to prepare new class material – multiferroic fibrous non-woven mat from basis material of Fe - Al -O. This new material consists of long fine fibres of $FeAl_2O_4$ mixed with $FeAlO_3$ their cross-section diameter being ≤1 μm. Two different blend solutions for electrospinning are used and the fibres prepared are similar in size and shape. Carefully controlled electrospinning conditions are selected in order to fabricate homogeneous and defectless non-woven mats. Further investigations with modified blend solutions for preparation of $FeAlO_3$ fibres only are in progress.

Acknowledgements

This study is supported by WBI – Belgium in the framework of the joint collaboration research programme with BAS, Bulgaria. The authors are grateful to Acad. Al. Petrov – Institute of Solid State Physics, BAS for the high-voltage DC supply grant. Many thanks also to Prof. D. Kovacheva – Institute of General and Inorganic Chemistry as well as to Prof. Dr. Al. Karamanov – Institute of Physical Chemistry, BAS for recording of XRD spectra and providing the thermal treatment of the as spun non woven mats correspondingly.

References

[1] Fiebig M, Lottermoser T, Frohlich D, Goltsev A V and Pisarev R V 2002 *Nature* **419** 818
[2] Eerenstein W, Nathur N D and Scott J F 2006 *Nature* **442** 759
[3] Schmid H 1994 *Ferroelectrics* **162** 317
[4] Devaux X, Rousset A, Broto J M, Rakoto H and Askenazy S 1990 *J. Mat. Science Letters* **9** 371
[5] Bhushan B, Mikherjee S, Bassumalick A, Bandopadhyay S K and Das D J 2008 *J. Hyperfine Interactions* **187** 1187
[6] Brandrup J and Immergut E H, *Polymer Handbook,* 3-rd edition, John Wiley & Sons, New York, 1989, Ch. 7
[7] Neumann E and Nadeau H 1963 *Anal. Chem.* **35**, 1454
[8] Starbova K, Petrov D, Starbov N and Lovchinov V 2012 *Ceram. International* **38** 4645
[9] PDF 34-192

Journal of Physics: Conference Series **398** (2012) 012051 doi:10.1088/1742-6596/398/1/012051

Synthesis of supported CeO_2 nanofibers via electrospinning

K Starbova[1], D Nihtianova[2], D Petrov[3], N Starbov[1] and V Lovchinov[1]
[1] Institute of Solid State Physics, Bulgarian Academy of Science, Sofia
[2] Institute of Mineralogy and Crystallography, Institute of General and Inorganic Chemistry, Bulgarian Academy of Science, Sofia
[3] Chair of Physical Chemistry, University of Plovdiv „Paisii Hilendarsky"

E-mail: ilka_05@yahoo.com

Abstract. Fibrous CeO_2 non-woven mats are synthesized via electrospinning. Homogeneous and stable aqueous polymer/cerium acetate blend spinning solutions are used. It is shown how the parameters of the solution, electrospinning, thermal post-processing can be successfully combined thus providing the synthesis of fibrous webs with mean diameters in the nanosized range. The morphologies of the samples are recorded under SEM and HRTEM while SAED is applied for studying their phase composition. The results obtained pave the way for the development of functional immobilized and self-supporting electrospun ceria ceramic materials.

1. Introduction

Cerium oxide could be reasonably classified as a strategic solid state material since it is extensively employed as a catalyst or as a promoter for supported metal or metal-oxide catalysts [1,2]. The so called structural promotion effect is usually attributed to the ability of the lattice defects of various crystalline cerium oxides to act as catalytically active sites whereas its textural promotion effect relates to excellent CeO_2 thermal and mechanical resistance [3]. Besides, non-stoichiometric cerium oxides, CeO_{2-x} exhibit greater catalytic activity than the stoichiometric one [4] the same behavior already been proved for oxygen-deficient crystalline phases of TiO_2 [5].

Similarly to all materials of applied interest, the contemporary progress in the field of advanced nanostructured CeO_2 ceramics is continuously stimulated by the development of novel synthesis techniques. In the last decade a revived preparation route, the electrospinning, is successfully applied for design of meso- micro- and nanostructured ceramic fibrous materials [6, 7]. The renewed interest to that relatively inexpensive, simple and exceptionally versatile fabrication technique could be primarily attributed to the synthesis of a broad set of low dimensional continuous fibres with high length-to-diameter ratio, specific surface area, and porosity and possibilities for their alignment and spooling [8 - 10]. Moreover, recently the electrospinning is combined with chemical and physical modification techniques, thus remarkably expanding the wealth of engineered fibrous materials over composites and ceramics [11]. Another peculiarity of the electrospun fibres is their relatively easy preparation in the form of supported or free-standing membranes and nonwoven or woven mats. Nowadays, the trend for synthesis of fibrous ceramics via electrospinning having novel chemical and phase composition and amazing morphologies is continuing. The overview of the opened literature shows, however, that the results on CeO_2 are on a small scale compared to electrospun nonwoven mats of other ceramics. For example using polyvinyl alcohol (PVA) as assisting polymer and $Ce(NO_3)_3$ ceramic precursors, CeO_2 nonwoven mat comprised of defect free fibres having diameters in the range 50-100 nm is synthesized [12]. The results presented are related, however, to both one and only polymer/ceramic precursor blend solution and electrospinning parameters. Further, an attempt for synthesis of CeO_2 is made using polyvinylbutiral (PVB) as assisting polymer [13]. The statement for hybrid mats synthesis built up of fibres of 200 to 500 nm diameters is supported, however, by an electron image of a couple CeO_2 fibres of high material density and low porosity. The specific surface area of those samples is 2,4 m^2/g is then not a surprising value. Several years later, using the same polymer and cerium precursor, PVB and $Ce(NO_3)_3$, hybrid as spun and crystalline CeO_2 nanofibers are synthesized [14]. It is established that by increasing the amount of $Ce(NO_3)_3$ in the spinning solution the average diameter of CeO_2 nanofibres increases while the opposite trend is observed on rising the calcination temperature. Since the micrographs of the final mats display small areas of poor fibre quality, the application of a complete set of imaging and analytical techniques is barely reasonable.

Summarizing the results discussed above it should be concluded, that PVA or PVB and $Ce(NO_3)_3$ could be applied as assisting polymers and cerium precursor in a narrow range of weigh ratios for preparation of CeO_2 nonwoven mats with low fibre yield and unsatisfactory fibre morphology.

In searching the basic rules governing the preparing of optimal spinning solutions, we continue to check a strategy with proven efficiency for synthesis of fibrous mats of other important nanostructured ceramics [15]. Most generally, the essence of the applied concept consists in using water soluble metal organic ceramic precursors which are able to preserve the viscoelastic properties of the assisting polymer in the blend solution in wide range of polymer/organic precursor weight ratios. On the basis of that concept we used cerium acetate $Ce(CH_3COO)_3$ and high molecular polymer. The main goal of the present study is to follow the opportunity for high yield synthesis of fibrous supported and free standing CeO_2 membranes via electrospinning for catalytic applications.

2. Experimental

The blend spinning solutions were prepared via gentle mixing aqueous solutions of 8,0 wt% polyethylene oxide (PEO) (M_v = 800 000) and saturated concentration of $Ce(CH_3COO)_3$ in weight ratios between 3:1 and 1:10. In that interval the blend solutions exhibited viscoelastic properties under electrospinning conditions. The applied field strength (AFS) between the capillary tip and counter electrode was kept 1,0 kV/cm while the flow rates were of the order of 0,02 - 0,1 mL/h. Pieces of soda lime glass and aluminium foil arranged over the collector, served as substrates. PEO free calcinated fibres are obtained via two-step non-isothermal procedure in a quartz tubular furnace under conditions of controlled PEO pyrolysis [16, 17]. Depending on the efficiency of emanating gases removal, self-supporting or immobilized fibrous membranes were obtained. The samples were imaged and recorded under scanning electron microscope Jeol T 200 and HRTEM Jeol 2100. The later was used for verifying the phase composition via SAED. The free-standing and supported membranes display no morphological and phase composition differences.

3. Results and discussions

In the present study a wide scale of PEO/$Ce(CH_3COO)_3$ weight combinations were prepared and tested under electrospinning conditions at a constant AFS of 1,0 kV/cm. In all cases dense nonwoven mats were accumulated on the substrate surfaces and the remaining parts of the collector. Further, the as spun samples were subjected to two stage thermal procedure aiming efficient polymer removal and calcination. Thus, supported and self-supported crystalline fibrous CeO_2 membranes are prepared.

Figure1 shows low and high magnification scanning electron micrographs of PEO free fibres. These were obtained from aqueous blend solution having PEO to $Ce(CH_3COO)_3$ weight ratio 1: 1 (a) and 1:10 (b). A synthesis of continuous defect free fibres in the first case is clearly seen. In addition, a pliability that is typical for pure PEO fibres is preserved thus evidencing polymer memory effect. At tenfold increase of $Ce(CH_3COO)_3$ weight concentration in the spinning solution, keeping the same electrospinning and thermal conditions, a dense fibre web is also synthesized as shown by the

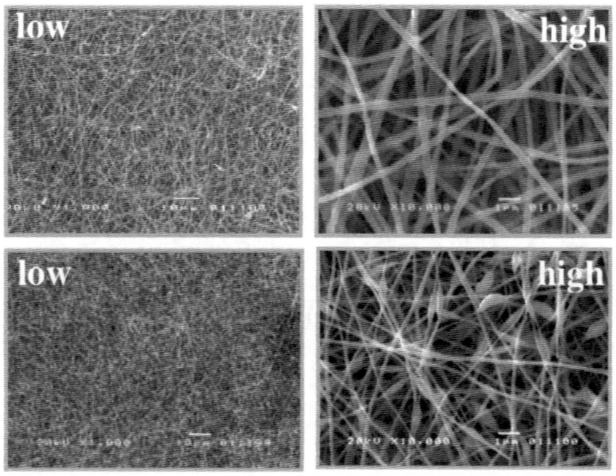

Figure 1. Scanning electron micrographs of PEO free calcinated fibres at low and high magnification: PEO / $Ce(CH_3COO)_3$ weight ratio 1:1 (a) and 1:10 (b); AFS 1,0 kV/cm

electron micrographs on figure 1b. Obviously, the increase of ceramic precursor weight part within the blend spinning solution is accompanied by a significant decrease of the mean fibre diameter. Thus, the results obtained demonstrate a trend opposite to that already observed earlier using PVB and $Ce(NO_3)_3$ based spinning solution [14]. Also, the electron micrographs on figure 1b shows that in addition to fibre mean diameter decrease, the high ceramic precursor concentration favours the formation bead defects. These experimental conditions represent the upper limit for the synthesis of defect free fibres, provided the concentration of PEO in the spinning solution is kept 8,0 wt%.

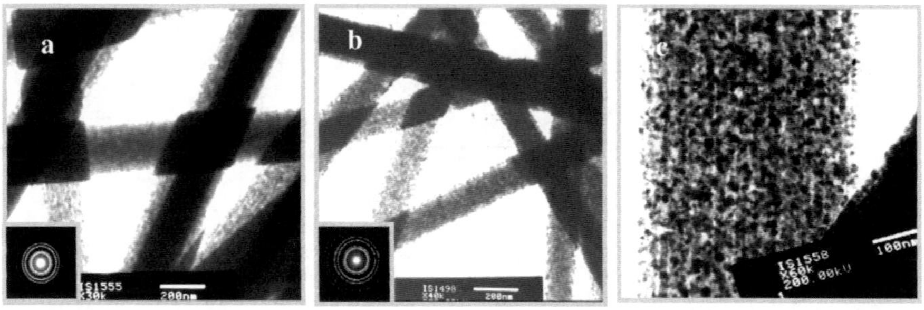

Figure 2 Transmission electron micrographs and SAED patterns of PEO free calcinated Ce-O fibres; AFS 1,0 kV/cm; PEO/$Ce(CH_3COO)_3$ weight ratio: 1:1 (a) and 1:4 (b); (c) magnified single fibre of (a).

Figure 2 presents transmission electron micrographs and corresponding SAED spectra of PEO free, calcinated at 500°C of samples electrospun from spinning solutions differing fourfold in PEO / $Ce(CH_3COO)_3$ solution weight ratio. Webs comprised of deffect free fibres built up of individual nanosized grains are seen on both micrographs. In those cases the trend of fiber mean diameter reduction from 220 nm to 140 nm with increasing the ceramic precursor concentration in the blend solution is also seen. Further, it was established that CeO_{2-x}: Cubic, a = 11.700Å, SG Ia3 [18a] , CeO_2 : Hexagonal, a = 8.360Å, c = 10.420Å, SG P6/mmm [18b], Ce_2O_3 : Rhomboedral (Hexagonal), a = 3.981 Å , c=6,063 Å , SG P321 [189c] and Ce_7O_{12} :Rhombohedral (Hexagonal) a = 10.370Å, c = 9.670Å, SG R [18d] are the basic phase constituents of the calcinated fibers as revealed by the diffraction analysis of the recorded polycrystalline SAED patters. Therefore, thermally stimulated either complete or partial reduction of the as spun hybrid mats takes place resulting in the evolution of crystalline grains of CeO_2, CeO_{2-x}, and traces of cerium oxides with lower oxygen content. Besides, the electrospun fibers are highly porous (figure 2c) and could supply nucleation sites for further loading of nanoclusters from various catalytically active elements or compounds. On this basis it is hard to avoid the conclusion that self-supporting or immobilized fibrous membranes of cerium oxides synthesized via elctrospinning could be successfully used as efficient catalyst supports. Moreover, similarly to the nanoparticles, the opportunity for nucleation and growth of nanosized grains having non-stoichiometric phase composition is a substantial prerequisite for an enhanced catalytic efficiency [4].

4. Conclusions

A simple experimental approach for high yield synthesis of immobilized and self-supporting electrospun cerium oxides based nonwoven mats is proposed. It relays on the preparation of reliable PEO/cerium three acetate blend spinning solutions. It is shown how the latter can be reproducible electrospun within a broad range of polymer/ceramic precursor weight ratios under same electrospinning conditions. The synthesized immobilized and self-supporting electrospun membranes refer to nanostructured crystalline fibrous materials having a significant porosity. These are very promising parameters for their further applications in the field of applied catalysis.

Acknowledgements
The authors are grateful to Prof. D Kovacheva – Institute of General and Inorganic Chemistry, Bulgarian Academy of Sciences for recording of XRD spectra.

Refferences

[1] Trovarelli A, Leitenburg C, Boaro M and Dolcetti C 1999 *Catal.Today* **50** 353
[2] Roggenbuck J, Schaeffer H, Tsoncheva and Minchev C 2007 *Microporous Mesoporous Mater* **101** 335
[3] Alves A, Berutti, Clemens F, Graule T and Bergmann C 2009 *Rev.Adv.Mater.Sci.* **21** 200
[4] Tschope A., Liu W, Stephanopoulos M and Ying J 1995 *J. Catal.* **157** 42
[5] Kobasa I et al 2008 *US Patent* 146441(A1)
[6] Reneker D and Chun I 1996 *Nanotechnology* **7** 216
[7] Li D and Xia Y 2004 *Advanced Materials* 16 1151
[8] Sigmund W, Yuh J, Park H, Maneeratana , Pyrgiotakis G, Daga A, Taylor K and Nino J 2006 *J. Am. Ceram. Soc.* **89** 395
[9] Formo E, Camargo P, Lim B, Jiang M and Xia Y 2009 *Chem. Phys. Lett.* **476** 56
[10] Formo E, Lee E, Campbell D and Xia Y 2008 *Nano Lett.* **8** 668
[11] Dai Y, Liu W, Formo E, Sun Y and Xia Y 2011 *Polym. Adv. Technol.* **22** 326
[12] Yang X., Shao C, Liu Y, Mu R and Guan H 2005 *Thin Solid Films* 478 228
[13] Azad A, Matthews T and Swary J 2005 *Materials Science and Engineering B* **123** 252
[14] Hwang A, Park J and Kang Y 2011 *Bull. Korean Chem. Soc.* **32** (9) 3338
[15] Starbova K, Petrov D, Starbov N and Lovchinov V 2012 *Ceram.International* **36** 4645
[16] Neumann E and Nadeau H, 1963 *Anal. Chem* 35 1454
[17] Starbova K, Krumov E, Starbov N, Spasova M and Rashkov I 2006 *Nanoscience & Nanotechnology* 6 Eds. E. Balabanova and I. Dragieva Heron Press Sofia p 237
[18] a) PDF 49 – 1415; b) PDF 44 – 1001; c) PDF 44 – 1086; d) PDF 71 – 0567

New yellow-emitting phosphorescent cyclometalated iridium(III) complex

P Ivanov [1,2,3], R Tomova[1], P Petrova[1], S Stanimirov[2], I Petkov[2]
[1]Institute of Optical Materials and Technolologies "Acad. J. Malinowski", Bulgarian Academy of Sciences, Acad. G. Bonchev str. bl. 109, 1113 Sofia, Bulgaria
[2]Laboratory of Organic Photochemistry, Faculty of Chemistry and Pharmacy, University of Sofia, 1 James Baurchier blvd., 1164 Sofia,

E-mail: ivanov.petar@mail.bg

Abstract. We have synthesized a new yellow iridium complex Iridium(III) bis[2-phenylbenzothiazolato-N,$C^{2'}$]-(1-phenylicosane-1,3-dionate) $(bt)_2Ir(bsm)$, based on the benzothiazole derivative. The synthesized molecule was identified by ^1H NMR and elemental analysis. The UV-Visible absorption and photoluminescence (PL) spectra of $(bt)Ir_2(bsm)$ in CH_2Cl_2 solution were found at 273 nm and 559 nm, respectively. The complex was used as a dopant into a hole-transporting layer (HTL) in a multilayered organic light emitting device (OLED) structure: ITO/doped-HTL/EL/ETL/M. ITO was a transparent anode of In_2O_3:SnO_2, M- a metallic Al cathode, HTL- 4,4'-bis(9H-carbazol-9-yl)biphenyl (CBP) incorporated in poly(N-vinylcarbazole) (PVK) matrix, EL- electroluminescent layer of bis(8-hydroxy-2-methylquinoline)-(4-phenylphenoxy)aluminum (BAlq) and ETL- electron-transporting layer of tris(8-hydroxyquinolinato)aluminum (Alq_3). The electroluminescent (EL) spectra of OLEDs were basically the sum of the emissions of BAlq at 496 nm and the emission of $(bt)_2Ir(bsm)$ at 559 nm. With increasing $(bt)_2Ir(bsm)$ concentration, the relative electroluminescent intensity of greenish-blue emission (at 496 nm) decreased, while the yellow (at 559 nm) – increased and CIE coordinates of the device shifted from (0.21, 0.33) at 0 wt % to (0.40, 0.48) at 8 wt % of the dopant. It was found that OLED with 0.5 wt % $(bt)_2Ir(bsm)$ had the best performance and stable color chromaticity at various voltages.

1. Introduction

Phosphorescent heavy metal complexes play a very important role in organic light-emitting diodes (OLEDs), because they can fully utilize both singlet and triplet excitons though the strong spin-orbital coupling caused by the heavy metal ions in complexes. Since Forrest and Thompson's group reported the first example of electrophosphorescence at room temperature [1], phosphorescent OLED materials and devices have been intensively studied due to their high external quantum efficiencies (EQEs) [2,3]. It has been widely accepted that the strong spin-orbit coupling induced by a heavy metal ion, such as Ir, Pt and Os promotes an efficient intersystem crossing (ISC) between the singlet and the triplet excited state manifold. Therefore, both singlet and triplet excitons can be harnessed and the strong electroluminescence with an internal efficiency theoretically approaching to 100% can be achieved. Of these phosphorescent materials, iridium complexes have been regarded as the most appropriate emitters because of their relative short lifetime and high quantum efficiency. The electronic transitions responsible for the phosphorescence in these complexes has been assigned to the mixture of metal-to-ligand charge transfer (^{1}MLCT, ^{3}MLCT) and ligand-centred (LC) $^3(\pi$-$\pi^*)$ lowest exited states [4, 5, 6, 7]. The holes and electrons in OLED are injected into opposite surfaces of a planar multilayer organic thin film. The holes and electrons migrate through the film, to a material interface, where they recombine to form radiative excited states or excitons. This electrically generated exciton can be either a singlet or a triplet. Both theoretical predictions and experimental measurements give a singlet/triplet ratio for these excitons of 1 to 3 [8]. The fluorescent materials typically used for OLEDs fabrication do not give detectable triplet emission (i.e. phosphorescence), nor is there evidence for significant intersystem crossing between the triplet and singlet manifolds at room temperature. The singlet/triplet ratio thus implies a limitation of 25% for the internal quantum efficiency for OLEDs based on the fluorescence. Baldo [1] and Thompson [9] have shown by doping OLEDs with heavy metal phosphors, that the singlet-triplet limitation can be eliminated. The excited states generated by electron-hole recombination are trapped at the phosphor, where strong spin-orbit coupling leads to singlet-triplet state mixing and, hence, efficient phosphorescent emission at room temperature. Both singlet and triplet excited states can be trapped at the phosphor. OLEDs with these heavy metal complexes are the most efficient OLEDs prepared to date, with internal quantum efficiencies exceeding 75% (photons/electrons) (>15% external efficiency) [10]. Furthermore, OLEDs have been prepared with $(C\text{^}N)_2Ir(LX)$ phosphor dopants, giving efficient green, yellow or red emission. The external quantum efficiencies for these devices vary from 5% to nearly 20%.

In this paper we have synthesized a new benzothiazole-based iridium complex, Iridium(III)bis[2-phenylbenzothiazolato-N,$C^{2'}$]-(1-phenylicosane-1,3-dionate), $(bt)_2Ir(bsm)$, as a solution processable triplet emitter for organic electrophosphorescent devices. Poly(N-vinylcarbazole) (PVK) was used as the polymeric host for the synthesized iridium complex and the optical and electroluminescence properties of the composite and OLEDs were studied.

2. Experimental

2.1. Instruments and measurements

^{1}H-NMR (600 MHz) spectra were recorded on a Bruker Avance II+ 600 at room temperature. The ^{1}H-NMR spectra were referenced to internal TMS. The elemental analyses were made on Elementar Vario EL III. The UV and fluorescence spectra of the complex in solution were recorded on a Thermo Spectronic Unicam 500 spectrophotometer and a Varian Cary Eclipse fluorescence spectrophotometer, respectively. The electroluminescent spectra (EL) were obtained with an Ocean Optics HR2000+ spectrometer. The current-voltage (I-V) curves were measured by programmable with Labview power supply. The luminescence (L) was measured in continuous DC mode and the light output was detected using a calibrated Hamamatsu silicon photodiode S2281-01. The electroluminescent and external quantum efficiency (η_L, η_{EQE}) were calculated by equations (1) and (2) and used for quantifying the properties of the OLEDs.

$$\eta_L = L / I \text{, cd/A} \tag{1}$$

$$\eta_{EQE} = (I\phi / I).100 \text{, \%} \tag{2}$$

where L is the luminescence (in cd/m^2), I and $I\phi$ are the densities of current and photocurrent (in A/m^2).

2.2. Synthesis of iridium complex

2.2.1. Synthesis of bis[di(2-phenylbenzothiazole)-μ-chloroiridium(III].Cyclometalated Ir(III) μ-chloro-bridged dimer of formula $(bt)_2Ir(\mu\text{-}Cl)_2Ir(bt)_2$ were synthesized by the method reported by Nonoyama [11]. 2-Phenylbenzothiazole (for synthesis see [12]) (2.2 equiv), iridium trichlorid hydrate (1 equiv),

2-ethoxyethanol : water = 3 : 1 were heated and stirred at 110 °C for 20 h under argon. After cooling to room temperature, the crude solution was filtrated to give an orange-yellow precipitate. Then the precipitate was washed with ethanol and hexane for several times. After dried of room temperature, a cyclometalated $(bt)_2Ir(\mu\text{-}Cl)_2Ir(bt)_2$ dimer was obtained.

2.2.2. Synthesis of iridium(III)bis[2-phenylbenzothiazolato-N,C$^{2'}$]-(1-phenylicosane-1,3-dionate). Dimer with 2.5 equiv of 1-phenylicosane-1,3-dione and 5 equiv Na_2CO_3 were heated and stirred at 110 °C in the presence of 2-ethoxyethanol under argon for 20 h (scheme 1). After cooling to room temperature, the crude product was filtrated to give an orange-yellow precipitate. Then the precipitate was washed with water, ethanol and hexane for several times. Finally, the crude product was purified using liquid chromatography (silica, dichloromethane) and dried. Yeld: 80%. ^{1}H NMR (600 MHz, $CDCl_3$), δ[ppm]: 1.06-1.09 (m, 3H), 1.19 (bs, 32H), 5.70 (s, 1H), 6.39-6.41 (m, 1H), 6.45 (d, 1H, J=7.2Hz), 6.57-6.60 (m, 2H), 6.79-6.83 (m, 2H), 7.09-7.12 (m, 1H), 7.17 (s, 1H), 7.26-7.33 (m, 5H), 7.58-7.63 (m, 4H), 7.77-7.79 (m, 2H), 7.96 (d, 1H, J=8.4Hz), 8.01-8.03 (m, 1H); $C_{52}H_{57}N_2O_2S_2Ir$. Elemental analysis: Calc. C 62.56%, H 5.75%, N 2.81%. Found: C 62.52%,H 5.76%, N 2.80%.

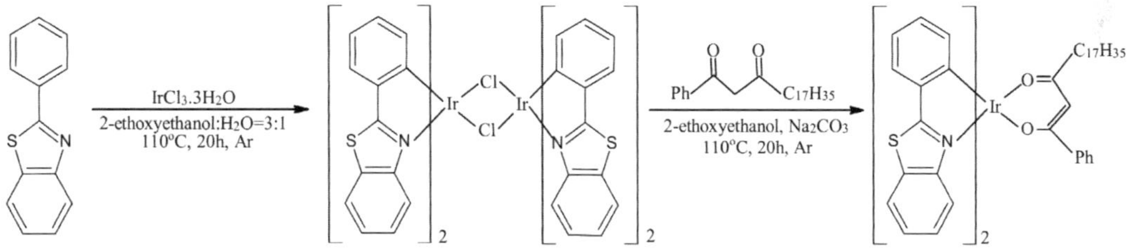

Scheme 1.

2.3. OLED fabrication

We investigated $(bt)_2Ir(bsm)$ as a dopant into a hole-transporting layer (HTL) in multilayered organic light emitting device (OLED) structure: ITO/doped-$HTL_{(30\ nm)}$/$EL_{(40\ nm)}$/$ETL_{(35\ nm)}$/$M_{(90\ nm)}$. ITO was a transparent anode of In_2O_3:SnO_2, M - a metallic Al cathode, HTL - 4,4'-Bis(9H-carbazol-9-yl)biphenyl (CBP) incorporated in poly(N-vinylcarbazole) (PVK) matrix, EL - electroluminescent layer of Bis(8-hydroxy-2-methylquinoline)-(4- phenylphenoxy) aluminum (BAlq) and ETL - electron-transporting layer of Tris (8-hydroxyquinolinato) Aluminum (Alq_3) or Bis[2-(2-benzothiazoly)phenolato]zinc(II) ($Zn(btz)_2$).

Devices with area 1cm^2 were prepared on commercial polyethylene terephtalate (PET) substrates coated with ITO (40Ω/sq). The layer (30nm) of PVK:$CBP_{10\%}$(10 wt% relatively to PVK) + $(bt)_2Ir(bsm)_{(X\%)}$ composite films were prepared by spin-coating from 0.75% solution in dichloroethane at 2000 rpm. BAlq, Alq_3 (or $Zn(btz)_2$) and metal cathode were deposited by thermal evaporation in vacuum better than 10^{-4} Pa at rates 2-5 Å/s.

3. Result and discussion

3.1. UV-VIS absorption and photoluminescence

The plot in figure 1 presents the UV-VIS absorption spectrum of $(bt)_2Ir(bsm)$ and photoluminescence spectra (PL) of: $(bt)_2Ir(bsm)$, PVK and CBP in dichloromethane. The absorption spectrum of the complex (curve 1) is a typical for that kind of complexes. It is not well resolved and shows complicate structure because of overlapping of various bands. It can be considered as composed of two parts. The region below 360 nm contains bands for electronic transitions, which lead to excitations in the bt-ligand. Most probably, the band around 330 nm is due to transition to lowest singlet ligand centred (^{1}LC) excited state. This statement is in accordance with published before articles discussing similar complexes. The shoulders above 360 nm are assigned to singlet and triplet metal-to-ligand charge transfer bands (^{1}MLCT, ^{3}MLCT), respectively. Effective spin-orbit coupling (SOC) in this complex leads to mixing of the singlet and triplet states making the band for resonant transition $^1S_0 \rightarrow {}^3$MLCT detectable in absorption spectrum (around 480 nm). The band for ^{1}MLCT transition is centred at 440 nm.

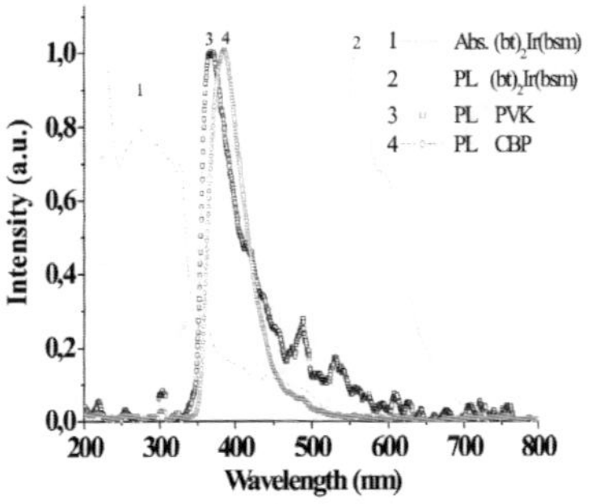

Figure 1. UV-VIS absorption spectrum of $(bt)_2Ir(bsm)$ and PL emission spectra of $(bt)_2Ir(bsm)$, PVK and CBP in dichloromethane.

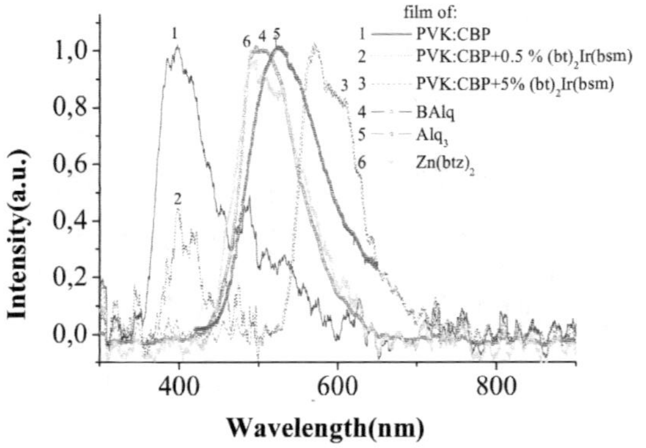

Figure 2. PL spectra of spin coated pure and doped with $(bt)_2Ir(bsm)$ PVK:CBP films and thermal evaporated BAlq, Alq_3 and $Zn(btz)_2$ films.

The complex showed emission in yellow region (table 1) with peak at 559 nm and shoulder around 590 nm (curve 2) while PVK (curve 3) and CBP (curve 4) – in blue region with peaks at 367 and 385 nm respectively. It can be expected that the efficient Förster and Dexter energy transfer from PVK and CBP to $(bt)_2Ir(bsm)$ would be possible since ^{1}MLCT and ^{3}MLCT absorption spectra of the dopant has broader spectral overlap with fluorescence spectra of PVK and CBP.

Table 1. Photophysical data for $(bt)_2Ir(bsm)$ complex in Ar saturated dichloromethane.

Complex	*Absorbtion λ[nm], (lgε)*	*Emission λ_{max} [nm]*	*QY*
$(bt)_2Ir(bsm)$	273(4.68), 298(4.66), 314(4.64), 327(4.63), 359(4.19), 400(3.98), 444(3.88), 485(3.75)	559	0.20

The PL spectra of spin coated pure and doped with $(bt)_2Ir(bsm)$ PVK:CBP films and thermal evaporated BAlq, Alq_3 and $Zn(btz)_2$ are shown in figure 2. It is seen that the blue emission at ~ 395 nm, attributed to PVK:CBP film, reduces significantly with increasing of dopant concentration from 0 to 5 wt %, indicating that energy transfer from PVK and CBP to $(bt)_2Ir(bsm)$ not only take place, but also increases with doping concentration and was completely brought to an end at 1.25 wt %. BAlq and Alq_3 were peaked at 496 nm (curve 4) and 525 nm (curve 5) respectively and $Zn(btz)_2$ demonstrated a peak at 490 nm and shoulder at 529 nm (curve 6).

3.2. Electroluminescence

The electroluminescence spectra (EL) of two types OLED with ETL of Alq_3 and $Zn(btz)_2$ at different concentrations of the dopant are presented in figures 3 and 4. In contrast to the discussed above PL spectra, in the EL spectra any PVK or CBP induced emission near ~400 nm was not observed, irrespective of the dopant concentration for two types of devices. The absence of major PVK related emission features has implied a charge trapping rather than an exciton transfer mechanism. The EL spectra were basically the sum of the emissions of $(bt)_2Ir(bsm)$ (at 559 nm) and BAlq (at 496 nm). With increasing of $(bt)_2Ir(bsm)$ concentration, the relative intensity of the greenish-blue emission (496 nm) decreased, while the yellow (559 nm) - increased and Commission Internationale de L'Eclairage (CIE) coordinates of OLEDs shifted from blue (0.21,0.33) at 0 wt % to yellow (0.40,0.48) at 8 wt % of the dopant for device with ETL of Alq_3 (table 2) and from greenish-blue (0.24,0.43) at 0 wt % to greenish-yellow(0.31,0.45) at 1 wt % of the dopant for device with ETL of $Zn(btz)_2$.

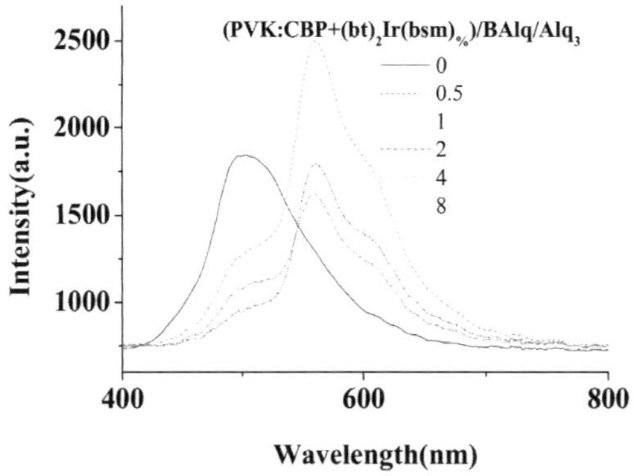

Figure 3. EL spectra of OLED with ETL of Alq_3 at different doping concentrations at DC 16 V.

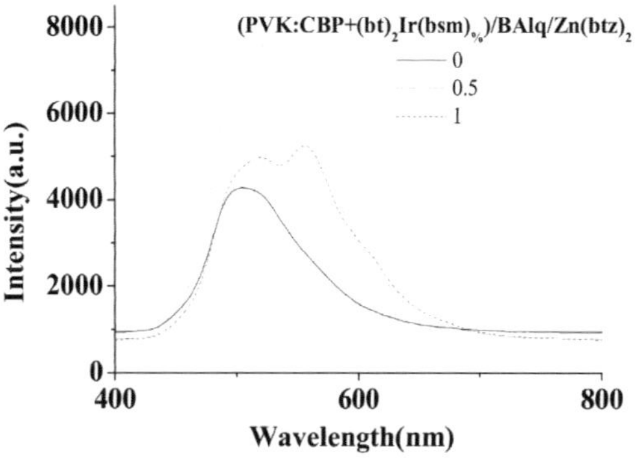

Figure 4. EL spectra of OLED with ETL of $Zb(btz)_2$ at different doping concentrations at DC 16 V.

All devices became visible between 10 - 12 V depending on the doping concentration. The luminescence-voltage characteristics of OLEDs are shown in figures 5 and 6 and results for electroluminescent and external quantum efficiencies are summarized in tables 2 and 3.

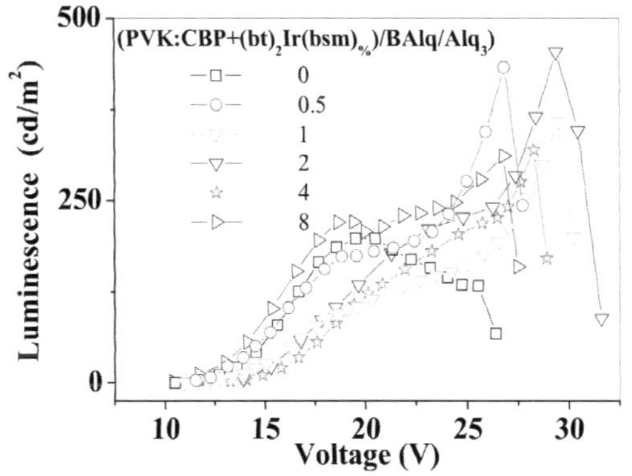

Figure 5. Luminescence-voltage characteristics of device with ETL of BAlq.

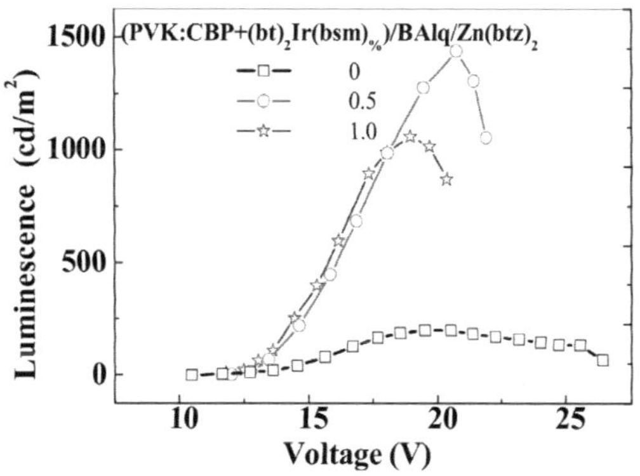

Figure 6. Luminescence-voltage characteristics of device with ETL of $Zn(btz)_2$.

Both type of devices showed maximum external quantum and current efficiency at 0.5 wt % of the dopant. The efficiencies were 2-3 times higher in comparison with undoped devices. The values of these parameters for device with ETL of Alq_3 (0.76 %, 1.46 cd/A) were nearly two times lower than that of the device with ETL of $Zn(btz)_2$ (1.58 %, 3.02 cd/A). We suppose that better performance of the device with ETL of $Zn(btz)_2$ is due to its better electron-transport properties [13].

Table 2. OLED characteristics of device ITO/PVK:CBP+$(bt)_2Ir(bsm)_{X\%}$/$BAlq_{(40nm)}$/$Alq_{3(35nm)}$/Al.

Doping concentrations	λ_{max} *(nm)*	L_{max} *(cd/m²)*	*Current efficiency(cd/A)*	*External QE (%)*	*CIE (x,y)*
0%	501	198	0.73	0.26	(0.21,0.33)
0.5%	558	434	1.46	0.76	(0.31,0.45)
1%	558	360	0.56	0.28	(0.34,0.47)
2%	559	455	0.56	0.29	(0.36,0.48)
4%	561	322	0.93	0.50	(0.41,0.50)
8%	562	313	0.53	0.27	(0.40,0.48)

Table 3. OLED characteristics of device ITO/PVK:CBP+$(bt)_2Ir(bsm)_{X\%}$/$BAlq_{(40nm)}$//$Zn(btz)_{2\ (35\ nm)}$/Al.

Doping concentrations	λ_{max} *(nm)*	L_{max} *(cd/m²)*	*Current efficiency(cd/A)*	*External QE (%)*	*CIE (x,y)*
0%	510	201	0.95	0.49	(0.24,0.43)
0.5%	555	1448	3.02	1.58	(0.28,0.45)
1%	556	1064	2.30	1.20	(0.31,0.45)

4. Conclusion

New yellow iridium complex $(bt)_2Ir(bsm)$ has been synthesizes as a triplet emitter for organic electro-phosphorescent devices. It was found that inserting of $(bt)_2Ir(bsm)$ in hole-transporting layer of electroluminescent devices increased the luminescence, tuned the color of illuminated light in yellow region, and raised more than twice the electroluminescent and external quantum efficiencies of the devices.

References

[1] Baldo M, O'Brien D, You Y, Shoustikou A, Thompson M and Forrest S 1998 *Nature* **395** 234
[2] Holder E, Langeveld B and Schubert U 2005 *Adv. Mater.* **17** 1109
[3] You Y and Park S 2009 *Dalton Trans.* 1267
[4] Colombo M, Brunold T, Riedener T and Gudel H 1994 *Inorg. Chem.* **33** 545
[5] Tamayo A, Alleyne B, Djurovich P, Lamansky S, Tsyba I, Bau N and Thompson M 2003 *J. Am. Chem. Soc.* 125 7377
[6] Lamansky S, Djurovich P, Murphy M, Abdel-Razaq F, Lee H, Adachi C, Burrows P, Forrest S and Thompson M 2001 *J. Am. Chem. Soc.* **123** 4304
[7] Brooks J, Babayan Y, Lamansky S, Djurovich P, Tsyba I, Bau R and Thompson M 2002 *Inorg. Chem.* **41** 3055
[8] Baldo M, O'Brien D, Thompson M and Forrest S 1999 *Phys. Rev.* **60** 14422
[9] Thompson M, Burrows P and Forrest S 1999 *Curr. Opin. Solid State Mater. Sci* **4** 369
[10] Adachi C, Baldo M, Forrest S and Thompson M 2000 *Appl. Phys. Lett.* **78** 1704
[11] Nonoyama M 1974 *Bull. Chem. Soc. Jpn.* **47** 767
[12] Deligeorgiev T, Kaloyanova S, Vasilev A and Vaquero J 2010 *Phosphorus Sulfur* **185** (11) 2292
[13] Yu G, Yin S, Liu Y, Shuai Z and Zhu D, *J. Am. Chem. Soc.* 2003, **125**, 14816–14824

Diffraction efficiency growth of nano-scale holographic recording produced in a corona discharge

I Bodurov[1], T Yovcheva[2], I Vlaeva[3], A Viraneva[2], R Todorov[1], G Spassov[4] and S Sainov[1]

[1]Institute of Optical Materials and Technologies "Acad. Jordan Malinowski", Bulgarian Academy of Sciences, "Acad. G. Bonchev" str., building 109, 1113 Sofia, Bulgaria
[2]Plovdiv University "Paisii Hilendarski", 24 "Tsar Asen" str., 4000 Plovdiv, Bulgaria
[3]University of Food Technologies, 26 "Maritsa" blvd., 4002 Plovdiv, Bulgaria
[4]College of Technical University of Sofia, 31 "Bulgaria" blvd., 2140 Botevgrad, Bulgaria

E-mail: bodurov@uni-plovdiv.bg

Abstract. The nano-scale holographic gratings ware recorded in 29 nm and 56 nm thick As_2S_3 films. The chalcogenide layers were deposited on a transparent chromium electrode with thickness 10 nm, produced on a glass substrate. Both chromium and chalcogenide films were deposited in one vacuum cycle by e-beam and thermal evaporation, respectively. The diode 532 nm diode laser was used as a light source in the present holographic experiments. The total internal reflection arrangement (Stetson-Nassenstein) was used in holographic recordings. The reference beam was totally reflected from the air-As_2S_3 boundary surface by an input glass prism. The object beam was normally incident on the recording medium. The corona charging was performed by a needle fixed at the distance of 1 cm from the holographic recording medium by applying a – 5 kV voltage. The diffraction efficiency increased from 9 to 30 times when the corona discharge was applied during the holographic recording, in comparison to the uncharged recording. The possible reason of the observed effect is discussed on the basis of the Franz-Keldysh effect and Moss rule.

1. Introduction

The holographic grating measuring methods are very useful for the rapidly developing field of nanoscience and nanotechnology. During the past years we have performed a series of experiments connected with holographic recording in nano-sized As_2S_3 films [1-4]. The results of these studies have demonstrated the possibilities of the total internal reflection holographic technique for grating and image recording in As_2S_3 with different nano-scale thicknesses. The main disadvantage of this recording is the low diffraction efficiency value, so the researchers are looking for different methods for resolving this problem. The application of strong electric field seems to be most promising in this direction. Recently, Nastas et al. [5-6] have reported diffraction efficiency enhancement of a holographic grating recorded in 1.0 μm and 0.3 μm thick film using corona discharge. Up to 3 times diffraction efficiency increase has achieved by 8 kV electric field applying [5-6]. Investigations are now being carried out to increase the diffraction efficiency by applying the electric field in nano-scale recording. In a recently published paper [7] spatial frequencies influence of the diffraction efficiency of holographic recording in As_2S_3 films with thicknesses of 29 nm, 15 nm and 10 nm has been presented. The authors found that the optimal thickness for low (310 mm^{-1}) and high (2700 mm^{-1}) spatial frequencies was about 30 nm.

In the present paper we reported the results obtained from the investigation of electric charge influence on the diffraction efficiency of holographic gratings recorded in 29 nm and 56 nm thick chalcogenide As_2S_3 film. This study is continuation of investigations reported earlier in [7].

2. Theoretical background

2.1. Holographic recording in nano-sized thin films

In the holographic scheme the plane object wave normally falls at the recording medium. The second reference wave (the evanescent wave) is created by total internal reflection (TIR).

The TIR of s-polarized plane wave with amplitude a_0 takes place at the prism-air interface in the case when the thickness of the registration medium is $d << \lambda$ (λ is the laser wavelength in the air) [8]. When the incident angle is $\varphi_1 > \varphi_c$ (φ_c is the critical angle) we have

$$\sin\varphi_c = 1/n_1 \tag{1}$$

where n_1 is the refractive index of the glass prism.

The evanescent wave amplitude is

$$a_e = \frac{2a_0 n_1 \cos\varphi_1}{\left(n_1^2 - 1\right)^{\frac{1}{2}}} \exp\left(-\frac{z}{z_0}\right) \tag{2}$$

where the penetration depth is defined by

$$z_0 = \frac{\lambda}{2\pi n_1 \left(\sin^2\varphi_1 - \sin^2\varphi_c\right)^{\frac{1}{2}}} \tag{3}$$

The plane wave amplitude is

$$a_P = a_{0_P} \exp\left(i\frac{2\pi}{\lambda} n_1 z\right) \tag{4}$$

In the case when $\frac{2\pi n_1 z}{\lambda} << 1$, $\frac{z}{z_0} << \left(\sin^2\varphi - \sin^2\varphi_c\right)^{\frac{1}{2}} << 1$, $\exp\left(-\frac{z}{z_0}\right) \approx 1$, the interference term, obtained after simple trigonometric transformations, is

$$I_{ep} = 8\frac{a_0 a_{0_p} \cos\varphi_1}{\left(n_1^2 - 1\right)^{\frac{1}{2}}} \cos^2\left(\frac{\pi n_1 x \sin\varphi_1}{\lambda}\right) \tag{5}$$

The grating step along x-axis is

$$\Lambda = 1/n_1 \sin\varphi_1 \tag{6}$$

2.2. Electric field influence on the diffraction efficiency

The Klein parameter Q [9] is low in the case of holographic recording in very thin photosensitive films. This parameter is used to distinguish thick (volume) from thin (plane) holographic grating and is defined as

$$Q = 2\pi\lambda d / n\Lambda^2 \tag{7}$$

where n is the refractive index of the recording medium; λ is the wavelength of the laser, using in recording; d is the thickness of the film; Λ is the diffraction grating's step.

The calculations for recording with 532 nm wavelength in 60 nm thick As_2S_3 film $(n_1 = 2.41)$ gives $Q \sim 0.5$, when grating step is 400 nm, corresponding to spatial frequency of 2500 mm^{-1}. This defined the grating as very thin. In spite of the low Q value, the coupled-wave theory, developed by H. Kogelnik [10] is accurate in the case of the thin grating diffraction efficiency, illuminated at Bragg – angle condition, when only one diffraction order is observed. This is thoroughly investigated in [11] for the low modulation condition, that is our case of nano-sized holographic recording.

According the coupled wave theory, the diffraction efficiency (η) is

$$\eta = \sin^2 \frac{\pi \Delta n d}{\lambda \cos\theta} \approx \left(\frac{\pi \Delta n d}{\lambda \cos\theta}\right)^2 \tag{8}$$

where Δn is the refractive index change (modulation), θ is the Bragg angle value, in the present case – the angle of incidence.

In order to calculate the refractive index change in corona discharge condition, one has to estimate the additional refractive index change under the applied high electric field F. For this purpose we shall use in the first place the connection between refractive index of semiconductors and energy gap (E_g) [12] and after that 1-D Franz-Keldysh effect [13]. The empirical Moss rule is defined as

$$n^4 E_g = 77 \tag{9}$$

We should mention that for As_2S_3 we have $n = 2.42$ for 532 nm (our measurements) according to [14] $n = 2.41$ and E_g is 2.32 eV [15]. The condition (9) in this case is satisfied very well – the difference is less than 1.5%.After some mathematical manipulation, we have

$$4\frac{\Delta n}{n} = -\frac{\Delta E_g}{E_g} \tag{10}$$

According to 1-D Franz-Keldysh effect, the parallel energy gap shift is [13]

$$-\Delta E_g = \frac{3}{2}\left[\frac{(e\hbar F)^2}{m^*}\right]^{\frac{1}{3}} \tag{11}$$

where m^* is the effective electron mass; e is the electron charge; F is the electric field intensity; $\hbar$ is the reduced Planck constant.

By using relations (8), (10) and (11), for the diffraction efficiency when electric field F is applied we have

$$\eta_F = \frac{9}{64}\frac{\pi^2 d^2 n^2}{\lambda^2 \cos^2\theta E_g^2}\left[\frac{(e\hbar F)^2}{m^*}\right]^{\frac{1}{3}} = 1.39\frac{d^2 n^2}{\lambda^2 \cos^2\theta E_g^2}\left[\frac{(e\hbar F)^2}{m^*}\right]^{\frac{1}{3}} \tag{12}$$

3. Experimental

3.1. Sample preparation

The samples were prepared in one vacuum cycle $(P \sim 10^{-3} Pa)$ used thin film's deposition system Leybold Haraeus A 702 Q. Firstly, thin transparent electrode from chromium with 10 nm thickness was deposited on suitably cleaned glass substrates by electron beam evaporation. The deposition rate was 0.2 nm/s. The transmission coefficient of the electrode in the spectral range 400 nm – 1500 nm was between 70 % and 80 % and the electrical resistivity was $\rho = 2$ mΩ.cm. Subsequently the investigated samples – thin layers from high purity As_2S_3 with thickness 29 nm and 56 nm were deposited on the chromium electrode by thermal evaporation.

Optical transmission and reflection measurement in the spectral range 400 nm – 1500 nm were carried out using UV – VIS – NIR spectrophotometer (Cary 05E, Varian, Australia).

3.2. Holographic recording

The high spatial frequency grating is recorded in the Stetson-Nassenstein regime, using the holographic setup presented in figure 1.

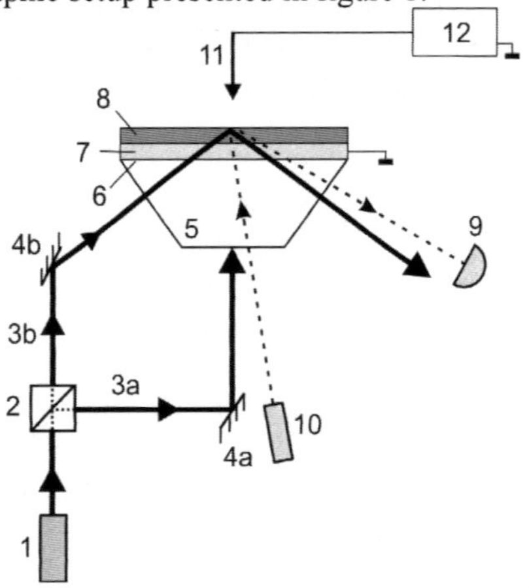

Figure 1. Holographic set-up for total internal reflection grating recording: 1 – recording laser, 532 nm; 2 – beam splitter; 3b – total internal reflecting beam; 3a – normally incident beam; 4a,b – mirrors; 5 – prism; 6 – matching liquid; 7 – chromium electrode; 8 – recording medium; 9 – power meter; 10 – monitoring He-Ne laser, 632.8 nm; 11 – corona electrode (needle); 12 - high voltage supply

The reference beam (3b) is totally reflected from the chalcogenide film – air interface at the 54° incidence angle. The object beam (3a) is incident normally on the chalcogenide film. The total laser beam intensity is 10 mW/cm^2. The grating's step Λ is 0.376 μm and the corresponding spatial frequency is 2660 mm^{-1}. A low intensity He-Ne laser (10) is employed as a read-out beam to monitor the build-up dynamics of the gratings during the holographic recording.

The holographic recording in chalcogenide As_2S_3 films was carried out simultaneously with their charging in a negative corona field. The corona charging was performed by the needle (11) fixed at a distance of 1 cm from the sample, applying a voltage of – 5 kV (12).

4. Results and discussion

The optical constants (refractive index n and extinction coefficient k) and thickness d of the thin films were calculated using transmittance and reflectance measurements and applying the triple method proposed by Konstantinov and Panayotov [16]. This method operated with three spectrophotometric measurements – (T, R_f, R_m) or (T, R_b, R_m), where T is the transmission, R_f, R_b and R_m – the reflections from the front and back side of the BK-7 substrate, and from the film deposited on Si – wafer substrate, respectively. The physically corrected solution for d is isolated from the condition $d = f(\lambda) = const$. Further, for a more precise estimation of refractive index (to an accuracy better than ± 0.05) and extinction coefficient, double (T, R_f) and (T, R_b) methods developed by Abeles and Theye were applied [17, 18].

The refractive indices of the As_2S_3 film with thickness 29 nm and 56 nm are presented in figure 2. The refractive index values are considerably different for the two films with different thicknesses. The above results are in agreement with previous reports, which claim the existence of a critical value for d, below which optical constants are thickness dependent [19]. It is seen that refractive indices of both layers increase after light exposition, which is a typical effect for the arsenic sulphide films. The observed refractive index change of Δn was about 0.15 for $\lambda = 532$ nm and it could be related with the photodarkening phenomena due to formation of localized electronic states in the tails of the condition and valance bands [20].

The diffraction efficiency dependences on the exposure for both 29 nm and 56 nm As_2S_3 films with and without corona charging are illustrated in figure 3.

The maximum value of diffraction efficiency for charged samples with thickness 29 nm is reached at exposure 250 mJ/cm^2, while for theose with thickness 56 nm - at about 550 mJ/cm^2. This is due to

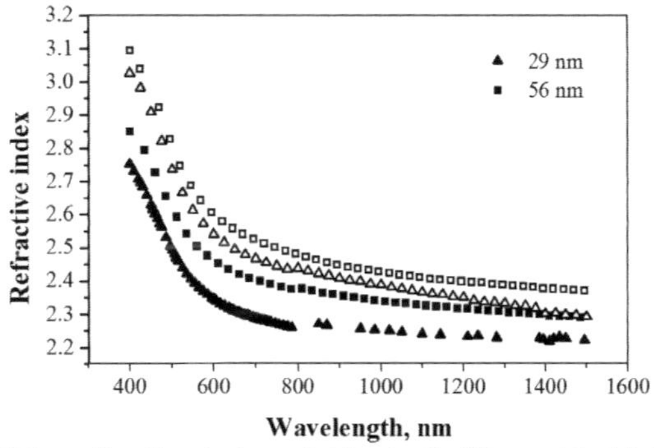

Figure 2. Dispersion of the refractive index of thin As_2S_3 films with thickness 29 nm and 56 nm before (solid symbols) and after illumination (open symbols)

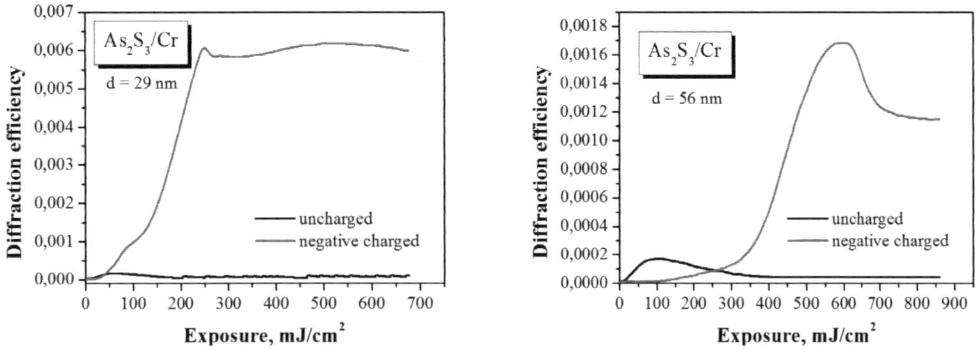

Figure 3. Diffraction efficiency as a function of the exposure for negatively charged As_2S_3 films with different thicknesses 29 nm (left) and 56 nm (right)

the fact that thicker samples require more exposition energy to reach maximum value and then to achieves saturation.

The experimental results clearly demonstrate that the corona charging during the holographic recording increases the diffraction efficiency of about one order of magnitude for 56 nm and 30 times for 29 nm thick As_2S_3 film. The possible reasons for the observed effect could be the following:

1. Additional refractive index modulation due to the Franz-Keldysh effect.
2. Diffusion of chromium oxide ions, having high refractive index, into chalcogenide film.

In our experiments, the average surface potential value measured by the vibrating electrode method is about 70 V. From equation (8) the refractive index change (modulation) without electric field is

$$\Delta n_0 = \sqrt{\eta}\lambda \cos\theta / \pi d \qquad (13)$$

For holographic recording in 29 nm and 56 nm As_2S_3 films Δn_0 are respectively 0.05 and 0.03.

The refractive index change when electric field F is applied is

$$\Delta n_F = \frac{3}{8}\frac{n}{E_g}\left[\frac{(e\hbar F)^2}{m^*}\right]^{\frac{1}{3}} \qquad (14)$$

In our calculations we have $n = 2.42$ for 29 nm thick film; $E_g = 2.32$; $F = 1.3\times10^9$ V.m^{-1} and $F = 2.4\times10^9$ V.m^{-1} for 56 nm and 29 nm respectively. According to Kolomiets [21] $m^* = 7.5m_0$ (m_0 is a free electron mass). The obtained diffraction efficiency ratio is

$$(\eta_F / \eta_0)_d = (\Delta n_F / \Delta n_0)^2 \qquad (15)$$

The calculated value of the ratio $(\eta_F / \eta_0)_d$ for film with thickness 56 nm is about 11, which is in satisfactory agreement with the experimentally obtained value 9. For 29 nm thick film, however, the calculated ratio is nearly the same – 12, when the experiment gives 30. Evidently, there is additional refractive index modulation related with chromium oxide ions diffusion. To elucidate this effect, the Auger Electron Spectroscopy (AES) analysis is performed with the analyzer energy resolution about 0.3 %. Auger analysis is carried out using 75 nA, 3 keV electron beam with a diameter 10 μm, directed normally to the surface. The CrO peak is clearly detected. The refractive index of CrO is 2.71 [22], so the corresponding modulation is 0.19. Thus

$$(\eta_F / \eta_0)_{29nm} = (\Delta n_{CrO} / \Delta n_0)^2 + (\Delta n_F / \Delta n_0)^2 = (0.19 / 0.005)^2 + (0.17 / 0.05)^2 = 28 \quad (16)$$

This result is in good agreement with the experiment.

5. Conclusion

This study shows that the corona discharge method is a promising tool for diffraction efficiency enhancement of nano-scale holographic recording and it is suitable for variety application of holographic technique in nano-optics. Significant increasing of the diffraction efficiency in the order of 10 – 30 times was observed when corona discharge was applied during the holographic recording. The possible reason of the observed effect is that the additional refractive index modulation could be due to the Franz-Keldysh effect and chromium oxide ions diffusion.

6. Acknowledgement

This paper was published with the financial support of project BG05PO001-3.3-005/0001 "Science & Business" branch, operational programme "Human Resource Development"

References

[1] Sainov S, Dikova J and Beev K 2006 *Opt. Commun.* **263** 163
[2] Sainov S and Stoycheva-Topalova R 2000 *J. Opt. A: Pure Opt.* **2** 117
[3] Sainov S, Dikova J and Topalova-Stoycheva R 2001 *J. Opt. Adv. Mat.* **3** 395
[4] Sainov S, Sainov V and Dikova J 2001 *J. Opt. Adv. Mat.* **3** 399
[5] Nastas A, Andriesh A, Prisakar A and Triduh G 2011 *"Holoexpo-2011"* 395
[6] Nastas A, Andriesh A, Bivol B, Prisakar A and Triduh G 2009 *J. Technol Phys.* 79
[7] Dikova J, Vlaeva I, Babeva Tz, Yovcheva T and Sainov S 2012 *OLEN* **50** 838
[8] Harick N 1967 *Internal reflection spectroscopy* (New York: Interscience)
[9] Smith H M 1975 *Principles of holography* (New York: John Wiley & Sons) p 109
[10] Kogelnik H 1969 *Bell System Tech. J.* **48** 2909
[11] Kaspar F G 1973 *J. Opt. Soc. Am.* **63** 37
[12] Moss T S 1959 *Optical properties of Semiconductors* (Oxford: Butterworth-Heinemann) p 48
[13] Vavilov V S 1962 *Soviet Physics Uspekhi* 761
[14] Kastner M 1973 *Phys. Rev. B* **7** 5237
[15] Mott N F and Davis E A 1979 *Electronic processes in Non-Crystalline Materials* (Oxford: Clarendon Press) p 291
[16] Panayotov V and Konstantinov I 1994 *Proc. SPIE* **2253** 1070
[17] Abeles F and Theye M L 1966 *Surf. Sci.* **5** 325
[18] Babeva T, Kitova S and Konstantinov I 1998 *Appl. Opt.* **37** 4260
[19] Todorov R, Paneva A and Petkov K 2010 *Thin Solid Films* **518** 3280
[20] Fritzsche H 2000 *Isulating and Semiconductor Glasses* ed P Boolchand (Singapore: World Scientific Publ.)
[21] Kolomiets B T, Mazets T F, Efendiev Sh M and Andriesh A M 1970 *J. Non-Crystalline Solids* **4** 45
[22] Malvern Instruments 2007 *Sample dispersion and refractive index guide Reference Manual* (Worcestershire UK)

Journal of Physics: Conference Series **398** (2012) 012054 doi:10.1088/1742-6596/398/1/012054

Contact angle analysis of corona treated polypropylene films

I Vlaeva[1], T Yovcheva[2], A Viraneva[2], S Kitova[3], G Exner[2], A Guzhova[4] and M Galikhanov[4]

[1] Department of Mathematics and Physics, University of Food Technologies, 26 Maritza Boulevard, 4000 Plovdiv, Bulgaria
[2] Department of Experimental Physics, Plovdiv University "P. Hilendarski", 24 Tzar Assen Str., 4000 Plovdiv, Bulgaria
[3] Institute of Optical Materials and Technologies "Acad. J. Malinowski", Bulgarian Academy of Science, 1113 Sofia, Bulgaria
[4] Kazan National Research Technological University, 68 Karl Marx Str., 420015 Kazan, Republic of Tatarstan, Russian Federation

E-mail: ivlaeva@yahoo.com

Abstract. In this work, the effect of the surface modification of polypropylene films via corona treatment was investigated. Polypropylene films were treated with negative and positive corona discharge, at atmospheric pressure, for 5 minutes, at two different temperatures - 25 °C and 90 °C. The changes in the surface free energy were investigated by means of contact angle measurements. The Bickerman's method was applied to determine the polar and dispersion components of the polymer surface free energy, on the basis of the theory of Owens, Wendt, Kaelble and Uy. Atomic force microscopy was used to analyze the polymer surface morphology changes of the films with temperature. According to the findings, in all cases the corona treatment increases the surface free energy of polypropylene films and its polar part, in comparison with the untreated samples. The effects of negative and positive corona polarities display some specific features which could be associated with different charged group introduced onto the film surface during the corona treatment. The total final effect depends on the simultaneous action of the two competing factors - temperature and corona polarity. The most pronounced effect was observed for high temperature negative corona treatment.

1. Introduction

Polypropylene (PP) is one of the most widely used polymers owing to its high transparency, small weight, thermal stability, chemical resistance, easiness to manufacture, and low cost [1, 2]. However, PP is known to be non-polar, low surface energy material, and its hydrophobic characteristics cause poor wettability. The wettability is very important characteristic for many industrial applications of PP films, for example, ink printing, paints and coatings, lamination and conventional adhesives, composites preparation [1, 2].

Different techniques, such as plasma, ultraviolet irradiation, and electron bombardment treatments have been used to overcome these problems by modifying the surface properties [1, 3-6]. Another possible technique for surface modification of polymers is corona discharge [1, 7]. Atmospheric-pressure corona discharge is very attractive for various industrial applications because of its relatively low-cost, in-line treatment, good process control, high speed, operation without vacuum, and independence of the shape of the parts. The ability to change the surface properties of the most external layers of the material without modifying its bulk characteristics is also an advantage. Hence, one can improve the adhesion without any loss of the good mechanical properties.

When polymers are subjected to corona discharge in air, various ion groups initiated by the corona are injected into sample surface, enhancing the concentration of polar groups on it [1, 7-9]. The process leads to surface free energy increase. The type of the inserted groups could be governed by changing the corona type, temperature, air humidity and others factors [8, 10, 11]. However, systematic studies on the influence of corona discharge conditions, and especially of the corona polarity, on surface modification are still lacking.

In the present paper, the effect of corona discharge polarity and temperature on the surface modification of PP films was investigated. The changes in the polar, γ_s^p, and dispersion, γ_s^d, components of surface free energy were used as a measure of the degree of surface modification. The Bickerman's method [12] for precise measurements of contact angles of very small liquid sessile drops in contact with substrates was used to determine γ_s^p and γ_s^d on the basis of the theory of Owens, Wendt, Kaelble and Uy [13, 14].

2. Materials and methods

2.1. Sample preparation

Isotactic polypropylene (i-PP) films with thickness of 20 μm were supplied from "Assenova Krepost" LTD, Bulgaria. The films were used as received. They were cleaned with alcohol in an ultrasonic bath for four minutes, washed in distilled water, and dried on filter paper under room conditions. Samples with dimension (3 x 3) cm^2 were cut from the cleaned films.

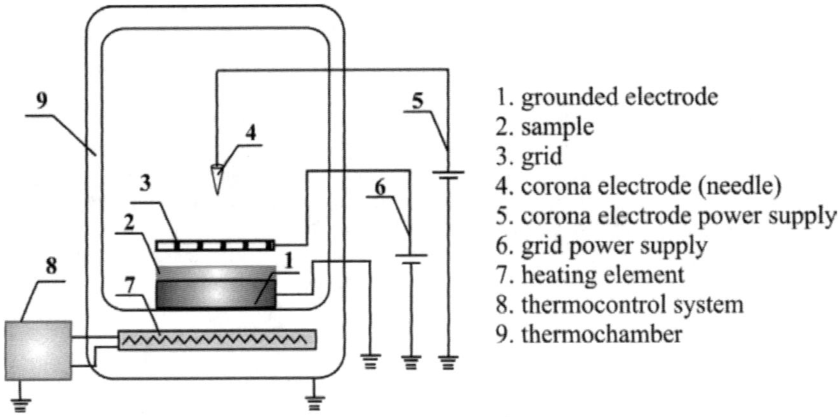

Figure 1. Experimental set-up for corona treatment.

2.2. Corona treatment

The samples corona treatment was carried out in a point-to-plane three-electrode corona discharge system [15]. The experimental scheme, presented in figure 1, consists of a corona electrode (needle), a grounded plate electrode, and a metal grid placed between them. Introducing a grid between the corona electrode and sample limits the sample surface potential to that of the grid and produces more uniform distribution of the charge on the sample surface. The distance between the grounded plate electrode and the grid was 10 mm and the distance from the grid to the corona electrode was 7 mm. A voltage of ± 5 kV was applied to the corona electrode and 1 kV with the same polarity, as that of the corona electrode, to the grid. The corona discharge system was placed into thermochamber with temperature control so that the samples, placed on the grounded plate electrode, were charged in positive or negative corona for 5 minutes, at two different temperatures (25 °C and 90 °C).

2.3. Morphology
i-PP surface morphology was investigated by means of Atomic force microscope (AFM NANOSURF FlexAFM). The samples were scanned with a standard silicon cantilever (type Tap 190Al-G) and the measurements were performed in ambient atmosphere, and dynamic operating mode. The applied force was always minimized, not to deform the samples. The surface topographic images of the films before and after temperature treatment were taken.

2.4. Contact angle measurements
For the measurements of the contact angle, θ, drops from distilled water and diiodomethane with varying volume were inserted onto the solid i-PP film surface, at room temperature, using a precise 10 µl micro syringe (Innovative Labor System GmbH, Germany) supplied with steel needle. The drop volume was from 2 µl to 6 µl. The drop diameters were measured with optical microscope (MBC-9, USSR) provided with a micrometer scale eyepiece. For each drop volume, and each testing liquid, 5 measurements were performed at different places onto the sample surface. The same procedure was repeated with six different samples. The final value of the contact angle is the mean value of the six samples measurements. The estimated error was less than 5 %.

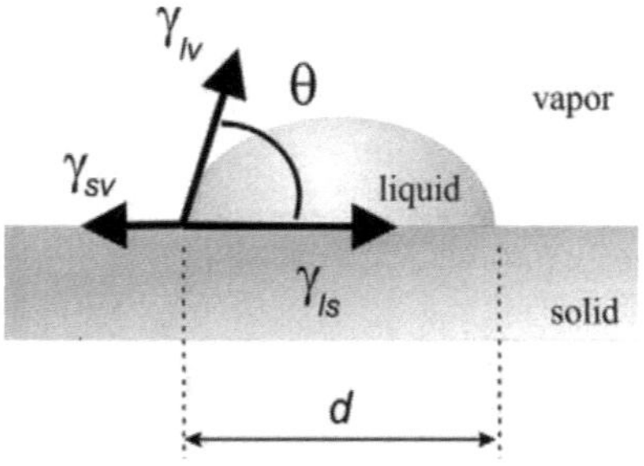

Figure 2. Scheme of the surface energy: γ_{SV} at solid-vapor interface, γ_{lS} at liquid-solid interface, and γ_{lV} at liquid-vapor interface; θ is the contact angle; d is the drop diameter.

Bickerman's method [4, 12], in which the effect of gravity distortion is negligible and the drop may be considered as a segment of a sphere, was applied to derive the contact angle:

$$\frac{d^3}{V} = \frac{24\sin^3\theta}{\pi\left(2 - 3\cos\theta + \cos^3\theta\right)} \tag{1}$$

where V is the drop volume, and d is the drop diameter.

Contact angle formation between a liquid drop and solid surface is presented in figure 2. The contact angle depends on the energies at the interfaces: solid-liquid (*sl*), solid-vapor (*sv*), and liquid-vapor (*lv*). The symbols γ with two indices describe the surface energy between the two phases in contact.

Following the theory of Owens and Wendt [13], and Kaelble and Uy [14], the surface free energy of a solid, γ_s, can be expressed as a sum of contributions from γ_s^d and γ_s^p components. Both can be determined from the contact angle data of polar and non-polar liquids with known dispersion, γ_{lv}^d, and polar, γ_{lv}^p, parts of their interfacial energy:

$$\gamma_{lv}\left(1 + \cos\theta\right) = 2\left(\gamma_s^d \gamma_{lv}^d\right)^{1/2} + 2\left(\gamma_s^p \gamma_{lv}^p\right)^{1/2} \tag{2}$$

where $\gamma_{lv} = \gamma_{lv}{}^{p} + \gamma_{lv}{}^{d}$.

3. Results and Discussion

In order to distinguish the effect of high temperature treatment from corona discharge, a preliminary AFM study of surface morphology was performed. As it is seen in figure 3, the morphology is insignificantly altered by the temperature increase. The difference between the lowest and the highest surface points is $\Delta h = 52.3\ nm$ at 25 °C and it displays minor change after temperature treatment at 90 °C ($\Delta h = 55.4\ nm$). The results indicate that the degree of mechanical interlocking remains the same and this mechanism does not play role in the adhesive properties increase. It has been already shown that the corona treatment provokes formation of nano- to micron-sided globular mounds but it does not alter the film surface roughness [7-9].

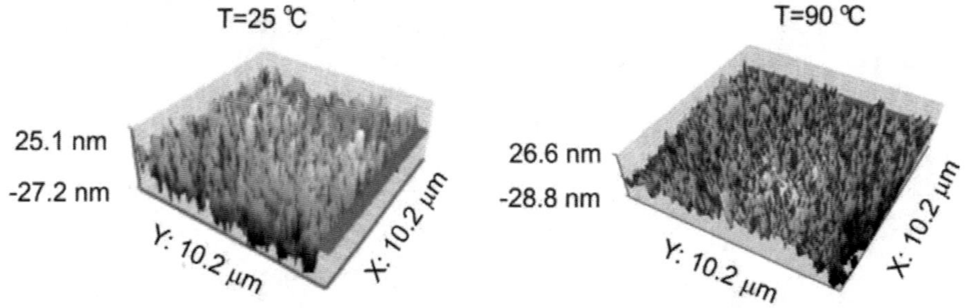

Figure 3. AFM images of 20 µm thick i-PP films, stored at room temperature (T = 25 °C), and after annealing at T=90 °C for 5 min.

The investigated samples were divided into three main groups – uncharged, positively charged, and negatively charged. In each group, two different series were prepared with different temperature treatment (25 °C and 90 °C). The obtained values of the distilled water, θ^{w}, and diiodomethane, θ^{DIM}, contact angles, γ_s of the films, and polarity $\left(P = \gamma_S{}^{p} / \left(\gamma_S{}^{p} + \gamma_S{}^{d}\right)\right)$ are given in table 1.

Table 1. θ^{w}, θ^{DIM}, γ_S and ***P*** of the untreated and corona treated i-PP films.

Sample type	*T* (°C)	θ^{w} (deg)	θ^{DIM} (deg)	γ_s (mJ.m^{-2})	*P*
Uncharged	25	97.0 ± 0.8	56.9 ± 0.9	30.4	0.03
	90	100.4 ± 0.5	61.4 ± 0.6	27.8	0.02
Positively charged	25	90.5 ± 0.8	49.4 ± 0.1	32.3	0.05
	90	91.0 ± 2.0	50.4 ± 0.7	34.2	0.05
Negatively charged	25	95.1 ± 0.7	56.6 ± 0.9	30.6	0.04
	90	90.0 ± 2.0	57.0 ± 3.0	30.3	0.12

The temperature treatment alone leads to an increase in the contact angles for both liquids and the corresponding decrease in the surface free energy. This effect could be explained by thermal annealing, where gradual structural change from smectic to monoclinic crystal structure takes place simultaneously with formation of a thin surface layer of low-molecular-weight compounds of i-PP [1].

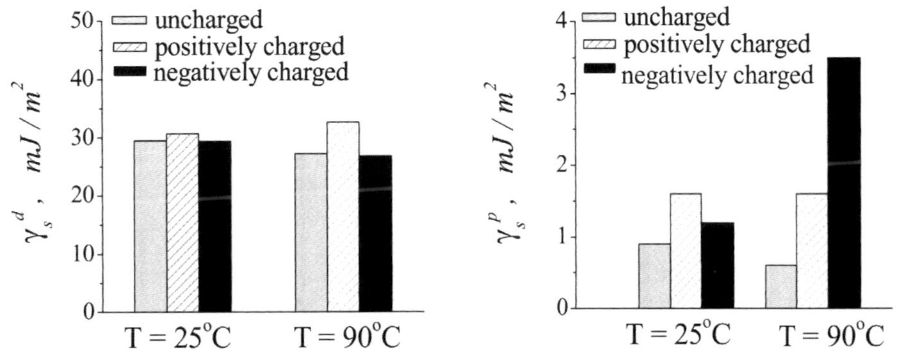

Figure 4. The dispersion, γ_s^d , and polar, γ_s^p , components of surface free energy of untreated and corona treated i-PP films at different temperatures.

For both experimental temperatures, θ^w decreases after the corona treatment and correspondingly the γ_s increases for both corona polarities, in agreement with the results of Cáceres at al. for negative corona treatment [16]. The values of γ_s^p and γ_s^d , depending on the treatment conditions, are shown in figure 4. As it can be seen, the changes in surface free energy of the corona treated samples are due to the increase of γ_s^p , leading further to enhanced polarity. The findings can be explained by introduction of oxygen containing polar groups such as hydroxyl, peroxy, hydroperoxide, carbonyl, carboxylic, carbonate, and ester onto the sample surface during the corona treatment [7, 8, 17, 18]. Our results (see table 1 and figure 4) indicate that the negative and positive corona polarities treatments display different specific features which could be explained by different kinds of polar groups, created during the corona treatment. Yovcheva at al. [19] based on X-ray photoelectron spectroscopy analysis, have found that the oxygen content differs in samples charged in different corona polarity, leading to the formation of different surface local levels. In negative corona polarity, where the most important ions are the CO_3^- , the electric field stimulates introduction of oxygen ions or oxygen-containing groups onto polymer surface. In positive corona polarity, where the prevailing ions are $(H_2O)_n H^+$ type, the electric field stimulates desorption of oxygen atoms from the surface. The presence of two different processes, oxygen adsorption and desorption during the sample charging, are most probably stimulated in different ways and/or have different rates in the positive and negative corona. The lower oxygen content in positively charged samples in comparison to the negatively charged samples [19] could explain the higher values of ν^d obtained for positive corona treated samples.

The observed changes of the surface energy are result for the interplay of the two simultaneously acting and competing effects: temperature and corona polarity. In general, the temperature increase leads to a decrease in surface free energy, while the corona treatment stimulates the surface free energy increase. According to our results, the most pronounced effect of increase in the surface polarity was observed for high temperature negative corona treatment. Obviously further detailed studies are needed to reveal the exact impact of the simultaneous action of these factors - temperature and corona polarity.

4. Conclusion

The obtained results indicate that corona discharge, applied in combination with high temperature leads to a decrease in the water contact angle, an increase of surface polarity, and consequently leads to an improved weattability of the films. The effect is observed for both corona polarities. The observations are believed to be due to introduction of different oxygen containing polar groups, such as hydroxyl, peroxy, hydroperoxide, carbonyl, carboxylic, carbonate, and ester onto the i-PP surface. Negative and positive corona polarities treatments display some specific features. The strongest increase in the surface polarity was observed for negative corona high temperature treatment. The differences in the results for both polarities could be associated with different type and amount of the charged group injected into the polymer surface during the corona treatment. The results show that the effect of the treatment was dependent on the simultaneous action of two treatment factors - temperature and corona polarity.

References

[1] Doğan F ed 2012 *Polypropylene* (Croatia: InTech)
[2] Karian H G ed 2003 *Handbook of Polypropylene and Polypropylene Composites – 2nd Edition, Revised and Expanded* (London: Taylor & Francis)
[3] Pandiyaraj K N, Selvarajan B, Deshmukh P and Gao C 2008 *Vacuum* **83** 332-9
[4] Kitova S, Minchev M and Danev G 2005 *J. Optoelectron. Adv. M.* **7(1)** 249-52
[5] Schnyder B, Lippert T, Kötz R, Wokaun A, Graubner V-M and Nuyken O 2003 *Surf. Sci.* **532–535** 1067-71
[6] Aronov D and Rosenman G 2007 *Surf. Sci.* **601** 5042-9
[7] O'Hare L-A, Leadley S and Parbhoo B 2002 *Surf. Interface Anal.* **33(4)** 335-42
[8] Strobel M, Jones V, Lyons C S, Ulsh M, Kushner M J, Dorai R and Branch M C 2003 *Polymers and Plasmas* **8(1)** 61-95
[9] Süzer S, Argun A, Vatansever O and Aral O 1999 *J. Appl. Polym. Sci.* **74(7)** 1846-50
[10] Yovcheva T, Mekishev G and Marinov A 2004 *J. Phys.: Condens. Mat.* **16** 455-64
[11] Yovcheva T, Mekishev G and Nedev S 2005 *J. Optoelectron. Adv. M.* **7(1)** 237-40
[12] Johnson Jr R E and Dettre R H 1969 *Surface and Colloid Science* vol 2 *Wettability and Contact Angles* ed E Matijevic (New York: Wiley–Interscience) 85-153
[13] Owens D K and Wendt R C 1969 *J. Appl. Polym. Sci.* **13(8)** 1741-7
[14] Kaelble D H and Uy K C 1970 *J. Adhesion* **2(1)** 50-60
[15] Yovcheva T A 2010 *Corona Charging of Synthetic Polymer Films* (New York: Nova Science Publishers, Inc.)
[16] Cáceres C A, Mazzola N, França M and Canevarolo S V 2012 *Polym. Test.* **31(4)** 505–11
[17] Strobel J M, Strobel M, Lyons C S, Dunatov C and Perron S J 1991 *J. Adhes Sci. Technol* **5(2)** 119-30
[18] Sellin N and de C Campos J S 2003 *Mat. Res.* **6(2)** 163-6
[19] Yovcheva T A, Avramova I A, Mekishev G A and Marinova T S 2007 *J. Electrostat.* **65(10-11)** 667-71

Heat – mechanically induced phase and relaxation transitions in amorphous polyester fibers

V Velev[1], A Popov[2], R Androsch[3], T Dimov[1], T Angelov[4] and H-J Radusch[3]

[1]Konstantin Preslavsky University, 115 Universitetska Street, 9712 Shumen, Bulgaria
[2]University "Prof. Dr. Assen Zlatarov", 1 "Prof. Jakimov" Street, 8000 Burgas, Bulgaria
[3]M LU, Halle-Wittenberg, Center of Engineering Sciences, Geusaer Straße, Gebäude 131, 06217 Merseburg, Germany
[4]Lukoil-Neftohim-Burgas AD, 8000 Burgas, Bulgaria

E-mail: v.velev@shu-bg.net

Abstract. In the current work we submit experimental data pertaining to the role of the heat-mechanical modification on the structural reorganization of poly (ethylene terephthalate) (PET) as-spun yarns. The thermal deformation experiments were realized using specialized device designed and produced in the author's laboratories. The structural characterizations of the studied fibers after the above described heat-mechanical treatments were realized using differential scanning calorimetry (DSC), wide-angle X-ray scattering (WAXS) and density measurements. Have been identified dependences between the heat-mechanical deformation parameters and the occurred in the studied PET fibers structural reorganization.

1. Introduction

It is well known that the flexible chain polymers are widely used in the production of fibers, thin films and more. Poly (ethylene terephthalate) (PET) is a crystallizable thermoplastic polymer, which is one of the basic fiber forming flexible chain polymers. The simultaneous heat and mechanical treatment is especially topical in the resolving a number of technical and technological problems in the attempts to improve the filaments strain-strength behavior. It is well known that the physical characteristics of the anisotropic polymer systems depends directly from their super molecular structural organization [1-4]. Most often the mechanical properties improving of highly anisotropic objects such as polymer fibers is realized technologically using a high temperature uniaxial orientation extending.

It is extremely important to study the influence of the parameters on different variants of heat mechanical fibers modification on their structure and structural transformation. The optimization of the conditions of the uniaxial extension depends on the fibers behavior during their processing.
The basic conditions of the heat mechanical fibers modification are as follows:

degree of the preliminary filament orientation;

temperature-time regime of the filaments extension;

values and ways of applying of the orientation stresses such as a gravitational loading or forced extension at a constant speed [4-6].

In this sense, a key problem is studying of the structural transformation mechanisms of the uniaxial orientation extending. Different experimental aspects of the super molecular structural reorganization and deformation behavior of PET fibers caused by thermal and mechanical treatments have been studied in recent years, including the authors of the paper. Were carried out heat-mechanical treatments of PET filaments at constant temperatures without application of stress [3, 4, 7, 8] and under strain stress [4, 9] as well in non-isothermal conditions [4, 10-12].

The optimal realization of the high-temperature orientation modification is a complicated and still not sufficiently well studied process. The role of the simultaneous heat-mechanical treratment on the super molecular structural development of amorphous PET yarns remains not fully understood. With purpose to clarify the complex effects of the temperature and tension stress values on the deformation behaviour and structure development in uniaxially oriented PET we carried out thermal deformation experiments with amorphous PET filaments under the following conditions. linear heating of the sample from room temperature to 170 ^0C, with heating speed of 3.5 degrees per minute and under precisely defined strain stresses from 0.02 *MPa* to 1.3 *MPa* with increment step of 0.1 *MPa*. The structural investigations of the heat-mechanically modified specimens were realized using differential scanning calorimetry (DSC), wide-angle X-ray scattering (WAXS) and density measurement.

2. Experimental

2.1. Materials

Amorphous PET as-spun multifilament yarn named B, have been selected as a precursor sample. The sample is prepared by melt spinning on the industrial spinning installation Furnet (France). Used for this purpose technological parameters are spinning speed of 1100 *m/min* and a melt flow of 50 *g/min*. The obtained complex yarn consists of 32 single filaments with single fiber diameter of 44 *μm*. The samples were characterized by measuring of their birefringence and flotation density and the received results are 0.008 for the birefringence and 1336 kg/m^3 for the density. The density show that the samples are practically amorphous with degree of crystallinity less than one percent and they are suitable for the realization of the purpose of this study.

2.2. Methods

The heat-mechanical treatments of the investigated filaments were realized using a specialized apparatus constructed and produced in our laboratory [13]. The gear involves a vertical cylindrical oven and a device for the sample deformation reading. One of the ends of the studied fibber bundle with initial length of 0,1 *m* was stationary attached in the oven to the device stand while the other one connected by an inextensible cord was attached to strain stress forming weight.

The thermal deformation experiment includes linear sample heating with a small heating rate of 3.5 *deg/min* from a room temperature up to temperature of 170^0C. During the heating the investigated PET bundle was simultaneously subjected to precise defined pulling tension in the interval from 0.02 *MPa* for the sample marked as B_1 up to 1.3 *MPa* for B_{13} with increments of 0.1 *MPa*. Should be noted that the tensile loading increments of 0.1 *MPa* allows examination of the tension role with enough good resolution. After reaching of the temperature of 170^0C simultaneously was taken off the applied tensile strain and the fiber bundle removed to a room temperature.

Structural changes in the investigated PET filaments occurred as a result of the above-described heat-mechanical treatments have been studied by differential scanning calorimetry (DSC) and wide-angle X-ray scattering (WAXS). The samples degree of crystallinity α was calculated using the data obtained by DSC. A Mettler-Toledo DSC-820 heat-flux module equipped with liquid nitrogen accessory was used for the calorimetric studies. The furnace was purged with nitrogen at a flow rate of 80 *ml/min*. Temperature calibration was done using the melting temperatures of indium and zinc, as well as the energy calibration is based on the heat of fusion of indium. The investigated PET fibers were cut into pieces of less than 1 *mm* and sealed in standard 40 μl aluminium pans. The value of the bulk heat of fusion of fully crystallized PET was taken from the literature [14].

Especially suitable for studying of the orientation processes in the arranged in a crystalline phase macromolecular chain segments are the various X-ray methods. The PET fiber structure

Journal of Physics: Conference Series **398** (2012) 012055 doi:10.1088/1742-6596/398/1/012055

development was studied as well by WAXS using a diffractometer HZG 4 (Freiberger Präzisionsmechanik, Germany) and *Ni*-filtered *CuKa* radiation with wavelength of 1.5418 $\dot{A}$. The fiber bundle were prepared as a layer with 2 *mm* thickness and 10 *mm* width, and mounted on the sample holder of the diffractometer. Equatorial scattering was monitored in transmission mode. The 2θ angle values are changed in the range from 10 up to 40 degrees with a step of 0.05 degrees and signal recording has been done in period of time 5 seconds. Were recorded the spectra of the studied PET bundle in two positions - when it is horizontal and vertical.

3. Results and discussion

For detailed identification of the occurred as a result of the processing structural changes in the PET fibers, they were analyzed using differential scanning calorimetry (DSC). Analysis of each sample involves three stages:

1) first heating in the temperature range from 25^0C to 300^0C with heating rate of 20 *deg/min*;
2) cooling from temperature 300^0C to temperature 25^0C with heating rate of 20 *deg/min*;
3) second sample heating in the same temperature range from 25^0C to 300^0C at the same heating rate of 20 *deg/min*.

The second heating after the sample cooling, under the same conditions (as in the first heating) is necessary in order to compare the two heating curves obtained for the sample under identical heating conditions but with a different background.

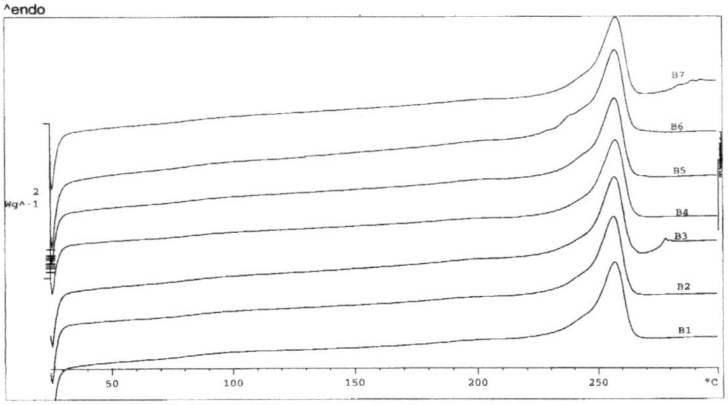

Figure 1. DSC curves from first heating of samples $B_1 \div B_7$.
The curves are shifted vertically for clarity.

The DSC curve of the amorphous untreated precursor sample B, show strongly expressed cold crystallization at a temperature of 123.94^0C and doublet melting peak at temperatures 256.3^0C and 262.3^0C. The doublet melting peaks reflect the presence of two phases with different degrees of perfection. Possibly the more perfect phase represents about 10 percent of the total crystallized phase and melts at a higher temperature. Is likely that these phases are formed in advance and have been further improved during the process of cold crystallization. DSC curves from the first heating of the treated sample B are present on the figures 1 and 2. From the shown figures it may be noted the following important features:

transition from glassy to rubbery state of PET fibers at about $75 \div 80^0C$;
absence of the typical for the amorphous PET specimens cold crystallization exopeak;
a small endopeak in the temperature range from 175^0C to 185^0C;
endopeak reflecting the melting process of the polymer in the temperature interval from 220^0C to 265^0C;
the melting peaks of some samples and namely B_7, B_8, B_{10} and B_{13} are slightly divided.

In addition, at all treated samples is observed endo effect of releasing of the frozen internal stresses and restoring of the polymer segmental mobility at a temperature about 80^0C. The observed attenuation and subsequent disappearance of this effect from sample B_1 to sample B_{13} is due to the increase of the applied tensile stress value.

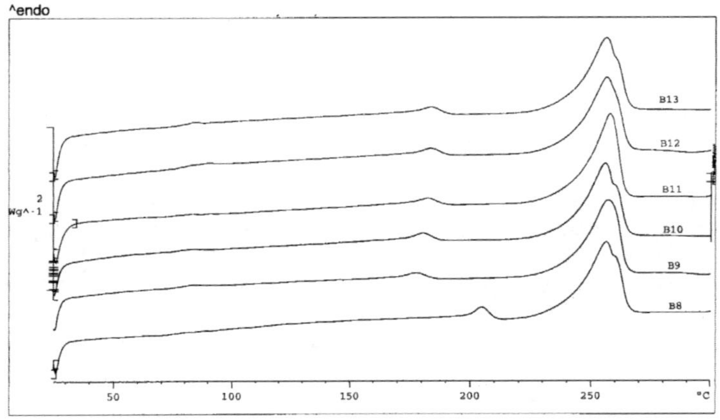

Figure 2. DSC curves from first heating of samples $B_8 \div B_{13}$.
The curves are shifted vertically for clarity.

It is interesting to note also that the small endopeak that appears at a temperature of 180^0C for sample B_1 and reflects the melting of imperfect components of the crystalline phase smoothly but steady sliding toward the higher temperatures with the increasing of the applied tensile stress value.

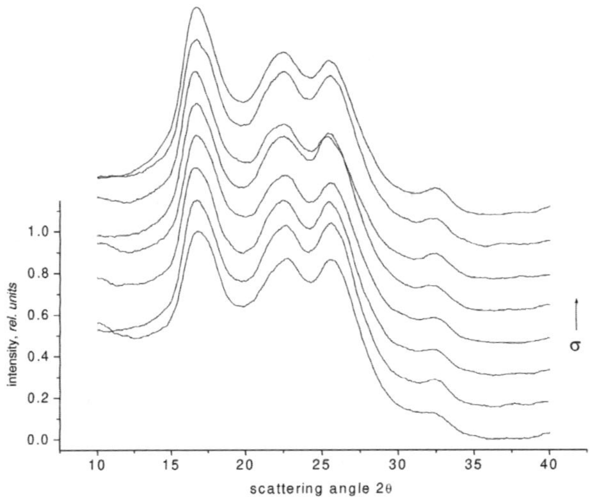

Figure 3. Representative WAXS curves of PET filaments from seria B.
The curves are shifted vertically for clarity.

The probable reason for appearance of the second endo peak on DSC curves is the melting of small crystal formations that appeared as a result of the preliminary fiber stretching, as an initial stage in the process of recrystallization of the polymer system. Obtained by differential scanning calorimetry results from the investigation of PET filaments series B can be summarized as follows:

the shape of DSC curves and especially the absence of crystallization peak, confirming the results from the X-ray analysis that all fibers of this series after the heat mechanical treatment are partially crystalline. The determined using the DSC data mass degree of crystallinity (α) is in the range from 26 to 30 percent;

most likely the appearance of a small endo peak in the temperature interval from $175^{0}C$ to $185^{0}C$ to be a consequence of the samples heat mechanical background. The probable reason for appearance of second endo peak on the DSC curve is the melting of small crystal formations that emerged as a result of the preliminary fiber stretching, as an initial stage in the process of the polymer recrystallization;

the melting peaks of some of the fibers subjected to relatively high tensile stress are divided. This fact can be explained by the influence of the applied mechanical stress during the heating of PET fibers in the range from glass transition temperature to the melting point.

Representative WAXS curves obtained for the studied fibers are shown in figure 3. From the figure we can see that regardless of the applied stress value during the heat mechanical experiments all PET fibers from the studied series B are partially crystalline with almost the same degree of crystallinity.

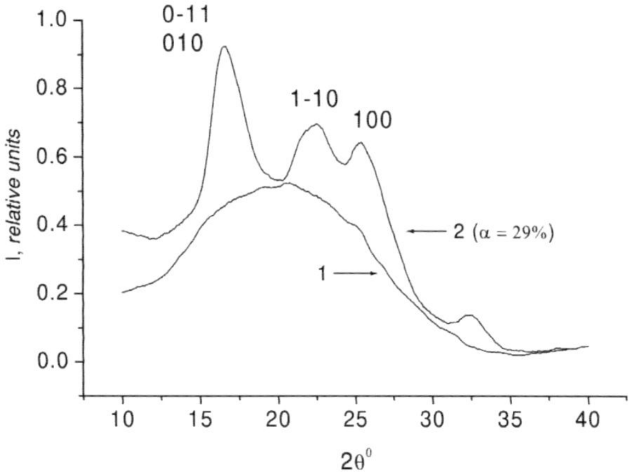

Figure 4. WAXS curves of amorphous untreated (1) and treated under srain stress of 0,8 *MPa* partially crystalline (2) PET filaments from seria B.

The received structural picture is expected taking into account that the temperature range in which is processed series B includes the cold crystallization process. Furthermore, figure 3 and the indexation of the main diffraction peaks of PET (figure 4) show that the increase of the tensile stress values leads to changes in the intensity of the main peaks. The intensity changes of the small angle doublet peak (0-11, 010) should be associated with deformation along the axis of the macromolecular chain segments and the their folds width. The existence and intensity of the other two peaks (1-10) and (100) can be related mainly with the distances between the planes of the folded segments. Variation of the peaks intensity depending on the values of the applied mechanical stress indicates that the deformation changes in the crystalline phase along the chains axis and the width of the folds are realized less and relatively monotonous. At the same time, with the mechanical loading increasing the distances between the folding planes change more significant and non-monotonicity. Figure 3 shows even a slight tendency for deviation of the diffraction reflection

(100) to the smaller angles. The observed effect can be interpreted as a slight increase in the distance between the planes of the folding of the macromolecular chain segments. With the tension increasing a similar trend can be accepted as logical. In support of this hypothesis is the strong non-monotonicity amending the intensity of peak (100) with increasing of the applied tensile stres. Increasing of the distance between the planes of folding determines the possibility for the intensity lowering of the diffraction effect whatever trend is observed for peak (100).

It should be noted that the experimental results obtained using DSC and WAXS are confirmed by the subsequent measurement of the studied PET fibers density.

4. Conclusions

The simultaneous application of tensile stress with linear heating has a strong impact on the course of the phase and relaxation transitions in amorphous PET filaments.

It was found that all investigated samples after the heat-mechanical treatment are partially crystalline which explains the absence of the typical for the amorphous PET specimens cold crystallization exopeak.

Was established relationship between strain stress values and the disappearance of the endo effect of releasing of the frozen internal stresses and restoring of the macromolecular segmental mobility at a temperature about 80^0C.

Appearance of a small endo peak at temperature from 175^0C to 185^0C and bifurcation of some melting peaks. These found effects can be explained with the role of the applied mechanical stress during the heating of PET fibers in the range from glass transition temperature to the melting point.

It was launched assumption concerning the intensity changes of the diffraction reflections (0-11, 010), (1-10) and (100).

Acknowledgement

Grants 5240, RD – 07-718/04.04.2012 from the University of Shumen, №D-002-61/2008 and NIH-244/2011 from University "Prof. Dr. Assen Zlatarov", Burgas, Bulgaria supported the present study.

References

[1] Ziabicki A and Jarecki L 2007 *J. Appl.Polym. Sci.* **105** 215 – 23
[2] Betchev Ch 1994 *Crystallization of Polymers* (M. Dosiere, Ed.) *Kluwer Academic Publ.*
[3] Keum J and Song K H H 2005 *Polymer* **46** 939–45
[4] Velev V, Popov A and Bogdanov B 2011, Moreno-Pirajan J C (ed.) *Thermodynamics* Rijeka 89
[5] Wu J, Schultz J M, Samon J M, Pangelinan A B, and Chuah H H 2001 *Polymer* **42** 7161-70
[6] Shabana H M 2004 *Polymer Testing* **23** 291 – 97
[7] Dimov T, Velev V, Iliev I and D. Betchev Ch 2007 *Journal of Optoelectronics and Advanced Materials* **9** 2 275 – 77
[8] Velev V, Dimov T, Popov A, Denev Y, Hristov H, Angelov T, Markova K, Zagortcheva M, Arhangelova N and Uzunov N 2010 *Journal of Physics Conference series* **253** 012026
[9] Llana P G and Boyce M C 1999 *Polymer* **40** 6729–51
[10] Velev V and Betchev Ch 2008 *Thermochimica Acta* **473** *50–53*
[11] Kong Y and Hay N J 2003 *Polymer* **44** 623-33
[12] Velev V, Androsch R, Radusch H-J and Betchev Ch 2001 *Macromol. Mater. Eng.* **286** 752-55
[13] Velev V and Betchev Ch 2000 *Polymer Testing* **19** 281-88
[14] Blundell D J and Osborn B N 1983 *Polymer* **24** 953

Thin polyphenylene vinylene electrophoretically and spin-coated films – photoelectrical properties

D Mladenova[1,2], I Zhivkov[1,2], I Ouzzane[1], M Vala[1], P Heinrichova[1], D Budurova[3], and M Weiter[1]

[1] Brno University of Technology, Faculty of Chemistry, Centre for Materials Research, Purkynova 118, 612 00 Brno, Czech Republic

[2] Institute of Optical Materials and Technologies "Acad. J. Malinowski", Bulgarian Academy of Sciences, Acad. G. Bonchev Str. bl. 101/109, 1113 Sofia, Bulgaria

[3] Institute of Polymers, Bulgarian Academy of Sciences, Acad. G. Bonchev Str. bl. 103A, 1113 Sofia, Bulgaria

E-mail: xcmladenova@fch.vutbr.cz

Abstract. Electrophoretic deposition (EPD) and spin coating (SC) techniques were compared on poly[2-methoxy-5-(3,7-dimethyloctyloxy)-1,4-phenylene vinylene] (MDMO-PPV) used as a model material. Absorption spectra measured in a toluene solution and a toluene/ acetonitrile suspension with the same MDMO-PPV concentration of 0.0033 g.l^{-1} were analysed. An observed broadening of the characteristic absorption peak could be related to the formation of tightly-folded polymer chains in the suspension. Thin films of about 300 nm thickness were prepared by EPD from suspension and SC from solution with a concentration of 0.0033 g.l^{-1} and 8.95 g.l^{-1}, respectively. The ITO|MDMO-PPV|Al structures with EPD MDMO-PPV films behave as a photovoltaic cell, while the same sample configuration with SC MDMO-PPV films acts more like a photoresistor. EPD and SC films exhibit the same charge-carrier photogeneration mechanism. It was clearly demonstrated that the EPD method uses the deposited material more efficiently, producing films with better photoelectrical properties.

1. Introduction

Conjugated polymers are a class of organic semiconductors that have attracted much interest in recent decades from both fundamental and practical points of view.[1] The π-electron system on their main chain provides unique electronic and optical properties that could not be predicted by conventional solid state theories.

Much effort has been devoted to developing electronic devices that use conjugated polymers as electroluminescent diodes, photovoltaic cells[2], gas sensors, and field-effect transistors.[3] In contrast to traditional inorganic semiconductors, which require high processing temperatures, conjugated polymers are synthesized at near room-temperature conditions. Their solubility might be significantly improved to deposit wet-process films under atmospheric pressure. As a result, large-area films could be obtained by relatively inexpensive (compared to vacuum-based methods) coating technologies – spin and dip coating, spray and Langmuir Blodgett (LB) deposition, ink-jet printing, electrophoretic deposition, etc.

The spin-coating (SC) technique is a widely used method to prepare high quality polymer films for electronic devices. However, this is one of the most uneconomical methods, because most of the solution dropped on the substrate is blown away during the spinning.[4]

Published under licence by IOP Publishing Ltd

Electrophoretic deposition (EPD) allows the preparation of films of several hundred nanometers from a low suspension concentration.[5] The particle size in the EPD suspension can be controlled by changing the ratio of solvent and non-solvent.[6] Deposition from a suspension with a high non-solvent concentration results in a rougher film structure. This effect seems to improve the performance of gas sensor devices[7], as the film porosity may enable the fast absorption/desorption of dopants and target molecules. In contrast, deposition from a suspension with a high solvent concentration results in smooth and uniform films, applicable as active layers in polymer electroluminescent devices or solar cells.[8]

EPD enables quick patterned deposition, because the covered area can be specified by the electrification of selected electrodes.[9]

The application of the EPD method in the development of thin layer based organic devices needs further investigation of the mechanism, kinetics of the electrophoretic process, and structure and properties of the films.

This paper presents a comparative study of the properties of poly[2-methoxy-5-(3,7-dimethyloctyloxy)-1,4-phenylene vinylene] (MDMO-PPV) films deposited electrophoretically and cast by SC. MDMO-PPV is a widely investigated organic semiconductor (HOMO and LUMO orbital positions at 5.4 and 3.2 eV, respectively)[10] with a potential application as a photoconductive or electroluminescent layers. The study seeks potential applications for these films in solar energy conversion technology and other branches of organic electronics.

2. Material and methods

A stock solution (concentration of 0.165 $g.l^{-1}$) was prepared by dissolving 3.3 mg MDMO-PPV (Sigma-Aldrich, catalogue number: 546461) in 20 ml toluene (Sigma-Aldrich Toluene, catalogue number: PLC22C11X), then heating for 5 min at 60°C, followed by ultrasonic treatment for 10 min. A homogeneous solution without clearly visible undissolved particles was obtained.

An EPD suspension with MDMO-PPV concentration of 0.0033 $g.l^{-1}$ and toluene/ acetonitrile ratio of 50 % v/v (Sigma-Aldrich Acetonitrile, catalogue number: 34851) was prepared. For a 20 ml suspension with a concentration of 0.0033 $g.l^{-1}$ and toluene/ acetonitrile ratio of 50 % v/v, 0.4 ml from the stock solution was diluted in 9.6 ml pure toluene, then 10 ml of acetonitrile was added. The suspension obtained clearly changes color[1], which could be related to the formation of (micro) nano-sized solid particles. The prepared suspension was immediately used for film deposition or measurement to prevent further particle coagulation.

A solution (0% v/v acetonitrile) with the same MDMO-PPV concentration of 0.0033 $g.l^{-1}$ was prepared for optical absorption measurements. The optical absorption spectra of the prepared suspension and solution were measured by a Varian Cary 50 UV-VIS dual beam spectrophotometer. The measurement was carried out in a quartz cuvette with an optical path of 10 mm. The baseline was taken from the cuvette, filled with pure toluene or toluene/ acetonitrile mixture in a 1:1 ratio (toluene/ acetonitrile). Then the spectra in the 650-270 nm range were measured.

The electrical parameters (voltage and current) of the EPD process were controlled by the Keithley 2410 SourceMeter (voltage range of 0-1100 V and current measurement from 10 pA to 1 A).

After applying voltage between the electrodes, the current continuously decreases. A deposition on the positive electrode starts when the current reaches about 80-40 μA. The film, about 300 nm thick, completely grows within 3-4 min.

A solution with a concentration of 8.95 $g.l^{-1}$ was prepared and SC was done on the KW-4A Spin Coater (Chemat Technology, Inc.). Spinning was applied at 1000 rpm for 3 sec. and 2600 rpm for 60 sec. to obtain about 300 nm-thick films.

The film thickness was determined by the chromatic distance method on the MicroProf® FRT optical surface measuring system.

Samples for electrical measurements of 25×15 mm were cut from commercial ITO-coated glass slides (Sigma-Aldrich, catalogue number: 636916). The top Al electrodes of about 100 nm thickness were deposited in a vacuum; copper wires were then bonded to the electrodes by a silver paste (Dotite® Silver Paint D-550).

Photoelectrical measurements were carried out in an oil-free vacuum of 2.2×10^{-5} Pa with a Keithley 6517A Electrometer. Monochromatic light was produced by a LOT-Oriel halogen lamp LSH502 and a LOT-Oriel monochromator MSH101. Light power was controlled by an iris diaphragm and measured by the Gigahertz-Optik - X97 Irradiance Radiometer.

Photoelectrical measurements start with spectral dependence of the photocurrent at zero applied voltage; *I-V* characteristics were then measured in both directions of the voltage scale in dark and exposed to light. Finally, the dependence of the photocurrent on the incidental light power (irradiance) was measured. The sample was exposed to monochromatic light at a wavelength of 560 nm.

3. Results and Discussion

3.1. UV-VIS spectra

The optical absorption spectra of the solution and the 50% v/v toluene/ acetonitrile suspension with the same MDMO-PPV concentration of 0.0033 $g.l^{-1}$ are presented in figure 1. The solution spectrum (figure 1, curve 2) consists of the characteristic MDMO-PPV absorption peak.[11] The peak is slightly broad and asymmetric due to the "red" shift with respect to the main peak position. This effect could be related to a slight interaction of the MDMO-PPV molecules.

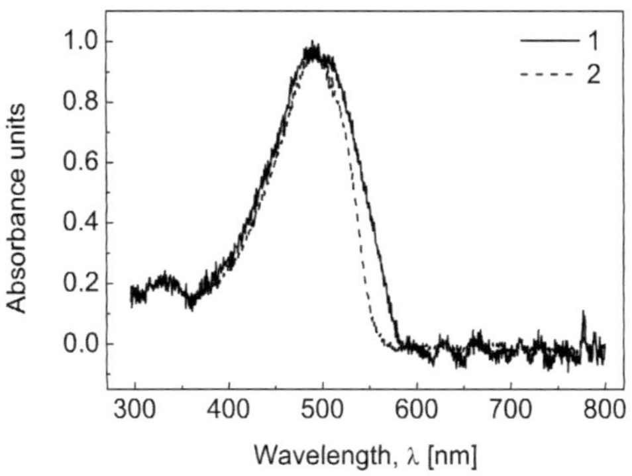

Figure 1. Optical absorption spectra of MDMO-PPV suspension (curve 1) and solution (curve 2).

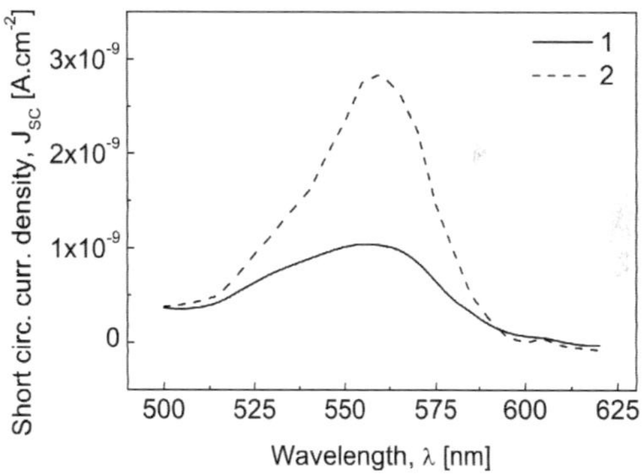

Figure 2. Spectral dependence of the photocurrent at zero applied voltage: EPD (curve 1) and SC (curve 2) films of about 300 nm thicknesses.

In the suspension spectrum (figure 1, curve 1), the "red" shoulder strength increases, which leads to a greater broadening of the peak. This effect could be related to the appearance of a precipitated solid phase during the suspension formation, caused by the polar precipitating acetonitrile. It is known that for more crystalline organic materials (e. g. poly(3-octylthiophene)), such precipitation can even lead to a "red" shift of the whole characteristic absorption peak.[12] The effect of the peak broadening could be used for a qualitative estimation of the suspension properties.

3.2. Electrical measurements

Spectral dependences of the photocurrent, measured at zero applied voltage between the electrodes for structures with EPD and SC MDMO-PPV films are presented in figure 2, curve 1, and figure 2, curve 2, respectively.

Comparing figures 1 and 2, it could be concluded that generally the MDMO-PPV photocurrent spectrum is similar to the absorption spectrum. There is a clear characteristic peak in the photocurrent spectrum at 560 nm. The peak position is "red" shifted (about 50 nm) with respect to the maximum of the absorption spectrum. This effect of energy gap reduction could be related to the formation of a solid state. The photocurrent peak obtained from EPD film (figure 2, curve 1) is slightly broader than the peak obtained from SC film (figure 2, curve 2). This result is similar to that obtained from absorption spectra measurement (Sec. 3.1). The photocurrent measured from structures with SC films is 3 times higher than that obtained from EPD samples.

I-V characteristics measured in the dark on the ITO|MDMO-PPV|Al structure with EPD and SC MDMO-PPV films of about 300 nm thickness are presented in figure 3. To estimate the electrical parameters of the samples, the data are plotted in a semilogarithmic scale (negative values of the current are multiplied by -1).

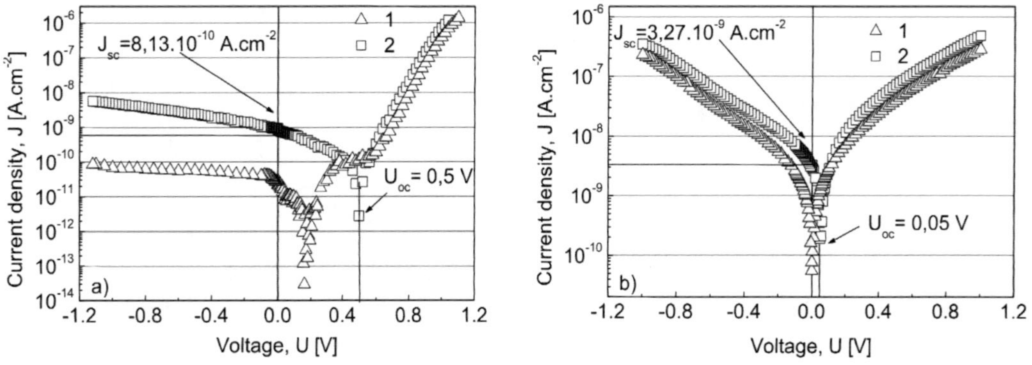

Figure 3. *I-V* characteristics measured in the dark (curve 1) and under illumination (curve 2) with monochromatic light (λ=560 nm, irradiance 0.59 mW.cm^{-2}) on ITO|MDMO-PPV|Al structures with a) EPD and b) SC MDMO-PPV film with about 300 nm thicknesses.

From the dark current measurements on a sample with EPD film (figure 3a, curve 1), a diode contact barrier could be observed at 0.16 V. Overcoming this barrier changes the sign but does not lead to a considerable increase of the current. Therefore this effect more probably results from the influence of charged defects in the depletion region of the diode. A similar but smaller contact barrier was observed on some of samples with SC films. Therefore it is probably not only connected with the EPD process, but more probably related to the metal (oxide)/organic interface.

Overcoming the second contact barrier at 0.6 V leads to an exponential increase of the current. Therefore, this value could be considered as the diode forward voltage drop (U_d). Applying reverse bias, the electrical current slightly grows. Comparing the *I-V* characteristics (figure 3a, curve 1) in forward and reverse directions (at ±1.1 V), a rectification ratio of 1.6×10^4 is determined. It could be concluded from the dark current measurements that the ITO|MDMO-PPV|Al structures with EPD MDMO-PPV films exhibit a clear diode behavior.

I-V characteristics were measured on the same sample under illumination (figure 3a, curve 2) with monochromatic light (λ=560 nm, irradiance 0.59 mW.cm^{-2}). For positive voltages applied to a sample with EPD film, the *I-V* characteristics measured are similar to those measured in dark (figure 3a, curve 1). For negative applied voltages, the photocurrent measured is more than 2 orders of magnitude higher than the dark current. This effect combined with the strongly asymmetric dark *I-V* characteristic (figure 3 curve 1) and the ability of photogeneration (figure 2) express a clear photovoltaic (PV) cell behaviour.

I-V characteristics measured on samples with SC MDMO-PPV films are nonlinear and symmetrical (figure 3b).

Processing the data for the EPD MDMO-PPV PV cell (figure 3a, curve 2) determined short circuit current $J_{SC}=8.1\times10^{-10}$ A.cm^{-2} and open circuit voltage $U_{OC}=0.5$ V. From the area confined by J_{SC} and U_{OC}, the dependence of the electrical power on the voltage is plotted on figure 4, right Y axis. The maximum electrical power found were $P_{MAX}=7.6\times10^{-8}$ mW.cm^{-2} at voltage $U_{MP}=0.2$ V and current $J_{MP}=3.6\times10^{-10}$ A.cm^{-2}.

The power conversion efficiency $\eta=1.3\times10^{-5}\%$ is calculated. The low value of the power conversion efficiency measured for organic semiconductors is not exceptional. A similar value of $4.2\times10^{-4}\%$ was reported in the literature for MEH-PPV/C_{60} donor-acceptor composite, where acceptor doping is used to improve the efficiency.[13]

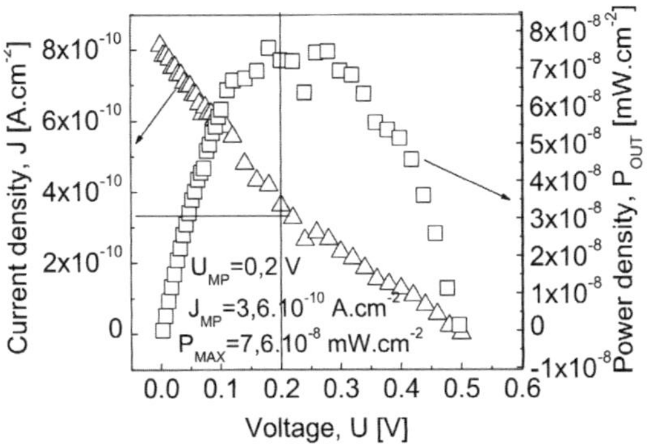

Figure 4. Dependence of the power (right Y axis) and current (left Y axis) on the voltage applied, as calculated from figure 3a, curve 2.

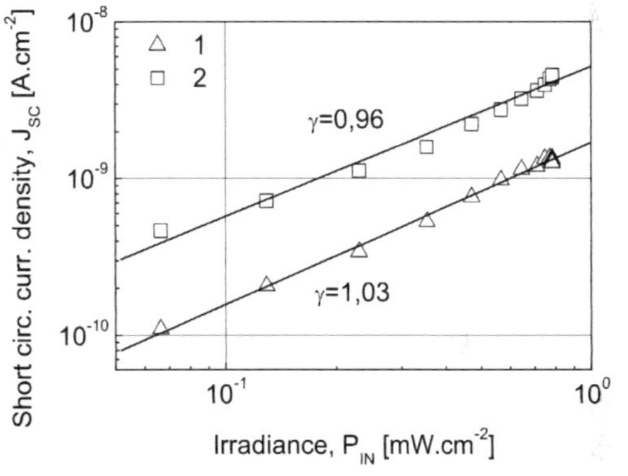

Figure 5. Dependence of the photocurrent on the light intensity measured at a fixed wavelength of $\lambda=560$ nm on samples with EPD (curve 1) and SC (curve 2) MDMO-PPV films.

The power conversion efficiency could be considerably increased in some of the following ways: decreasing the film thickness; using donor-acceptor composite; and preparing multilayer structures to reduce the electrode/semiconductor contact barrier. Such investigations are beyond the scope of this work. It should be noted here that our previous work demonstrated the advantage of the EPD method for constructing a multilayer structure from films generally dissolved in the same solvent.[14] Similar multilayer structures could be used further for improving the PV properties.

The dependences of the photocurrent on the incidental light power (irradiance) for EPD and SC MDMO-PPV films are plotted on figure 5. In both cases, the photocurrent is proportional to G^{γ}, where G is the photogeneration rate and γ is the slope in the presented graphs. For the EPD and SC devices, a slope of 0.96 and 1.03 is found, respectively. This result agrees with the literature data[15] and represents a monomolecular photogeneration mechanism. It might be concluded that the EPD process of film preparation compared to the SC method does not change the photogeneration mechanism.

The short circuit photocurrent measured from samples with SC films (figure 5, curve 2) follows the tendency observed in figure 2 and is about 1.5 times higher than EPD films (figure 5, curve 1). This result demonstrates the good photoconductive properties of the MDMO-PPV films, which could be observed at low contact barrier. It confirms that the key point for increasing the MDMO-PPV power efficiency is the optimization of the PV cell, thus increasing the Uoc. This could be achieved better with EPD (compared to SC) technique.

4. Conclusion

Absorption spectra of solution and suspension with the same MDMO-PPV concentration of 0.0033 $g.l^{-1}$ were measured. An observed broadening of the characteristic peak could be related to the formation of a solid phase in the suspension.

Thin films of about 300 nm thicknesses were prepared by EPD from a suspension and SC from a solution with a concentration of 0.0033 $g.l^{-1}$ and 8.95 $g.l^{-1}$, respectively. The EPD method produces films with the same film thickness as SC method from more than 2500 times lower concentrations.

ITO|MDMO-PPV|Al structures with EPD MDMO-PPV films behave as photovoltaic devices; the same sample configuration with SC MDMO-PPV films acts more like a photoresistor. EPD and SC films exhibit the same mechanism of charge-carrier photogeneration.

Acknowledgements

This work was supported by the South Moravian Region and the Seventh Framework Programme for Research and Development (grant SIGA 885), by the IGA Brno University of Technology via FCH/FEKT-S-11-2 project, Grant Agency of the Czech Republic project No. P205/10/2280, project "Centre for Materials Research at FCH BUT" No. CZ.1.05/2.1.00/01.0012 supported by ERDF and the National Fund of the Ministry of Education and Science, Bulgaria (grant DO 02-254).

References

[1] Heeger A J 2001 *Rev. Mod. Phys.* **73** 681
[2] Nielsen T D, Cruickshank C, Foged S, Thorsen J and Krebs F C 2010 *Solar Energy Materials and Solar Cells* **94** 1553
[3] Burroughs J H, Bradley D D C, Brown A R, Marks R N, Mackay K, Burns P L and Holmes A B 1990 *Nature* **55347** 539
[4] Tada K and Onoda M 2009 *J. Phys. D: Appl. Phys.* **42** 172001 (5pp)
[5] Tada K and Onoda M 2009 *Thin Solid Films* **518** 711
[6] Tada K and Onoda M 2003 *Jpn. J. Appl. Phys.* **42** L1279
[7] Miasik J J, Hooper A and Tofield B C 1986 *J. Chem. Soc., Faraday Trans. 1* **82** 1117
[8] Piok T, Gemarith S, Gadermaier C, Plank H, Wenzl F P, Patil S and Neher D 2003 *Adv. Mater.* **15** 800
[9] Saitoh M, Yuasa J, Tada K, Onoda M, Nakashima T and Kawai T 2011 *Physics Procedia* **14** 52
[10] Thompson B C and Frechet J M J 2008 *Angew. Chem. Int. Ed.* **47** 58
[11] van Hal P, Wienk M, Kroon J, Verhees W, Slooff L, van Gennip W, Jonkheijm P and Nanesen R 2003 *Adv. Mater.* **15** 118
[12] Trung Vu Quoc 2006 Dissertation: *Electrophoretic deposition of semiconducting polymer metal/oxide nanocomposites and characterization of the resulting films* (Dresden) 109
[13] Mo X, Mizokuro T, Mochizuki H, Tanigaki N and Hiraga T 2005 *Jpn. J. Appl. Phys.* **44** 1B 656
[14] Mladenova D, Dimov D, Karashanova D, Boyadzhiev S, Dobreva T, Budurova D, Sinigersky V and Zhivkov I 2010 *J. Optoelectron. Adv. Mater.* **12** 1952
[15] Mihailetchi V D, Wildeman J and Blom P W M 2005 *Phys. Rev. Lett.* **94** 126602

The Study of the Influence of Deposition Method on Electrical and Optical Properties of PPV Polymer with High Glass Temperature

P Heinrichová[a], I Zhivkov[a,b], D Mladenova[a,b], J David[a], M Vala[a], and M Weiter[a]

[a] Brno University of Technology, Faculty of Chemistry, Materials Research Centre, Purkynova 118, 612 00 Brno, Czech Republic
[b] Institute of Optical Materials and Technologies "Acad. J. Malinowski", Bulgarian Academy of Sciences, Acad. G. Bonchev Str. bl. 101/109, 1113 Sofia, Bulgaria

E-mail: xcheinrichova@fch.vutbr.cz

Abstract. The paper is focused on comparison of electrical and optical properties of thin layers of copolymer with high glass transition temperature Tg-PPV (poly[2-methoxy-5-(3′,7′-dimethyloctyloxy)-1,4-phenylenevinylene], poly{2-[4′-(3″,7″-dimethyloctyloxy)-1′,4′-phenyl]-1,4-phenylenevinylene} and poly{2-[3′-(3″,7″-dimethyloctyloxy)-1′,4′-phenyl]-1,4-phenylenevinylene)}) prepared by two different deposition methods: by spin-coating and by electrophoretic deposition. We found several differences in absorption and emission spectra using those two preparation techniques. On the other hand, the photoelectrical properties were similar.

1. Introduction

Study of organic electronics opens the way to develop new electronic applications, which are not achievable using classical semiconducting inorganic materials as silicon, gallium arsenide, etc. Thin layers of organic semiconductors could be easily deposited by simple "wet" techniques. Key role for the control of the film structure and properties plays the choice of the deposition method. This paper is focused on a comparison of two "wet" deposition methods of organic functional materials. One is the standard laboratory deposition method spin-coating (SC) and second method is the electrophoretic deposition (EPD).

While SC is a method for deposition from solution, EPD works with colloidal particles dispersed in poor solvents of deposed material. The suspension is prepared by precipitation of deposed material solution by adding poor solvents. During the process of particle formation they are charged due to the requirement that an electrochemical equilibrium with solvent has to be achieved. Then the charged particles are deposed on the electrode by applying electrical field. Whole mechanism of the EPD is not completely known yet due to the complexity of the process, but it is used in the industry for various applications [1].

In comparison with SC [2], [3] the properties of the EPD layers are more strongly influenced by the preparation conditions. On the other hand EPD allows controlled preparation of nanostructured layers [4], and a formation of multilayer stacks from materials soluble in the same solvent [5]. In comparison

Journal of Physics: Conference Series **398** (2012) 012057 doi:10.1088/1742-6596/398/1/012057

with SC the EPD uses more effectively the deposed material, but consumes more of high quality solvents.

This work is focused on the study of thin layer preparation of polymeric materials for organic electronics. Poly(*p*-phenylenevinylene) derivatives are one group of polymeric materials with a potential application in the organic electronics. One of the well-known and described derivate is the MDMO-PPV (poly[2-methoxy-5-(3′,7′-dimethyloctyloxy)-1,4-phenylenevinylene]), which serves as model electron donor material for organic electronics. Structure of MDMO-PPV is shown on figure 1. MDMO-PPV has relatively low glass transition temperature T_g = 45 °C [6], which influences negatively the thermal stability of the active layer. Triblock copolymer of poly[2-methoxy-5-(3′,7′-dimethyloctyloxy)-1,4-phenylenevinylene], poly{2-[4′-(3″,7″-dimethyloctyloxy)-1′,4′-phenyl]-1,4-phenylenevinylene} and poly{2-[3′-(3″,7″-dimethyloctyloxy)-1′,4′-phenyl]-1,4-phenylenevinylene)}, denoted here as Tg-PPV (see figure 1) with T_g = 138 °C [6, 7] seems to be a more suitable alternative. Tg-PPV is generally synthesized for use in the organic light emitting diodes. It is distributed by Merck KGaA, Darmstadt, Germany under the commercial name PDO-123 [8].

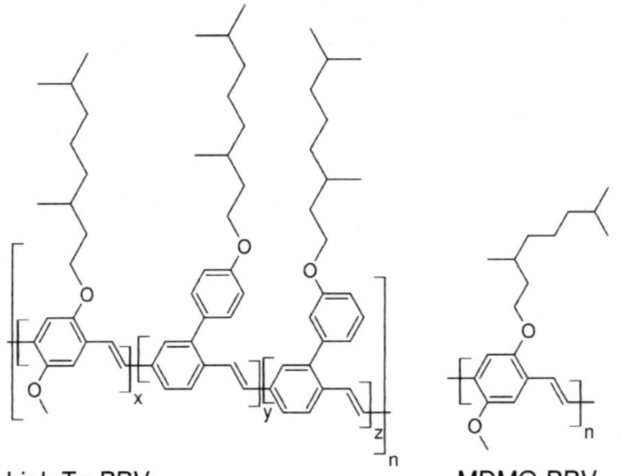

Figure 1. Structure of the used here high Tg-PPV and more studied in the literature MDMO-PPV derivate.

2. Experimental

Films of Tg-PPV were prepared on glass substrates commercially covered with ITO (indium thin oxide) electrode (Sigma–Aldrich Co., Steinheim, Germany).

SC thin layers of polymer were deposed from chlorobenzene solution of concentration 6 g/l of Tg-PPV. The deposition was performed at rotation speed of 1000 RPM for a time of 1 min on a spin-coater KW-4A (Chemat Technology, Inc., CA, U.S.A.).

EPD was carried out from Tg-PPV suspension with a concentration of $7.74 \cdot 10^{-3}$ g/l. Suspension was prepared from solution of Tg-PPV in toluene by precipitation with acetonitrile. Optimal ratio of toluene and acetonitrile of 42.5 % was determined by light absorption measurements on Varian Cary 50 double-beam spectrometer. Deposition was realized under constant current condition in the range of 0.060 to 0.150 mA. Source of electrical current was Keithley 2410 Source Meter (Keithley Instruments Inc., Cleveland, OH, U.S.A.). The value of the current controls the rate of deposition. Time of deposition varied between 30 and 60 s and influenced the final film thickness.

Optical properties for quality assurance of thin layers and for suspension optimization were studied by a measurement of UV-VIS absorption spectra on Varian Cary Probe 50 UV–VIS spectrometer and fluorescence spectra on Horiba Jobin Yvon Fluorolog (Horiba Jobin Yvon S.A.S., Chilly Mazarin,

France). Liquid samples were measured in Hellma QS 10 mm quartz cuvette (Hellma GmbH, Mühlheim, Germany).

Thicknesses of layers were characterized by optical profilometer MicroProf FRT (Fries Research & Technology GmbH).

Samples for electrical characterization with semiconductor Tg-PPV thin layers were subsequently covered with aluminium electrodes by vacuum deposition. Deposition was realized in the deposition chamber U311B Tesla (Tesla, Opočno, Czech Republic).

The samples were electrically characterized in cryostat VPF-475 (Janis Research Company Inc., Wilmington, MA, U.S.A.) at 10^{-4} Pa at laboratory temperature by picoampermeter Keithley 6478. Current voltage characteristics were measured on samples both in dark and irradiated by LSH502 LOT Oriel xenon lamp with emission spectrum similar to the sun light (LOT-Oriel Co., Darmstadt, Germany). Intensity of the incident light of 1 mW·cm^{-2} was controlled by irradiance meter X97 Gigahertz-Optik with radiometric detector head RW-3703-4 for UV-VIS range (Gigahertz-Optik GmbH, Türkenfeld, Germany).

3. Results and discussion

3.1. Optical study of suspension

A suspension concentration is a key parameter for optimal electrophoretic deposition. Suitable volume ratio of acetronitrile/toluene for suspension preparation was found by UV-VIS absorption spectrometry. Figure 2 A) shows absorption and emission spectra of Tg-PPV solution in toluene (0% acetonitrile) and after addition of 42 and 90 % volume ratio of acetronitrile. Noticeable bathochromic shift of absorption band from 476 nm to 494 nm with the increase of acetonitrile volume ratio was observed. The emission spectra behave similar, but they are more sensitive to acetonitrile addition. Bathochromic shift of emission maximum is from 549 nm to 590 nm. Generally the evaluation of the emission spectra is more difficult due to effects as fluorescence quenching, reabsorption of emitted light, changes in vibrational structure of spectra and others. On the other hand fluorescence spectra provide valuable additional information.

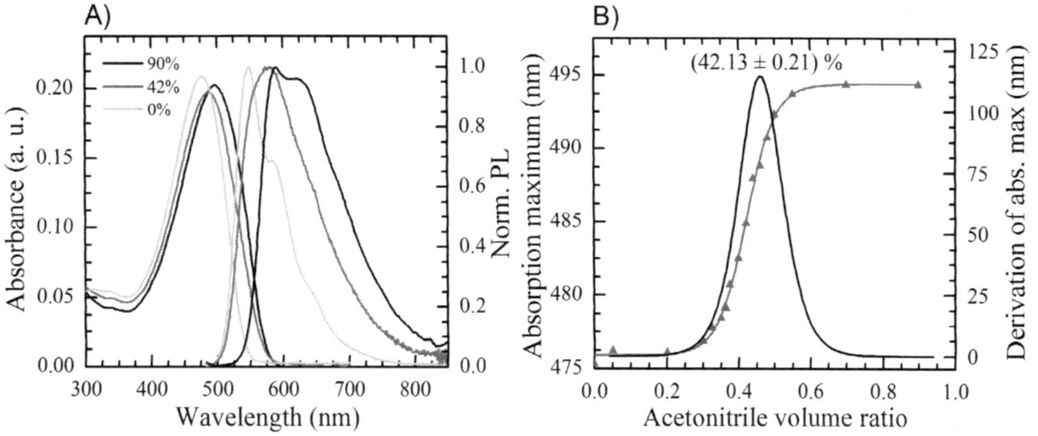

Figure 2. A) Absorption and emission spectra of toluene solution of Tg-PPV after addition of 0, 42 and 90 % volume ratio of acetonitrile. B) A dependence of absorption maxima position on the acetronitrile volume ratio *w*. Right axis is related to the first derivative of the curve. Maximum at 42.13 % of acetonitrile corresponds to the inflection point.

The position of the absorption maxima were precisely determined by taking the first derivative of the absorption spectra. The obtained values are plotted as a function of the acetonitrile volume ratio in figure 2 B). Sigmoid curve shows range (from 30 % to 55), where acetonitrile volume ratio influences the suspension concentration. A first derivative of the sigmoid curve was also taken and inflection point at (42.13 ± 0.21) % was found. This acetonitrile volume ratio was used for EPD of thin layers.

3.2. Optical study of thin layers

Absorption and emission spectra of Tg-PPV: solution in toluene; solution in chlorobenzene; toluene/acetonitrile suspension with 90% acetonitrile; thin EPD film produced from a toluene/acetonitrile suspension with 42.5 % acetonitrile; thin SC films produced from a solution in chlorobenzene are presented in figure 3.

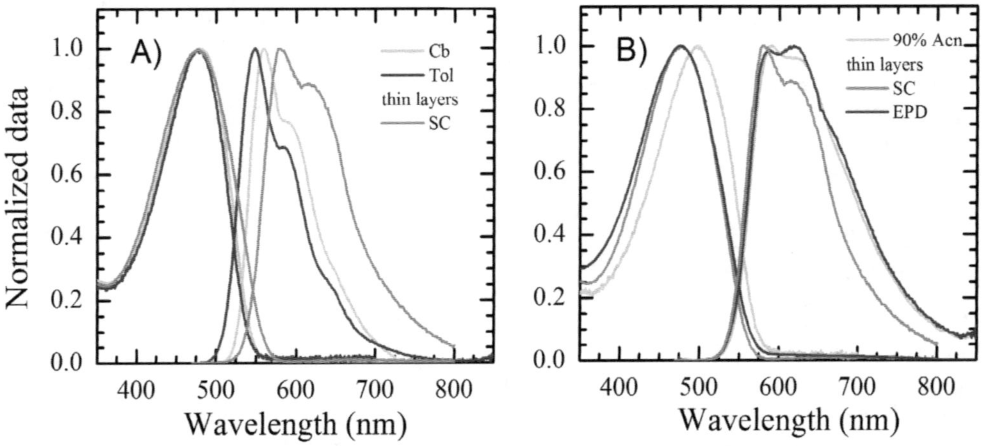

Figure 3. Comparison of absorption and emission spectra of Tg-PPV plot: A) solutions in chlorobenzene (Cb) and toluene (Tol) and thin layer prepared by SC, plot B) suspension prepared by addition of 90% volume acetronitrile (90% Acn) to a toluene solution, and thin SC and EPD layers.

Absorption spectra of chlorobenzene solution and thin film prepared by SC from chlorobenzene solution don't have resolute vibrational structure. This effect is usually related to the disordered structure of polymer [9]. Both spectra showed the same position of absorption band, only thin layer absorption edge is 20 nm red shifted. This effect is related to the closer arrangement of polymer chains in the layer due to aggregation [9]. Emission spectrum of SC layer is found fully bathochromically shifted compared to the spectrum of chlorobenzene solution [10].

More differences in spectra taken from toluene solution, acetonitrile suspension and EPD layer are seen. EPD layer performs an absorption edge at 580 nm like suspension, prepared by adding of 90% volume of acetronitrile and like SC layer, but absorption maximum lies at 476 nm like in spectra, measured from solutions and SC layer. Emission spectrum of suspension and EPD layers are overlapped, only vibrational structure of spectrum is changed. Perhaps there was particularly preserved nanostructure of suspension present and particularly spill in uniform film [11]. Strict interpretation needs more investigations to be done. However, the spectra measured don't indicate a degradation of the polymer after EPD, which may have been caused by electrical current during deposition.

3.3. Electrical study of thin layers

Several thicknesses of thin films were prepared by EPD. The aim was to prepare layers comparable to the 120 nm layer prepared by SC as a reference.

Figure 4 shows samples with best current voltages curves of SC and EPD layers. Layers were characterized in dark and under illumination with light of intensity 1 mW/cm^2. In dark, the EPD layers were less conductive. The threshold voltage V_T, when current start strongly grow, for SC layers was around 1.1 V, whereas for SPD layers are located at much higher voltages (higher than 2 V); see figure 4. It is caused by an energy barrier for charge carries. Origin of barriers may be different: morphology of layer, bad adhesion of layers and other. Current voltage characteristics of EPD layers under illumination were on average comparable with those, taken from SC layers.

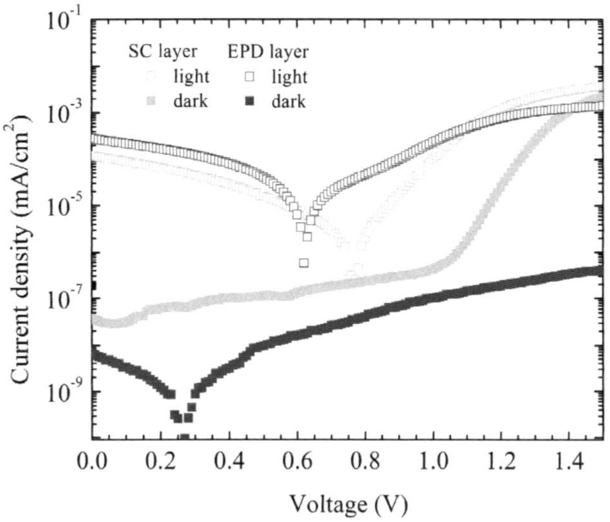

Figure 4. Comparison of current voltage characteristics measured on samples with SC and EPD layers.

Table 1. Average photovoltaic conversion parameters of three EPD layers with different thicknesses and one SC layer.

sample	tickness (nm)	j_{SC} (mA/cm2)	V_{OC} (U)	*FF*	μ(%)
SP	120 ± 20	$(1.2 \pm 0.7)\cdot 10^{-4}$	0.6 ± 0.2	0.22 ± 0.08	$(1.8 \pm 0.3)\cdot 10^{-3}$
EPD 1	≈ 100	$1.75\cdot 10^{-4}$	0.52	0.22	$2.02\cdot 10^{-3}$
EPD 2	290 ± 80	$(1.8 \pm 0.6)\cdot 10^{-4}$	0.57 ± 0.05	0.26 ± 0.03	$(2.7 \pm 0.9)\cdot 10^{-3}$
EPD 3	540 ± 80	$0.09\cdot 10^{-4}$	0.59	0.21	$0.11\cdot 10^{-3}$

Parameters as short current density j_{SC}, open circuit voltageV_{OC}, fill factor *FF* and maximal photovoltaic efficiency μ were calculated from current voltages measurements taken under light exposure. Results are compared by plotting diagrams on figure 5 and summarized in table 1. Thin EPD layers (sample EPD 1 and EPD 2) have similar parameters as SC layers within the measured uncertainty. Thick layers EPD 3 (540 nm) has a reduced value of the short current density. This is an expected result, because the conductivity of the organic layers is driven by the electric field, which depends on the distance between the electrodes [12, 13].

4. Conclusion

Electrical and optical properties of thin Tg-PPV layers prepared by SC and EPD were compared. The differences in the emission spectra could be related to a nanostructure of the EPD films, while the SC films are more structureless.

Similar photovoltaic parameters of SC and EPD layers were found, only the dark-conductivity of EPD layers is lower, due to presence of higher barrier observed in current voltage curves.

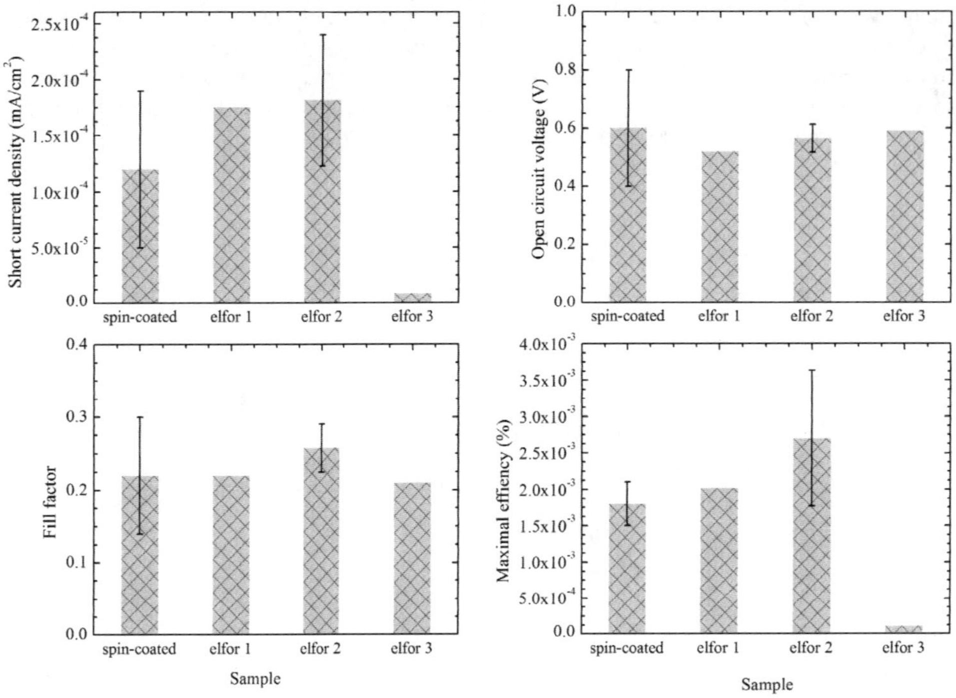

Figure 5. Diagrams of average photovoltaic conversion parameters of three EPD layers with different thicknesses (see table 1) and one SC layer.

Acknowledgement

This work was supported by the South Moravian Region and the Seventh Framework Programme for Research and Development (grant SIGA 885), Grant Agency of the Czech Republic project No. P205/10/2280, project "Centre for Materials Research at FCH BUT" No. CZ.1.05/2.1.00/01.0012 supported by ERDF and the National Fund of the Ministry of Education and Science, Bulgaria (grant DO 02-254).

References

[1] Van Tassel J J and Randall C A 2006 *Key Engineering Materials* **314** 167-174
[2] Norrman K, Ghanbari-Siahkali A and Larsen N B 2005 *Annu. Rep. Prog. Chem., Sect. C,* **101** 174–201
[3] Besra L and Liu M 2007 *Progress in Materials Science* **52** 1–61
[4] Tada K and Onoda M 2006 *Thin Solid Films* **499** 19-22
[5] Mladenova D, Sinigersky V et al. 2010 *Journal of Physics: Conference Series* **253** 012041
[6] Vandewal K, Gadisa A et al. 2008 *Advanced Functional Materials* **18** 2064–2070
[7] Bertho S, Haeldermans I et al. 2006 *Spie-Int Society Optical Engineering* DOI: 10.1117/12.663256.
[8] *Innovative Materials for Next Generation Electronics*, Merck. Electronical document URL <http://www.merck-chemicals.com/showBrochure/201004.144.ProNet.pdf> [cit. 2012-08-02].
[9] Lemmer U and Göbel E O 1997 *Primary Photoexcitations In Conjugated Polymers: Molecular Exciton Versus Semiconductor Band Model* ed Sariciftci N. S. (World scientific, Singapur) Chapter 9
[10] Traiphol R and Charoenthai N 2008 *Synthetic Metals* **158** 135–142

[11] Tada K and Onoda M 2008 *J. Phys. D: Appl. Phys.* **41** 4
[12] Blom P W M and Vissenberg M C J M 2000 *Materials Science and Engineering* **27** 53–94
[13] Deibel C, Dyakonov V 2010 *Rep. Prog. Phys.* **73** 096401

Journal of Physics: Conference Series **398** (2012) 012058 doi:10.1088/1742-6596/398/1/012058

Electrical conductivity, photoconductivity and gas sensitivity of Ge-Se-Te thin films

D Nesheva[1], Z Levi[1], Y S Raptis[2], C Raptis[2], K Petkov[3] and V Vassilev[4]
[1] Institute of Solid State Physics, Bulgarian Academy of Sciences, 72 Tzarigradsko chaussee blvd, 1784 Sofia, Bulgaria
[2] Department of Physics, National Technical University of Athens, GR-15780 Athens, Greece
[3] Institute of Optical Materials and Technologies, Bulgarian Academy of Sciences, Acad. G. Bonchev str., bl. 109, 1113 Sofia, Bulgaria
[4] Department of Non-Ferrous Metals Metallurgy and Semiconductors Technology, University of Chemical Technology and Metallurgy, 8 Kl. Ohridski blvd, 1756 Sofia, Bulgaria

E-mail: nesheva@issp.bas.bg

Abstract. Films of $Ge_xSe_yTe_z$ ($x \approx 33$) with a thickness of 0.8 μm and four different compositions were prepared by thermal evaporation of preliminary synthesized glasses. The glass and film compositions were determined by Energy-Dispersive X-ray Spectroscopy. X-ray diffraction measurements have confirmed the amorphous structure of both bulk and thin film samples. The temperature dependence of the dark conductivity has been measured in the range 20 – 110 °C. No significant change of the room temperature conductivity (around $5x10^{-7}$ S/cm) and the dark current activation energy (around 0.7 eV) have been observed with increasing *z*. Spectral photocurrent measurements have revealed that the increase of Te content results in: (i) a photoconductivity decrease and (ii) a red shift of the low-energy side of the spectrum which indicates an optical band gap decrease. Resistive chemical sensing tests carried out at room temperature with ammonia, acetone, ethanol or water vapours in air as carrying gas have shown that the $Ge_{32}Se_{55}Te_{13}$ films are sensitive to ammonia but not to water vapour and hence they are suitable for selective ammonia sensors operating at room temperature. The observed sensitivity is related to the specific lattice structure of the region nearest to the film surface.

1. Introduction

Amorphous chalcogenides show a unique set of photoinduced effects which makes them very interesting for optical fabrication and performance [1,2]. Besides, these glasses have attracted much attention for the development of active and passive infrared devices. A great variety of alloys such as Ge-Sb-Te, In-Sb-Te, Te-Ge-Sb-S, Ge-Sb-Se-Te and many others have been suggested for use as the active material in non-volatile memories [3-5]. Intensive investigations in the field of ion-selective electrodes on the basis of chalcogenide glass materials have been carried out during the last two decades [6-8].

Today, there is an increasing demand for gas sensors with selectivity for specific gases in different fields of applications, e.g. industrial tasks like process control, challenges in homeland security, detection of chemical or biological hazardous substances, pollution control etc. Compared to metal oxide devices, chalcogenide-based sensors appear superior as they can be readily operated at ambient temperatures. Also, in comparison to their crystalline counterparts, amorphous chalcogenides exhibit useful advantages which originate from their good long-term stability and resistance to chemical reaction. In a series of works [9,10] a class of gas sensors using thin films from pure tellurium or its alloys from the As–Ge–Te system has been proposed to detect some pollutants in ambient air. Such films show very good sensing properties to NO_2 even at room temperature. Results on the interaction of Te-based thin films with different gas species such as propylamine, carbon oxide [11] and ammonia [12] have also been reported. Also, a relevant study has been performed on $Te_{13}Ge_x$-based binary, ternary and quaternary alloys of Te, Ge, Se, As, Mg, I and S [13] in an attempt to find appropriate material compositions for use in future NO sensors operating at room temperature. All these results have motivated further studies to find appropriate materials for each separate analyte. A way to obtain suitable sensing properties is the development, preparation and characterization of complex chalcogenide glasses.

In this study, $Ge_xSe_yTe_z$ thin films with four different compositions have been prepared by thermal vacuum evaporation of preliminary synthesized glassy materials. Their structure, electrical and photoelectrical properties have been studied and their sensitivity to various vapours such as ammonia, alcohol, acetone, water and isopropanol has been tested.

2. Experimental details

Selected complex glasses of the $(GeSe_2)_x(GeTe)_y(PbTe)_z$ system ($x = 54 - 90$, $y = 9 - 36$, $z = 0, 10$) at four different compositions (see table 1) were synthesized by the conventional melt quenching technique. The synthesis was performed with step-wise heating up to 773, 1073 and 1248 K, at rates 3 - 4, 2 - 3 and 2 – 3 $K.min^{-1}$, respectively. Homogeneous annealing of the melts was carried out at these temperatures for 0.5, 0.5 and 2.0 h, respectively, combined with continuous vibration stirring at the highest temperature. The samples were quenched in melting ice and at a cooling rate of 15 $K.sec^{-1}$. The glassy state was verified by XRD and electron microscopy analyses [14].

Thin films (of around 0.8 μm thickness) were prepared by thermal evaporation of the synthesized and powdered glasses in a high vacuum of 6×10^{-6} Torr at evaporation rate of 0.8 - 1.0 $nm.sec^{-1}$. The films were deposited on crystalline silicon or glass substrates at room temperature for investigations of composition/surface morphology and for electrical measurements, respectively.

The composition of both bulk samples and thin films was determined by a scanning electron microscope with an X-ray microanalyzer Jeol Superprobe 733 (Japan) at an electron accelerating voltage of 20 kV and a current of 1.4 nA. The results from the X-ray microanalysis show (table 1) that no Pb has been detected in the films produced from source material with 10 mol % of PbTe. Thus ternary $Ge_xSe_yTe_z$ films have been produced with near stoichimetric Ge content (the atomic concentration has been determined with an accuracy of ± 1%) and four different Se/Te compositions.

Table 1. Source glass and thin film compositions determined by X-ray microanalysis.

Sample No	Source materials (mol %)	Glasses (at %)	Thin films (at %)
1.	$(GeSe_2)_{90}(GeTe)_{10}$	$Ge_{34}Se_{62}Te_4$	$Ge_{31}Se_{66}Te_3$
2.	$(GeSe_2)_{81}(GeTe)_9(PbTe)_{10}$	$Ge_{32}Se_{57}Te_7Pb_4$	$Ge_{32}Se_{64}Te_4$
3.	$(GeSe_2)_{67.5}(GeTe)_{22.5}(PbTe)_{10}$	$Ge_{34}Se_{50}Te_{12}Pb_4$	$Ge_{33}Se_{59}Te_8$
4.	$(GeSe_2)_{54}(GeTe)_{36}(PbTe)_{10}$	$Ge_{35}Se_{43}Te_{18}Pb_4$	$Ge_{32}Se_{55}Te_{13}$

The surface morphology of the films was studied by means of a Scanning electron microscope Philips SEM 515 operating in a regime of secondary electrons and at an accelerating voltage of 20 kV. The angle between the electron beam and the normal to the film surface was varied between 0^0 and

40^0. The obtained images showed quite smooth film surface with no distinct grains larger than 100 nm (that is the image resolution).

Stokes Raman spectra in a pseudo-backscattering geometry were recorded using a system of a SPEX 1403 double monochromator (operated at a spectral slit width of ~ 3 cm^{-1}) and a cooled photomultiplier connected to photon counting equipment. The 488 nm (2.54 eV) line of an Ar+ laser was used for the excitation in a line focusing configuration (cylindrical lens) in order to avoid degradation of the films. The Raman spectra were obtained at 293 K in air.

Dark current dc measurements were carried out by means of a Keithley 6487 Picoammeter using co-planar aquadag contacts on the top surface of the layers; the contacts produce a gap cell with an active area of ~ (1×10) mm^2. Current-voltage (*I-V*) characteristics were measured in dark at applied electric fields in the range 1 - 10^2 V cm^{-1} which showed Ohmic-like behaviour of the contacts. The temperature dependences of the dark and photocurrent were determined in the 293 – 383 K range at a heating rate of 0.1 $K.sec^{-1}$. Prior to each measurement, the samples were heated in dark up to 383 K in order to achieve charged carrier equilibrium and to desorb water molecules captured on the film surface. The applied temperatures are well below the glass transition temperatures we determined (> 500 K for all compositions studied) and the repetition of the described measuring cycle did not cause changes in the film conductivity and dark current activation energy.

The spectral dependence of the photocurrent was measured in air at 293 K in the wavelength range 500-1000 nm under illumination with monochromatic light (2 - 7×10^{15} $photons.s^{-1}.cm^{-2}$) from a diffraction grating monochromator MDR 2 at a spectral bandpass per slit width of 2 $nm.mm^{-1}$. The films were illuminated with chopped light (2 Hz, ON/OFF ratio = 1) and thus only the contribution of photoexcited charge carriers in the total current through the films was detected. The as-measured photocurrent spectra were corrected for the spectral sensitivity of the system (to constant photon flux) and then normalized.

A preliminary test of the films sensitivity to various gases has been performed by measuring the dc dark current (I_d) at room temperature before (in a vacuum cryostat evacuated to ~ 10^{-2} Torr) and after exposure to water vapour, ammonia, ethanol, acetone and isopropanol vapours in air as the carrier gas [15]. The vapour content in the air which entered the cryostat upon opening the entrance valve was not strictly controlled. The level of this content was determined by the vapour drawn into the cryostat from a soaked piece of cloth fixed to the valve entrance. The exposure procedure described was repeated several times on each sample for each analyte and a good reproducibility has been observed. Normally, the evacuation of the cryostat for 3 - 5 minutes was enough to completely return the samples in the initial state (i.e. to recover the initial I_d values) and sample heating was not necessary. Although these measurements are qualitative, they can be useful to distinguish compositions suitable for sensor applications.

3. Results and Discussion

3.1. Dark conductivity

Current-voltage characteristics measured at room temperature in air are depicted in figure 1. The linear fit performed shows that the characteristics are Ohmic-like ($I_d \sim V^m$, $m \approx 1$), which ensures correct determination of both film conductivity and its thermal activation energy. In order to eliminate the effect of water vapour and other molecules adsorbed on the film surface, the conductivity values were calculated using dark current values measured in vacuum of ~ 10^{-2} Torr. It has been obtained that the room temperature dark conductivity varies within less than an order of magnitude (between 1 and 5×10^{-7} $S.cm^{-1}$). The weak conductivity variations observed upon increasing the Te content from 3 to 13 at % implies that the films are most likely amorphous. This implication is confirmed by the X-ray diffraction data given below (figure 2). Probably, Te atoms replace Se ones, but do not create high free carrier concentration. On the other hand, in amorphous chalcogenides a disorder increase results in an increase of the density of gap states trapping holes and thus diminishes the density of free holes in the valence band. Hence, the overall dark conductivity decrease observed with increasing Te content (figure 1) may be connected with some increase of the level of network disorder. This suggestion is in agreement with the results of other authors [16].

Figure 2 shows an X-ray diffraction spectrum of a $Ge_{32}Se_{64}Te_4$ film. A broad band is seen that is centred at $2\Theta \sim 24^{\circ}$; no strong narrow bands are observed. Similar spectra have been obtained for the other three compositions. They show that films of all composition are amorphous, thus confirming the assumption made on the basis of the electrical measurements.

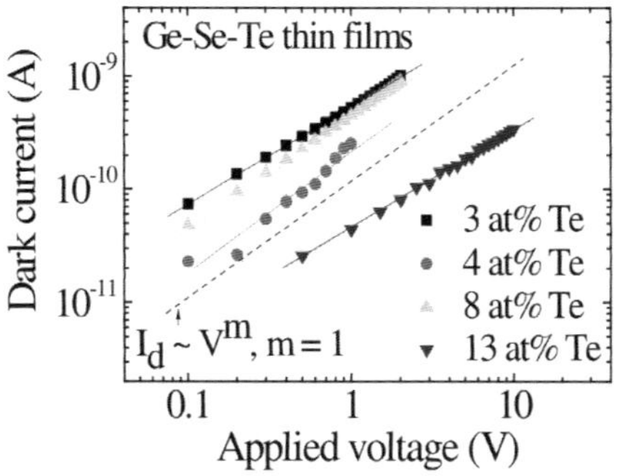

Figure 1. Current-voltage characteristics of four Ge-Se-Te thin films of various compositions.

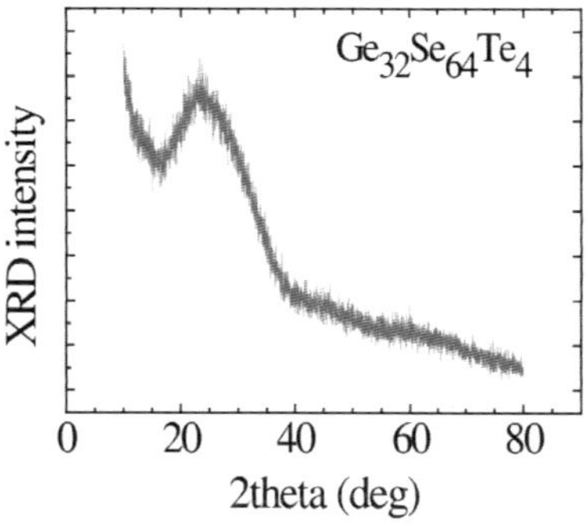

Figure 2. XRD spectrum of a $Ge_{32}Se_{64}Te_4$ thin film.

A series of dark current temperature dependences measured on $Ge_xSe_yTe_z$ films with various compositions are shown in figure 3 in semi-log coordinates. The observed linearity of all dependences indicates that they obey the relation

$$I_d \sim \exp(-E_d/kT) \tag{1}$$

which corresponds to a thermally activated process. In eq.(1) E_d is the dark current activation energy, k the Boltzman constant and T the measuring temperature. The obtained values for the activation energy are in the range (0.67-0.70) ± 0.03 eV i.e. the activation energy practically does not change with the composition. Activation energies of 0.8 - 1.1 eV [17,18] and ~ 0.35 eV [19] have been reported for amorphous $GeSe_2$ and GeTe, respectively. The obtained values of 0.67 - 0.7 eV are smaller but close to the activation energy of amorphous $GeSe_2$.

At high temperatures (normally $T > 300$ K), the electrical conductivity of most undoped chalcogenide glasses is due to transport of free carriers in the extended states of the valence bands and the activation energy E_d represents the energy distance between the Fermi level E_F and the valence band. Taking into account that the measuring temperatures in this study are $T \geq 293$ K one can relate the electrical conductivity of the Ge-Se-Te films to transport of holes via the extended state of the valence band.

3.2. Photoconductivity

Using the maximum current in the as-measured photocurrent spectral dependences, the relative film photoconductivity has been calculated. Comparison of the values obtained has shown that the film photoconductivity decreases approximately 100 times with increasing Te content. This decrease is most likely due to the above assumed network disorder increase which may cause a decrease of both lifetime and mobility of the photoexcited carriers.

Normalized spectral dependences of the photocurrent are depicted in figure 4. Some difference is observed in the high energy part of the photocurrent spectra which is related to light absorption close to the film surface. This difference may be due to the effect of gas molecules absorbed on the films surface. It is also seen from figure 4 that the increase of Te from 3-4 at.% to 8-13 at.% causes a "red"

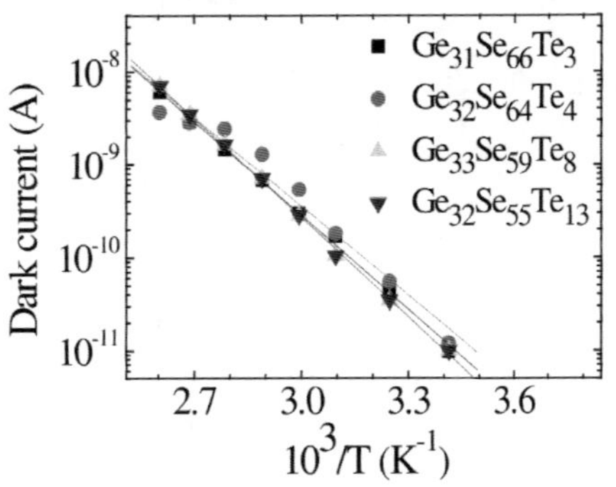

Figure 3. Temperature dependences of the dc dark current through Ge-Se-Te thin films of compositions denoted in the figure. The solid lines represent a linear fit of the experimental points.

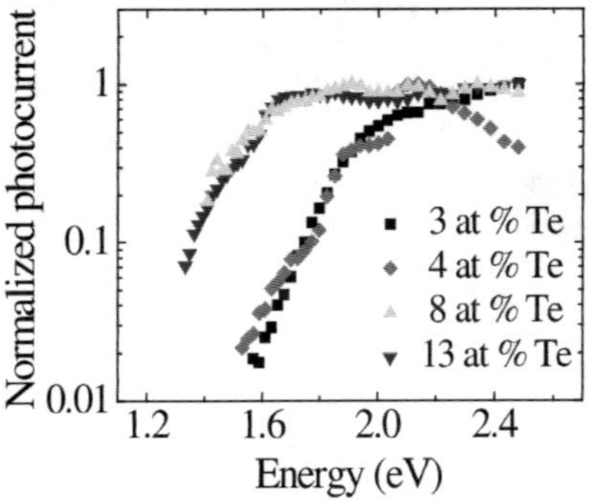

Figure 4. Normalized photocurrent of $Ge_xSe_yTe_z$ films with increasing Te content. The fine structure seen in the spectra is due to light interference.

shift of the low-energy part of the spectrum and thus an extension of the range of film spectral sensitivity is obtained. This observation indicates that the optical band gap E_g^o of the films decreases with increasing Te content. The "red" shift corresponds to formation of ternary Ge-Se-Te alloys whose band gap decreases with increasing Te content. Similarly, short wavelength absorption edge shift towards longer wavelength has been registered as a result of Te → Se substitution in the $Ge_{20}Se_{80-x}Te_x$ glass system [20]. A rough estimation of the optical band gap has been made based on the energy corresponding to the photocurrent value of ($I_p^{max}/2$) at the low energy side of the curves. Reasonable values of ~ 1.9 eV and ~ 1.55 eV have been obtained for samples NNo 1, 2 and NNo 3, 4, respectively (the room temperature optical gap of $GeSe_2$ is $E_g^o = 2.1$ eV [21] and that of GeTe $E_g^o = 0.7$ eV [19]). The E_g^o values we obtained from optical transmission measurements also show a gradual decrease with increasing Te content; values of 1.86, 1.58, 1.49 and 1.33 eV have been determined. The photoelectrically and optically determined E_g^o values are in good agreement for the lowest Te content but not for the higher ones. The observed discrepancy may be understood assuming that the films are not completely homogeneous and the main contribution to the photocurrent comes from regions with lower Te content which displayed higher photoconductivity. The inhomogeneity assumed could also be the reason for the E_d constancy with composition though the optical band gap decreases with increasing Te content.

3.3. Chemical sensitivity

The results from the test for the chemical sensitivity of the films are summarized in table 2. The films response time, i.e. the time for achieving the maximum dark current alteration, was quite short, between 5 and 40 sec. The films do not show any systematic change of their sensitivity to any vapour with changing Se/Te content. Additionally, films with very similar Te content of 3-4 at % show quite different behaviour. The most interesting result obtained in these experiments concerns the $Ge_{32}Se_{55}Te_{13}$ films which show high sensitivity to ammonia at room temperature and a response time of 20 sec. It is important that $Ge_{32}Se_{55}Te_{13}$ films are practically insensitive to water vapour.

In order to understand better the results of table 2, one should keep in mind that the sensitivity is mainly related to changes in the conductivity of the near surface region caused by the adsorbed vapour molecules. The observed dispersion of the results indicates that the initial surface state of the films is quite different, which implies different compositions or/and different structure. As mentioned in the introduction, tellurium plays an important role in the gas sensitivity of chalcogenide films. Because of the application of thermal evaporation for the film preparation, one could expect [22] that the near

Table 2. Chemical sensitivity of Ge-Se-Te thin films to five different vapours in carrying gas air. The ratio of the dark current under film exposure to the respective vapour, I_{exp}, to the equilibrium dark current in vacuum, I_{eq}, is taken as measure of the film sensitivity.

Film composition	ammonia	ethanol	isopropanol	acetone	water
	I_{exp}/I_{eq}	I_{exp}/I_{eq}	I_{exp}/I_{eq}	I_{exp}/I_{eq}	I_{exp}/I_{eq}
$Ge_{31}Se_{66}Te_3$	15	400	20	7.5	300
$Ge_{32}Se_{64}Te_4$	28	22	28	42	40
$Ge_{33}Se_{59}Te_8$	1.8	3.3	4.4	8.9	1.8
$Ge_{32}Se_{55}Te_{13}$	750	-	-	-	1.5

surface region will be slightly Te-enriched when compared with the interior of the film. The enrichment depends on the film deposition conditions and therefore it may vary from batch to batch thus causing variations in the film response. Precise control of the deposition conditions is required for obtaining reproducible results. From the SEM images the surface microstructure of all films was very similar and hence one could not expect that it will significantly affect the film sensitivity.

We have obtained structural (Raman) data for the near-surface (100-150 nm) layer of the films corresponding to the penetration depth of the 488 nm exciting line. Figure 5 shows the normalized spectra of two films, namely those with 4 and 13 at% Te. The spectra of the other two films with 3 and 8 at% Te are similar to that of the 4 at% film. Three main bands are observed, with two of them attributed [23] to symmetric vibrations in corner-sharing (CS, ~200 cm^{-1}) and edge-sharing (ES, ~215 cm^{-1}) $Ge\text{-}Se_4$ tetrahedra. The third band, at ~ 175 cm^{-1}, has also been observed in ternary $Ge_{20}Se_{80-x}Te_x$ glasses [24] showing a similar intensity increase for Te-rich samples. This, Te-content dependent, band has been related [24] to Te-Te short chains bridging $Ge\text{-}Se_4$ tetrahedra. Existence of Ge-Ge bonds may also have some contribution to this band. The spectrum of the 13 at% Te film shows a well expressed additional band at ~90 cm^{-1} (figure 5) which may be attributed to bending vibrations in $Ge\text{-}Te_4$ tetrahedra [25]. The Raman scattering results confirm that the introduction of Te in the glassy network of $Ge\text{-}Se_4$ tetrahedra results in an increase of disorder and this is compatible with the photocurrent reduction observed upon increasing the Te-content. Hence, the structure of the near surface region of the 13 at% Te films is quite different from the structure of the other films and one can reasonably assume that this difference is responsible for the observed good room temperature sensitivity and selectivity of films with this composition.

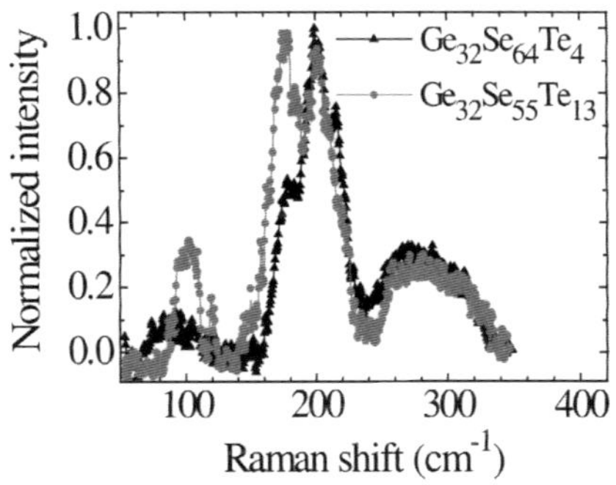

Figure 5. Raman scattering spectra of two $Ge_xSe_yTe_z$ thin films whose composition is given in the figure.

Nowadays, the sensors using metal oxides hold a leading position in the group of chemical gas sensors [26]. However they have some drawbacks such as low selectivity, a relatively long response and recovery times, and a strong effect of the water vapour on their response. Besides, they operate at high temperatures (423 K – 623 K) and this requires power consumption of ~ 1 W. Therefore, the observed high selectivity, good sensitivity and fast response make the $Ge_{32}Se_{55}Te_{13}$ films good candidates for the preparation of ammonia sensor operating at room temperature in standard conditions (in humid air).

4. Conclusions

It has been shown that the investigated $Ge_xSe_yTe_z$ thin films are amorphous. No considerable changes of both the room temperature conductivity (varying between 1×10^{-7} and 5×10^{-7} $S.cm^{-1}$) and the dark current activation energy (0.67 - 0.7 eV) have been observed with increasing Te content from 3 - 4 to 13 at % which has been related to the amorphous structure of the films. However, a strong photoconductivity decrease has been observed with increasing Te content which has been assigned to a disorder increase of the amorphous network. A "red" shift of the low-energy side of the photocurrent spectra has been seen with increasing Te content. It indicates that the optical band gap decreases as a result of Te $\rightarrow$ Se substitution in the films. It has been found that the $Ge_{32}Se_{55}Te_{13}$ films are sensitive to ammonia vapours, but not to air humidity. The observed sensitivity and selectivity have been connected with the specific network structure of the region nearest to the film surface. These properties, together with their short response time, make the $Ge_{32}Se_{55}Te_{13}$ films interesting for the preparation of ammonia sensors operating at room temperature in standard conditions (in humid air).

Acknowledgments

The publication of this paper is supported under the Human Resources Developments Programme of the European Social Fund though grant scheme № BG051PO001-3.3-05/0001 "Science and business", contract D02-743/02.08.2012. The authors thank the undergraduate student M Christov for making a part of the electrical measurements.

References

[1] Zakery A and Elliott S R 2003 *J. Non-Cryst. Solids* **330** 1
[2] Mehta N 2006 *J Scientific and Industrial Research* **65** 777
[3] Yamada N, Nagata K, Nishiuchi K and Furukawa Sh 1998 *EP patent* 0899725 A1
[4] Liu J 2009 *US patent* 7638357 B2
[5] Lung H-L 2011 *US patent* 7932101 B2
[6] Schubert J et al 1999 *Appl. Phys. A* **69** [Suppl.] S803
[7] Kloock J P, Mourzina Y G, Ermolenko Y, Doll Th, Schubert J and Schöning M J 2004 *Sensors* **4** 156
[8] Vassilev V S and Boycheva S V 2005 *Talanta* **67** 20
[9] Marian S, Tsiulyanu D, Marian T and Liess H-D 2001 *Pure Appl. Chem.* **73** 2001
[10] Tsiulyanu D, Stratan I, Tsiulyanu A, Liess H-D and Eisele I 2007 *Sensors and Actuators* B **121** 406
[11] Tsiulyanu D, Marian S and Liess H D 2002 *Sensors and Actuators* B **85** 232
[12] Sen S, Muthe K P, Joshi N, Gadkari S C, Gupta S K, M Roy Jagannath, Deshpande S K and Yakhami J V 2004 *Sensors and Actuators* B **98** 154
[13] Wüsten J and Potje-Kamloth K 2010 *Sensors and Actuators* B **145** 216
[14] Aljihmani L Petkov K and Vassilev V 2012 *J. Non-Cryst. Solids* **358** 364
[15] Nesheva D, Aneva Z, Reynolds S, Main C and Fizgerald A G 2006 *J. Optoel .Adv. Mat.* **8** 2120
[16] Kalužný J, Ležal D, Mariani E and Zavadil J 2004 *Chalcogenide Letters* **1** 41
[17] Bakr N A, Aziz M S and Hammam M. 2000 *Egypt. J. Sol.* **23** 45
[18] Qamhieh M and Adriaenssens G J 2001 *J. Non-Cryst. Solids* **292** 80
[19] Tsu R, Howard W E and Esaki L 1970 *J. Non-Cryst. Solids* **4** 322
[20] Lezal D, Pedlikova J and Zavadil J 2004 *J. Optoel. Adv. Mat.* **6** 133
[21] Sleeckx E, Tich L, Nagels P and Callaerts R 1996 *J. Non-Cryst. Solids* **198-200** 723
[22] Kasap S O, 1991 *Handbook of Imaging Materials,* (New York: Marcel Dekker Inc) p 329
[23] Sugai S 1987 *Phys. Rev.* B **35** 1345

Journal of Physics: Conference Series **398** (2012) 012058 doi:10.1088/1742-6596/398/1/012058

[24] Moharram A H, Hefni M A and Abdel-Baset 2010 *J Appl Phys* **108** 073505
[25] Andrikopoulos K S, Yannopoulos S N, Kolobov A V, Fons P and Tominaga J 2007 *J. Phys. Chem Solids* **68** 1074
[26] Baltes H, Göpel W and Hesse J (eds.) 1996 *Sensors Update* (Weinheim: Wiley VCH) vol 2